Annals of Mathematics Studies

Number 129

The Admissible Dual of GL(N) via Compact Open Subgroups

by

Colin J. Bushnell and Philip C. Kutzko

PRINCETON UNIVERSITY PRESS

PRINCETON, NEW JERSEY

1993

Princeton University Press books are printed on acid-free paper and meet the guidelines for permanence and durability of the Committee on Production Guidelines for Book Longevity of the Council on Library Resources

Printed in the United States of America
by Princeton University Press, 41 William Street
Princeton, New Jersey

Library of Congress Cataloging-in-Publication Data

Bushnell, Colin J. (Colin John), 1947-
The admissible dual of GL(N) via compact open subgroups / by Colin J. Bushnell & Philip C. Kutzko.
p. cm.—(Annals of mathematics studies ; no. 129)
Includes bibliographical references and index.
ISBN 0-691-03256-4—ISBN 0-691-02114-7 (pbk.)
1. Representations of groups. 2. Nonstandard mathematical analysis. I. Kutzko, Philip C., 1946- . II. Title.
III. Series.
QA171.B978 1993
512′.2—dc20 92-33614

To Lesley and David

Contents

This work gives a full description of a method for analysing the admissible complex representations of the general linear group $G = GL(N, F)$ of a non-Archimedean local field F in terms of the structure of these representations when they are restricted to certain compact open subgroups of G. We define a family of representations of these compact open subgroups, which we call *simple types*. The first example of a simple type, the "trivial type", is the trivial character of an Iwahori subgroup of G. The irreducible representations containing a given simple type are then classified, via an isomorphism of Hecke algebras, by the irreducible representations of some $H = GL(M, K)$, where K is some finite extension of F and M some divisor of N, which contain the trivial type in H. This leads to a complete classification of the irreducible smooth representations of G, including an explicit description of the supercuspidal representations as induced representations.

Acknowledgements: The first-named author was supported in part by SERC grant GR/E47650. The second-named author was supported in part by NSF grant DMS-8704194 and by SERC grant GR/F73366. Both authors wish to thank the Institute for Advanced Study for its hospitality during their visit in the Academic Year 1988–89. This visit was supported in part by NSF grant DMS-8610730. They also wish to acknowledge the hospitality of the Center for Advanced Studies of the University of Iowa.
This document was typeset by $\mathcal{A}\mathcal{M}\mathcal{S}$-TEX.

The Admissible Dual of GL(N)
via Compact Open Subgroups

INTRODUCTION

In this book, we give a new and effectively complete classification of the irreducible smooth complex representations of the general linear group $G = GL(N, F)$ over a non-Archimedean local field F. The study of the representation theory of G and other reductive groups over a local field arose from several sources, notably general questions of harmonic analysis on locally compact groups, and connections with arithmetic and automorphic forms:— see, among others, the influential accounts **[GGP-S]**, **[HC]**, **[JL]**, **[S]**. The subject has been largely driven by a wide-ranging family of conjectures of Langlands unified in the "Principle of Functoriality". This was first enunciated in **[La1]** and subsequently refined:— see **[La2]** and the expositions in **[BoCs]**. Progress with these conjectures will presumably require a clear description of the irreducible representations of these reductive groups, of which $GL(N)$ is the most accessible example. The methods of this work have been developed in the hope that they will prove amenable both to generalisation and application in this program.

When discussing the representation theory of a reductive algebraic group, in virtually any context, the one technique which immediately suggests itself is that of induction from parabolic subgroups. If G is a reductive algebraic group (identified here with its group of points over some fixed field of definition) and P is a proper parabolic subgroup of G with unipotent radical U and Levi decomposition $P = MU$, then $M \cong P/U$ is a reductive group with smaller semisimple rank than G, and therefore presumably of simpler structure. An irreducible representation of M can be inflated to P and then induced to G, the composition factors of this representation yielding irreducible representations of G. (Such induced representations always seem to have finite composition length, so this procedure can, in principle, be carried out.) The representation theory of G can thus be approached via a clear strategy:

(a) *classify the irreducible representations of Levi factors of proper parabolic subgroups of G;*

(b) *describe the decomposition of representations of G induced from proper parabolic subgroups;*

(c) *describe the representations of G which cannot be obtained from steps (a) and (b).*

In some circumstances, for example in appropriate categories of representations for groups over connected fields, step *(c)* is effectively empty.

For smooth representations of reductive groups over a non-Archimedean local field, considerable progress has been made along these lines. In this context, it is known that a representation of G induced from an irreducible representation of a parabolic subgroup has finite composition length **[Cs]**. Representations which do not occur as factors of such "parabolically induced" representations are here called *supercuspidal*. Given an irreducible smooth representation π of G which is not supercuspidal, there is a proper parabolic subgroup P of G, with unipotent radical U, and an irreducible supercuspidal representation σ of P/U such that π occurs as a factor of the representation $\mathrm{Ind}(\sigma)$ induced from σ. Further, σ is uniquely determined up to the action of an appropriate Weyl group in G:— see **[Cs]**. By choosing σ correctly in its Weyl orbit, one can even arrange for the given π to appear as a G-subspace (or, if preferred, quotient) of $\mathrm{Ind}(\sigma)$. To work out explicitly the structure of these induced representations seems, on present evidence, to be a subtle and difficult problem. It has been solved in **[BZ]**, **[Z]** for the case $G = GL(N, F)$, and some recent progress for groups of symplectic type has been announced in **[T]**, but there are very few general results on the subject.

Thus, for $G = GL(N)$, one has only to consider step *(c)*, which amounts to classifying the irreducible supercuspidal representations of $GL(M, F)$, for $M \leq N$. These are of crucial importance in that all the other representations are, so to speak, built from supercuspidal ones. This is particularly clear in the context of the local Langlands Conjectures:— irreducible supercuspidal representations of $GL(N, F)$ should parametrise the *irreducible* N-dimensional representations of the Weil group of F. On this basis alone, supercuspidal representations would appear, in all probability, to be intrinsically arithmetical in nature and so not approachable through any essentially geometric method in the spirit of the Zelevinsky classification of the nonsupercuspidal representations. It has long been conjectured (see **[K3]** for the history) that the supercuspidal representations can all be obtained by induction from open, compact mod centre subgroups of G. We shall prove this conjecture, as a consequence of a completely different strategy for understanding the admissible dual (i.e. the set of equivalence classes of irreducible smooth representations) of G.

There are reasons, in the present context, for seeking a description of the admissible dual of G which is more uniform than one would get by following the standard strategy given by steps *(a)*–*(c)* above. For, such an approach inevitably divides the dual into two parts, supercuspidal and nonsupercuspidal, and these parts have to be described by totally different, effectively incompatible methods. Such a situation would be, to say the least, technically inconvenient in the context of the Langlands program, where one envisages operations on representations (e.g. change of base or group) which could not respect this distinction.

To motivate our approach, let us return for a moment to the general situation, where G is the group of F-points of some reductive algebraic group defined over the non-Archimedean local field F. There is one case which has attracted particular attention, and which is easier to work with. This concerns the decomposition of representations $\mathrm{Ind}(\sigma)$ as above, but where P is a Borel subgroup of G and σ is an unramified quasicharacter of P/U (unramified here meaning that σ is trivial on a maximal compact subgroup). In the seminal paper **[Bo]**, it is shown, for a wide class of groups G, that the irreducible components of such representations are *characterised* by the property of containing the trivial character of an Iwahori subgroup $\mathcal{I}$ of G. Irreducible representations of G of this form are then parametrised by the simple modules over the Hecke algebra $\mathcal{H}(G, \mathbf{1}_\mathcal{I})$ of compactly supported functions on G which are bi-invariant under translation by $\mathcal{I}$. (In this notation, $\mathbf{1}_\mathcal{I}$ is the trivial character of $\mathcal{I}$.) This algebra can be explicitly described in terms of generators and relations, in which case it is sometimes referred to as an "affine Hecke algebra". The simple modules over an affine Hecke algebra are classified, in terms of L-group data and under very general hypotheses, in **[KL]**. Thus the classification of irreducible representations with an Iwahori-fixed vector (it seems reasonable to call these representations *arithmetically unramified*) is completely known in this generality.

We describe here, in the context of $G = GL(N, F)$, an extension of this method. The approach is, on the one hand, naturally adapted to the description of supercuspidal representations and, on the other, shows that the classification of the nonsupercuspidal representations, along with the problem of describing the decomposition of parabolically induced representations, can be reduced in an entirely natural manner to the arithmetically unramified case.

So, take $G = GL(N, F)$, and let $(\pi, \mathcal{V})$ denote an irreducible smooth complex representation of G. Thus, by definition, $\mathcal{V}$ is a complex vector space, and π is a homomorphism $G \to \mathrm{Aut}_\mathbb{C}(\mathcal{V})$ such that every $v \in \mathcal{V}$ is fixed by the image of some open subgroup of G. Let K be an open compact subgroup of G. When we view $\mathcal{V}$ as a K-space, we get a discrete (i.e. algebraic) direct sum decomposition

$$\mathcal{V} = \sum_\sigma \mathcal{V}^\sigma,$$

where σ ranges over the (equivalence classes of) irreducible smooth (therefore finite-dimensional) representations of K, and $\mathcal{V}^\sigma$ denotes the σ-isotypic component of $\mathcal{V}$. Since the representation $(\pi, \mathcal{V})$ is admissible (by **[J2]**), we know that the isotypic components $\mathcal{V}^\sigma$ are all finite-dimensional. We say that π *contains* σ if $\mathcal{V}^\sigma \neq \{0\}$. Our method rests on the construction of certain pairs (K, σ), which we call *types*. We shall see that every irreducible smooth representation $(\pi, \mathcal{V})$ of G contains some type, and we can classify those π which contain a fixed type.

These types come in two sorts, *simple* ones and *split* ones. We first discuss the simple ones. The structure of these is rather involved, having both an arithmetical component and a geometrical one:— this is to be expected, given the function they are to perform. They have to be constructed at length rather than defined axiomatically, but we can give here a rough outline of their properties. A simple type is a pair (J, λ) consisting of an open compact subgroup J of G and a smooth irreducible representation λ of J, both J and λ being of a very special kind. Attached to (J, λ) is an integer $e \geq 1$ and a field extension K/F, contained in the matrix algebra $A = \mathbb{M}(N, F)$ and of degree N/e. Write C for the A-centraliser of K, so that $C \cong \mathbb{M}(e, K)$. We have the properties:

(i) *$J \cap C^\times$ is an Iwahori subgroup of $C^\times$;*
(ii) *the G-intertwining of λ is $JC^\times J$.*

In *(ii)*, as usual, the G-intertwining of λ is the set of $g \in G$ such that λ and the conjugate representation λ^g of K^g have a component in common when restricted to $K \cap K^g$. Our Main Theorem is a finer version of *(ii)*. Let $\mathcal{H}(G, \lambda)$ be the usual Hecke algebra of compactly supported λ-spherical functions on G:— see §**4** for the definition. In particular, $\mathcal{H}(G, \lambda)$ is an associative $\mathbb{C}$-algebra with identity whose simple modules are in canonical bijection (up to equivalence) with the irreducible representations of G which contain λ. Our Main Theorem asserts:

(iii) *Let $\mathbf{1}$ denote the trivial character of the Iwahori subgroup $J \cap C^\times$ of $C^\times$. Then there is an isomorphism $\mathcal{H}(G, \lambda) \cong \mathcal{H}(C^\times, \mathbf{1})$ of $\mathbb{C}$-algebras with* 1.

There is not a canonical isomorphism here in general, but there is a canonical *family* of isomorphisms which differ, one from the other, in a fairly trivial way.

It follows that the set of irreducible smooth representations of G which contain λ is parametrised by the set of simple modules over the ring $\mathcal{H}(C^\times, \mathbf{1})$ (up to equivalence). As we have observed, these modules are known from **[KL]** (which amounts here to a very special case of **[Z]**). While we never need explicitly the classification of these modules, they are in fact in bijection with the set of conjugacy classes of pairs (s, u), where $s, u \in GL(e, \mathbb{C})$, s is semisimple, u is unipotent and $sus^{-1} = u^q$. Here, $q = q_K$ is the cardinality of the residue class field of

K. This parametrisation of the irreducible representations of G containing λ does indeed depend on the choice of algebra isomorphism in *(iii)*, but changing the isomorphism corresponds to tensoring representations with an unramified one-dimensional representation. Moreover, if the irreducible representation π of G does contain a simple type (J, λ), then (J, λ) is uniquely determined up to G-conjugacy.

The next point to remark on is that the algebra isomorphisms of *(iii)* above preserve support of functions, in a certain sense. It follows easily that the induced correspondence, between irreducible representations of G which contain λ and arithmetically unramified irreducible representations of $C^\times$, takes supercuspidal representations to supercuspidal representations. However, the group $C^\times \cong GL(e, K)$ has an arithmetically unramified supercuspidal representation if and only if $e = 1$, in which case all irreducible representations are supercuspidal. It follows, for our simple type (J, λ), that λ occurs in a supercuspidal representation of G if and only if every irreducible representation of G containing λ is supercuspidal, and that this is equivalent to the condition $e = 1$. (We call the simple type (J, λ) *maximal* if it has the property $e = 1$, or, equivalently, K/F is a maximal subfield of A.) Moreover, if (J, λ) is maximal, and π contains λ, then there is a uniquely determined extension of λ to a representation Λ of $K^\times J$ such that $\pi \cong \mathrm{Ind}(\Lambda)$, the representation smoothly induced from Λ. It also follows that two irreducible supercuspidal representations π_1, π_2 of G contain the same simple type if and only if $\pi_2 \cong \pi_1 \otimes \chi \circ \det$, for some unramified quasicharacter χ of $F^\times$.

We next have the task of determining, in real-life terms, those irreducible smooth representations $(\pi, \mathcal{V})$ of G which contain a fixed simple type (J, λ). Assuming that (J, λ) is not maximal, there is a proper parabolic subgroup P of G, with unipotent radical U, which has the following properties. Write $\mathcal{V}_U$ for the *Jacquet module* of $\mathcal{V}$ relative to P. Thus $\mathcal{V}_U = H_1(U, \mathcal{V})$ is the maximal quotient of $\mathcal{V}$ on which U acts trivially. Let π_U denote the natural representation of P/U on $\mathcal{V}_U$, and let e, K be the integer and the field attached to (J, λ) as above. Suppose that $(\pi, \mathcal{V})$ contains λ. Then we have:

(iv)(a) $P/U \cong GL(N/e, F)^e$;

(iv)(b) $\mathcal{V}_U \neq \{0\}$ *and there is a unique maximal simple type* (J', λ') *in* $GL(N/e, F)$ *such that every composition factor of* π_U *is of the form* $\sigma_1 \otimes \ldots \otimes \sigma_e$, *with each* σ_i *supercuspidal and containing* λ'.

In other words, there is a maximal simple type (J', λ') attached invariantly to (J, λ) (which we call the maximal simple type *associated* to (J, λ)) and, to use the terminology of **[BZ]**, any irreducible representation of G containing λ has *supercuspidal support* consisting of representations containing λ'. This amounts to saying that the supercuspidal support consists of unramified twists $\pi' \otimes \chi \circ \det$ of a single supercuspidal π' which contains λ'. Indeed the converse holds here:— if the support of π consists of supercuspidals all of which contain λ', then π contains λ.

Besides simple type, there is a complementary notion, at present rather more primitive, of *split type*. This is again a pair (K, ϑ) consisting of an open compact subgroup K of G and an irreducible smooth representation ϑ of K. These split types are constructed to have the following properties:

(v) *Let $(\pi, \mathcal{V})$ be an irreducible smooth representation of G which contains no simple type. Then π contains a split type.*

(vi) *Let (K, ϑ) be a split type in G. There exists a proper parabolic subgroup subgroup P of G, with unipotent radical U, such that for any irreducible smooth representation $(\pi, \mathcal{V})$ of G which contains ϑ, we have $\mathcal{V}_U \neq \{0\}$.*

It follows that an irreducible supercuspidal representation of G must contain some simple type, and is hence induced from an open compact mod centre subgroup of G. Moreover, *(iv)* and its converse now imply that an irreducible representation of G contains some simple type if and only if its supercuspidal support consists of unramified twists of a single supercuspidal representation of some $GL(N/e, F)$, $e \geq 1$.

If π' denotes an irreducible supercuspidal representation of some $GL(M, F)$, we write $\mathfrak{I}(\pi')$ for the set of equivalence classes of representations $\pi' \otimes \chi \circ \det$, where χ ranges over the unramified quasicharacters of $F^\times$, and call this the *inertial equivalence class of* π'. To complete our classification, we have to account for those irreducible representations π of G which contain no simple type. We know that π contains no simple type only if it contains a split type, and that this is equivalent to the supercuspidal support of π meeting at least two distinct inertial equivalence classes of supercuspidal representations. Here the situation is very simple. Write Ind^u for the "normalised" smooth induction functor (the one which takes unitarisable representations to unitarisable representations). Let P be a parabolic subgroup of G, with unipotent radical U, and suppose that

$$P/U \cong GL(N_1, F) \times \ldots \times GL(N_t, F).$$

Let π_i be an irreducible smooth representation of $GL(N_i, F)$, $1 \leq i \leq t$, and view $\pi_1 \otimes \ldots \otimes \pi_t$ as a representation of P via inflation. Following the standard practice, we put

$$\pi_1 \times \ldots \times \pi_t = \mathrm{Ind}^u(\pi_1 \otimes \ldots \otimes \pi_t).$$

Using this notation, we get

(vii) Let π be an irreducible smooth representation of G which contains no simple type. Then there exist integers $r \geq 2$, $N_1, N_2, \ldots N_r > 0$, satisfying $\sum_i N_i = N$, and an irreducible representation π_i of $GL(N_i, F)$, $1 \leq i \leq r$, such that

(a) π_i contains some simple type, $1 \leq i \leq r$;

(b) if π_i' denotes some element of the supercuspidal support of π_i, then $\mathfrak{I}(\pi_i') \cap \mathfrak{I}(\pi_j') = \emptyset$, whenever $i \neq j$;

(c) $\pi \cong \pi_1 \times \pi_2 \times \ldots \times \pi_r$.

Moreover, the expression in (c) is unique up to permutation of the factors, any permutation being permissible here.

Thus, overall, an irreducible smooth representation of G is described by a set of simple types, and for each of these simple types, a Kazhdan-Lusztig invariant. While we are not here concerned with such matters, this mirrors exactly the structure of "continuous ϕ-semisimple Deligne representations" of the Weil group of F, as one would expect in the context of the local Langlands Conjectures. Further, the structure of simple types is extremely well-adapted to computation of Godement-Jacquet local constants $\varepsilon(\pi, s)$ (see **[GJ]**) via Gauss sums, as in **[BF2]**. If necessary, one could even contemplate using the explicit computational techniques of **[BF1]**.

Historically, it was R. E. Howe who originally suggested (in **[H2]** and **[HM1]**) that one should study representations of groups like G by first considering their restrictions to suitable compact open subgroups of G and then giving a classification in terms of Hecke algebra isomorphisms. In his paper **[H2]** on $GL(N, F)$, he proposed the notion of "essential K-type". This is a particular sort of abelian character of a principal congruence subgroup of the maximal compact subgroup $GL(N, \mathfrak{o}_F)$ of $GL(N, F)$ (where $\mathfrak{o}_F$ is the discrete valuation ring in F). He showed that any irreducible smooth representation of $G = GL(N, F)$ contains a finite, nonzero number of essential K-types, which he was able to estimate. As a first application, he gave a proof, completely different from the standard one, that the process of induction from parabolic subgroups takes irreducible representations to representations of finite length. A more precise notion, that of *fundamental stratum*, was proposed by A. Moy in a conjecture in **[Mo2]** (although he called it a "minimal K-type").

To describe this conjecture, it is convenient to introduce some of the language of the body of the paper. Let $\mathfrak{o}_F$ denote the discrete valuation ring in F, and $\mathfrak{p}_F$ its maximal ideal. Let $\mathfrak{A}$ be a hereditary $\mathfrak{o}_F$-order in the algebra $A = \mathbb{M}(N, F)$. (We recall the definition in **(1.1)**.) In particular, $\mathfrak{A}$ is a compact open subring of A, and its unit group $U(\mathfrak{A})$ is a compact open subgroup of G. The Jacobson radical $\mathfrak{P}$ of $\mathfrak{A}$ is an invertible two-sided ideal of $\mathfrak{A}$, which gives rise to a filtration of $U(\mathfrak{A})$ by open normal subgroups $U^n(\mathfrak{A}) = 1+\mathfrak{P}^n$, $n \geq 1$. If $n > m \geq n/2$, the quotient $U^m(\mathfrak{A})/U^n(\mathfrak{A})$ is abelian, indeed isomorphic to $\mathfrak{P}^m/\mathfrak{P}^n$. Its Pontrjagin dual has a very convenient description, which leads us to the language of strata. A *stratum* in A is a 4-tuple $[\mathfrak{A}, n, m, b]$ consisting of a hereditary order $\mathfrak{A}$ in A, integers $n > m$, and an element $b \in \mathfrak{P}^{-n}$. We say that two strata $[\mathfrak{A}, n, m, b]$, $[\mathfrak{A}, n, m, b']$ are *equivalent* if $b \equiv b' \pmod{\mathfrak{P}^{-m}}$. We now fix a continuous character ψ_F of the additive group of F assumed, for convenience, to be nontrivial on $\mathfrak{o}_F$ but trivial on $\mathfrak{p}_F$. We put ψ_A (or just ψ) $= \psi_F \circ \mathrm{tr}$, where tr denotes the trace from A to F. Suppose next that $n > m \geq [n/2] \geq 0$, where $[\ ]$ denotes the greatest integer function. Then, for $b \in \mathfrak{P}^{-n}$, the map $\psi_b : u \mapsto \psi(b(u-1))$, $u \in U^{m+1}(\mathfrak{A})$, is a character of $U^{m+1}(\mathfrak{A})/U^{n+1}(\mathfrak{A})$, and the map $b \mapsto \psi_b$ establishes a bijection between the set of equivalence classes of strata $[\mathfrak{A}, n, m, b]$ and the dual group $(U^{m+1}(\mathfrak{A})/U^{n+1}(\mathfrak{A}))^\wedge$. A stratum of the form $[\mathfrak{A}, n, n-1, b]$, with $n \geq 1$, is called *fundamental* if the coset $b + \mathfrak{P}^{1-n}$ contains no nilpotent element of the algebra A. Moy's conjecture then amounts to

> *Let π be an irreducible smooth representation of G. There exists a fundamental stratum $[\mathfrak{A}, n, n-1, b]$ such that $\pi \mid U^n(\mathfrak{A})$ contains ψ_b.*

(There is, in fact, another possibility:— π might contain the trivial character of $U^1(\mathfrak{A}')$, for some hereditary order $\mathfrak{A}'$. We ignore this in the present discussion.) This conjecture was proved in **[Bu1]** (see also **[HM2]**). As described in **[Bu1]**, it implies the existence of an explicit formula for the Godement-Jacquet local constant $\varepsilon(\pi, s)$ when π is supercuspidal, and gives a quick local proof of Carayol's Theorem **[Ca]** that every supercuspidal representation of G is induced from an open compact mod centre subgroup in the case where the dimension N is prime:— a convincing display of the power of the method.

The fundamental stratum in π is by no means unique, but it does have useful uniqueness properties. The hereditary order $\mathfrak{A}$ has a "central ramification index" $e = e(\mathfrak{A}|\mathfrak{o}_F)$ defined by $\mathfrak{p}_F\mathfrak{A} = \mathfrak{P}^e$. Then any two fundamental strata $[\mathfrak{A}, n, n-1, b]$ occurring in the same irreducible representation π have the same "normalised level" $n/e(\mathfrak{A}|\mathfrak{o}_F)$. Moreover, the equivalence class of $[\mathfrak{A}, n, n-1, b]$ gives rise to a sort of characteristic polynomial $\phi_b(X) \in \mathsf{k}_F[X]$, where $\mathsf{k}_F = \mathfrak{o}_F/\mathfrak{p}_F$ is the residue class field of F. The stratum $[\mathfrak{A}, n, n-1, b]$ is fundamental if and only if $\phi_b(X)$ is not a power of X, and any two fundamental strata appearing in the same irreducible π have the same characteristic polynomial.

On the basis of this, a finer result was given in **[K4]**. Call the fundamental stratum $[\mathfrak{A}, n, n-1, b]$ *split* if the polynomial $\phi_b(X)$ has at least two distinct monic irreducible factors in $\mathrm{k}_F[X]$. It was then shown in **[K4]** that an irreducible π which contains a split fundamental stratum has a matrix coefficient which is not compactly supported mod centre, hence that such a π is not supercuspidal. This yields a second quick local proof of Carayol's theorem. Another argument in **[K4]** gives a refinement of the notion of nonsplit fundamental stratum. For a hereditary order $\mathfrak{A}$ as above, define $\mathfrak{K}(\mathfrak{A}) = \{g \in G : g\mathfrak{A}g^{-1} = \mathfrak{A}\}$. This is an open compact mod centre subgroup of G which normalises the unit groups $U(\mathfrak{A})$, $U^n(\mathfrak{A})$, as above. A stratum $[\mathfrak{A}, n, m, \alpha]$, $m < n$, is called *alfalfa* if

(a) $\alpha \notin \mathfrak{P}^{1-n}$;
(b) the algebra $E = F[\alpha]$ *is a field, and* $E^\times \subset \mathfrak{K}(\mathfrak{A})$;
(c) α *is* minimal *over* F.

Here, *minimal* means first that the normalised valuation $\nu(\alpha) = \nu_E(\alpha)$ is relatively prime to the ramification index $e(E|F)$ and also that the residue class field k_E is generated over k_F by the residue class of the element $\alpha^{e(E|F)}\pi_F^{-\nu(\alpha)}$, where π_F is any prime element of F. In particular, any alfalfa stratum is fundamental. (Alfalfa strata are the first examples of what we now call simple strata, but we retain the original term in this historical survey.) The other result in **[K4]** then shows that, given a nonsplit fundamental stratum $[\mathfrak{A}, n, n-1, b]$, there exists an alfalfa stratum $[\mathfrak{A}', n', n'-1, \alpha]$ with the property that it is contained in any irreducible representation of G which contains $[\mathfrak{A}, n, n-1, b]$.

Thus, at least in the search for supercuspidal representations, one is reduced to investigating irreducible representations of G which contain some alfalfa stratum. In certain cases, there is a shortcut, outlined in **[HM1]** and **[HM3]**. Suppose that the dimension N (as in $G = GL(N, F)$) is not divisible by the residual characteristic p of F (the so-called "tamely ramified" case). Let $[\mathfrak{A}, n, n-1, \alpha]$ be an alfalfa stratum in A. Because of the tame ramification condition, one can find a compact open subgroup U' of G, and an irreducible representation ψ'_α of U', such that $U' \supset U^n(\mathfrak{A})$, and $\psi'_\alpha | U^n(\mathfrak{A})$ is a multiple of ψ_α, with the following properties. First, any irreducible representation of G which contains ψ_α also contains ψ'_α. Second, if H denotes the G-centraliser of α (so that $H \cong GL(M, F[\alpha])$, where we put $M = N/[F[\alpha] : F]$), there is an algebra isomorphism $\mathcal{H}(G, \psi'_\alpha) \cong \mathcal{H}(H, \mathbf{1})$, where $\mathbf{1}$ denotes the trivial character of $U^n(\mathfrak{A}) \cap H$. Since $U^n(\mathfrak{A}) \cap H$ is the group of n-units of a hereditary $\mathfrak{o}_{F[\alpha]}$-order in $\mathbb{M}(M, F[\alpha])$, one has an effective inductive step. Assuming one can maintain the hypothesis of tame ramification (e.g. by assuming $p > N$), this process continues to give a classification in terms of affine Hecke algebras, along with a determination of the supercuspidal representations (this latter being valid whenever p does not divide N). The method is also effective, for rather different reasons, in the case $p = N$. A similar approach, when all ramification is tame, yields results for certain other groups of small rank:— see **[Mo3]**, **[Mo4]**, **[Mo5]**. However, it should be noted that this procedure is not susceptible to further generalisation. At a basic technical level, it depends on the splitting of certain exact sequences, on which we comment further after (1.4.10) below. These sequences do not split in the presence of wild ramification. This simple fact forces us to adopt an entirely different scheme, leading to quite different Hecke algebra isomorphisms.

That one must expect to work harder in the general case is already clear from **[KM2]**, which proves the induction conjecture for the supercuspidal representations of G in the case where N is the product of two (not necessarily distinct) primes. That paper contains the seeds of several of the ideas of the present one. One starts there with an irreducible representation π of G which, we may assume, contains an alfalfa stratum $[\mathfrak{A}, n, n-1, \alpha]$, i.e., it contains the character ψ_α of $U^n(\mathfrak{A})$. In particular, it contains the character ψ_α of $U^{m+1}(\mathfrak{A})/U^{n+1}(\mathfrak{A})$ for some stratum $[\mathfrak{A}, n, m, \alpha]$, $m < n$, such that $[\mathfrak{A}, n, n-1, \alpha]$ is alfalfa. We therefore say it contains the stratum $[\mathfrak{A}, n, m, \alpha]$. We choose such a stratum so as to minimise the quantity $m/e(\mathfrak{A})$. Assume that $m > [n/2]$, so that π contains some character $\psi_{\alpha+b}$ of $U^m(\mathfrak{A})$ with $b \in \mathfrak{P}^{-m}$. Write $E = F[\alpha]$, $B =$ the A-centraliser of E, and $\mathfrak{B} = \mathfrak{A} \cap B$. Then $\mathfrak{B}$ is a hereditary $\mathfrak{o}_E$-order in B, and $U^t(\mathfrak{A}) \cap B^\times = U^t(\mathfrak{B})$, $t \geq 0$. The restriction of ψ_b to $U^m(\mathfrak{B})$ then defines a stratum $[\mathfrak{B}, m, m-1, \beta]$ in B. The main labour of **[KM2]** is to show that one has here a situation similar to that of **[Bu1]** and **[K4]**:— one can arrange for $[\mathfrak{B}, m, m-1, \beta]$ to be fundamental. If it is split fundamental, one has a nonsupercuspidal representation, while otherwise one can arrange for it to be alfalfa. It then follows that the element $\alpha + b$ generates a field which normalises $\mathfrak{A}$. One can compute the intertwining of $\psi_{\alpha+b}$. If it is compact mod centre (equivalently, if the field $F[\alpha+b]$ is a maximal subfield of A), any irreducible π which contains $\psi_{\alpha+b}$ is supercuspidal and induced from a compact-mod-centre subgroup. If not, then one can achieve $m = [n/2]$, and one must consider, instead of ψ_α, extensions to $U^m(\mathfrak{B})U^{[n/2]+1}(\mathfrak{A})$ of characters of the group $U^{m+1}(\mathfrak{B})U^{[n/2]+1}(\mathfrak{A})$, for some $m \leq [n/2]$, which agree with ψ_α on $U^{[n/2]+1}(\mathfrak{A})$ and factor through the determinant $\det_B$ on $U^{m+1}(\mathfrak{B})$. Such an extension again defines a stratum $[\mathfrak{B}, m, m-1, \beta]$ in B, and one proceeds as in the first case. Either way, one is finished in the case of N a product of two primes, as soon as one achieves $[F[\alpha, \beta] : F] > [F[\alpha] : F] > 1$. One might expect that it would be possible, in the general case, to continue this process by working in B

over the (now) alfalfa stratum $[\mathfrak{B}, m, m-1, \beta]$. However, this is not the case:— one has to return to the algebra A and work with the stratum $[\mathfrak{A}, n, m-1, \alpha+b]$. This feature is intrinsic to the general (i.e. possibly wildly ramified) case. It is to deal with this situation that we need the notion of simple stratum, on which the entire work is founded.

It is now appropriate to start reviewing the layout of the present paper, although we shall make further comments on the literature below. §**1** is technical and contains algebraic apparatus and computational techniques which we shall use throughout the work. First among these is the *tame corestriction*. Suppose we have a subfield E/F of A with centraliser B. In the standard way, we can identify A, B with their Pontrjagin duals. The tame corestriction is a (suitably normalised) map $s : A \to B$ which realises restriction of additive characters. When $\mathfrak{A}$ is a hereditary $\mathfrak{o}_F$-order in A with $E^\times \subset \mathfrak{K}(\mathfrak{A})$, the map s has particular pleasant properties relative to $\mathfrak{A}$ and the hereditary $\mathfrak{o}_E$-order $\mathfrak{B} = \mathfrak{A} \cap B$. It generalises the "derivative map" S_α of **[KM1]**. We then introduce the notion of a *simple stratum*. A stratum $[\mathfrak{A}, n, m, \beta]$ is simple if, roughly speaking, the algebra $F[\beta]$ is a field which normalises $\mathfrak{A}$, and β is chosen within its $\mathfrak{P}^{-m}$-coset to minimise the field degree $[F[\beta] : F]$. In this situation, there is a close interplay between a tame corestriction s on A relative to $F[\beta]/F$ and the adjoint map $a_\beta : A \to A$ given by $x \mapsto \beta x - x\beta$. This enables us to compute conjugacy classes and centralisers very efficiently, and we exploit this technique repeatedly throughout the paper. The first application is a *formal intertwining theorem* for simple strata. One says that $x \in G$ intertwines the stratum $[\mathfrak{A}, n, m, \beta]$ if the intersection of $\beta + \mathfrak{P}^{-m}$ and $x^{-1}(\beta + \mathfrak{P}^{-m})x$ is nonempty. When $n > m \geq [n/2] \geq 0$, this is the same as the intertwining of the character ψ_β of $\boldsymbol{U}^{m+1}(\mathfrak{A})$. The main result of §**1** computes the formal intertwining of a simple stratum, and shows that it is, in some sense, as small as possible.

§**2** gives the basic structure theorems for simple strata. It starts by showing that they have the refinement properties we will need in the representation theory. More precisely, let $[\mathfrak{A}, n, m, \beta]$ be a simple stratum. Set $E = F[\beta]$ (which is a field), $B =$ the A-centraliser of E, $\mathfrak{B} = \mathfrak{A} \cap B$, and let s be a tame corestriction on A relative to E/F. Take an element $b \in \mathfrak{P}^{-m}$, and suppose that the stratum $[\mathfrak{B}, m, m-1, s(b)]$ in B is simple (i.e. alfalfa in this context). The first result of §**2** shows that $[\mathfrak{A}, n, m-1, \beta+b]$ is equivalent to a simple stratum. This gives a method of constructing simple strata in A from alfalfa strata. The main result of the section then shows that all simple strata are obtained this way, and gives various invariance properties of the construction. The section concludes with a "rigidity" property of obvious importance for representation theory. We say that two strata $[\mathfrak{A}_i, n_i, m_i, b_i]$ *intertwine* in G if there exists $x \in G$ such that

$$x^{-1}(b_1 + \mathfrak{P}_1^{-m_1})x \cap (b_2 + \mathfrak{P}_2^{-m_2}) \neq \emptyset,$$

where $\mathfrak{P}_i$ is the radical of $\mathfrak{A}_i$. Then two simple strata $[\mathfrak{A}, n, m, \beta_i]$ in A intertwine if and only if one is equivalent to a $\mathfrak{K}(\mathfrak{A})$-conjugate of the other.

In §**3**, we construct a family of abelian characters attached to a simple stratum $[\mathfrak{A}, n, m, \beta]$, $m \geq 0$. We start by constructing a pair $\mathfrak{H}(\beta, \mathfrak{A}) \subset \mathfrak{J}(\beta, \mathfrak{A})$ of $\mathfrak{o}_F$-orders in A. These are defined in terms of the structure of the stratum, and are related by a collection of exact sequences generalising those of §**1**. They define families of compact open subgroups $H^m(\beta, \mathfrak{A})$, $J^m(\beta, \mathfrak{A})$ of G, $m \geq 0$, by $H^m = \mathfrak{H} \cap \boldsymbol{U}^m(\mathfrak{A})$, $J^m = \mathfrak{J} \cap \boldsymbol{U}^m(\mathfrak{A})$. When $m \geq [n/2]$, we get $H^m = J^m = \boldsymbol{U}^m(\mathfrak{A})$. We then proceed to define, for $m \geq 0$, a finite family $\mathcal{C}(\mathfrak{A}, m, \beta)$ of abelian characters θ (which we call *simple characters*) of the group $H^{m+1}(\beta, \mathfrak{A})$. These generalise the extended ψ_α's of **[KM2]**, described above. Indeed, when $m \geq [n/2]$, the only element of $\mathcal{C}(\mathfrak{A}, m, \beta)$ is the character ψ_β of $\boldsymbol{U}^{m+1}(\mathfrak{A}) = H^{m+1}(\beta, \mathfrak{A})$. §**3** is devoted largely to computing the intertwining of these simple characters θ, and checking invariance properties of their construction. In **(3.4)** we show that $J^{m+1}(\beta, \mathfrak{A})$ is a "Heisenberg group" relative to the subgroup $H^{m+1}(\beta, \mathfrak{A})$ and the character θ, for any $\theta \in \mathcal{C}(\mathfrak{A}, m, \beta)$. This is simply a reflection of the exact sequence calculus alluded to earlier. In **(3.5)**, we show that the simple characters have a rigidity property generalising that of simple strata. **(3.6)** investigates another surprising property. Given a simple stratum $[\mathfrak{A}, n, 0, \beta]$, there will in general be many hereditary $\mathfrak{o}_F$-orders $\mathfrak{A}'$ in A for which $[\mathfrak{A}', n', 0, \beta]$ is a simple stratum (for some n'). When this is the case, the sets $\mathcal{C}(\mathfrak{A}, 0, \beta)$, $\mathcal{C}(\mathfrak{A}', 0, \beta)$ *are in canonical bijection.* Indeed, the same applies even when the algebra spanned by $\mathfrak{A}'$ is not A itself, but some algebra $\mathbb{M}(N', F)$ containing a copy of the field $F[\beta]$. Thus the simple characters θ have, in some measure, an existence independent of the objects used to define them. We see later, in §**7**, that this is a reflection of the properties of the Jacquet functor.

§**4** is a short section, collecting together a number of basically well known general results on Hecke algebras. We have gathered these together, with proofs, for convenience of reference.

It is in §**5** that we finally come to the central notion of *simple type.* Given a simple stratum $[\mathfrak{A}, n, 0, \beta]$ in A and a character $\theta \in \mathcal{C}(\mathfrak{A}, 0, \beta)$, there is a unique irreducible representation η of $J^1(\beta, \mathfrak{A})$ which contains θ. We first show that η can be extended to a representation κ of $J(\beta, \mathfrak{A})$. This is a tricky process, which has led to some difficulties with the literature (see **[Wa]** and **[KM2a]** for more details). We produce this extension κ (which is not unique) via a method, inspired in part by

[Wa], which exploits the relations between the character sets $\mathcal{C}(\mathfrak{A}, 0, \beta)$, when $\mathfrak{A}$ is allowed to vary over hereditary orders in A normalised by the field $F[\beta]$. Not all possible extensions are of interest to us:— we only use what we call "β-extensions" of η. These are characterised by being intertwined by the whole of the G-centraliser of β. This construction occupies **(5.1)**, **(5.2)**. At this stage, we therefore have a canonical family of representations κ of the group $J(\beta, \mathfrak{A})$. The quotient $J(\beta, \mathfrak{A})/J^1(\beta, \mathfrak{A})$ is canonically isomorphic to $\boldsymbol{U}(\mathfrak{B})/\boldsymbol{U}^1(\mathfrak{B})$, where $\mathfrak{B}$ is the $\mathfrak{A}$-centraliser of β, and so is a direct product of groups $GL(n_i, \mathrm{k}_E)$, for various integers $n_i \geq 1$, $1 \leq i \leq e = e(\mathfrak{B}|\mathfrak{o}_E)$, k_E being the residue field of $E = F[\beta]$. We now insist that the n_i are all equal, to some integer f, say, which amounts to demanding that the order $\mathfrak{A}$ be principal. We take an irreducible cuspidal representation $\boldsymbol{\sigma}_0$ of $GL(f, \mathrm{k}_E)$, form the e-fold tensor product $\boldsymbol{\sigma}_0 \otimes \ldots \otimes \boldsymbol{\sigma}_0$, and inflate this to a representation $\boldsymbol{\sigma}$ of $J(\beta, \mathfrak{A})$. We can then form the irreducible representation $\lambda = \kappa \otimes \boldsymbol{\sigma}$, and (J, λ) is what we call a simple type. There is a significant special case which only fits uncomfortably into this definition:— this is given by a principal order $\mathfrak{A}$, with $\boldsymbol{U}(\mathfrak{A})/\boldsymbol{U}^1(\mathfrak{A})$ isomorphic to $GL(f, \mathrm{k}_F)^e$. We get a simple type $(\boldsymbol{U}(\mathfrak{A}), \boldsymbol{\sigma})$ by taking an irreducible cuspidal representation $\boldsymbol{\sigma}_0$ of $GL(f, \mathrm{k}_F)$, taking the e-fold tensor product of $\boldsymbol{\sigma}_0$ with itself, and inflating to get a representation $\boldsymbol{\sigma}$ of $\boldsymbol{U}(\mathfrak{A})$. In practice however, we can always treat the two cases simultaneously. The next step is to estimate the intertwining of a simple type (J, λ) attached to the simple stratum $[\mathfrak{A}, n, 0, \beta]$. This has to be done in terms of a certain affine Weyl group $\widetilde{\boldsymbol{W}}(\mathfrak{B})$. Explicitly, we write $e = e(\mathfrak{B}|\mathfrak{o}_E)$, $ef = N/[E : F]$ (in fact these are the same integers given by $\boldsymbol{U}(\mathfrak{B})/\boldsymbol{U}^1(\mathfrak{B}) \cong GL(f, \mathrm{k}_E)^e$ in the notation above). The elements of this affine Weyl group $\widetilde{\boldsymbol{W}}(\mathfrak{B})$ parametrise the canonical basis of a certain affine Hecke algebra $\mathcal{H}(e, q^f)$, where $q = q_E$ denotes the cardinality of the residue class field of E. (In the terms we used earlier in *(i)–(iii)*, if K denotes an unramified field extension of E of degree f with $K^\times \subset \mathfrak{K}(\mathfrak{A})$, and C the A-centraliser of K, then $J \cap C^\times$ is an Iwahori subgroup of $C^\times$, and $\mathcal{H}(e, q^f)$ is canonically isomorphic to the Hecke algebra of $C^\times$ relative to the trivial character of $J \cap C^\times$. Moreover, $\widetilde{\boldsymbol{W}}(\mathfrak{B})$ can be identified with the affine Weyl group of $C^\times$ relative to $J \cap C^\times$.) We get a canonical embedding $\widetilde{\boldsymbol{W}}(\mathfrak{B}) \to J\backslash G/J$ whose image contains (in fact equals) the G-intertwining of λ. We then prove our Main Theorem, which asserts that the $\mathbb{C}$-algebras (with 1) $\mathcal{H}(G, \lambda)$, $\mathcal{H}(e, q^f)$ are isomorphic.

This isomorphism $\mathcal{H}(G, \lambda) \cong \mathcal{H}(e, q^f)$ generalises the Hecke algebra isomorphism of Waldspurger **[Wa]**. The viewpoint of **[Wa]** is, however, rather different. It starts with an irreducible supercuspidal representation of $GL(M, F)$, for some M, given explicitly as an induced representation. (The construction used in **[Wa]** gives all supercuspidals in the case of M prime (known from **[Ca]**) or M not divisible by the residual characteristic of F (known from **[H1]** and **[Mo]**). In the general case, however, it does not give them all.) **[Wa]** proceeds to construct a representation of a certain compact open subgroup of $GL(nM, F)$, for any $n \geq 1$, which is essentially a simple type (it is a special case of the representation we denote by λ_P in §**7**), and proves the Hecke algebra isomorphism. While we cannot appeal to it directly, this work has frequently inspired us in this area.

With the Hecke algebra isomorphism established, we can start our applications to representation theory. In §**6**, we classify the supercuspidal representations which contain a simple type and show, in particular, that they are all induced from open, compact mod centre subgroups. We then go on, in §**7**, to classify the irreducible representations of G which contain a fixed simple type λ. This involves building up a fairly elaborate machinery of Iwahori decompositions, but leads to a maximal simple type (J', λ') in $GL(N/e, F)$ attached invariantly to (J, λ). This, together with the Hecke algebra isomorphism, puts us in the position of being able to follow quite closely the arguments of **[Cs]** to analyse the irreducible representations containing λ. It turns out that the irreducible $\boldsymbol{\pi}$ contains λ if and only if its supercuspidal support consists of representations which contain λ'. We refine this to deal with the admissible representations of G which are generated by their λ-isotypic components, via a categorical equivalence generalising those of **[Bo]** and **[Wa]**. We then use this equivalence to show that our Hecke algebra isomorphisms commute with (normalised) induction from appropriate parabolic subgroups. Further, the Hecke algebra isomorphism, when suitably normalised, preserves various canonical involutions. This implies that it preserves the discrete series. Using the results of **[Bo]**, we thus get a determination of the discrete series representations which contain λ (and we see later that any discrete series representation contains some λ). We also give an explicit formula for the formal degrees. All of this takes up §**7**.

We thus come to §**8** with a complete account of representations which contain a simple type. It remains only to describe those which contain no simple type. We start by defining a "split type", and show that a representation which contains no simple type must contain a split type. Then we show that a representation containing a split type has a nonzero Jacquet module relative to a proper parabolic subgroup of G which is determined by that split type. It follows that an irreducible supercuspidal representation of G must contain some simple type. Therefore all the supercuspidal representations have already been described in §**6**, and this completes the proof of the induction conjecture. Moreover, an irreducible representation of G contains some simple type if and only if its supercuspidal support consists of unramified twists of a single representation, and such representations are classified via the Hecke algebra isomorphism. From there, as we mentioned above, it is a comparatively simple step to get the classification of the "atypical representations".

Our theory of simple types (and split types) has been developed exclusively in the context of $GL(N)$, and it is inevitable that one should ask whether there are analogous structures in more general reductive groups over F. In fact there is a similar theory of simple types and Hecke algebra isomorphisms in the special linear group $SL(N, F)$, an account of which will appear elsewhere. In another direction, the papers **[M1-4]** of Morris give substantial preliminary evidence for the existence of similar structures in p-adic classical groups. Further, though we are here only really concerned with $GL(N)$, we have taken care to use only those auxiliary results (i.e. ones not involving analysis of restriction to compact open subgroups) which are known in great generality.

There is another remark to be made in this connection. We have not fully checked the details, but it seems clear to us that one could replace the standard results on, in particular, supercuspidal support, by taking a more general approach to the rigidity theorem for simple strata (and simple characters). This would give a more consistent and self-contained treatment, based purely on restriction and intertwining arguments on compact open subgroups. The partial results we obtained in an earlier phase of the work (as reported at the Bowdoin conference, but excluded from the write-up **[BK]**) indicated that this would be an extremely laborious process which might also entail some loss of generality. However, the possibility of doing this should not be forgotten.

For numerous informative and stimulating conversations, we have to thank our colleagues D. C. Manderscheid, L. E. Morris, C. D. Keys, and especially G. Henniart, who also read parts of an earlier draft of the manuscript. We are happy to acknowledge the contribution of an anonymous referee, whose helpful and detailed comments stimulated us to make extensive revisions and corrections to the original version. The foundations of this work were laid during the academic year 1988–9 while the authors were Members of the Institute for Advanced Study. Preliminary accounts of some of the material were delivered in seminars there:—the long-suffering audience deserves a word of appreciation here. A great debt of gratitude is owed to R. P. Langlands for arranging our visit to the Institute, for allowing us to so use his seminar, and for his unfailing encouragement. Finally, we acknowledge the indulgent support of our home institutions King's College London and the University of Iowa, along with the superb facilities and the hospitality of the University

House (now the Center for Advanced Studies) of the University of Iowa where the final stages of this project were carried out.

Department of Mathematics,
King's College,
Strand,
London WC2R 2LS.

Department of Mathematics,
University of Iowa,
Iowa City,
IA 52242.

Comments for the reader

We ask of the reader a knowledge of the basic theory of representations of finite groups (although we shall invariably use this in the context of profinite groups), and some facility in working with the algebra and topology of non-Archimedean local fields. As for the smooth representation theory of groups like $G = GL(N, F)$, sufficient general background can be extracted from the first few pages of Cartier's survey article **[Cr]** (but we have different ideas as to what should be called Frobenius Reciprocity). Those substantial results which we need from time to time are generally quoted in full, and can be found in, especially, Casselman's notes **[Cs]**.

The work is based on detailed knowledge of the intertwining properties of certain representations of certain compact subgroups of G. This requires some intricate and unfamiliar machinery. The first half of the book is concerned exclusively with this, but it has been arranged so that the essence of the material can be extracted from a short list of definitions and results. The reader wishing to reach the representation theory of G with the minimum of delay can therefore move quite rapidly through this part, absorbing only selected definitions and statements as follows:

1: An understanding of the basic definitions in **(1.1–2)** is probably essential. The key definitions (tame corestriction, k_0, $\mathfrak{N}$) and properties at the beginnings of the next couple of sections will be adequate for a first reading, while the definition (1.5.5) and the main result (1.5.8) of the next section are essential. (1.6) is technical, and may be omitted at a first reading.

2: Only the statements of the main theorems (2.2.8), (2.4.1), (2.6.1) need be read at first.

3: In **(3.1)**, the definitions and properties of the rings $\mathfrak{H}$, $\mathfrak{J}$ (and thus the groups H, J) (through to (3.1.10)) are basic. Likewise the definition of simple character in **(3.2)** (i.e. (3.2.1/3)). In **(3.3)**, there is only the main theorem (3.3.2) and the corollaries (3.3.18)–(3.3.21). In the remaining sections, the significant results are (3.4.1), (3.5.11), (3.6.1) and (3.6.14).

4: This is an exposition of some general properties of Hecke algebras, to be consulted only at need.

5: The content of the first two sections is subsumed in the definition of the representation η in (5.1.1) and the definition and existence of β-extensions (5.2.1), (5.2.2). **(5.3)** is purely technical, and **(5.4)** is a review of standard material concerning affine Weyl groups. From the definition (5.5.10) of simple type, one can skip straight to the Main Theorem (5.6.6), its Corollary (5.6.17), and then (5.7.1).

The representation theory of G starts in §**6**. Even within this some shortcuts are possible. For example, the reader only interested in supercuspidal representations can read §**6** and then pass straight to §**8** (where only the first three sections are relevant). The principal applications of the theory are summarised in **(8.4)**.

An overview of the material and its relations to some other parts of the subject is given in Henniart's Bourbaki talk **[He]**. It might also be helpful to read **[K4]** and the early parts of **[Bu1]**, where some of the basic ideas of the present work can be seen in action in a considerably more straightforward context.

1. EXACTNESS AND INTERTWINING

This section introduces most of the algebraic apparatus that we shall need, at least to the extent of definitions and basic properties. All of this is associated with a pair $(\mathfrak{A}, E)$ consisting of a hereditary order $\mathfrak{A}$ in an algebra $A \cong \mathbb{M}(N, F)$ and a subfield E/F of A which *normalises* $\mathfrak{A}$ in the sense that $x\mathfrak{A}x^{-1} = \mathfrak{A}$ for all $x \in E^\times$.

After some introductory reminiscences in **(1.1)**, we produce in **(1.2)** a family of tensor decompositions (the "(W, E)-decompositions") of various objects associated with $(\mathfrak{A}, E)$. **(1.3)** introduces the notion of tame corestriction. This is a bimodule map s from A to the centraliser B of the field E which, after suitable identifications, realises restriction of additive characters from A to B. In **(1.4)**, we fix a generator β of E/F, i.e. $E = F[\beta]$, and consider the adjoint map $a_\beta : x \mapsto \beta x - x\beta$ on A. Examination of a_β relative to the filtration of A given by powers of the radical of $\mathfrak{A}$ gives rise to the lattices $\mathfrak{N}_k(\beta, \mathfrak{A})$ and a "critical exponent" $k_0(\beta, \mathfrak{A})$. There is a close interplay between adjoint maps and tame corestrictions, which we express via a family of exact sequences describing the lattices $\mathfrak{N}_k(\beta, \mathfrak{A})$. These exact sequences find many applications throughout the paper, usually in computation of centralisers and intertwining.

In **(1.5)**, we use the exponent $k_0(\beta, \mathfrak{A})$ to define *simple strata*, which will be a major object of study in what follows. The machinery of the preceding subsections then comes together to prove a formal intertwining theorem for simple strata. This, and indeed all of the concepts and machinery developed here have repeated applications throughout the paper.

(1.1) Hereditary orders

We start by recalling some basic terminology and facts, mainly concerned with hereditary orders. We also take the opportunity to introduce some notation which will remain standard throughout the paper:

F = *a non-Archimedean local field;*

$\mathfrak{o}_F$ = *the discrete valuation ring in* F*;*

$\mathfrak{p}_F$ = *the maximal ideal of* $\mathfrak{o}_F$*;*

$\mathsf{k}_F = \mathfrak{o}_F/\mathfrak{p}_F$, *the residue class field of* F*;*

ψ_F = *some fixed continuous character of the additive group of* F, *with conductor* $\mathfrak{p}_F$ *(i.e.* ψ_F *is null on* $\mathfrak{p}_F$ *but not on* $\mathfrak{o}_F$*);*

V = *an* F*-vector space of finite dimension* $N \geq 1$*;*

$A = \mathrm{End}_F(V)$*;*

$G = \mathrm{Aut}_F(V)$;
$\psi_A = \psi = \psi_F \circ \mathrm{tr}_{A/F}$, *where* tr *denotes trace.*

If W is some finite-dimensional F-vector space, a *lattice* in W is a compact open subgroup of W. Such is an *$\mathfrak{o}_F$-lattice* if it is also an $\mathfrak{o}_F$-module. Equivalently, an $\mathfrak{o}_F$-lattice in W is a finitely generated $\mathfrak{o}_F$-submodule of W containing an F-basis of W. If W happens to be an F-algebra with 1, an *$\mathfrak{o}_F$-order* in W is an $\mathfrak{o}_F$-lattice in W which is also a subring of W (with the same 1). If $\mathfrak{A}$ is an $\mathfrak{o}_F$-order, an *$\mathfrak{A}$-lattice* is an $\mathfrak{A}$-module which is also an $\mathfrak{o}_F$-lattice.

Now let $\mathfrak{A}$ be an $\mathfrak{o}_F$-order in the algebra $A = \mathrm{End}_F(V)$ as above. Then $\mathfrak{A}$ is *(left) hereditary* if every (left) $\mathfrak{A}$-lattice is $\mathfrak{A}$-projective. For a full account of hereditary orders, see **[Re]**, and the supplementary summaries in **[BF2]** and **[Bu1]**. We here recall, without proofs, the structure theory of hereditary orders, concentrating on those aspects which will be most useful to us.

An *$\mathfrak{o}_F$-lattice chain in V* is a nonempty set $\mathcal{L}$ of $\mathfrak{o}_F$-lattices in V which is linearly ordered under inclusion and such that $xL \in \mathcal{L}$ for all $L \in \mathcal{L}$, $x \in F^\times$. A lattice chain $\mathcal{L}$ is thus a disjoint union of orbits

$$F^\times.L = \{\pi_F^n L : n \in \mathbb{Z}\},$$

for various $L \in \mathcal{L}$, and where π_F is some prime element of F. The linear ordering implies that, given $L, L' \in \mathcal{L}$, there exists $m \in \mathbb{Z}$ such that $\pi_F^{m+1} \subset L' \subset \pi_F^m$. This implies that the set $\mathcal{L}$ is countable, so we may index it by $\mathbb{Z}$, $\mathcal{L} = \{L_i : i \in \mathbb{Z}\}$, so that $L_i \supsetneq L_{i+1}$, $i \in \mathbb{Z}$. The prime element π_F then acts as an order-preserving automorphism of $\mathbb{Z}$, i.e. as a translation. Thus an $\mathfrak{o}_F$-lattice chain in V just a sequence $\{L_i : i \in \mathbb{Z}\}$ of $\mathfrak{o}_F$-lattices in V such that

(1.1.1) *(i)* $L_i \supsetneq L_{i+1}$, $i \in \mathbb{Z}$;
(ii) there exists $e \in \mathbb{Z}$ *such that* $L_{i+e} = \mathfrak{p}_F L_i$, $i \in \mathbb{Z}$,

which is the definition commonly used elsewhere. This integer $e = e(\mathcal{L})$ is uniquely determined, and is called the *$\mathfrak{o}_F$-period* of $\mathcal{L}$.

Given an $\mathfrak{o}_F$-lattice chain $\mathcal{L}$ in V, we define a sequence of $\mathfrak{o}_F$-lattices in A by

$$\textbf{(1.1.2)} \qquad \mathrm{End}^n_{\mathfrak{o}_F}(\mathcal{L}) = \{x \in A : xL_i \subset L_{i+n}, i \in \mathbb{Z}\}$$

for each $n \in \mathbb{Z}$. In particular, $\mathrm{End}^0_{\mathfrak{o}_F}(\mathcal{L}) = \mathfrak{A} = \mathfrak{A}(\mathcal{L})$ is an $\mathfrak{o}_F$-order in A, and indeed a hereditary order. Every hereditary $\mathfrak{o}_F$-order in A is of this form, for some lattice chain $\mathcal{L}$. We can recover the lattice chain $\mathcal{L}$ from the order $\mathfrak{A}(\mathcal{L})$, up to a shift in the index:— $\mathcal{L}$ is precisely the set of all $\mathfrak{A}$-lattices in V. In other words, $\mathcal{L} \mapsto \mathfrak{A}(\mathcal{L})$ gives a bijection, which reverses inclusions, between the set of lattice chains in V (modulo shift in index) and the set of hereditary orders in A. We recall one special case: the hereditary order $\mathfrak{A}(\mathcal{L})$ is called *principal* if $(L_i : L_{i+1}) = (L_j : L_{j+1})$, for all $i, j \in \mathbb{Z}$. See, especially, **[BF2]** for an account of the special properties of these orders.

Above, the lattices $\mathrm{End}^n_{\mathfrak{o}_F}(\mathcal{L})$ are $(\mathfrak{A}, \mathfrak{A})$-bimodules. In particular, $\mathrm{End}^1_{\mathfrak{o}_F}(\mathcal{L})$ is the Jacobson radical of $\mathfrak{A}$. We denote it

$$\mathfrak{P} = \mathrm{rad}(\mathfrak{A}) = \mathrm{End}^1_{\mathfrak{o}_F}(\mathcal{L}), \quad \textit{where } \mathfrak{A} = \mathfrak{A}(\mathcal{L}).$$

As a fractional ideal of $\mathfrak{A}$, the radical $\mathfrak{P}$ is invertible, and we have

$$\mathfrak{P}^n = \mathrm{End}^n_{\mathfrak{o}_F}(\mathcal{L}), \quad n \in \mathbb{Z}.$$

It will also be useful to remember the property

$$\mathfrak{P}^n L_i = L_{i+n}, \quad i, n \in \mathbb{Z}.$$

The period e of $\mathcal{L}$ is also a function of $\mathfrak{A} = \mathfrak{A}(\mathcal{L})$, so we often write $e = e(\mathcal{L}) = e(\mathfrak{A}) = e(\mathfrak{A}|\mathfrak{o}_F)$. In these terms, it is given by

$$\mathfrak{p}_F \mathfrak{A} = \mathfrak{P}^e.$$

Let $\mathfrak{A}$ be a hereditary order in A as above, with $\mathfrak{P} = \mathrm{rad}(\mathfrak{A})$. We define a sequence of compact open subgroups of G by

$$\begin{aligned} \boldsymbol{U}^0(\mathfrak{A}) &= \boldsymbol{U}(\mathfrak{A}) = \mathfrak{A}^\times, \\ \boldsymbol{U}^n(\mathfrak{A}) &= 1 + \mathfrak{P}^n, \ n \geq 1. \end{aligned}$$

Also, if $\mathfrak{A} = \mathfrak{A}(\mathcal{L})$, we set

$$\mathfrak{K}(\mathfrak{A}) = \mathrm{Aut}(\mathcal{L}) = \{x \in G : xL_i \in \mathcal{L}, \ i \in \mathbb{Z}\}.$$

This is an open compact-mod-centre subgroup of G, and the $\boldsymbol{U}^n(\mathfrak{A})$, for $n \geq 0$, are normal subgroups of it. In particular, $\mathfrak{A}^\times$ is the unique maximal compact subgroup of $\mathfrak{K}(\mathfrak{A})$.

Since the order $\mathfrak{A}$ determines the chain $\mathcal{L}$, we can equivalently define the group $\mathfrak{K}(\mathfrak{A})$ by

$$\mathfrak{K}(\mathfrak{A}) = \{x \in G : x^{-1}\mathfrak{A}x = \mathfrak{A}\}.$$

In fact, $\mathfrak{K}(\mathfrak{A})$ is precisely the G-normaliser of $\boldsymbol{U}^n(\mathfrak{A})$, for any given $n \geq 0$. To see this in the case $n \geq 1$, we note that the normaliser of $\boldsymbol{U}^n(\mathfrak{A})$ is just $\{x \in G : x\mathfrak{P}^n x^{-1} = \mathfrak{P}^n\}$. Any x which normalises $\mathfrak{P}^n$ must also normalise the order $\{a \in A : \mathfrak{P}^n a \subset \mathfrak{P}^n\}$. However, since $\mathfrak{P}^n$ is an invertible fractional ideal of $\mathfrak{A}$, we have $\mathfrak{P}^{-n}\mathfrak{P}^n = \mathfrak{A}$. Thus $\mathfrak{P}^n a \subset \mathfrak{P}^n$ implies $\mathfrak{P}^{-n}\mathfrak{P}^n a \subset \mathfrak{P}^{-n}\mathfrak{P}^n$, or $\mathfrak{A}a \subset \mathfrak{A}$. This says $a \in \mathfrak{A}$, whence any $x \in G$ which normalises $\boldsymbol{U}^n(\mathfrak{A})$ lies in $\mathfrak{K}(\mathfrak{A})$. For the case $n = 0$, see (the proof of) **[BF2]** (1.3.10).

There is also a "valuation" map associated with the hereditary order $\mathfrak{A}$. Define

(1.1.3) $$\nu_{\mathfrak{A}}(x) = \max\{n \in \mathbb{Z} : x \in \mathfrak{P}^n\}, \quad x \in A,$$

with the understanding that $\nu_{\mathfrak{A}}(0) = \infty$. In particular, if $x \in \mathfrak{K}(\mathfrak{A})$, it is easy to see that $\nu_{\mathfrak{A}}(x) = n$, where n is given by

$$x\mathfrak{A} = \mathfrak{A}x = \mathfrak{P}^n.$$

(See also **[BF2]** §1.1.)

Now let ψ be the additive character of A defined above. For a subset S of A, define

(1.1.4) $$S^* = \{x \in A : \psi(xs) = 1, \ s \in S\}.$$

If S is a lattice in A, then so is S^*, and this "star" operation preserves module structures (while reversing their orientation). In particular, we get

$$(\mathfrak{P}^n)^* = \mathfrak{P}^{1-n}, \quad n \in \mathbb{Z},$$

(see **[Bu1]** p. 190).

Let r, n be integers satisfying

$$n > r \geq \left[\tfrac{n}{2}\right] \geq 0,$$

where $[x]$ denotes the greatest integer $\leq x$, for $x \in \mathbb{R}$. We then have a canonical isomorphism

$$\boldsymbol{U}^{r+1}(\mathfrak{A})/\boldsymbol{U}^{n+1}(\mathfrak{A}) \xrightarrow{\approx} \mathfrak{P}^{r+1}/\mathfrak{P}^{n+1}$$

given by $x \mapsto x - 1$. This leads to an isomorphism

(1.1.5) $$(\boldsymbol{U}^{r+1}(\mathfrak{A})/\boldsymbol{U}^{n+1}(\mathfrak{A}))^\wedge \xrightarrow{\approx} \mathfrak{P}^{-n}/\mathfrak{P}^{-r},$$

where "hat" $^\wedge$ denotes Pontrjagin dual. Explicitly, this is given by

(1.1.6) $$\begin{gathered} b + \mathfrak{P}^{-r} \mapsto \psi_{A,b} = \psi_b, \quad b \in \mathfrak{P}^{-n}, \quad \textit{where} \\ \psi_b(1+x) = \psi(bx), \quad x \in \mathfrak{P}^{r+1}. \end{gathered}$$

We return to the $\mathfrak{o}_F$-lattice chain $\mathcal{L}$ as above, with $\mathfrak{A} = \mathfrak{A}(\mathcal{L})$. An *$\mathfrak{o}_F$-basis of $\mathcal{L}$* is an F-basis $\{v_1, v_2, \ldots, v_N\}$ of V such that

(1.1.7) *$\{v_1, v_2, \ldots, v_N\}$ is an $\mathfrak{o}_F$-basis of some $L_i \in \mathcal{L}$ and*

$$L_j = \coprod_{k=1}^{N} \mathfrak{p}_F^{a(j,k)} v_k, \quad j \in \mathbb{Z},$$

for integers $a(j,1) \leq a(j,2) \leq \ldots \leq a(j,N)$.

Observe that this definition specifies the integers $a(j,r)$ quite closely. The containment relation $L_{j+1} \subset L_j$ gives us $a(j+1,r) \geq a(j,r)$ for all r and j. Further, periodicity implies $a(j+e,r) = a(j,r)+1$ for all j, r, where $e = e(\mathcal{L})$. Therefore, given $j \in \mathbb{Z}$, there exists $r(j)$, $1 \leq r(j) \leq N$, such that

(1.1.8)
$$a(j,r) = \begin{cases} a(j,1) \text{ if } 1 \leq r \leq r(j), \\ a(j,1) - 1 \text{ if } r(j)+1 \leq r \leq N. \end{cases}$$

Any $\mathfrak{o}_F$-lattice chain $\mathcal{L}$ has an $\mathfrak{o}_F$-basis $\{v_1, \ldots, v_N\}$. Moreover, if we are given $L \in \mathcal{L}$, we may choose the basis $\{v_1, \ldots, v_N\}$ to span L over $\mathfrak{o}_F$. If we use a basis of $\mathcal{L}$ to identify A with the matrix algebra $\mathbb{M}(N, F)$, then $\mathfrak{A}(\mathcal{L})$ becomes identified with an order of matrices over $\mathfrak{o}_F$ which are upper triangular (in blocks) modulo $\mathfrak{p}_F$:— see **[Bu1]** (1.8).

In the other direction, if we are given a basis $\mathcal{V} = \{v_1, v_2, \ldots, v_N\}$ of V, we can define a lattice chain $\overline{\mathcal{L}} = \overline{\mathcal{L}}(\mathcal{V})$ as the set of all $\mathfrak{o}_F$-lattices of the form

$$\mathfrak{p}_F^m(\mathfrak{o}_F v_1 + \mathfrak{o}_F v_2 + \ldots + \mathfrak{o}_F v_r + \mathfrak{p}_F v_{r+1} + \ldots \mathfrak{p}_F v_N),$$

where $1 \leq r \leq N$ and $m \in \mathbb{Z}$. Then $e(\overline{\mathcal{L}}) = N$, and any lattice chain $\mathcal{L}$ which has $\mathcal{V}$ as a basis is contained in $\overline{\mathcal{L}}$. In terms of the associated hereditary orders, we have $\mathfrak{A}(\overline{\mathcal{L}}) \subset \mathfrak{A}(\mathcal{L})$. Thus a choice of basis $\mathcal{V}$ of V specifies a minimal hereditary $\mathfrak{o}_F$-order in A, namely $\mathfrak{A}(\overline{\mathcal{L}}(\mathcal{V}))$.

(1.1.9) Proposition: *Let $\mathcal{L}^1$, $\mathcal{L}^2$ be $\mathfrak{o}_F$-lattice chains in V, and put $\mathfrak{A}_i = \mathrm{End}^0(\mathcal{L}^i)$, $i = 1, 2$. The following conditions are equivalent:*

(i) the $\mathfrak{o}_F$-lattice chains $\mathcal{L}^i$ have a common $\mathfrak{o}_F$-basis;

(ii) the sets $\mathcal{L}^1 \cap \mathcal{L}^2$, $\mathcal{L}^1 \cup \mathcal{L}^2$ of $\mathfrak{o}_F$-lattices are $\mathfrak{o}_F$-lattice chains;

(iii) there exist hereditary $\mathfrak{o}_F$-orders $\mathfrak{A}_3$, $\mathfrak{A}_4$ in A such that $\mathfrak{A}_3 \supset \mathfrak{A}_i \supset \mathfrak{A}_4$, $i = 1, 2$.

Proof: Suppose first that *(i)* holds. Let $\mathcal{V} = \{v_1, \ldots, v_N\}$ be a common basis of the lattice chains $\mathcal{L}^i$. If L denotes the $\mathfrak{o}_F$-span of the v_i, then $L \in \mathcal{L}^1 \cap \mathcal{L}^2$ and the hereditary (indeed maximal) $\mathfrak{o}_F$-order $\mathrm{End}_{\mathfrak{o}_F}(L)$

contains both $\mathfrak{A}_1$ and $\mathfrak{A}_2$. On the other hand, the $\mathfrak{A}_i$ both contain $\mathfrak{A}(\overline{\mathcal{L}}(\mathcal{V}))$, so *(i)⇒(iii)*.

Assuming *(iii)* holds, write $\mathcal{L}^i$ for the lattice chain defining $\mathfrak{A}_i$, $1 \le i \le 4$. Each $L \in \mathcal{L}^1 \cup \mathcal{L}^2$ is, in particular, an $\mathfrak{A}_4$-lattice, so $\mathcal{L}^1 \cup \mathcal{L}^2 \subset \mathcal{L}^4$. Therefore the set $\mathcal{L}^1 \cup \mathcal{L}^2$ of lattices is linearly ordered by inclusion. It is surely stable under multiplication by $F^\times$, so $\mathcal{L}^1 \cup \mathcal{L}^2$ is indeed a lattice chain. On the other hand, the set $\mathcal{L}^1 \cap \mathcal{L}^2$ is linearly ordered and $F^\times$-stable. However, any $\mathfrak{A}_3$-lattice in V is an $\mathfrak{A}_i$-lattice, for $i = 1, 2$. Thus $\mathcal{L}^1 \cap \mathcal{L}^2 \supset \mathcal{L}^3$. In particular, $\mathcal{L}^1 \cap \mathcal{L}^2$ is nonempty, and so a lattice chain. This proves *(iii)⇒(ii)*.

Finally, assume that *(ii)* holds. Fix $L \in \mathcal{L}^1 \cap \mathcal{L}^2$, and choose a basis of the lattice chain $\mathcal{L}^1 \cup \mathcal{L}^2$ whose $\mathfrak{o}_F$-span is L. This is the common basis required for *(i)*. ■

Remarks: *(i)* In (1.1.9), we can replace *(ii)* by the superficially weaker statement that *$\mathcal{L}^1 \cap \mathcal{L}^2 \neq \emptyset$ and the set $\mathcal{L}^1 \cup \mathcal{L}^2$ is linearly ordered by inclusion*. Indeed, this is what we used in the proof.

(ii) If $\mathcal{L}^1$, $\mathcal{L}^2$ satisfy the conditions of (1.1.9), we get various incidence relations connecting the orders $\mathfrak{A}_i$ and their radicals. For example, set $\mathfrak{P}_i = \text{rad}(\mathfrak{A}_i)$, $1 \le i \le 4$, and let L be an $\mathfrak{A}_4$-lattice in V. Write $\mathcal{L}^1 = \{L_j^1 : j \in \mathbb{Z}\}$. We can then find $j \in \mathbb{Z}$ such that $L_j^1 \supset L \supset L_{j+1}^1$. We have $\mathfrak{P}_1 L \subset \mathfrak{P}_1 L_j^1 = L_{j+1}^1 \subset L$, hence $\mathfrak{P}_1 L \subset L$ for all $\mathfrak{A}_4$-lattices L. It follows that $\mathfrak{P}_1 \subset \mathfrak{A}_4 \subset \mathfrak{A}_2$. By symmetry, we also get $\mathfrak{P}_2 \subset \mathfrak{A}_1$. There are many relations of this sort.

(1.2) Hereditary orders relative to subfields

Now suppose that, in addition to our hereditary order $\mathfrak{A} = \text{End}^0_{\mathfrak{o}_F}(\mathcal{L})$ in A, we have a subfield E/F of A. Thus we may view V as an E-vector space, via the given inclusion $E \to A$.

(1.2.1) Proposition: *Let $\mathfrak{A}$ be a hereditary order in A, with $\mathfrak{A} = \text{End}^0_{\mathfrak{o}_F}(\mathcal{L})$, for some $\mathfrak{o}_F$-lattice chain $\mathcal{L} = \{L_i\}$ in V. Let E/F be some subfield of A. The following conditions are equivalent:*

(i) $E^\times \subset \mathfrak{K}(\mathfrak{A})$ (i.e. E normalises $\mathfrak{A}$);

(ii) each L_i is an $\mathfrak{o}_E$-lattice and there is an integer e' such that $\mathfrak{p}_E L_i = L_{i+e'}$ for all i (i.e. $\mathcal{L}$ is an $\mathfrak{o}_E$-lattice chain in the E-vector space V).

Proof: Suppose that $E^\times \subset \mathfrak{K}(\mathfrak{A})$. We show first that each L_i is an $\mathfrak{o}_E$-module, hence $\mathfrak{o}_E$-lattice. The group $\boldsymbol{U}^0(\mathfrak{A})$ is the unique maximal compact subgroup of $\mathfrak{K}(\mathfrak{A})$, so $\mathfrak{o}_E^\times \subset \boldsymbol{U}^0(\mathfrak{A})$. In particular, each L_i is a module for the ring $R = \mathfrak{o}_F[\mathfrak{o}_E^\times]$ generated over $\mathfrak{o}_F$ by $\mathfrak{o}_E^\times$. It is therefore enough to show that $R = \mathfrak{o}_E$. First, we have $R \supset \boldsymbol{U}^1(\mathfrak{o}_E)$, so $\mathfrak{p}_E \subset R$. Further, R contains a complete set of representatives of $\mathfrak{o}_E$ mod $\mathfrak{p}_E$, namely 0 together with the group of roots of unity in E of order relatively prime to the residual characteristic of E. The assertion follows.

Now take a prime element π_E of E. Then $\pi_E \in \mathfrak{K}(\mathfrak{A})$, and it therefore acts as an automorphism of $\mathcal{L}$. That is to say, there is an integer e' such that $L_{i+e'} = \pi_E L_i = \mathfrak{p}_E L_i$ for all i.

This proves *(i)*$\Rightarrow$*(ii)*. The converse is similar. ■

Now suppose that the equivalent conditions of (1.2.1) hold. Define

$$B = \mathrm{End}_E(V) = \text{ the } A\text{-centraliser of } E, \tag{1.2.2}$$

$$\mathfrak{B} = \mathfrak{A} \cap B, \quad \mathfrak{Q} = \mathfrak{P} \cap B. \tag{1.2.3}$$

(1.2.4) Proposition: *With the notation above, $\mathfrak{B}$ is a hereditary $\mathfrak{o}_E$-order in B, and $\mathfrak{Q}$ is the Jacobson radical of $\mathfrak{B}$. Indeed, we have $\mathfrak{Q}^n = \mathfrak{P}^n \cap B$ for all $n \in \mathbb{Z}$. Further, the integer e' of* (1.2.1)*(ii) is given by*

$$e' = e(\mathfrak{B}) = e(\mathfrak{B}|\mathfrak{o}_E) = \frac{e(\mathfrak{A}|\mathfrak{o}_F)}{e(E|F)}.$$

Proof: Immediately from the definitions we have $\mathfrak{B} = \mathrm{End}^0_{\mathfrak{o}_F}(\mathcal{L}) \cap B$, which is exactly $\mathrm{End}^0_{\mathfrak{o}_E}(\mathcal{L})$. Further, the intersection of B with $\mathrm{End}^n_{\mathfrak{o}_F}(\mathcal{L})$ is $\mathrm{End}^n_{\mathfrak{o}_E}(\mathcal{L})$ for any n, and all the assertions follow. ■

We now forget about our order $\mathfrak{A}$ for the moment, and concentrate on the pair A, E. Let W be an F-subspace of V such that the canonical map

$$E \otimes_F W \to V,$$

induced by the inclusion of W in V and the given action of E on V is an isomorphism (i.e. W is the F-span of an E-basis of V). This induces an isomorphism of F-algebras

$$A \cong \mathrm{End}_F(E) \otimes_F \mathrm{End}_F(W).$$

It is convenient to abbreviate

$$A(E) = \mathrm{End}_F(E).$$

Thus the choice of W induces an embedding of algebras

$$\iota_W : A(E) \to A, \tag{1.2.5}$$

which extends the given embedding $E \to A$, viewing E as canonically embedded in $A(E)$. We view $A(E)$ as a (E, E)-bimodule, and this gives us the trivial identity

$$A(E) = A(E) \otimes_E E.$$

On the other hand, the isomorphism $E \otimes_F W \cong V$ of left E-spaces induces an algebra isomorphism $E \otimes_F \mathrm{End}_F(W) \cong B$. In all, we get an *isomorphism*

$$A(E) \otimes_E B \cong A \tag{1.2.6}$$

of $(A(E), B)$*-bimodules.* Explicitly, this is given by the canonical inclusion of B in A, the embedding of ι_W of $A(E)$ in A determined by W, and multiplication in A. We refer to the isomorphism (1.2.6) as *the* (W, E)*-decomposition of* A.

The algebra $A(E)$ contains the hereditary order

$$\mathfrak{A}(E) = \mathrm{End}^0_{\mathfrak{o}_F}(\{\mathfrak{p}_E^i : i \in \mathbb{Z}\}). \tag{1.2.7}$$

This is the unique hereditary $\mathfrak{o}_F$-order in $A(E)$ which is normalised by E. It is, of course, a *principal* order.

(1.2.8) Proposition: *Let* $\mathfrak{A} = \mathrm{End}^0_{\mathfrak{o}_F}(\mathcal{L})$ *be a hereditary order in* A*, and suppose it is normalised by the subfield* E/F *of* A*. Define* $\mathfrak{B}$*,* $\mathfrak{Q}$ *by* (1.2.3) *and* $\mathfrak{A}(E)$ *by* (1.2.7). *Let* W *be the* F*-span of an* $\mathfrak{o}_E$*-basis of the* $\mathfrak{o}_E$*-lattice chain* $\mathcal{L}$*. Then the* (W, E)*-decomposition* (1.2.6) *of* A *restricts to an isomorphism*

$$\mathfrak{A} \cong \mathfrak{A}(E) \otimes_{\mathfrak{o}_E} \mathfrak{B} \tag{1.2.9}$$

of $(\mathfrak{A}(E), \mathfrak{B})$*-bimodules.*

Proof: Let $\{w_1, w_2, \dots, w_m\}$ be some $\mathfrak{o}_E$-basis of $\mathcal{L} = \{L_i\}$, numbered so that the $\mathfrak{o}_E$-span of the w_j is L_0. Then each lattice L_i is a sum of groups of the form $\mathfrak{p}_E^j.w_k$, for various pairs (j, k). These groups are all stable under the action of $\iota_W(\mathfrak{A}(E))$, and surely $\mathfrak{B} \subset \mathfrak{A}$, so (1.2.6) maps $\mathfrak{A}(E) \otimes_{\mathfrak{o}_E} \mathfrak{B}$ into $\mathfrak{A}$. It clearly preserves the required module structures. We have therefore only to show that the image of $\mathfrak{A}(E) \otimes_{\mathfrak{o}_E} \mathfrak{B}$ is the whole of $\mathfrak{A}$.

Suppose first that $\mathfrak{B}$ is a *maximal* $\mathfrak{o}_E$-order in B. Thus $\mathfrak{B} = \mathrm{End}_{\mathfrak{o}_E}(L_0)$. Let M_0 be the $\mathfrak{o}_F$-span of the $\{w_j\}$. Then $\mathcal{L} = \{\mathfrak{p}_E^i.M_0 : i \in \mathbb{Z}\}$, and

$$\mathfrak{A} \cong \mathfrak{A}(E) \otimes_{\mathfrak{o}_F} \mathrm{End}_{\mathfrak{o}_F}(M_0)$$

as $\mathfrak{o}_F$*-algebras.* Further, $\mathfrak{B} = \mathfrak{o}_E \otimes_{\mathfrak{o}_F} \mathrm{End}_{\mathfrak{o}_F}(M_0)$, and the assertion follows in this case.

In general, let $\mathfrak{B}_0$ be the maximal $\mathfrak{o}_E$-order $\mathrm{End}_{\mathfrak{o}_E}(L_0)$, and let $\mathfrak{A}_0$ be the hereditary $\mathfrak{o}_F$-order in A defined by the same lattice chain $\{\mathfrak{p}_E^i L_0\}$ as $\mathfrak{B}_0$. Note that our space W contains an $\mathfrak{o}_E$-basis of L_0, so the first case applies to show that $\mathfrak{A}(E)\otimes_{\mathfrak{o}_E}\mathfrak{B}_0$ maps onto $\mathfrak{A}_0$. We have $\mathfrak{A}_0\supset\mathfrak{A}$, and it is enough to check that

$$(\mathfrak{A}(E)\otimes_{\mathfrak{o}_E}\mathfrak{B}_0:\mathfrak{A}(E)\otimes_{\mathfrak{o}_E}\mathfrak{B})=(\mathfrak{A}_0:\mathfrak{A}).$$

To do this, first observe that $\mathfrak{A}(E)$ is a free right $\mathfrak{o}_E$-module of rank $[E:F]$, so $(\mathfrak{A}(E)\otimes_{\mathfrak{o}_E}\mathfrak{B}_0:\mathfrak{A}(E)\otimes_{\mathfrak{o}_E}\mathfrak{B})=(\mathfrak{B}_0:\mathfrak{B})^{[E:F]}$. To compute $(\mathfrak{B}_0:\mathfrak{B})$, we put $m_i=\dim_{\mathsf{k}_E}(L_i/L_{i+1})$, for $0\le i\le e'-1$, where we write $e'=e(\mathfrak{B}|\mathfrak{o}_E)$. Thus, in particular, $\sum_{i=0}^{e'-1}m_i=m=\dim_E(V)$. The block matrix description **[Bu]** (1.8) (see also **(2.5)** below) of hereditary orders then gives

$$(\mathfrak{B}_0:\mathfrak{B})=q_E^{(m^2-m')/2},$$

where $m'=\sum_i m_i^2$ and $q_E=\#\mathsf{k}_E$. Likewise, if $\mathfrak{M}$ is any maximal $\mathfrak{o}_F$-order in A, we have

$$(\mathfrak{M}:\mathfrak{A})=q_F^{(N^2-N')/2},$$

where we put $e=e(\mathfrak{A}|\mathfrak{o}_F)$, $n_i=\dim_{\mathsf{k}_F}(L_i/L_{i+1})$, and $N'=\sum_{i=0}^{e-1}n_i^2$. Of course, we have $N=m[E:F]$, $e=e'e(E|F)$, $q_E=q_F^{f(E|F)}$. Further, the sequence of integers $(n_0,n_1,\dots,n_{e-1})$ is $(f(E|F)m_0,\dots,f(E|F)m_{e'-1})$ repeated $e(E|F)$ times. We therefore get

$$(\mathfrak{M}:\mathfrak{A})=q_F^{(N^2-e(E|F)f(E|F)^2m')/2}.$$

Next, let $\mathcal{L}^0=\{L_i^0:i\in\mathbb{Z}\}$ be the $\mathfrak{o}_F$-lattice chain defining $\mathfrak{A}_0$. We have $e(\mathfrak{A}_0|\mathfrak{o}_F)=e(E|F)$, and $\dim_{\mathsf{k}_F}(L_i^0/L_{i+1}^0)=N/e(E|F)=mf(E|F)$ for all i. Thus

$$(\mathfrak{M}:\mathfrak{A}_0)=q_F^{(N^2-m^2f(E|F)^2e(E|F))/2},$$

which gives us

$$(\mathfrak{A}_0:\mathfrak{A})=q_F^{f(E|F)^2e(E|F)(m^2-m')/2}=(q_E^{(m^2-m')/2})^{[E:F]}$$

as required. ∎

Of course, we extend our terminology and refer to (1.2.9) as the *(W,E)-decomposition of $\mathfrak{A}$.*

(1.2.10) Corollary: *The isomorphism* (1.2.9) *identifies*

$$\mathfrak{A}(E)\otimes_{\mathfrak{o}_E}\mathfrak{Q}^n\cong\mathfrak{P}^n,\quad n\in\mathbb{Z}.$$

Proof: We have $\mathfrak{A}(E)\otimes_{\mathfrak{o}_E}\mathfrak{Q}^n=\mathfrak{A}(E)\otimes_{\mathfrak{o}_E}\mathfrak{B}.\mathfrak{Q}^n$, which has image $\mathfrak{A}.\mathfrak{Q}^n\subset\mathfrak{P}^n$. However, $\mathfrak{A}.\mathfrak{Q}^n\subset\mathfrak{P}^n=\mathfrak{P}^n\mathfrak{B}=\mathfrak{P}^n\mathfrak{Q}^{-n}\mathfrak{Q}^n\subset\mathfrak{A}.\mathfrak{Q}^n$, so $\mathfrak{A}\mathfrak{Q}^n=\mathfrak{P}^n$, and the corollary follows. ∎

It is worth pointing out another consequence of (the proof of) (1.2.8).

(1.2.11) Corollary: *In the situation of* (1.2.8), *let* $x \in \mathfrak{K}(\mathfrak{A}(E))$. *Then* $\iota_W(x) \in \mathfrak{K}(\mathfrak{A})$.

Proof: Since $x \in \mathfrak{K}(\mathfrak{A}(E))$, there exists $t \in \mathbb{Z}$ such that $x\mathfrak{p}_E^i = \mathfrak{p}_E^{i+t}$, $i \in \mathbb{Z}$. On the other hand, if $L \in \mathcal{L}$, then

$$L = \sum_{i=1}^{m} \mathfrak{p}_E^{a(i,L)}.w_i,$$

for various integers $a(i, L)$. It follows that

$$\iota_W(x).L = \sum_{i=1}^{m} \mathfrak{p}_E^{a(i,L)+t}.w_i = \mathfrak{p}_E^t.L \in \mathcal{L}.$$

Thus $\iota_W(x) \in \mathfrak{K}(\mathfrak{A})$, as required. ■

(1.3) Tame corestriction

Again let E/F be a subfield of A, with centraliser B. We shall need an auxiliary subspace C of A defined by:

(1.3.1) $C =$ *the orthogonal complement of* B *under the symmetric bilinear form* $A \times A \to F$ *given by* $(x, y) \mapsto \mathrm{tr}_{A/F}(xy)$.

In the present circumstances (that is, where F is a non-Archimedean local field), the extension E/F has a primitive element, $E = F[\delta]$, for some δ. The bilinear form defined by the matrix trace is nondegenerate so, comparing dimensions, we have:

$$C = \{\delta x - x\delta : x \in A\}.$$

We now consider A and B as (B, B)-bimodules.

(1.3.2) Proposition: *(i) Let* f *be a nonzero* (B, B)*-homomorphism* $A \to B$. *Then* f *is surjective, with kernel* C.

(ii) The group $\mathrm{Hom}_{(B,B)}(A, B)$ *is an* E*-vector space of dimension at most* 1.

(iii) $f(E) \subset E$, *for any* $f \in \mathrm{Hom}_{(B,B)}(A, B)$.

Remark: We shall see in (1.3.4) below that the space $\mathrm{Hom}_{(B,B)}(A, B)$ is nonzero.

Proof: Since B is a central simple E-algebra, it is a simple (B, B)-bimodule, and $\mathrm{End}_{(B,B)}(B) = E$. Thus any nonzero bimodule homomorphism $A \to B$ is surjective. The kernel certainly contains C, hence equals C by comparing dimensions. Therefore any two such maps differ by a (B,B)-automorphism of B, i.e. an element of $E^\times$. To prove *(iii)*, let $x \in E$ and $b \in B$. Then

$$bf(x) = f(bx) = f(xb) = f(x)b,$$

so $f(x)$ lies in the centre E of B. ∎

(1.3.3) Definition: *A tame corestriction on A (relative to E/F) is a (B,B)-bimodule homomorphism $s : A \to B$ such that $s(\mathfrak{A}) = \mathfrak{A} \cap B$ for any hereditary $\mathfrak{o}_F$-order $\mathfrak{A}$ in A which is normalised by $E^\times$.*

A tame corestriction, if such exists, is therefore uniquely determined up to multiplication by a unit $u \in \mathfrak{o}_E^\times$.

(1.3.4) Proposition: *(i) Let ψ_E, ψ_F be continuous additive characters of E, F with conductors $\mathfrak{p}_E$, $\mathfrak{p}_F$ respectively. Define additive characters ψ_B, ψ_A of B, A by $\psi_B = \psi_E \circ \mathrm{tr}_{B/E}$, $\psi_A = \psi_F \circ \mathrm{tr}_{A/F}$ respectively. There exists a unique map $s : A \to B$ such that*

$$\psi_A(ab) = \psi_B(s(a)b), \quad a \in A,\ b \in B. \tag{1.3.5}$$

This map s is a tame corestriction on A relative to E/F.

(ii) Let $\mathfrak{A}$ be any hereditary order in A normalised by E, and put $\mathfrak{P} = \mathrm{rad}(\mathfrak{A})$. Then, for any tame corestriction s on A relative to E/F and any $n \in \mathbb{Z}$, we have $s(\mathfrak{P}^n) = \mathfrak{P}^n \cap B$.

Proof: The pairing $(b_1, b_2) \mapsto \psi_B(b_1b_2)$, $b_i \in B$, identifies B with its Pontrjagin dual. Further, for $a \in A$, the map $b \mapsto \psi_A(ab)$ is a continuous character of the additive group of B. Therefore there exists a unique element $s(a) \in B$ such that $\psi_A(ab) = \psi_B(s(a)b)$ for all $b \in B$. This gives the existence and uniqueness of the map s satisfying (1.3.5). It is clearly a (B,B)-bimodule homomorphism from A to B.

Now let $\mathfrak{A}$ be a hereditary $\mathfrak{o}_F$-order in A which is normalised by E. Set $\mathfrak{P} = \mathrm{rad}(\mathfrak{A})$, $\mathfrak{B} = \mathfrak{A} \cap B$ and $\mathfrak{Q} = \mathrm{rad}(\mathfrak{B}) = \mathfrak{P} \cap \mathfrak{B}$. To prove all the remaining assertions, we need only show that $s(\mathfrak{P}^n) = \mathfrak{Q}^n$, for all $n \in \mathbb{Z}$. By analogy with the operation "star" defined by (1.1.4), we put

$$S^\dagger = \{y \in B : \psi_B(ys) = 1,\ s \in S\} \tag{1.3.6}$$

for any subset S of B. In particular, we get $(\mathfrak{Q}^n)^\dagger = \mathfrak{Q}^{1-n}$. Take $a \in \mathfrak{P}^n$. Then $\psi_A(ay) = 1$ for all $y \in \mathfrak{P}^{1-n}$, especially all $y \in \mathfrak{P}^{1-n} \cap B = \mathfrak{Q}^{1-n}$. Thus $s(a) \in (\mathfrak{Q}^{1-n})^\dagger$, i.e. $s(\mathfrak{P}^n) \subset \mathfrak{Q}^n$. To prove the opposite inclusion, take $b \in \mathfrak{Q}^n$. The map $y \mapsto \psi_B(by)$ is then a character of B which is null on $\mathfrak{Q}^{1-n} = \mathfrak{P}^{1-n} \cap B$. It can therefore be extended to a character of A which is null on $\mathfrak{P}^{1-n}$. This extension is of the form $x \mapsto \psi_A(ax)$ for some $a \in \mathfrak{P}^n$, so that $\psi_A(ax) = \psi_B(bx)$ if $x \in B$. This means that $s(a) = b$, as required. ∎

The characters ψ_E, ψ_F of the proposition are uniquely determined up to multiplication by units of $\mathfrak{o}_E$, $\mathfrak{o}_F$ respectively. Therefore:

(1.3.7) *Let s' be some tame corestriction on A relative to E/F. There exist characters ψ_E, ψ_F of E, F defining characters ψ_B, ψ_A as in* (1.3.4), *so that*

$$\psi_A(ab) = \psi_B(s'(a)b), \quad a \in A,\ b \in B.$$

(1.3.8) Remarks: *(i)* The notion of tame corestriction generalises the maps S_α introduced in **[KM1]**. The applications we have in mind also generalise those of **[KM1]**.

(ii) If E/F is separable, the orthogonal projection $\mathrm{pr} : A \to B$ (relative to the pairing $(x, y) \mapsto \mathrm{tr}_{A/F}(xy)$ on A) is a nonzero (B, B)-homomorphism. (Indeed, this property characterises separable extensions E/F.) For $b, b' \in B$, we have $\mathrm{pr}(b) = b$ and

$$\psi_A(bb') = \psi_F(\mathrm{tr}_{A/F}(bb')) = \psi_F(\mathrm{tr}_{E/F}(\mathrm{tr}_{B/E}(bb'))).$$

The character $\psi_F \circ \mathrm{tr}_{E/F}$ has conductor $\mathfrak{p}_F \mathfrak{D}_{E/F}^{-1}$, where $\mathfrak{D}_{E/F}$ is the different of E/F. Thus, when E/F is separable, the map pr is a tame corestriction if and only if $\mathfrak{p}_F \mathfrak{D}_{E/F}^{-1} = \mathfrak{p}_E$, i.e. if and only if the extension E/F is tamely ramified.

(iii) In general, the tame corestriction s induces a surjective map $\mathfrak{P}^r/\mathfrak{P}^{r+1} \to \mathfrak{Q}^r/\mathfrak{Q}^{r+1}$, for any $r \in \mathbb{Z}$, and hence a map $\mathfrak{Q}^r/\mathfrak{Q}^{r+1} \to \mathfrak{Q}^r/\mathfrak{Q}^{r+1}$. We have just seen that this last map is an isomorphism when E/F is tamely ramified. By contrast, when E/F is not tamely ramified, we get $s(\mathfrak{Q}^r) \subset \mathfrak{Q}^{r+1}$. For, take $a \in \mathfrak{Q}^r$ and $b \in B$. The same computation as before yields $\psi_A(ab) = \psi_F \circ \mathrm{tr}_{E/F}(\mathrm{tr}_{B/E}(ab))$. The character $\psi_F \circ \mathrm{tr}_{E/F}$ of E has conductor $\mathfrak{p}_F \mathfrak{D}_{E/F}^{-1}$ again, and $\mathfrak{p}_F \mathfrak{D}_{E/F}^{-1} \supset \mathfrak{o}_E$ when E/F is wildly ramified. Thus $\psi_A(ab) = \psi_B(s(a)b) = 1$ when $b \in \mathfrak{Q}^{-r}$, and it follows that $s(a) \in (\mathfrak{Q}^{-r})^\dagger = \mathfrak{Q}^{r+1}$.

These tame corestriction maps also respect the notion of (W, E)-decomposition introduced above.

(1.3.9) Proposition: *In the situation of* (1.2.8), *let s_0 be a tame corestriction on $A(E)$ relative to E/F. Then, in terms of the (W, E)-decomposition* (1.2.6), *the map $s = s_0 \otimes 1_B$ is a tame corestriction on A relative to E/F.*

Proof: Fix the character ψ_F, and set $\psi_{A(E)} = \psi_F \circ \mathrm{tr}_{A(E)/F}$. The centraliser of E in $A(E)$ is E itself, and there exists a character ψ_E of E which satisfies

$$\psi_{A(E)}(ab) = \psi_E(s_0(a)b), \quad b \in E,\ a \in A(E),$$

by (1.3.7). For $x \in A(E)$, $y \in B$, we have

$$\mathrm{tr}_{A/F}(x \otimes y) = \mathrm{tr}_{A(E)/F}(x.\mathrm{tr}_{B/E}(y)).$$

Setting $a = x \otimes y$ in this formula, we get

$$\begin{aligned}\psi_A(a.b) = \psi_F(\mathrm{tr}_{A/F}(x \otimes yb)) &= \psi_F(\mathrm{tr}_{A(E)/F}(x.\mathrm{tr}_{B/E}(yb)) \\ = \psi_{A(E)}(x.\mathrm{tr}_{B/E}(yb)) &= \psi_E(s_0(x).\mathrm{tr}_{B/E}(yb)) \\ = \psi_E(\mathrm{tr}_{B/E}(s_0(x)yb)) &= \psi_B(s(x \otimes y)b), \quad b \in B.\end{aligned}$$

The assertion now follows from (1.3.4). ■

We now fix a tame corestriction s on A relative to E/F (but everything we do will be independent of this choice). Let L be some (say left) $\mathfrak{o}_E$-lattice in A.

(1.3.10) Definition: *The lattice L is E-*exact *if* $s(L) = L \cap B$.

This definition is certainly independent of the choice of s.

(1.3.11) Proposition: *Let L be a left $\mathfrak{o}_E$-lattice in A. Then L is E-exact if and only if $(L \cap B)^\dagger = (L^* \cap B)$ (notation* (1.3.6), (1.1.4)*).*

Proof: Take s, ψ_A, ψ_B as in (1.3.5). For $x \in B$, we have $x \in s(L)^\dagger$ if and only if $\psi_B(s(L).x) = \{1\}$. However, for $y \in L$, $\psi_B(s(y).x) = \psi_A(yx)$ which is null for $y \in L$ if and only if $x \in L^* \cap B$. This shows that

$$s(L)^\dagger = L^* \cap B$$

for any $\mathfrak{o}_E$-lattice L in A. In the case to hand, we therefore have $s(L) = L \cap B$ if and only if $(L \cap B)^\dagger = s(L)^\dagger = L^* \cap B$, as required. ■

For example, if $\mathfrak{A}$ is a hereditary order in A normalised by E, and $\mathfrak{P} = \mathrm{rad}(\mathfrak{A})$, we have seen in (1.3.4)*(ii)* that the lattices $\mathfrak{P}^n$ are all E-exact. We now explore some sufficient conditions for E-exactness. In practice, the importance of this technique is that it allows us to compute $L \cap B$ for many lattices L which arise naturally in our applications. It replaces the cumbersome "block matrix" arguments which have hitherto been necessary in this area.

(1.3.12) Proposition: *Fix a (W, E)-decomposition of A, and view $A(E)$ (hence $\mathfrak{A}(E)$) as a subring of A via the map ι_W. Let L be an $(\mathfrak{A}(E), \mathfrak{o}_E)$-bilattice in A which is invariant under conjugation by $E^\times$. Then L is E-exact.*

Remark: The class of $(\mathfrak{A}(E), \mathfrak{o}_E)$-bilattices in A invariant under conjugation by $E^\times$ is closed under the operations of sum and intersection. Thus, if L_1, L_2 satisfy the conditions of (1.3.12), then the lattices L_1,

L_2, $L_1 + L_2$, $L_1 \cap L_2$ are all E-exact. We can use this fact to produce larger classes of E-exact lattices:— see (1.3.16) below.

Proof of (1.3.12): This is based on the following lemma.

(1.3.13) Lemma: *Let U be a finite-dimensional E-vector space, viewed as an (E, E)-bimodule by $u\alpha = \alpha u$, $\alpha \in E$, $u \in U$. View $A(E) \otimes_E U$ as an (E, E)-bimodule via the natural left action of E on $A(E)$ and the right action of E on U. The maps*

$$\upsilon : M \mapsto \mathfrak{A}(E) \otimes_{\mathfrak{o}_E} M, \qquad \delta : L \mapsto L \cap U = L \cap (1 \otimes U)$$

are mutually inverse bijections between the following sets:

(a) $\mathfrak{o}_E$-lattices M in U;

(b) $\mathrm{Ad}(E^\times)$-invariant $(\mathfrak{A}(E), \mathfrak{o}_E)$-bilattices L in $A(E) \otimes_E U$.

Before proving this lemma, we deduce (1.3.12). Write $L = \mathfrak{A}(E) \otimes_{\mathfrak{o}_E} M$, for some $\mathfrak{o}_E$-lattice M in B. By (1.3.13), $1 \otimes M = \mathfrak{o}_E \otimes_{\mathfrak{o}_E} M = L \cap B$. However, by (1.3.9),

$$s(\mathfrak{A}(E) \otimes_{\mathfrak{o}_E} M) = s_0(\mathfrak{A}(E)) \otimes_{\mathfrak{o}_E} M = \mathfrak{o}_E \otimes_{\mathfrak{o}_E} M,$$

as required. ∎

Now let us prove (1.3.13). Let M be an $\mathfrak{o}_E$-lattice in U. We assert that $(\mathfrak{A}(E) \otimes_{\mathfrak{o}_E} M) \cap U = M$. To see this, we choose a right $\mathfrak{o}_E$-basis $\{x_1 (= 1), x_2, \dots, x_r\}$ of $\mathfrak{A}(E)$ and an $\mathfrak{o}_E$-basis $\{m_1, \dots, m_s\}$ of M. Then $\{x_i \otimes m_j\}$ is a right $\mathfrak{o}_E$-basis of $\mathfrak{A}(E) \otimes M$, and also a right E-basis of $A(E) \otimes_E U$. The subspace spanned by the $1 \otimes m_j$ is just U, and our assertion is immediate.

Therefore $\delta \circ \upsilon$ is the identity map, whence υ is injective and δ is surjective. We need only show that δ is injective. Suppose we have a pair L_1, L_2 of $\mathrm{Ad}(E^\times)$-invariant $(\mathfrak{A}(E), \mathfrak{o}_E)$-bilattices in $A(E) \otimes U$ with $L_1 \neq L_2$ and $\delta(L_1) = \delta(L_2)$. Then $L_1 \cap L_2$ is also an $\mathrm{Ad}(E^\times)$-invariant $(\mathfrak{A}(E), \mathfrak{o}_E)$-bilattice, and $\delta(L_1 \cap L_2) = \delta(L_1) = \delta(L_2)$. Either $L_2 \supsetneq L_1$ or else $L_1 \supsetneq L_1 \cap L_2$. Therefore we need only treat the following special case:

(1.3.14) Lemma: *Let $L_1 \supset L_2$ be $\mathrm{Ad}(E^\times)$-invariant $(\mathfrak{A}(E), \mathfrak{o}_E)$-bilattices in $A(E) \otimes_E U$ with $L_1 \cap U = L_2 \cap U$. Then $L_1 = L_2$.*

Proof: We prove this by induction on $\dim_E(U)$. So, suppose first that $\dim_E(U) = 1$. If $\mathfrak{P}(E)$ denotes the radical of $\mathfrak{A}(E)$, we have $\mathfrak{P}(E)^j \cap E = \mathfrak{p}_E^j$, so the case $\dim(U) = 1$ of (1.3.14) now follows from:

(1.3.15) Lemma: *Let L be an $(\mathfrak{A}(E), \mathfrak{o}_E)$-bilattice in $A(E)$ which is invariant under conjugation by $E^\times$. Then $L = \mathfrak{P}(E)^j$, for some $j \in \mathbb{Z}$.*

Proof: To prove this, we note that the relevant properties of L are unchanged by scaling on the right with an element of $E^\times$. Assuming that $L \neq \mathfrak{P}(E)^j$ for any j, we can therefore arrange that L is contained (strictly) in $\mathfrak{A}(E)$, but not contained in $\mathfrak{P}(E)$. Thus L is contained in some maximal left ideal of $\mathfrak{A}(E)$. The lattice $L_1 = L + \mathfrak{P}(E)$ is then contained in the same maximal left ideal, and is an $\mathrm{Ad}(E^\times)$-invariant $(\mathfrak{A}(E), \mathfrak{o}_E)$-bilattice. In other words, we can assume that L lies strictly between $\mathfrak{A}(E)$ and $\mathfrak{P}(E)$. Consider the image $\overline{L}$ of L in the ring $\mathfrak{A}(E)/\mathfrak{P}(E) \cong \mathbb{M}(s, \mathrm{k})^r$, where $s = f(E|F)$ (residue class degree), $r = e(E|F)$ (ramification index) and $\mathrm{k} = \mathrm{k}_F$ is the residue field of F. These matrix factors are permuted transitively under conjugation by a prime element of E. However, $\overline{L}$ is a proper left ideal of $\mathfrak{A}(E)/\mathfrak{P}(E)$, so $\overline{L} = \coprod \mathfrak{a}_i$, where $\mathfrak{a}_i$ is a left ideal of the i-th matrix factor $\mathbb{M}(s, \mathrm{k})$. The $\mathfrak{a}_i$'s must all have the same dimension, so they are all proper left ideals. The residue field k_E of E acts on each matrix factor as a maximal subfield, so $\mathfrak{a}_i$ is a proper left ideal of $\mathbb{M}(s, \mathrm{k})$ which is right invariant under the action of a maximal subfield of $\mathbb{M}(s, \mathrm{k})$. However, the $\mathbb{M}(s, \mathrm{k})$-endomorphism ring of $\mathfrak{a}_i$ is isomorphic to $\mathbb{M}(l, \mathrm{k})$, where l is the length of $\mathfrak{a}_i$ as $\mathbb{M}(s, \mathrm{k})$-module. We therefore have a k-algebra homomorphism $\mathrm{k}_E \to \mathbb{M}(l, \mathrm{k})$. This implies that l is divisible by $[\mathrm{k}_E : \mathrm{k}] = s$, which in turn implies $\mathfrak{a}_i = \mathbb{M}(s, \mathrm{k})$ or $\{0\}$ and $\overline{L} = \mathfrak{A}(E)$ or $\mathfrak{P}(E)$. This contradiction proves (1.3.15). ■

Now we return to the proof of (1.3.14) with the assumption that $\dim(U) \geq 2$. Fix a nonzero $u \in U$, and consider the lattices

$$(L_1 \cap (A(E) \otimes Eu)) \supset (L_2 \cap (A(E) \otimes Eu)).$$

If we have a strict containment here, we get $L_1 \cap Eu \neq L_2 \cap Eu$ by the first case of the induction (i.e. (1.3.15)). This would imply $L_1 \cap U \neq L_2 \cap U$, contrary to hypothesis. Therefore we can assume we have equality here. The assumption $L_1 \neq L_2$ now implies $\widetilde{L}_1 \neq \widetilde{L}_2$, where $\widetilde{L}_i$ denotes the image of L_i in $(A(E) \otimes_E U)/(A(E) \otimes_E Eu) = A(E) \otimes U/Eu$. The result follows by induction. ■

This also completes the proofs of (1.3.13) and (1.3.12).

Now let $\mathfrak{A}$ be some hereditary $\mathfrak{o}_F$-order in A, and let E/F be a subfield of A such that $E^\times \subset \mathfrak{K}(\mathfrak{A})$. As before, put $B = \mathrm{End}_E(V)$, and let $\mathfrak{P}$ denote the radical of $\mathfrak{A}$. Then, for any $n \in \mathbb{Z}$ and any $h \in B^\times$, the lattice $\mathfrak{P}^n h$ is $\mathrm{Ad}(E^\times)$-invariant, it is a right $\mathfrak{o}_E$-lattice, and it is also a left $\mathfrak{A}(E)$-lattice *for any* (W, E)-decomposition of A. Thus the set of lattices generated, using the operations of sum and intersection, by the set of right $B^\times$-translates of powers of $\mathfrak{P}$ consists of E-exact lattices.

By duality (or applying symmetry to (1.3.12)), the same applies using left $B^\times$-translates of powers of $\mathfrak{P}$, and so we get another class of

E-exact lattices. It is not clear, however, that we can mix these two classes, using sum and intersection, and still get E-exact lattices. There is, however, one important case in which we can do this.

(1.3.16) Example: *Let $h_1, h_2 \in B^\times$, and $r, s, t, u \in \mathbb{Z}$. Let $L = \mathfrak{P}^r \cap (h_1\mathfrak{P}^s + \mathfrak{P}^t h_2 + \mathfrak{P}^u)$. Then L is E-exact.*

Proof: Write $\mathfrak{B} = \mathfrak{A} \cap B$. Note that, for $v_1 \in \mathfrak{B}^\times$, the lattice L is E-exact if and only if $v_1 L$ is E-exact. Also, $h_1\mathfrak{P}^r = h_1 v_2 \mathfrak{P}^r$ for any $v_2 \in \mathfrak{B}^\times$. In other words, we can replace h_1 by $v_1 h_1 v_2$ for any $v_1, v_2 \in \mathfrak{B}^\times$. However, the group $\mathfrak{B}^\times$ contains an Iwahori subgroup of $B^\times$ (which is nothing other than the unit group of a minimal hereditary $\mathfrak{o}_E$-order $\mathfrak{B}'$ in B contained in $\mathfrak{B}$), so we may take h_1 to belong to the affine Weyl group of $B^\times$ relative to this Iwahori subgroup. (These concepts are recalled in detail in **(5.4)** below.) We let W denote the F-linear span of some common $\mathfrak{o}_E$-basis $\{w_i\}$ of the lattice chains defining $\mathfrak{B}$ and $\mathfrak{B}'$ (*cf.* (1.1.9)), and take the corresponding (W, E)-decomposition of $\mathfrak{A}$. We show that, in this situation, we have

$$h_1^{-1}\mathfrak{A}(E)h_1 \subset \mathfrak{A}.$$

This will imply that $h_1\mathfrak{P}^s$ is an $\mathrm{Ad}(E^\times)$-invariant $(\mathfrak{A}(E), \mathfrak{o}_E)$-bilattice. The remaining terms $\mathfrak{P}^r$, $\mathfrak{P}^t h_2$, $\mathfrak{P}^u$ certainly have these properties. (1.3.16) will then follow from (1.3.12) and the succeeding remark. To prove this assertion, we note that $h_1(1 \otimes w_i) = \pi_E^k \otimes w_j$, for some preassigned prime element π_E of E and some integers j, k. Thus, if $x \in \mathfrak{A}(E)$ and $n \in \mathbb{Z}$, $h_1^{-1}xh_1$ maps $\mathfrak{p}_E^n \otimes w_i$ to itself. Since each lattice in $\mathcal{L}$ is a sum of such groups, it follows that $h_1^{-1}xh_1$ lies in $\mathfrak{A}$, as required. ∎

Remark: This proof shows that, given $h_1 \in B^\times$, there exists a (W, E)-decomposition of $\mathfrak{A}$ for which $h_1\mathfrak{P}^s$ is an $\mathrm{Ad}(E^\times)$-invariant $(\mathfrak{A}(E), \mathfrak{o}_E)$-bilattice, and hence L (as in (1.3.16) has the same properties.

We need one more technical result in this section, for which we recall the subspace C of A defined by (1.3.1).

(1.3.17) Proposition: *Fix a (W, E)-decomposition of A, and view $A(E)$ as a subring of A via the map ι_W. Let L_1, L_2 be $\mathrm{Ad}(E^\times)$-invariant $(\mathfrak{A}(E), \mathfrak{o}_E)$-bilattices in A. Then*

$$(L_1 + L_2) \cap C = L_1 \cap C + L_2 \cap C.$$

Proof: We start with an elementary fact. Given an abelian group G, together an endomorphism f and subgroups H, K, there is a canonical isomorphism

$$\frac{C \cap (H + K)}{(C \cap H) + (C \cap K)} \xrightarrow{\approx} \frac{f(H) \cap f(K)}{f(H \cap K)},$$

where $C = \mathrm{Ker}(f)$. Explicitly, this isomorphism is given by the map $x + y \mapsto f(x)$, where $x \in H$, $y \in K$, and $x + y \in C \cap (H + K)$.

We apply this to the endomorphism of $A = \mathrm{End}_F(V)$ given by a tame corestriction s relative to E/F. The assertion of the Proposition is thus equivalent to

$$s(L_1 \cap L_2) = s(L_1) \cap s(L_2).$$

However, by (1.3.12), the lattices L_1, L_2, $L_1 \cap L_2$ are all E-exact, so

$$s(L_1 \cap L_2) = L_1 \cap L_2 \cap B = (L_1 \cap B) \cap (L_2 \cap B) = s(L_1) \cap s(L_2),$$

as required. ■

(1.4) Adjoint maps

We continue in the same situation:— $A = \mathrm{End}_F(V)$, E/F is a subfield of A normalising a hereditary order $\mathfrak{A} = \mathrm{End}^0_{\mathfrak{o}_F}(\mathcal{L})$, for some $\mathfrak{o}_F$-lattice chain $\mathcal{L}$ in V, with $\mathfrak{P} = \mathrm{rad}(\mathfrak{A})$. We again write B = the A-centraliser of E, $\mathfrak{B} = \mathfrak{A} \cap B$, $\mathfrak{Q} = \mathrm{rad}(\mathfrak{B}) = \mathfrak{P} \cap B$. We also fix a tame corestriction s on A relative to E/F.

We work relative to a fixed element $\beta \in E$ satisfying

$$E = F[\beta],$$

and define a map $a_\beta : A \longrightarrow A$ by

(1.4.1) $$a_\beta(x) = \beta x - x\beta, \quad x \in A.$$

Then a_β is a (B, B)-bimodule homomorphism $A \longrightarrow A$ with kernel B. Its image is certainly contained in C (see (1.3.1)) and, comparing dimensions, we see that $a_\beta(A) = C$. We therefore get an infinite exact sequence

$$\dots \xrightarrow{s} A \xrightarrow{a_\beta} A \xrightarrow{s} A \xrightarrow{a_\beta} \dots$$

When we wish to specify the dependence on the algebra A, we write $a_\beta = a_{\beta,A}$. This dependence on A is quite straightforward.

(1.4.2) Proposition: *Let $A = A(E) \otimes_E B$ be a (W, E)-decomposition of A. Then $a_{\beta,A} = a_{\beta,A(E)} \otimes 1_B$.*

Proof: For $x \in A(E)$, $y \in B$, we have

$$(\beta \otimes 1)(x \otimes y) - (x \otimes y)(\beta \otimes 1) = \beta x \otimes y - x\beta \otimes y,$$

since, in particular, the identification $A(E) \otimes B = A$ preserves (E, E)-bimodule structures. The result is now obvious. ■

The main object of this section is the investigation of the above infinite exact sequence connecting a_β and s when we restrict to the hereditary order $\mathfrak{A}$.

(1.4.3) Definition: *Let $\beta \in E$ as above, and $k \in \mathbb{Z}$. Define*

$$\mathfrak{N}_k = \mathfrak{N}_k(\beta, \mathfrak{A}) = \{x \in \mathfrak{A} : a_\beta(x) \in \mathfrak{P}^k\}.$$

Then $\mathfrak{A} \supset \mathfrak{N}_k \supset \mathfrak{P}^{-\nu_\mathfrak{A}(\beta)} \cap \mathfrak{A}$, so $\mathfrak{N}_k$ is a lattice in A. Further, it is a $(\mathfrak{B}, \mathfrak{B})$-bilattice, and $\mathfrak{N}_k \cap B = \mathfrak{B}$.

(1.4.4) Lemma: *With the notation above, we have*

(i) $\bigcap_{k \in \mathbb{Z}} \mathfrak{N}_k(\beta, \mathfrak{A}) = \mathfrak{B}$;

(ii) $\mathfrak{N}_k(\beta, \mathfrak{A}) \subset \mathfrak{B} + \mathfrak{P}$, for all sufficiently large k.

Proof: *(i)* If $x \in \mathfrak{N}_k$ for all k, then $a_\beta(x) \in \mathfrak{P}^k$ for all $k \in \mathbb{Z}$. This implies $a_\beta(x) = 0$, whence $x \in \mathfrak{B}$.

(ii) Suppose the contrary. For each $k \in \mathbb{Z}$, we can then find $x_k \in \mathfrak{N}_k$ with $x_k \notin \mathfrak{B} + \mathfrak{P}$. The sequence $\{x_k\}$ is contained in the compact set $\mathfrak{A} \setminus (\mathfrak{B} + \mathfrak{P})$, so it has a subsequence converging in $\mathfrak{A} \setminus (\mathfrak{B} + \mathfrak{P})$. However, the limit of this subsequence lies in the intersection of the $\mathfrak{N}_k$, thereby contradicting *(i)*. ■

In the other direction, we have $\mathfrak{N}_k = \mathfrak{A}$ for k sufficiently small. Indeed, suppose that $\nu_\mathfrak{A}(\beta) = -n$ (in the notation of (1.1.3)). Then $a_\beta(\mathfrak{A}) \subset \mathfrak{P}^{-n}$, whence $\mathfrak{N}_k = \mathfrak{A}$ for $k \leq -n$.

(1.4.5) Definition: *Suppose that $E \neq F$. Define*

$$k_0 = k_0(\beta, \mathfrak{A}) = \max\{k \in \mathbb{Z} : \mathfrak{N}_k \not\subset \mathfrak{B} + \mathfrak{P}\}.$$

Of course, if $E = F$, we have $\mathfrak{A} = \mathfrak{B} + \mathfrak{P} = \mathfrak{N}_k$ for all k. It is best to treat this case as exceptional and set

$$k_0(\beta, \mathfrak{A}) = -\infty \quad \textit{when } F[\beta] = F. \tag{1.4.6}$$

(1.4.7) Proposition: *With the notation above, let k, $r \in \mathbb{Z}$ and suppose that $k \geq k_0(\beta, \mathfrak{A})$, $r \geq 1$. Then the following sequences are exact:*

$$0 \longrightarrow \mathfrak{N}_k/\mathfrak{N}_{k+r} \xrightarrow{a_\beta} \mathfrak{P}^k/\mathfrak{P}^{k+r} \xrightarrow{s} \mathfrak{Q}^k/\mathfrak{Q}^{k+r} \longrightarrow 0,$$
$$0 \longrightarrow \mathfrak{N}_k/\mathfrak{B} \xrightarrow{a_\beta} \mathfrak{P}^k \xrightarrow{s} \mathfrak{Q}^k \longrightarrow 0.$$

Proof: In the first sequence, exactness at the first place follows from the definition of $\mathfrak{N}_k$, while exactness at the last is implied by (1.3.4)*(ii)*. Likewise, in the second sequence, exactness at the outside places is immediate. We start by fixing $k \geq k_0$, and proving exactness of the first sequence for infinitely many values of r. Taking the limit as $r \to \infty$, we get the exactness of the second sequence. Exactness of the first, for *all* values of $r \geq 1$ will then follow.

(1.4.8) Lemma: *(i) We have* $\mathfrak{Q}\mathfrak{N}_k = \mathfrak{N}_k\mathfrak{Q} = \mathfrak{N}_{k+1} \cap \mathfrak{P}$, *and* $\mathfrak{N}_k = \mathfrak{Q}^{-1}\mathfrak{N}_{k+1} \cap \mathfrak{A} = \mathfrak{N}_{k+1}\mathfrak{Q}^{-1} \cap \mathfrak{A}$, *for all* $k \in \mathbb{Z}$.

(ii) For $k \in \mathbb{Z}$, *the following are equivalent:*

(a) $k \geq k_0$;

(b) $\mathfrak{N}_{k+1} = \mathfrak{B} + \mathfrak{Q}\mathfrak{N}_k$.

Proof: In *(i)*, we have $\mathfrak{Q}\mathfrak{N}_k \subset \mathfrak{Q}\mathfrak{A} = \mathfrak{P}$, while $a_\beta(\mathfrak{Q}\mathfrak{N}_k) = \mathfrak{Q}a_\beta(\mathfrak{N}_k) \subset \mathfrak{Q}\mathfrak{P}^k = \mathfrak{P}^{k+1}$. Thus $\mathfrak{Q}\mathfrak{N}_k \subset \mathfrak{N}_{k+1} \cap \mathfrak{P}$. In the other direction, take $x \in \mathfrak{N}_{k+1} \cap \mathfrak{P}$. Then $\mathfrak{Q}^{-1}x \subset \mathfrak{Q}^{-1}\mathfrak{P} = \mathfrak{A}$, while $a_\beta(\mathfrak{Q}^{-1}x) \subset \mathfrak{Q}^{-1}\mathfrak{P}^{k+1} = \mathfrak{P}^k$. Thus $\mathfrak{Q}^{-1}x \subset \mathfrak{N}_k$, and so $x \in \mathfrak{B}x = \mathfrak{Q}\mathfrak{Q}^{-1}x \subset \mathfrak{Q}\mathfrak{N}_k$. Symmetrically, we get $\mathfrak{N}_k\mathfrak{Q} = \mathfrak{N}_{k+1} \cap \mathfrak{P}$.

Similarly, we have $a_\beta(\mathfrak{Q}^{-1}\mathfrak{N}_{k+1}) \subset \mathfrak{P}^k$, so $\mathfrak{Q}^{-1}\mathfrak{N}_{k+1} \cap \mathfrak{A} \subset \mathfrak{N}_k$. On the other hand, if $x \in \mathfrak{N}_k$, we have just seen that $\mathfrak{Q}x \subset \mathfrak{N}_{k+1}$, and thus $x \in \mathfrak{Q}^{-1}\mathfrak{N}_{k+1}$, as required. The other equality is proved symmetrically, so this completes the proof of *(i)*.

In *(ii)*, we observe that $\mathfrak{B}+\mathfrak{Q}\mathfrak{N}_k \subset \mathfrak{B}+\mathfrak{P}$, so certainly *(b)*$\Rightarrow$*(a)*. By part *(i)*, we have $\mathfrak{B}+\mathfrak{Q}\mathfrak{N}_k \subset \mathfrak{N}_{k+1}$ for all k. So, assume $k \geq k_0$ and take $x \in \mathfrak{N}_{k+1}$. Thus $x \in \mathfrak{B} + \mathfrak{P}$, by the definition of k_0, so write $x = y + z$, $y \in \mathfrak{B}$, $z \in \mathfrak{P}$, so that, in particular, $a_\beta(x) = a_\beta(z)$. Then $\mathfrak{Q}^{-1}z \subset \mathfrak{A}$ and $a_\beta(\mathfrak{Q}^{-1}z) = \mathfrak{Q}^{-1}a_\beta(x) \subset \mathfrak{P}^k$. In other words, $\mathfrak{Q}^{-1}z \subset \mathfrak{N}_k$, whence $z \in \mathfrak{Q}\mathfrak{N}_k$ and $x \in \mathfrak{B} + \mathfrak{Q}\mathfrak{N}_k$, as required. ■

Iterating, (1.4.8)*(ii)* gives

(1.4.9) $$\mathfrak{N}_{k+r} = \mathfrak{B} + \mathfrak{Q}^r\mathfrak{N}_k, \quad \text{for } k \geq k_0,\ r \geq 1.$$

We can now prove the exactness of the first sequence in (1.4.7) taking $r = et$, where $e = e(\mathfrak{A})$ and $t \geq 1$. We only have to prove exactness in the middle, and we know that $s \circ a_\beta$ is null, so it is enough to show

$$(\mathfrak{N}_k : \mathfrak{N}_{k+et}).(\mathfrak{Q}^k : \mathfrak{Q}^{k+et}) = (\mathfrak{P}^k : \mathfrak{P}^{k+et}), \quad k \geq k_0,\ t \geq 1.$$

Take a prime element π_F of F, so that $\pi_F^t\mathfrak{A} = \mathfrak{P}^{et}$, $\pi_F^t\mathfrak{B} = \mathfrak{Q}^{et}$. By (1.4.9), $\mathfrak{N}_k/\pi_F^t\mathfrak{N}_k$ maps onto $\mathfrak{N}_k/\mathfrak{N}_{k+et}$ with kernel $\mathfrak{B}/\pi_F^t\mathfrak{B}$. The index $(M : \pi_F^t M)$ is the same for all lattices M in A, so

$$\begin{aligned}(\mathfrak{P}^k : \mathfrak{P}^{k+et}) &= (\mathfrak{P}^k : \pi_F^t\mathfrak{P}^k) = (\mathfrak{N}_k : \pi_F^t\mathfrak{N}_k) \\ &= (\mathfrak{N}_k : \mathfrak{N}_{k+et}).(\mathfrak{B} : \pi_F^t\mathfrak{B}) = (\mathfrak{N}_k : \mathfrak{N}_{k+et}).(\mathfrak{Q}^k : \mathfrak{Q}^{k+et}),\end{aligned}$$

since $\mathfrak{Q}^{k+et} = \pi_F^t\mathfrak{Q}^k$. This proves the index equality above, and with it the Proposition. ■

(1.4.10) Corollary: *For $m, k \in \mathbb{Z}$, $k \geq k_0(\beta, \mathfrak{A})$, the following sequences are exact:*

$$0 \longrightarrow \mathfrak{Q}^m \mathfrak{N}_k / \mathfrak{Q}^m \xrightarrow{a_\beta} \mathfrak{P}^{m+k} \xrightarrow{s} \mathfrak{Q}^{m+k} \longrightarrow 0,$$
$$0 \longrightarrow \mathfrak{Q}^m \mathfrak{N}_k / \mathfrak{Q}^m \mathfrak{N}_{k+1} \longrightarrow \mathfrak{P}^{m+k} / \mathfrak{P}^{m+k+1} \longrightarrow \mathfrak{Q}^{m+k} / \mathfrak{Q}^{m+k+1} \longrightarrow 0.$$

Proof: To get the first one here, we tensor the second sequence of (1.4.7) on the left with $\mathfrak{Q}^m$ over $\mathfrak{B}$, to get a sequence

$$0 \longrightarrow \mathfrak{Q}^m \otimes (\mathfrak{N}_k / \mathfrak{B}) \xrightarrow{1 \otimes a_\beta} \mathfrak{Q}^m \otimes \mathfrak{P}^k \xrightarrow{1 \otimes s} \mathfrak{Q}^m \otimes \mathfrak{Q}^k \longrightarrow 0.$$

This sequence is exact, since $\mathfrak{Q}^m$ is $\mathfrak{B}$-projective (as $\mathfrak{B}$ is hereditary). Further, for any left $\mathfrak{B}$-lattice M in any finitely generated B-module W, the multiplication map $\mathfrak{Q}^m \otimes_{\mathfrak{B}} M \to \mathfrak{Q}^m M$ is an isomorphism, because again M is $\mathfrak{B}$-projective. Further, when applied to our last sequence, this multiplication map carries $1 \otimes a_\beta$ to a_β and $1 \otimes s$ to s, since a_β and s are (B, B)-homomorphisms. We have $\mathfrak{Q}^m \mathfrak{P}^k = \mathfrak{P}^{m+k}$, and this gives the desired exactness of the first sequence. The exactness of the second follows immediately. ∎

Remark: The second sequence of (1.4.7), or more generally the first one of (1.4.10), invariably splits as an exact sequence of, say, left $\mathfrak{B}$-modules:— this follows from the fact that $\mathfrak{Q}^{m+k}$ is a projective $\mathfrak{B}$-module. However, these sequences are also exact sequences of $(\mathfrak{B}, \mathfrak{B})$-*bimodules* and, as such, split if and only if the extension E/F is *tamely ramified*. For, if E/F is tamely ramified, we can take s to be the orthogonal projection $A \to B$ (see (1.3.8)), and this is split by the inclusion $\mathfrak{Q}^k \to \mathfrak{P}^k$, for any k. Conversely, suppose we have a splitting $\mathfrak{Q}^k \to \mathfrak{P}^k$ of $s : \mathfrak{P}^k \to \mathfrak{Q}^k$ as $(\mathfrak{B}, \mathfrak{B})$-homomorphism. This extends, by F-linearity, to a splitting $B \to A$ of the (B, B)-homomorphism $s : A \to B$. It is therefore of the form $b \mapsto ub$, for some $u \in E^\times$. Since $u\mathfrak{Q}^k \subset \mathfrak{P}^k$, it follows that $u \in \mathfrak{o}_E$. If $u \in \mathfrak{p}_E$, we would have $s(u\mathfrak{Q}^k) \subset \mathfrak{Q}^{k+1}$, so we deduce that $u \in \mathfrak{o}_E^\times$. We may adjust s by any unit of E, so we assume that $u = 1$. In other words, s is split by the inclusion $B \to A$. Thus s acts as the identity on B. If E/F were not separable, we would have $s(B) = 0$, so E/F is separable. Now it follows from (1.3.8) that E/F is tamely ramified.

We now establish a few formal properties of the lattices $\mathfrak{N}_k = \mathfrak{N}_k(\beta, \mathfrak{A})$.

(1.4.11) Proposition: *(i) For all $k \in \mathbb{Z}$, the lattice $\mathfrak{N}_k(\beta, \mathfrak{A})$ is an $\mathfrak{o}_F$-order in A, and a $(\mathfrak{B}, \mathfrak{B})$-bimodule.*

(ii) We have $\mathfrak{Q}^r \mathfrak{N}_k(\beta, \mathfrak{A}) = \mathfrak{N}_k(\beta, \mathfrak{A}) \mathfrak{Q}^r$, for all $r, k \in \mathbb{Z}$. In particular, $\mathfrak{N}_k(\beta, \mathfrak{A})$ is invariant under conjugation by $\mathfrak{K}(\mathfrak{B})$.

(iii) Let $k_0 = k_0(\beta, \mathfrak{A})$. Then $a_\beta(\mathfrak{A}) \supset \mathfrak{P}^{k_0} \cap C$, where $C = a_\beta(A)$ as in (1.3.1), and k_0 is the least integer with this property.

Proof: Of course, this result holds trivially if $E = F[\beta] = F$, with $k_0 = -\infty$, since then $\mathfrak{N}_k = \mathfrak{A}$ for all k. We therefore ignore this case.

In general, assertion *(i)* follows immediately from the definitions. In *(ii)*, it is enough to treat the cases $r = \pm 1$, and the case $r = 1$ is already given by (1.4.8). For $r = -1$, we have $\mathfrak{N}_k = \mathfrak{B}\mathfrak{N}_k = \mathfrak{Q}^{-1}\mathfrak{Q}\mathfrak{N}_k = \mathfrak{Q}^{-1}\mathfrak{N}_k\mathfrak{Q}$, so $\mathfrak{N}_k\mathfrak{Q}^{-1} = \mathfrak{Q}^{-1}\mathfrak{N}_k\mathfrak{Q}\mathfrak{Q}^{-1} = \mathfrak{Q}^{-1}\mathfrak{N}_k$, as required. For the second assertion, if $x \in \mathfrak{K}(\mathfrak{B})$, then $x\mathfrak{B} = \mathfrak{B}x = \mathfrak{Q}^n$, where $n = \nu_{\mathfrak{B}}(x)$ (see **(1.1)**), so $x\mathfrak{N}_k = \mathfrak{Q}^n\mathfrak{N}_k = \mathfrak{N}_k\mathfrak{Q}^n = \mathfrak{N}_k x$, as required.

In *(iii)*, we have $a_\beta(\mathfrak{A}) \supset a_\beta(\mathfrak{N}_{k_0}) = \operatorname{Ker}(s \mid \mathfrak{P}^{k_0})$ (by (1.4.7)), which is just $\mathfrak{P}^{k_0} \cap \operatorname{Ker}(s \mid A) = \mathfrak{P}^{k_0} \cap C$. Now suppose that $k < k_0$ while $a_\beta(\mathfrak{A}) \supset \mathfrak{P}^k \cap C$. Then

$$a_\beta(\mathfrak{P}) = \mathfrak{Q}.a_\beta(\mathfrak{A}) \supset \mathfrak{P}^{k+1} \cap C \supset \mathfrak{P}^{k_0} \cap C.$$

Since the last term here is $a_\beta(\mathfrak{N}_{k_0})$, it follows that $\mathfrak{N}_{k_0} \subset \mathfrak{B} + \mathfrak{P}$, which is nonsense. ■

We therefore have

$$\textbf{(1.4.12)} \qquad \mathfrak{P}^{\nu_{\mathfrak{A}}(\beta)} \supset a_\beta(\mathfrak{A}) \supset \mathfrak{P}^{k_0} \cap C, \quad k_0 = k_0(\beta, \mathfrak{A}).$$

The non-negative integer $k_0 - \nu_{\mathfrak{A}}(\beta)$ therefore measures how the map a_β "spreads" the filtration of A given by the powers of $\mathfrak{P}$ (assuming, of course, that $E \neq F$). Note, however, that it is possible in (1.4.12) to have $a_\beta(\mathfrak{A})$ contained in $\mathfrak{P}^{1+\nu_{\mathfrak{A}}(\beta)}$. (To get an example of this, replace β by $\beta' = \alpha + \beta$, where $\alpha \in F^\times$ and $\nu_{\mathfrak{A}}(\alpha) = e(\mathfrak{A}|\mathfrak{o}_F).\nu_F(\alpha) < \nu_{\mathfrak{A}}(\beta)$. Then $\nu_{\mathfrak{A}}(\beta') = \nu_{\mathfrak{A}}(\alpha)$, while $a_{\beta'} = a_\beta$.)

We now investigate how the objects $\mathfrak{N}_k(\beta, \mathfrak{A})$, $k_0(\beta, \mathfrak{A})$ depend on the order $\mathfrak{A}$. Since the element β will be fixed throughout the discussion, we tend to omit it from the notation at this particular point. We also fix a (W, E)-decomposition

$$\mathfrak{A} = \mathfrak{A}(E) \otimes_{\mathfrak{o}_E} \mathfrak{B}$$

of $\mathfrak{A}$, as in (1.2.8).

(1.4.13) Proposition: *Let $e = e(\mathfrak{B}|\mathfrak{o}_E)$ be the $\mathfrak{o}_E$-period of the lattice chain defining $\mathfrak{A}$. Then*

(i) $\mathfrak{N}_k(\beta, \mathfrak{A}(E)) \otimes_{\mathfrak{o}_E} \mathfrak{B} = \mathfrak{N}_{ek}(\beta, \mathfrak{A})$, for $k \in \mathbb{Z}$;

(ii) $k_0(\beta, \mathfrak{A}) = ek_0(\beta, \mathfrak{A}(E))$.

Proof: We start by proving both statements when $\mathfrak{B}$ is a *maximal* order in B, so that $e = 1$ here. (1.4.2) shows

$$a_\beta(\mathfrak{N}_k(\mathfrak{A}(E)) \otimes \mathfrak{B}) = a_\beta(\mathfrak{N}_k(\mathfrak{A}(E))) \otimes \mathfrak{B} \subset \mathfrak{P}(E)^k \otimes \mathfrak{B} = \mathfrak{P}^k, \quad k \in \mathbb{Z},$$

where $\mathfrak{P}(E)$ denotes the radical of $\mathfrak{A}(E)$. Thus $\mathfrak{N}_k(\mathfrak{A}(E)) \otimes \mathfrak{B} \subset \mathfrak{N}_k(\mathfrak{A})$ for all $k \in \mathbb{Z}$.

Now write $C(E) = a_\beta(A(E))$. Then $C = C(E) \otimes_E B$. Since $\mathfrak{B}$ is a free $\mathfrak{o}_E$-module, we have $\mathfrak{P}^k \cap C = (\mathfrak{P}(E)^k \otimes \mathfrak{B}) \cap C = (\mathfrak{P}(E)^k \cap C(E)) \otimes \mathfrak{B}$. On the other hand, $a_\beta(\mathfrak{A}) = (a_\beta(\mathfrak{A}(E)) \otimes \mathfrak{B}$, so $a_\beta(\mathfrak{A}) \supset \mathfrak{P}^k \cap C$ if and only if $a_\beta(\mathfrak{A}(E)) \supset \mathfrak{P}(E)^k \cap C(E)$. (1.4.11)*(iii)* now shows that $k_0(\beta, \mathfrak{A}) = k_0(\beta, \mathfrak{A}(E))$.

It follows that $\mathfrak{N}_k(\mathfrak{A})$, $\mathfrak{N}_k(\mathfrak{A}(E)) \otimes \mathfrak{B}$ have the same image under a_β, namely $\mathfrak{P}^k \cap C$, provided $k \geq k_0(\mathfrak{A})$. These two lattices surely have the same intersection (namely $\mathfrak{B}$) with the kernel of a_β, so they are equal.

To extend this to all values of k, we recall from (1.4.8) that $\mathfrak{N}_{k-1}(\mathfrak{A}) = \mathfrak{N}_k \mathfrak{Q}^{-1} \cap \mathfrak{A}$ for all $k \in \mathbb{Z}$. We apply this in the present context, with the inductive assumption that $\mathfrak{N}_k(\mathfrak{A}) = \mathfrak{N}_k(\mathfrak{A}(E) \otimes \mathfrak{B}$:

$$
\begin{aligned}
\mathfrak{N}_{k-1}(\mathfrak{A}) &= \mathfrak{N}_k \mathfrak{Q}^{-1} \cap \mathfrak{A} = (\mathfrak{N}_k(\mathfrak{A}(E)) \otimes \mathfrak{Q}^{-1}) \cap \mathfrak{A} \\
&= (\mathfrak{N}_k(\mathfrak{A}(E))\mathfrak{p}_E^{-1} \otimes \mathfrak{B}) \cap \mathfrak{A} = (\mathfrak{N}_k(\mathfrak{A}(E))\mathfrak{p}_E^{-1} \cap \mathfrak{A}(E)) \otimes \mathfrak{B} \\
&= \mathfrak{N}_{k-1}(\mathfrak{A}(E)) \otimes \mathfrak{B},
\end{aligned}
$$

on applying (1.4.8) to $\mathfrak{A}(E)$. This completes the proof of the proposition when $\mathfrak{B}$ is a maximal order.

In the general case, let $\mathfrak{B}_i$, $1 \leq i \leq e$, be the maximal $\mathfrak{o}_E$-orders in B which contain $\mathfrak{B}$. For each i, let $\mathfrak{A}_i$ be the $\mathrm{Ad}(E^\times)$-invariant hereditary order in A with $\mathfrak{A}_i \cap B = \mathfrak{B}_i$, so that $\mathfrak{A}$ is the intersection of the $\mathfrak{A}_i$. Our space W, which is the F-span of an $\mathfrak{o}_E$-basis of the chain defining $\mathfrak{A}$, also contains a basis of the chain defining $\mathfrak{B}_i$, $1 \leq i \leq e$, by (1.1.9). Therefore our chosen (W, E)-decomposition gives

$$\mathfrak{A}_i = \mathfrak{A}(E) \otimes_{\mathfrak{o}_E} \mathfrak{B}_i,$$

for all i. Since $\mathfrak{N}_k(\mathfrak{A}(E))$ is at least a flat $\mathfrak{o}_E$-module, we have

$$\mathfrak{N}_k(\mathfrak{A}(E)) \otimes_{\mathfrak{o}_E} \mathfrak{B} = \bigcap_i \mathfrak{N}_k(\mathfrak{A}(E)) \otimes_{\mathfrak{o}_E} \mathfrak{B}_i,$$

while $\mathfrak{N}_k(\mathfrak{A}(E)) \otimes \mathfrak{B}_i = \mathfrak{N}_k(\mathfrak{A}_i)$ by the first case. However,

$$\bigcap_i \mathfrak{N}_k(\mathfrak{A}_i) = \{x \in \bigcap_i \mathfrak{A}_i : a_\beta(x) \in \bigcap_i \mathfrak{P}_i^k\},$$

where $\mathfrak{P}_i$ denotes the radical of $\mathfrak{A}_i$. The intersection of the $\mathfrak{A}_i$ is $\mathfrak{A}$, and the intersection of the $\mathfrak{P}_i^k$ is $\mathfrak{P}^{ek}$, so we have

$$\mathfrak{N}_k \otimes_{\mathfrak{o}_E} \mathfrak{B} = \bigcap \mathfrak{N}_k(\mathfrak{A}_i) = \mathfrak{N}_{ek}(\mathfrak{A}),$$

as required.

For *(ii)*, we write $C(E) = C \cap A(E) = a_\beta(A(E))$ as above. Then, inside A, we have $C = C(E).\mathfrak{B}$. Also

$$a_\beta(\mathfrak{A}) = a_\beta(\mathfrak{A}(E) \otimes \mathfrak{B}) \supset (\mathfrak{P}(E)^{\bar{k}_0} \cap C(E)) \otimes \mathfrak{B} = \mathfrak{P}^{e\bar{k}_0} \cap C,$$

where $\bar{k}_0 = k_0(\beta, \mathfrak{A}(E))$. Thus $e\bar{k}_0 \geq k_0 = k_0(\beta, \mathfrak{A})$. If, however, $e\bar{k}_0 > k_0$, we would have $\mathfrak{N}_{e\bar{k}_0}(\mathfrak{A}) \subset \mathfrak{B} + \mathfrak{P}$, whence

$$\begin{aligned}\mathfrak{N}_{\bar{k}_0}(\mathfrak{A}(E)) \subset (\mathfrak{N}_{\bar{k}_0}(\mathfrak{A}(E)) \otimes \mathfrak{B}) \cap A(E) &= \mathfrak{N}_{e\bar{k}_0}(\mathfrak{A}) \cap A(E)\\ &\subset (\mathfrak{B} + \mathfrak{P}) \cap A(E) \subset \mathfrak{o}_E + \mathfrak{P}(E),\end{aligned}$$

contrary to the definition of $\bar{k}_0$. Thus $k_0 = e\bar{k}_0$, and this completes the proof of *(ii)*. ∎

The quantity $k_0(\beta, \mathfrak{A}(E))$ depends only on β (as an element of the field E). It seems worth seeking a purely field-theoretic interpretation of it:— this would seem to be related to the phenomenon of "Sprungstellen" discussed in **[Ko]**.

We now recall a basic definition from **[KM1]**. Let E/F be a field extension as above, with $E = F[\beta]$. Let ν_E be the normalised additive valuation on E, put $\nu = \nu_E(\beta)$, and let π_F denote some prime element of F. Then β is *minimal over* F if it satisfies

(1.4.14) *(i)* $\gcd(\nu, e(E|F)) = 1$, *and*
(ii) $\pi_F^{-\nu}.\beta^{e(E|F)} + \mathfrak{p}_E$ *generates the residue class field extension* $\mathsf{k}_E/\mathsf{k}_F$.

(1.4.15) Proposition: *Suppose, in the situation above, that* $E \neq F$. *Then we have* $\nu_{\mathfrak{A}}(\beta) \leq k_0(\beta, \mathfrak{A})$, *with equality if and only if* β *is minimal over* F.

Proof: We have already remarked (after (1.4.4)) on the first inequality. By part *(ii)* of (1.4.13), we may as well assume that E is a maximal subfield of A, so that $A = A(E)$ and $\mathfrak{A} = \mathfrak{A}(E)$. In particular, $\mathfrak{A}$ is a principal order and $\mathfrak{B} = \mathfrak{o}_E$. Put $\nu_{\mathfrak{A}}(\beta) = -n$, and consider the map $a'_\beta : \mathfrak{A}/\mathfrak{B} \to \mathfrak{A}/\mathfrak{B}$ induced by $x \mapsto \beta x \beta^{-1} - x$, $x \in \mathfrak{A}$. Then $k_0(\beta, \mathfrak{A}) = -n$ if and only if $\operatorname{Ker}(a'_\beta) \subset \mathfrak{o}_E + \mathfrak{P}$. However, in this situation, $\mathfrak{o}_E + \mathfrak{P}$ is the set of fixed points for the natural conjugation action of $E^\times$ on $\mathfrak{A}/\mathfrak{P}$, and, further, $\mathfrak{o}_E + \mathfrak{P}/\mathfrak{P} = \mathfrak{o}_E/\mathfrak{p}_E$ is a maximal subfield of the semisimple k_F-algebra $\mathfrak{A}/\mathfrak{P}$. The primitive central idempotents of $\mathfrak{A}/\mathfrak{P}$ are permuted transitively by the cyclic group $E^\times/F^\times\mathfrak{o}_E^\times$. (1.4.14)*(i)* holds (for β) if and only if the only such idempotents commuting with β are 0 and 1. If we assume that (1.4.14)*(i)* holds, then (1.4.14)*(ii)* holds if and only if $(\pi_F^{-\nu}.\beta^{e(E|F)} + \mathfrak{p}_E)$ has the same centraliser in $\mathfrak{A}/\mathfrak{P}$ as $(\mathfrak{o}_E/\mathfrak{p}_E)^\times$. The proposition follows. ∎

The "if" implication in (1.4.15) is due to Carayol **[Ca]**.

Remark: Again suppose that $E = F[\beta]$ with β minimal over F, $E \neq F$. Put $\nu_{\mathfrak{A}}(\beta) = -n$. Then $\mathfrak{N}_{-n}(\beta, \mathfrak{A}) = \mathfrak{A}$ by definition, while (1.4.15) and (1.4.8)*(ii)(b)* imply $\mathfrak{N}_{1-n}(\beta, \mathfrak{A}) = \mathfrak{B} + \mathfrak{P}$. Indeed, (1.4.9) implies

$$\mathfrak{N}_{r-n}(\beta, \mathfrak{A}) = \mathfrak{B} + \mathfrak{P}^r, \quad r \geq 0.$$

This also holds, in a rather trivial way, when $E = F$ since then $\mathfrak{B} = \mathfrak{A}$ and $\mathfrak{N}_k(\beta, \mathfrak{A}) = \mathfrak{A}$ for all k.

We conclude this section with a more elaborate exact sequence, generalising that of (1.4.7).

(1.4.16) Proposition: *Put $k = k_0(\beta, \mathfrak{A})$, $\mathfrak{N} = \mathfrak{N}_k(\beta, \mathfrak{A})$, and let $i, j, m, n \in \mathbb{Z}$. Let $h_1, h_2 \in B^\times$. Define*

$$L = \mathfrak{P}^i \cap (h_1\mathfrak{P}^j + \mathfrak{P}^m h_2 + \mathfrak{P}^n),$$
$$M = \mathfrak{Q}^{i-k}\mathfrak{N} \cap (h_1\mathfrak{Q}^{j-k}\mathfrak{N} + \mathfrak{Q}^{m-k}\mathfrak{N}h_2 + \mathfrak{Q}^{n-k}\mathfrak{N}).$$

Then the sequence

$$M \xrightarrow{a_\beta} L \xrightarrow{s} A$$

is exact.

Proof: The assertion amounts to showing that $L \cap C = a_\beta(M)$, where C is the space defined by (1.3.1). Define $L_2 = h_1\mathfrak{P}^j + \mathfrak{P}^m h_2 + \mathfrak{P}^n$, $M_2 = h_1\mathfrak{Q}^{j-k}\mathfrak{N} + \mathfrak{Q}^{m-k}\mathfrak{N}h_2 + \mathfrak{Q}^{n-k}\mathfrak{N}$. Fix a (W, E)-decomposition of $\mathfrak{A}$. Arguing as in (1.3.16), we can adjust h_1 so that $h_1\mathfrak{P}^j$ is an $\mathrm{Ad}(E^\times)$-invariant left $\mathfrak{A}(E)$-lattice. We can now apply (1.3.17) and (1.4.10) to show that $L_2 \cap C = a_\beta(M_2)$. Put $L_1 = \mathfrak{P}^i$, $M_1 = \mathfrak{Q}^{i-k}\mathfrak{N}$, so that $L_1 \cap C = a_\beta(M_1)$, by (1.4.10) again.

We have $L \cap C = (L_1 \cap C) \cap (L_2 \cap C)$, so we have to show

$$a_\beta(M_1 \cap M_2) = a_\beta(M_1) \cap a_\beta(M_2).$$

As in the proof of (1.3.17), this is equivalent to showing

$$(M_1 + M_2) \cap B = (M_1 \cap B) + (M_2 \cap B).$$

Here, we certainly have containment in the direction $\supset$. On the other hand,

$$M_1 + M_2 \subset \mathfrak{P}^{i-k} + (h_1\mathfrak{P}^{j-k} + \mathfrak{P}^{m-k}h_2 + \mathfrak{P}^{n-k}),$$

while $(\mathfrak{P}^{i-k} + (h_1\mathfrak{P}^{j-k} + \mathfrak{P}^{m-k}h_2 + \mathfrak{P}^{n-k})) \cap B = \mathfrak{Q}^{i-k} + (h_1\mathfrak{Q}^{j-k} + \mathfrak{Q}^{m-k}h_2 + \mathfrak{Q}^{n-k})$ by (1.3.16). Therefore

$$(M_1 + M_2) \cap B \subset \mathfrak{Q}^{i-k} + (h_1\mathfrak{Q}^{j-k} + \mathfrak{Q}^{m-k}h_2 + \mathfrak{Q}^{n-k})$$
$$= (M_1 \cap B) + (M_2 \cap B),$$

appealing to (1.3.16) again. We conclude that

$$(M_1 + M_2) \cap B = M_1 \cap B + M_2 \cap B,$$

as required. ■

(1.5) Simple strata and intertwining

Let $A = \mathrm{End}_F(V)$ as before. A *stratum in* A is a 4-tuple $[\mathfrak{A}, n, r, b]$ consisting of a hereditary $\mathfrak{o}_F$-order $\mathfrak{A}$ in A, integers $n > r$, and an element $b \in A$ such that $\nu_{\mathfrak{A}}(b) \geq -n$. If $[\mathfrak{A}_i, n_i, r_i, b_i]$ are strata in A, $i = 1, 2$, and $\mathfrak{P}_i = \mathrm{rad}(\mathfrak{A}_i)$, we say they are *equivalent*, denoted

$$\text{(1.5.1)} \qquad \begin{aligned} [\mathfrak{A}_1, n_1, r_1, b_1] \sim [\mathfrak{A}_2, n_2, r_2, b_2], \quad \textit{if} \\ b_1 + \mathfrak{P}_1^{-r_1} = b_2 + \mathfrak{P}_2^{-r_2}. \end{aligned}$$

(1.5.2) Proposition: *Let* $[\mathfrak{A}_i, n_i, r_i, b_i]$, $i = 1, 2$, *be strata in* A, *as above, and assume that they are equivalent. Then* $\mathfrak{P}_1 = \mathfrak{P}_2$, *and* $r_1 = r_2$. *Further, if* $\nu_{\mathfrak{A}_i}(b_i) = -n_i$ *for* $i = 1, 2$, *then also* $n_1 = n_2$.

Proof: (1.5.1) implies $\mathfrak{P}_1^{-r_1} = \mathfrak{P}_2^{-r_2}$. However, we have for example, $\mathfrak{A}_1 = \{x \in A : x\mathfrak{P}_1^m \subset \mathfrak{P}_1^m\}$, for any $m \in \mathbb{Z}$. To see this, we note that this set certainly contains $\mathfrak{A}_1$ while, if $x\mathfrak{P}_1^m \subset \mathfrak{P}_1^m$, then $x\mathfrak{P}_1^m\mathfrak{P}_1^{-m} \subset \mathfrak{P}_1^m\mathfrak{P}_1^{-m}$. However, since $\mathfrak{P}_1$ is an invertible fractional ideal of $\mathfrak{A}_1$, we have $\mathfrak{P}_1^m\mathfrak{P}_1^{-m} = \mathfrak{A}_1$. Thus $x\mathfrak{A}_1 \subset \mathfrak{A}_1$, whence $x \in \mathfrak{A}_1$.

We deduce that $\mathfrak{A}_1 = \mathfrak{A}_2$, hence $\mathfrak{P}_1 = \mathfrak{P}_2$ and $r_1 = r_2$. The remaining statement is now immediate. ■

(1.5.3) Remark: Let ψ_A be the additive character of A used in **(1.3)**. Let $[\mathfrak{A}, n, r, b]$ be a stratum in A, and assume that

$$\text{(1.5.4)} \qquad n > r \geq [\tfrac{n}{2}] \geq 0,$$

where $[x]$ is the greatest integer function. The character ψ_b of $\boldsymbol{U}^{r+1}(\mathfrak{A})/\boldsymbol{U}^{n+1}(\mathfrak{A})$ is defined as in (1.1.6). This depends only on the equivalence class of the stratum and, provided the condition (1.5.4) is satisfied, we see that an equivalence class of strata $[\mathfrak{A}, n, r, b]$ is the same as a character of $\boldsymbol{U}^{r+1}(\mathfrak{A})/\boldsymbol{U}^{n+1}(\mathfrak{A})$. Moreover, we have $\nu_{\mathfrak{A}}(b) = -n$ if and only if ψ_b is non-null on $\boldsymbol{U}^n(\mathfrak{A})$, as in (1.1.5).

(1.5.5) Definition: *Let* $[\mathfrak{A}, n, r, \beta]$ *be a stratum in* A. *It is* pure *if*

(i) *the algebra* $E = F[\beta]$ *is a field,*

(ii) $E^\times \subset \mathfrak{K}(\mathfrak{A})$,

(iii) $\nu_{\mathfrak{A}}(\beta) = -n$.

It is called simple *if, in addition,*

(iv) $r < -k_0(\beta, \mathfrak{A})$.

The basic example of a simple stratum is provided by the minimal elements recalled in (1.4.14). If $E = F[\alpha]$ is a field, with α minimal over F, and such that $E^\times \subset \mathfrak{K}(\mathfrak{A})$ and $\nu_{\mathfrak{A}}(\alpha) = -n$, then the stratum $[\mathfrak{A}, n, r, \alpha]$ is simple for any $r < n$, by (1.4.15). The situation is actually very straightforward in this case:

(1.5.6) Exercise: *Let $\alpha \in G$, and put $E = F[\alpha]$. The following conditions are equivalent:*

(i) E is a field and α is minimal over F;

(ii) there exists a hereditary order $\mathfrak{A}$ in A with $\alpha \in \mathfrak{K}(\mathfrak{A})$, and for any such $\mathfrak{A}$, we have $E^\times \subset \mathfrak{K}(\mathfrak{A})$.

Simple strata of the form $[\mathfrak{A}, n, r, \alpha]$, with α minimal over F, are discussed extensively in **[KM1]**, and their special role in the representation theory of G is discussed in **[K4]**. The main result of the present section generalises to arbitrary simple strata the intertwining theorem of **[KM1]**.

For the moment, let $[\mathfrak{A}, n, r, b]$ be any stratum in A. The *formal G-intertwining of* $[\mathfrak{A}, n, r, b]$ is the set

(1.5.7) $$\mathcal{I}_G[\mathfrak{A}, n, r, b] = \{x \in G : x^{-1}(b + \mathfrak{P}^{-r})x \cap (b + \mathfrak{P}^{-r}) \neq \emptyset\},$$

where $\mathfrak{P}$ is the radical of $\mathfrak{A}$. This clearly only depends on the equivalence class of the stratum, that is, only on the coset $b + \mathfrak{P}^{-r}$. We therefore sometimes refer to $\mathcal{I}_G[\mathfrak{A}, n, r, b]$ as the *intertwining of the coset $b + \mathfrak{P}^{-r}$*. Moreover, if (1.5.4) is satisfied, then the formal G-intertwining of the stratum $[\mathfrak{A}, n, r, b]$ is identical with the G-intertwining, in the ordinary sense (recalled in **(3.3)** below), of the character ψ_b of $\boldsymbol{U}^{r+1}(\mathfrak{A})$:— see **[H2]** pp. 484–5 for a discussion of this.

(1.5.8) Theorem: *Let $[\mathfrak{A}, n, r, \beta]$ be a simple stratum in A. Let B denote the A-centraliser of β, $\mathfrak{P} = \mathrm{rad}(\mathfrak{A})$, $\mathfrak{B} = \mathfrak{A} \cap B$, $\mathfrak{Q} = \mathrm{rad}(\mathfrak{B})$. Write $k = k_0(\beta, \mathfrak{A})$, $\mathfrak{N} = \mathfrak{N}_k(\beta, \mathfrak{A})$. Then*

$$\mathcal{I}_G[\mathfrak{A}, n, r, \beta] = (1 + \mathfrak{Q}^{-(r+k)}\mathfrak{N})B^\times(1 + \mathfrak{Q}^{-(r+k)}\mathfrak{N}).$$

Proof: Observe to start with that the integer $d = -(r + k)$ is positive, by the definition of simple stratum.

Abbreviate $\mathcal{I} = \mathcal{I}_G[\mathfrak{A}, n, r, \beta]$. We show first that

(1.5.9) $$(1 + \mathfrak{Q}^d\mathfrak{N})\mathcal{I}(1 + \mathfrak{Q}^d\mathfrak{N}) = \mathcal{I}.$$

For example, take $x \in \mathcal{I}$, so that

$$\beta x \equiv x\beta \pmod{\mathfrak{P}^{-r}x + x\mathfrak{P}^{-r}},$$

and let $y \in \mathfrak{Q}^d\mathfrak{N}$. Then $x(1+y) \in \mathcal{I}$ if and only if

$$\beta x \equiv x(1+y)\beta(1+y)^{-1} \pmod{\mathfrak{P}^{-r}x + x\mathfrak{P}^{-r}},$$

since $(1+y) \in \mathfrak{A}^\times$. However,

$$(1+y)\beta(1+y)^{-1} = \beta - a_\beta(y)(1+y)^{-1},$$

and we have $a_\beta(y) \in \mathfrak{P}^{-r}$ (by (1.4.10)). Thus $x(1+y)\beta(1+y)^{-1} \equiv x\beta \pmod{x\mathfrak{P}^{-r}}$, and $x(1+y) \in \mathcal{I}$, as required. By symmetry, we have (1.5.9).

So, take $x \in \mathcal{I}$. We have to show that the double coset $(1+\mathfrak{Q}^d\mathfrak{N})x(1+\mathfrak{Q}^d\mathfrak{N})$ meets $B^\times$ nontrivially. Put $t = \nu_{\mathfrak{A}}(x)$. Then

$$a_\beta(x) = \beta x - x\beta \in \mathfrak{P}^{-r}x + x\mathfrak{P}^{-r} \subset \mathfrak{P}^{t-r},$$

so, by (1.4.10), there exists $y \in \mathfrak{Q}^{t+d}\mathfrak{N}$ such that $a_\beta(y) = a_\beta(x)$. Put $\gamma_1 = x - y \in B$; indeed, we can arrange $\gamma_1 \in B^\times$, since $B^\times$ is dense in B. Thus

$$x \equiv \gamma_1 \pmod{\mathfrak{Q}^{t+d}\mathfrak{N}}, \quad \gamma_1 \in B^\times.$$

We show, by induction, that for $j \geq 1$, there exists $x_j \in (1+\mathfrak{Q}^d\mathfrak{N})x(1+\mathfrak{Q}^d\mathfrak{N})$ and $\gamma_j \in B^\times$ such that

$$x_j \equiv \gamma_j \pmod{\mathfrak{Q}^{t+jd}\mathfrak{N}}. \tag{1.5.10}$$

Granting this for the moment, we note that the coset $\mathcal{C} = (1+\mathfrak{Q}^d\mathfrak{N})x(1+\mathfrak{Q}^d\mathfrak{N})$ is compact open (this follows from (1.4.11)). If j is sufficiently large, $\mathcal{C}$ therefore contains $z + \mathfrak{Q}^{t+jd}\mathfrak{N}$ for any $z \in \mathcal{C}$. The theorem follows.

So, we are reduced to proving (1.5.10). The case $j = 1$ has already been done. So we assume (1.5.10). Dropping the subscript j, we have

$$x = \gamma + y, \quad \gamma \in B^\times, \ y \in \mathfrak{Q}^{t+jd}\mathfrak{N}.$$

Substituting this into the defining relation

$$x\beta \equiv \beta x \pmod{x\mathfrak{P}^{-r} + \mathfrak{P}^{-r}x},$$

we find

$$a_\beta(y) \in (\gamma+y)\mathfrak{P}^{-r} + \mathfrak{P}^{-r}(\gamma+y) \subset \gamma\mathfrak{P}^{-r} + \mathfrak{P}^{-r}\gamma + \mathfrak{P}^{t+jd-r}.$$

However, $y \in \mathfrak{Q}^{t+jd}\mathfrak{N}$, so $a_\beta(y) \in \mathfrak{P}^{t+jd+k}$, and therefore

$$a_\beta(y) \in \mathfrak{P}^{t+jd+k} \cap (\gamma\mathfrak{P}^{-r} + \mathfrak{P}^{-r}\gamma + \mathfrak{P}^{t+jd-r}).$$

By (1.4.16), we have $a_\beta(y) = a_\beta(z)$ where $z \in \mathfrak{Q}^{t+jd}\mathfrak{N} \cap (\gamma\mathfrak{Q}^d\mathfrak{N} + \mathfrak{Q}^d\mathfrak{N}\gamma + \mathfrak{Q}^{t+(j+1)d}\mathfrak{N})$. We can thus write z in the form $z = \gamma a + b\gamma + c$, where $a, b \in \mathfrak{Q}^d\mathfrak{N}$, $c \in \mathfrak{Q}^{t+(j+1)d}\mathfrak{N}$, and $\gamma a + b\gamma \in \mathfrak{Q}^{t+jd}\mathfrak{N}$. Therefore $y = \delta + z$, for some $\delta \in \mathfrak{Q}^{t+jd}$. We can replace γ by $\gamma + \delta$ throughout, the effect of this being that $x = \gamma + z$, with z as above. Moreover, as before, we may continue to assume $\gamma \in B^\times$. Consider the element

$$x_{j+1} = (1-b)x(1-a)(1-a^2)^{-1} \in (1+\mathfrak{Q}^d\mathfrak{N})x(1+\mathfrak{Q}^d\mathfrak{N}).$$

Expanding, we get

$$x_{j+1} = (\gamma + \gamma a + b\gamma + c - b\gamma - b\gamma a - b^2\gamma - bc)(1-a)(1-a^2)^{-1}.$$

By definition, the term $b\gamma a + b^2\gamma$ lies in $b.\mathfrak{Q}^{t+jd}\mathfrak{N} \subset \mathfrak{Q}^{t+(j+1)d}\mathfrak{N}$. Further, the lattice $\mathfrak{Q}^{t+(j+1)d}\mathfrak{N}$ is invariant under multiplication by $1+\mathfrak{Q}^d\mathfrak{N}$. So, misappropriating the O-notation in the obvious way, we get

$$\begin{aligned} x_{j+1} &= (\gamma + \gamma a + O(\mathfrak{Q}^{t+(j+1)d}\mathfrak{N})).(1-a)(1-a^2)^{-1} \\ &= (\gamma + \gamma a - \gamma a - \gamma a^2 + O(\mathfrak{Q}^{t+(j+1)d}.\mathfrak{N}))(1-a^2)^{-1} \\ &= \gamma + O(\mathfrak{Q}^{t+(j+1)d}\mathfrak{N}). \end{aligned}$$

This completes our induction, and the proof of the theorem. ■

It is possible to prove more general results of this kind. First, suppose we have strata $[\mathfrak{A}_i, n_i, r_i, b_i]$ in A, for $i = 1, 2$. Define their formal intertwining by

$$\textbf{(1.5.11)} \quad \mathcal{I}_G([\mathfrak{A}_1, n_1, r_1, b_1], [\mathfrak{A}_2, n_2, r_2, b_2]) = \{x \in G : x^{-1}(b_1 + \mathfrak{P}_1^{-r_1})x \cap (b_2 + \mathfrak{P}_2^{-r_2}) \neq \emptyset\},$$

where $\mathfrak{P}_i$ denotes the radical of $\mathfrak{A}_i$.

Now suppose that $[\mathfrak{A}_i, n_i, r_i, \beta]$ are simple, for $i = 1, 2$ (with the same β). Then, with some effort, it is possible to show that

$$\textbf{(1.5.12)} \quad \mathcal{I}_G([\mathfrak{A}_1, n_1, r_1, \beta], [\mathfrak{A}_2, n_2, r_2, \beta]) = (1 + \mathfrak{Q}_1^{-(r_1+k_1)}\mathfrak{N}_1)B^\times(1 + \mathfrak{Q}_2^{-(r_2+k_2)}\mathfrak{N}_2),$$

where B is the A-centraliser of β, $\mathfrak{P}_i$ is the radical of $\mathfrak{A}_i$, $\mathfrak{Q}_i = \mathfrak{P}_i \cap B$, $k_i = k_0(\beta, \mathfrak{A}_i)$, $\mathfrak{N}_i = \mathfrak{N}_{k_i}(\beta, \mathfrak{A}_i)$.

(1.6) The simple intersection property

We conclude this section with another result which we shall not need for some time, but the methods used to prove it belong in the present chapter. We fix a hereditary $\mathfrak{o}_F$-order $\mathfrak{A}$ in A, with radical $\mathfrak{P}$, and a subfield E/F of A such that $E^\times$ normalises $\mathfrak{A}$. We write B for the A-centraliser of E, $\mathfrak{B} = \mathfrak{A} \cap B$, $\mathfrak{Q} = \mathfrak{P} \cap B$.

(1.6.1) Theorem: *With the notation above, let $x \in B^\times$, and let $n \geq 1$. Then*

$$\boldsymbol{U}^n(\mathfrak{A})x\boldsymbol{U}^n(\mathfrak{A}) \cap B^\times = \boldsymbol{U}^n(\mathfrak{B})x\boldsymbol{U}^n(\mathfrak{B}).$$

Proof: Let $t = \nu_{\mathfrak{A}}(x)$, and let $\bar{x} \in \boldsymbol{U}^n(\mathfrak{A})x\boldsymbol{U}^n(\mathfrak{A}) \cap B^\times$. We proceed by induction on the hypothesis:

(1.6.2) *Given $m \geq 0$, there exists $\bar{x}_m \in \boldsymbol{U}^n(\mathfrak{B})\bar{x}\boldsymbol{U}^n(\mathfrak{B})$ such that*

$$\bar{x}_m \equiv x \pmod{\mathfrak{P}^{(m+1)n+t}}.$$

Granting this for the moment, we see that the sequence $\{\bar{x}_m\}$ converges to x. The double coset $\boldsymbol{U}^n(\mathfrak{B})\bar{x}\boldsymbol{U}^n(\mathfrak{B})$ is compact, so we conclude that $x \in \boldsymbol{U}^n(\mathfrak{B})\bar{x}\boldsymbol{U}^n(\mathfrak{B})$, or $\bar{x} \in \boldsymbol{U}^n(\mathfrak{B})x\boldsymbol{U}^n(\mathfrak{B})$, as required.

The case $m = 0$ in the induction is immediate: we take $\bar{x}_0 = \bar{x}$. So take $m \geq 0$, and assume we have found the required element $\bar{x}_m$. Thus $\bar{x}_m = (1+a)x(1+b) = x + ax + xb + axb$, for some $a, b \in \mathfrak{P}^n$ with $ax + xb + axb \in \mathfrak{P}^{(m+1)n+t}$. Put

$$b' = b - b^2 + b^3 - \ldots + (-1)^m b^{m+1}.$$

We show first that $ax + xb + axb \equiv ax + xb' \pmod{\mathfrak{P}^{(m+2)n+t}}$. Again we proceed inductively. Take $r \leq m$ and assume we have the congruence modulo $\mathfrak{P}^{(r+1)n+t}$ (the case $r = 0$ being trivial):

$$ax + xb + axb \equiv ax + x(b - b^2 + \ldots - (-b)^r) \pmod{\mathfrak{P}^{(r+1)n+t}}.$$

Since $r \leq m$, we have $ax + xb + axb \in \mathfrak{P}^{(r+1)n+t}$, whence $ax + x(b - b^2 + \ldots - (-b)^r) \in \mathfrak{P}^{(r+1)n+t}$. It follows that

$$axb + xb^2 - xb^3 + \ldots + x(-b)^{r+1} \in \mathfrak{P}^{(r+2)n+t}.$$

This gives

$$ax + xb + axb \equiv ax + x(b - b^2 + \ldots - (-b)^{r+1}) \pmod{\mathfrak{P}^{(r+2)n+t}},$$

as desired. So indeed $ax + xb + axb = ax + xb' + b''$, for some $b' \in \mathfrak{P}^n$, $b'' \in \mathfrak{P}^{(m+2)n+t}$. However, we have $ax + xb + axb \in B$, so we can use (1.3.16) to write $ax + xb + axb = ax + xb' + b'' = cx + xd + e$, with $c, d \in \mathfrak{Q}^n$, $e \in \mathfrak{Q}^{(m+2)n+t}$, $cx + xd \in \mathfrak{Q}^{(m+1)n+t}$. Now consider the element $\bar{x}_{m+1}$ of $\boldsymbol{U}^n(\mathfrak{B})\bar{x}\boldsymbol{U}^n(\mathfrak{B})$ given by

$$\begin{aligned}
\bar{x}_{m+1} &= (1-c)\bar{x}_m(1+d)^{-1} \\
&= (1-c)(x + cx + xd + e)(1+d)^{-1} \\
&\equiv (x + xd)(1+d)^{-1} \pmod{\mathfrak{P}^{(m+2)n+t}} \\
&\equiv x \pmod{\mathfrak{P}^{(m+2)n+t}}.
\end{aligned}$$

This completes our induction and the proof of the theorem. ∎

2. THE STRUCTURE OF SIMPLE STRATA

This section is devoted to giving a thorough account of the structure of simple strata. The basic plan is to build them all from the ones given by minimal elements, which we treat as the "atoms" of the theory. In **(2.2)**, we give the procedure for constructing more general simple strata from minimal elements, as another application of the machinery of §**1**. Indeed, it is in this section (in fact in (2.2.1)) that another aspect of the exact sequences of **(1.4)** emerges, as the first appearance of an "equivalence on centraliser implies conjugacy" principle.

(2.3) is a technical section. It recalls and, to some extent, reworks material from **[Bu1]** and **[K4]**. While many of the results there are not new, it is important to have these ideas to hand in a form consistent with our present setup. The main result is (2.3.12), which functions as a lemma for use in **(2.4)**. The principal result of the whole section is (2.4.1), which gives us effective control over simple strata and their equivalence classes. **(2.5)** is again technical, reworking another part of **[K4]**. It is only mildly more general than the original, but a number of crucial details need to be filled in. It gives a "standard form" for equivalence classes of nonsplit fundamental strata which eases subsequent, necessarily explicit, computations involving these objects. Its immediate purpose is to complete a couple of proofs left over from **(2.3)** and **(2.4)**.

(2.6) is comparatively straightforward, but the importance of its main result, the "intertwining implies conjugacy" theorem for simple strata, amply justifies the preceding labour.

Unexplained notations in this section are all to be found in §**1**.

(2.1) Equivalence of pure strata

We start by collecting together a few useful identities concerning pure or simple strata. First, let $[\mathfrak{A}, n, r, \beta]$ be some pure stratum in the algebra $A = \mathrm{End}_F(V)$. As before, we write B for the A-centraliser of β, $\mathfrak{B} = B \cap \mathfrak{A}$, $\mathfrak{Q} = \mathrm{rad}(\mathfrak{B}) = \mathfrak{P} \cap B$.

(2.1.1) Lemma: *For $t, m \in \mathbb{Z}$, we have*

$$\mathfrak{Q}^t\mathfrak{N}_m(\beta, \mathfrak{A}) = \{x \in \mathfrak{P}^t : a_\beta(x) \in \mathfrak{P}^{t+m}\}.$$

Proof: We certainly have $\mathfrak{Q}^t\mathfrak{N}_m(\beta, \mathfrak{A}) \subset \mathfrak{Q}^t\mathfrak{A} = \mathfrak{P}^t$ and $a_\beta(\mathfrak{Q}^t\mathfrak{N}_m(\beta, \mathfrak{A})) \subset \mathfrak{P}^{t+m}$, so let $x \in \mathfrak{P}^t$ satisfy $a_\beta(x) \in \mathfrak{P}^{t+m}$. Then $\mathfrak{Q}^{-t}x \subset \mathfrak{A}$ and $a_\beta(\mathfrak{Q}^{-t}x) \subset \mathfrak{Q}^{-t}\mathfrak{P}^{t+m} = \mathfrak{P}^m$. Thus $\mathfrak{Q}^{-t}x \subset \mathfrak{N}_m(\beta, \mathfrak{A})$, whence $x \in \mathfrak{B}x = \mathfrak{Q}^{-t}\mathfrak{Q}^t x \subset \mathfrak{Q}^t\mathfrak{N}_m(\beta, \mathfrak{A})$, as required. ■

We now take two pure strata $[\mathfrak{A}, n, r, \beta_i]$, $i = 1, 2$, and assume that

(2.1.2) $$[\mathfrak{A}, n, r, \beta_1] \sim [\mathfrak{A}, n, r, \beta_2],$$

in the notation of (1.5.1). We write B_i for the A-centraliser of β_i, $\mathfrak{B}_i = \mathfrak{A} \cap B_i$, $\mathfrak{Q}_i = \mathfrak{P} \cap B_i$. Since the order $\mathfrak{A}$ will be fixed throughout, we often abbreviate $\mathfrak{N}_m(\beta_i, \mathfrak{A}) = \mathfrak{N}_m(\beta_i)$.

(2.1.3) Proposition: *In the above situation we have*

$$\mathfrak{Q}_1^t \mathfrak{N}_m(\beta_1) = \mathfrak{Q}_2^t \mathfrak{N}_m(\beta_2), \quad m, t \in \mathbb{Z},\ m \leq -r.$$

Proof: If $m \leq -r$ and $x \in \mathfrak{P}^t$, then $a_{\beta_1}(x) \equiv a_{\beta_2}(x) \pmod{\mathfrak{P}^{t+m}}$, and the assertion follows from (2.1.1). ■

(2.1.4) Proposition: *Suppose that the strata $[\mathfrak{A}, n, r, \beta_i]$, $i = 1, 2$, are simple and equivalent to each other. Set $E_i = F[\beta_i]$ and use the other notation above. Then*

(i) $\mathfrak{B}_1 + \mathfrak{P} = \mathfrak{B}_2 + \mathfrak{P}$;

(ii) $k_0(\beta_1, \mathfrak{A}) = k_0(\beta_2, \mathfrak{A})$;

(iii) $e(E_1|F) = e(E_2|F)$ *and* $f(E_1|F) = f(E_2|F)$.

Proof: Set $k_i = k_0(\beta_i)$, so that $k_i < -r$, $i = 1, 2$. Applying (2.1.3) (with $t = 0$) we get

(2.1.5) $$\mathfrak{N}_m(\beta_1) = \mathfrak{N}_m(\beta_2), \quad m \leq -r.$$

On the other hand, the definition (1.4.5) gives

(2.1.6) $$\mathfrak{N}_m(\beta_i) + \mathfrak{P} = \mathfrak{B}_i + \mathfrak{P}, \quad m > k_i.$$

Combining these (with $m = -r$) we get

$$\mathfrak{B}_1 + \mathfrak{P} = \mathfrak{N}_{-r}(\beta_1) + \mathfrak{P} = \mathfrak{N}_{-r}(\beta_2) + \mathfrak{P} = \mathfrak{B}_2 + \mathfrak{P},$$

which proves *(i)*.

Now, (1.4.5) further says that $\mathfrak{N}_m(\beta_i) + \mathfrak{P} \supset \mathfrak{B}_i + \mathfrak{P}$ with equality if and only if $m > k_i$. (2.1.5) therefore shows that, for $m \leq -r$, we have $\mathfrak{B}_1 + \mathfrak{P} = \mathfrak{N}_m(\beta_1) + \mathfrak{P}$ if and only if $\mathfrak{B}_2 + \mathfrak{P} = \mathfrak{N}_m(\beta_2) + \mathfrak{P}$. Assertion *(ii)* now follows.

To prove *(iii)*, we start by observing that $(\mathfrak{B}_i + \mathfrak{P})/\mathfrak{P} \cong \mathfrak{B}_i/\mathfrak{Q}_i$ as k_F-algebras. Part *(i)* therefore implies that the semisimple k_F-algebras $\mathfrak{B}_1/\mathfrak{Q}_1$, $\mathfrak{B}_2/\mathfrak{Q}_2$ are isomorphic. However, the quantities $e(E_i|F)$ and $f(E_i|F)$ can be recovered by comparing the Wedderburn structure constants of the k_F-algebras $\mathfrak{B}_i/\mathfrak{Q}_i$, $\mathfrak{A}/\mathfrak{P}$. Explicitly, we have (*cf.* (1.2.4))

$$e(E_i|F) = \frac{e(\mathfrak{A}|\mathfrak{o}_F)}{e(\mathfrak{B}_i|\mathfrak{o}_{E_i})},$$

and, for example, $e(\mathfrak{A}|\mathfrak{o}_F)$ is the number of simple components of $\mathfrak{A}/\mathfrak{P}$. Further, the residue field k_{E_i} is the centre of any simple factor of $\mathfrak{B}_i/\mathfrak{Q}_i$, and $f(E_i|F) = [\mathsf{k}_{E_i} : \mathsf{k}_F]$. ■

(2.2) Refinements of simple strata

Let $[\mathfrak{A}, n, r, \beta]$ be a simple stratum in the algebra $A = \mathrm{End}_F(V)$. As before, we write E for the field $F[\beta]$, B for the A-centraliser of E (or β), $\mathfrak{B} = \mathfrak{A} \cap B$, and $\mathfrak{Q} = \mathfrak{P} \cap B$, the Jacobson radical of the hereditary $\mathfrak{o}_E$-order $\mathfrak{B}$ in B. We also fix a tame corestriction s on A relative to E/F, as in **(1.3)**. A *refinement* of our given simple stratum is a stratum of the form $[\mathfrak{A}, n, r-1, \beta + b]$, where $b \in \mathfrak{P}^{-r}$. We can then form the *derived stratum* $[\mathfrak{B}, r, r-1, s(b)]$, which is a stratum in B. Of course, this depends on the choice of s, but only in a rather trivial way.

Remark: Suppose, for the moment, that we have $n > r > [\frac{n}{2}] \geq 0$, so that the characters

$$\psi_\beta \in (\boldsymbol{U}^{r+1}(\mathfrak{A})/\boldsymbol{U}^{n+1}(\mathfrak{A}))\hat{}, \quad \psi_{\beta+b} \in (\boldsymbol{U}^{r}(\mathfrak{A})/\boldsymbol{U}^{n+1}(\mathfrak{A}))\hat{}$$

are defined (see (1.5.3), (1.1.6)). Of course, $\psi_{\beta+b}|\boldsymbol{U}^{r+1}(\mathfrak{A}) = \psi_\beta$ as $b \in \mathfrak{P}^{-r}$. Consider the restriction of $\psi_{\beta+b}$ to $\boldsymbol{U}^r(\mathfrak{B}) = \boldsymbol{U}^r(\mathfrak{A}) \cap B$. Let ψ_B be the additive character of B related to s and ψ_A by (1.3.5). This restriction is then $\psi_{B,s(\beta+b)}$. The factor $\psi_{B,s(\beta)}$ is insignificant for most purposes: the element $s(\beta)$ lies in E (by (1.3.2)), so $\psi_{B,s(\beta)}$ is the restriction of a one-dimensional character of $B^\times$. (Equivalently, it factors through the determinant map $\det_B : B^\times \to E^\times$.) Thus the equivalence class of $[\mathfrak{B}, r, r-1, s(b)]$ determines the interesting component $\psi_{B,s(b)}$ of the restriction.

(2.2.1) Proposition: *Let $[\mathfrak{A}, n, r, \beta]$ be a simple stratum in A, as above, and let $[\mathfrak{A}, n, r-1, \beta+b]$, $[\mathfrak{A}, n, r-1, \beta+b']$ be refinements of it. Write $k = k_0(\beta, \mathfrak{A})$, $\mathfrak{N} = \mathfrak{N}_k(\beta, \mathfrak{A})$. The derived strata $[\mathfrak{B}, r, r-1, s(b)]$, $[\mathfrak{B}, r, r-1, s(b')]$ in B are then equivalent if and only if there exists $y \in \mathfrak{Q}^{-(r+k)}\mathfrak{N}$ such that*

$$[\mathfrak{A}, n, r-1, (1+y)^{-1}(\beta+b)(1+y)] \sim [\mathfrak{A}, n, r-1, \beta+b'].$$

Proof: We have $[\mathfrak{B}, r, r-1, s(b)] \sim [\mathfrak{B}, r, r-1, s(b')]$ if and only if $s(b-b') \in \mathfrak{Q}^{1-r}$. By (1.4.10), this is equivalent to the existence of $y \in \mathfrak{Q}^{-(r+k)}\mathfrak{N}$ with $b - b' \equiv a_\beta(y) \pmod{\mathfrak{P}^{1-r}}$. However, since $-(r+k) \geq 1$,

$$(1+y)^{-1}(\beta+b)(1+y) \equiv \beta + b + a_\beta(y) \pmod{\mathfrak{P}^{1-r}},$$

and the result follows. ∎

Therefore, up to an essentially trivial conjugation (which does not affect the equivalence class of the original stratum $[\mathfrak{A}, n, r, \beta]$), the equivalence class of the refinement $[\mathfrak{A}, n, r-1, \beta+b]$ only depends on that of the derived stratum $[\mathfrak{B}, r, r-1, s(b)]$.

Our next result is well known:— it is essentially due to Carayol (see **[Ca]**, especially (3.3)). However, it provides a useful illustration of our machinery in action in a comparatively straightforward situation, so we have chosen to give a complete proof. It is an unusual result in that it yields exact conclusions starting from data which are approximate in nature, being given in terms of congruences. It enables us to obtain very precise results in certain circumstances (like (2.2.3) below).

To start with, recall that a stratum of the form $[\mathfrak{A}, n, n-1, \alpha]$ is simple if and only if the algebra $F[\alpha]$ is a field, α is minimal over F and $\alpha \in \mathfrak{K}(\mathfrak{A})$ (see (1.4.14/15) and (1.5.6)).

(2.2.2) Proposition: *Let* $[\mathfrak{A}, n, n-1, \alpha]$ *be a simple stratum in* A *such that* $F[\alpha]$ *is a maximal subfield of* A. *Let* $[\mathfrak{A}, n, n-1, \alpha']$ *be a stratum equivalent to* $[\mathfrak{A}, n, n-1, \alpha]$. *Then* $[\mathfrak{A}, n, n-1, \alpha']$ *is simple and* $F[\alpha']$ *is a maximal subfield of* A.

Proof: The case $F[\alpha] = F$ is completely trivial here, so we exclude this possibility. Since $F[\alpha]$ is a maximal subfield of A, the A-centraliser of α is just $F[\alpha]$. Thus, by (1.5.8), the G-intertwining of $[\mathfrak{A}, n, n-1, \alpha]$ is $\boldsymbol{U}^1(\mathfrak{A})F[\alpha]^\times \boldsymbol{U}^1(\mathfrak{A})$, which equals $F[\alpha]^\times \boldsymbol{U}^1(\mathfrak{A})$, since $F[\alpha]^\times \subset \mathfrak{K}(\mathfrak{A})$ and $\mathfrak{K}(\mathfrak{A})$ normalises $\boldsymbol{U}^1(\mathfrak{A})$. In particular, this intertwining set (call it $\mathcal{I}$) is compact mod centre (in G) and contained in $\mathfrak{K}(\mathfrak{A})$. However, the equivalent strata $[\mathfrak{A}, n, n-1, \alpha]$, $[\mathfrak{A}, n, n-1, \alpha']$ have the same intertwining, and the G-centraliser of α' surely intertwines $[\mathfrak{A}, n, n-1, \alpha']$. We deduce that this centraliser is compact mod centre, and this implies (using, for example, the Jordan decomposition) that $F[\alpha']$ is a maximal subfield of A. Further, $F[\alpha']^\times \subset \mathcal{I} \subset \mathfrak{K}(\mathfrak{A})$, so the stratum $[\mathfrak{A}, n, n-1, \alpha']$ is at least pure.

Now consider the set $\mathcal{I} \cap \boldsymbol{U}(\mathfrak{A})$. This equals $\mathfrak{o}_{F[\alpha]}^\times \boldsymbol{U}^1(\mathfrak{A})$, and certainly contains $\mathfrak{o}_{F[\alpha']}^\times \boldsymbol{U}^1(\mathfrak{A})$. Since $F[\alpha]$, $F[\alpha']$ are fields, we have

$$\mathfrak{o}_{F[\alpha']} + \mathfrak{P} = \mathfrak{o}_{F[\alpha']}^\times \boldsymbol{U}^1(\mathfrak{A}) \,\dot{\cup}\, \mathfrak{P} \subset \mathfrak{o}_{F[\alpha]}^\times \boldsymbol{U}^1(\mathfrak{A}) \,\dot{\cup}\, \mathfrak{P} = \mathfrak{o}_{F[\alpha]} + \mathfrak{P}.$$

The index $(\mathfrak{o}_{F[\alpha]} + \mathfrak{P} : \mathfrak{P})$ is $(\mathfrak{o}_{F[\alpha]} : \mathfrak{p}_{F[\alpha]}) = \#\mathsf{k}_{F[\alpha]}$, and likewise for α'. On the other hand, we know that $F[\alpha]$, $F[\alpha']$ are maximal subfields of A which normalise $\mathfrak{A}$, so $f(F[\alpha]|F) = f(F[\alpha']|F) = \dim(V)/e(\mathfrak{A})$. We deduce

$$\mathfrak{o}_{F[\alpha']} + \mathfrak{P} = \mathfrak{o}_{F[\alpha]} + \mathfrak{P}.$$

Now consider the endomorphism of $\mathfrak{A}/\mathfrak{P}$ induced by $x \mapsto \alpha x \alpha^{-1} - x$. By definition, this has kernel $\mathfrak{N}_{1-n}(\alpha, \mathfrak{A}) + \mathfrak{P}/\mathfrak{P}$, which equals $\mathfrak{o}_{F[\alpha]} + \mathfrak{P}/\mathfrak{P}$ by (1.4.15). However, we have

$$\alpha x \alpha^{-1} \equiv \alpha' x \alpha'^{-1} \pmod{\mathfrak{P}}, \quad x \in \mathfrak{A}.$$

We deduce that

$$\mathfrak{N}_{1-n}(\alpha',\mathfrak{A})+\mathfrak{P}=\mathfrak{o}_{F[\alpha]}+\mathfrak{P}=\mathfrak{o}_{F[\alpha']}+\mathfrak{P}.$$

This implies $\mathfrak{N}_{1-n}(\alpha',\mathfrak{A})\subset\mathfrak{o}_{F[\alpha']}+\mathfrak{P}$, and hence $k_0(\alpha',\mathfrak{A})<1-n$. Since $F[\alpha']$ is a maximal subfield of A, we have $\alpha'\notin F$ and hence $k_0(\alpha',\mathfrak{A})\geq -n$. Therefore $k_0(\alpha',\mathfrak{A})=-n$, and the result follows from (1.4.15). ■

We now return to our simple stratum $[\mathfrak{A},n,r,\beta]$ above, and prove a special case of the main result of this section.

(2.2.3) Proposition: *Let $b\in\mathfrak{P}^{-r}$ and suppose that the stratum $[\mathfrak{B},r,r-1,s(b)]$ is simple. Suppose also that $E_1=E[s(b)]$ $(=F[\beta,s(b)])$ is a maximal subfield of B. Then*

(i) the stratum $[\mathfrak{A},n,r-1,\beta+b]$ is simple.

(ii) the field $K=F[\beta+b]$ satisfies $e(K|F)=e(E_1|F)$ (ramification indices) and $f(K|F)=f(E_1|F)$ (residue class degrees). In particular, K is a maximal subfield of A.

(iii) We have $k_0(\beta+b,\mathfrak{A})=\max\{k_0(\beta,\mathfrak{A}),k_0(s(b),\mathfrak{B})\}$, i.e.

$$k_0(\beta+b,\mathfrak{A})=\begin{cases}-r=k_0(s(b),\mathfrak{B}) & \text{if } s(b)\notin E,\\ k_0(\beta,\mathfrak{A}) & \text{if } s(b)\in E.\end{cases}$$

Remark: The hypothesis here is independent of the choice of the tame corestriction s. Also, the case $\beta\in F$ is trivial, as we may take s to be the identity map here. We therefore ignore this possibility in the proof.

Proof: We start by bounding the G-centraliser of the element $\beta+b$. Let $x\in G$ commute with $\beta+b$, and write $t=\nu_{\mathfrak{A}}(x)$. Then

$$0=(\beta+b)x-x(\beta+b)\equiv a_\beta(x)\pmod{\mathfrak{P}^{t-r}}.$$

Put $k=k_0(\beta,\mathfrak{A})$. (1.4.10) shows that

$$a_\beta(A)\cap\mathfrak{P}^{t-r}=a_\beta(\mathfrak{Q}^{t-r-k}\mathfrak{N}_k(\beta,\mathfrak{A}))\subset a_\beta(\mathfrak{P}^{t-r-k}).$$

In other words, there exists $y\in\mathfrak{P}^{t-r-k}$ such that $a_\beta(y)=a_\beta(x)$. The simplicity of $[\mathfrak{A},n,r,\beta]$ implies $t-r-k>t$, whence $x-y\in\mathfrak{P}^t\cap\operatorname{Ker}(a_\beta)=\mathfrak{Q}^t$. Therefore we may write $x=\gamma+y$, $\gamma\in\mathfrak{Q}^t$, $y\in\mathfrak{P}^{t-r-k}$, to get

$$0=(\beta+b)(\gamma+y)-(\gamma+y)(\beta+b)\equiv a_\beta(y)+b\gamma-\gamma b\pmod{\mathfrak{P}^{t-2r-k}}.$$

Applying s, we deduce

(2.2.4) $$a_{s(b)}(\gamma)\equiv 0\pmod{\mathfrak{Q}^{t-2r-k}}.$$

The next step is to show that (2.2.4) implies, for our element x,

$$x \in \mathfrak{p}_{E_1}^t + \mathfrak{P}^{t+1}.$$

For this, it is enough to show $\gamma \in \mathfrak{p}_{E_1}^t + \mathfrak{Q}^{t+1}$, since, by definition, $y = x - \gamma \in \mathfrak{P}^{t-r-k} \subset \mathfrak{P}^{t+1}$. Suppose first that we are in the case $s(b) \notin E$. Since E_1 is a maximal subfield of A (or B), the B-centraliser of $s(b)$ is precisely E_1. We have $E_1^\times \subset \mathfrak{K}(\mathfrak{A})$, so (1.2.4) gives $\mathfrak{A} \cap E_1 = \mathfrak{B} \cap E_1 = \mathfrak{o}_{E_1}$, since $\mathfrak{o}_{E_1}$ is the unique hereditary order in E_1. Further, $\mathfrak{Q}^m \cap E_1 = \mathfrak{p}_{E_1}^m$, for all $m \in \mathbb{Z}$. Since $s(b) \notin E$, (1.4.15) gives $k_0(s(b), \mathfrak{B}) = -r < -2r - k$, so (2.2.4) implies (via (1.4.10), as above), that $\gamma \in \mathfrak{p}_{E_1}^t + \mathfrak{Q}^{t-2r-k} \subset \mathfrak{p}_{E_1}^t + \mathfrak{Q}^{t+1}$, as required. In the case $s(b) \in E$, we have $E = E_1$, this is a maximal subfield of A, and $\mathfrak{Q} = \mathfrak{p}_E = \mathfrak{p}_{E_1}$. The relation $\gamma \in \mathfrak{p}_{E_1}^t + \mathfrak{Q}^{t+1}$ therefore holds trivially.

Thus, in all cases, we have $x \in \mathfrak{p}_{E_1}^t + \mathfrak{P}^{t+1}$, and $x \notin \mathfrak{P}^{t+1}$ by definition. Since $E_1^\times \subset \mathfrak{K}(\mathfrak{A})$, the set $\mathfrak{p}_{E_1}^t + \mathfrak{P}^{t+1} \setminus \mathfrak{P}^{t+1}$ is contained in $\mathfrak{K}(\mathfrak{A})$, and we conclude that the G-centraliser of $\beta + b$ is contained in $\mathfrak{K}(\mathfrak{A})$. In particular, this centraliser is compact mod centre, which implies that the algebra $K = F[\beta + b]$ is a field, and indeed a maximal subfield of A with $K^\times \subset \mathfrak{K}(\mathfrak{A})$. This fact that K is a maximal subfield of A has several other consequences:

(a) *$\mathfrak{A}$ is the unique hereditary $\mathfrak{o}_F$-order in A which is normalised by K, and so $\mathfrak{A} \cong \mathfrak{A}(K) = \mathrm{End}_{\mathfrak{o}_F}^0(\{\mathfrak{p}_K^i : i \in \mathbb{Z}\})$;*

(b) $e(K|F) = e(\mathfrak{A}|\mathfrak{o}_F) = e(E|F)e(\mathfrak{B}|\mathfrak{o}_E) = e(E|F)e(E_1|E) = e(E_1|F)$;

(c) $[K : F] = [E_1 : F] = N = \dim_F(V)$.

This proves *(ii)*. We have also shown that the stratum $[\mathfrak{A}, n, r-1, \beta+b]$ is pure.

It therefore remains only to compute $k_0(\beta+b, \mathfrak{A})$. We have $\beta \equiv \beta+b \pmod{\mathfrak{P}^{-r}}$, so $\mathfrak{N}_m(\beta, \mathfrak{A}) = \mathfrak{N}_m(\beta + b, \mathfrak{A})$ for $m \leq -r$, by (2.1.3). If $k_0(\beta + b, \mathfrak{A}) < -r$, we can apply (2.1.4) to get $k_0(\beta + b, \mathfrak{A}) = k_0(\beta, \mathfrak{A})$. Thus either $k_0(\beta + b, \mathfrak{A}) = k_0(\beta, \mathfrak{A})$ or else $k_0(\beta + b, \mathfrak{A}) \geq -r$.

Writing $k = k_0(\beta, \mathfrak{A})$, we have $k_0(\beta + b, \mathfrak{A}) = k$ if and only if $\mathfrak{N}_{k+1}(\beta + b, \mathfrak{A}) \subset \mathfrak{o}_K + \mathfrak{P}$ by the definition of k_0, and, since $\mathfrak{o}_K \subset \mathfrak{N}_k(\beta+b, \mathfrak{A})$, this is the same as saying that $\mathfrak{N}_{k+1}(\beta+b, \mathfrak{A})+\mathfrak{P} = \mathfrak{o}_K+\mathfrak{P}$. Since $k+1 \leq -r$, we have $\mathfrak{N}_{k+1}(\beta+b, \mathfrak{A})+\mathfrak{P} = \mathfrak{N}_{k+1}(\beta, \mathfrak{A})+\mathfrak{P} = \mathfrak{B}+\mathfrak{P}$ by (2.1.3). Thus $k_0(\beta + b, \mathfrak{A}) = k_0(\beta, \mathfrak{A})$ if and only if $\mathfrak{B} + \mathfrak{P} = \mathfrak{o}_K + \mathfrak{P}$. By *(ii)*, this can hold if and only if E is a maximal subfield of A, i.e. $s(b) \in E$, and $\mathfrak{B} = \mathfrak{o}_E$.

So, we now assume that $s(b) \notin E$ and show that $\mathfrak{N}_{1-r}(\beta + b, \mathfrak{A}) \subset \mathfrak{o}_K + \mathfrak{P}$. This will finish the proof of the Proposition. To do this, take $x \in \mathfrak{N}_{1-r}(\beta + b, \mathfrak{A})$, so that

$$(\beta + b)x - x(\beta + b) \equiv 0 \pmod{\mathfrak{P}^{1-r}}.$$

Since $x \in \mathfrak{A}$, this implies $a_\beta(x) \in \mathfrak{P}^{-r}$. Arguing as before, (1.4.10) now implies $x = \gamma + y$, for some $\gamma \in \mathfrak{B}$, $y \in \mathfrak{P}^{-r-k}$, whence

$$0 \equiv (\beta + b)(\gamma + y) - (\gamma + y)(\beta + b) \equiv a_\beta(y) + b\gamma - \gamma b \pmod{\mathfrak{P}^{1-r}}.$$

(The terms yb, by lie in $\mathfrak{P}^{-2r-k} \subset \mathfrak{P}^{1-r}$, as $k < -r$.) Therefore $a_{s(b)}(\gamma) \in \mathfrak{Q}^{1-r}$ and, as $-r = k_0(s(b), \mathfrak{B})$, this means $\gamma \in \mathfrak{o}_{E_1} + \mathfrak{Q}$ and so $x \in \mathfrak{o}_{E_1} + \mathfrak{P}$. Altogether, we get

$$\textbf{(2.2.5)} \qquad \mathfrak{o}_K + \mathfrak{P} \subset \mathfrak{N}_{1-r}(\beta + b, \mathfrak{A}) + \mathfrak{P} \subset \mathfrak{o}_{E_1} + \mathfrak{P}.$$

We have $(\mathfrak{o}_K + \mathfrak{P} : \mathfrak{P}) = (\mathfrak{o}_K : \mathfrak{p}_K)$, and $(\mathfrak{o}_{E_1} + \mathfrak{P} : \mathfrak{P}) = (\mathfrak{o}_{E_1} : \mathfrak{p}_{E_1})$ which equals $(\mathfrak{o}_K : \mathfrak{p}_K)$ by *(ii)*. The containments (2.2.5) are therefore equalities, and we deduce that $\mathfrak{N}_{1-r}(\beta + b, \mathfrak{A}) \subset \mathfrak{o}_K + \mathfrak{P}$, as required. ■

(2.2.6) Warning: (2.2.3) (and also the following theorem (2.2.8)) says nothing whatsoever about the element b. In general, it will not generate a field, nor will it commute with β. Indeed, suppose that the field extension $F[\beta]/F$ is not tamely ramified, and take a refinement $[\mathfrak{A}, n, r-1, \beta + b]$ of $[\mathfrak{A}, n, r, \beta]$, where b commutes with β. This says $b \in \mathfrak{Q}^{-r}$. By (1.3.8)*(iii)*, we have $s(b) \in \mathfrak{Q}^{1-r}$, so the derived stratum $[\mathfrak{B}, r, r-1, s(b)]$ is equivalent to $[\mathfrak{B}, r, r-1, 0]$. (2.2.1) now says that $[\mathfrak{A}, n, r-1, \beta + b]$ is equivalent to a conjugate of $[\mathfrak{A}, n, r-1, \beta]$. So, in the presence of wild ramification, we can get no interesting new fields from this refinement process by using a "perturbation" b which commutes with β.

By contrast, if $F[\beta]/F$ is tamely ramified, we can assume that s is the identity map on B (see (1.3.8) again). In this case, up to conjugation and equivalence, all refinements of $[\mathfrak{A}, n, r, \beta]$ arise from elements b which commute with β, and, in the situation above, there is no distinction between the fields $F[\beta + b]$, $F[\beta, s(b)]$.

(2.2.7) Remark: The proof of (2.2.3) gives more general results of the following sort. Take $[\mathfrak{A}, n, r, \beta]$ as before, and $b \in \mathfrak{P}^{-r}$. Assume that $[\mathfrak{B}, r, r-1, s(b)]$ is pure, that $E_1 = E(s(b))$ is a maximal subfield of B, and that

$$-2r > k_0(\beta, \mathfrak{A}) + k_0(s(b), \mathfrak{B}).$$

Then again $F[\beta + b]$ is a maximal subfield of A normalising $\mathfrak{A}$, $e(F[\beta + b]|F) = e(E_1|F)$, and $k_0(\beta + b, \mathfrak{A}) = \max\{k_0(\beta, \mathfrak{A}), k_0(s(b), \mathfrak{B})\}$. A similar remark holds in the more general situation of (2.2.8) below.

We now give a more general version of (2.2.3) without the assumption that various subfields are maximal. This leads inevitably to some loss of precision, because of the failure of (2.2.2) in this situation.

(2.2.8) Theorem: *Let $[\mathfrak{A}, n, r, \beta]$ be a simple stratum in A. Let B be the A-centraliser of $E = F[\beta]$, and $\mathfrak{B} = B \cap \mathfrak{A}$. Let $b \in A$ with $\nu_{\mathfrak{A}}(b) = -r$, and let s be a tame corestriction on A relative to $F[\beta]/F$. Suppose that the stratum $[\mathfrak{B}, r, r-1, s(b)]$ is equivalent to some simple stratum $[\mathfrak{B}, r, r-1, c]$ in B. Then $[\mathfrak{A}, n, r-1, \beta + b]$ is equivalent to a simple stratum $[\mathfrak{A}, n, r-1, \beta_1]$. Moreover, if $E_1 = F[\beta, c]$, $K = F[\beta_1]$, we have*

(i) $e(K|F) = e(E_1|F)$, $f(K|F) = f(E_1|F)$;

(ii) $k_0(\beta_1, \mathfrak{A}) = \max\{k_0(\beta, \mathfrak{A}), k_0(c, \mathfrak{B})\}$.

Remark: By (2.1.4), everything here is independent of the choices of the elements β_1, c, subject to the stated conditions.

Proof: Let B_1 denote the A-centraliser of the field E_1, and put $\mathfrak{B}_1 = \mathfrak{A} \cap B_1$. We apply (1.2.8) relative to the extension E_1/F to get a decomposition

$$\mathfrak{A} = \mathfrak{A}(E_1) \otimes_{\mathfrak{o}_{E_1}} \mathfrak{B}_1. \tag{2.2.9}$$

The next step is to show that our tame corestriction s on A relative to E/F takes the form $s' \otimes 1_{B_1}$, for some tame corestriction s' on $A(E_1)$ relative to E/F. We have a decomposition $A(E_1) = A(E) \otimes_E \operatorname{End}_E(E_1)$, so $A = A(E) \otimes_E \operatorname{End}_E(E_1) \otimes_{E_1} B_1$, which collapses to $A(E) \otimes_E B$. A tame corestriction s' on $A(E_1)$ relative to E/F is of the form $s'' \otimes 1_{\operatorname{End}_E(E_1)}$, for a tame corestriction s'' on $A(E)$ relative to E/F, by (1.3.9). Thus $s' \otimes 1_{B_1} = s'' \otimes 1_B$, which is a tame corestriction on A relative to E/F, by (1.3.9). We can therefore adjust s' (by a unit of $\mathfrak{o}_E$) to get $s' \otimes 1 = s$.

Next, let $\mathfrak{C}$ denote the hereditary $\mathfrak{o}_E$-order $\mathfrak{A}(E_1) \cap \operatorname{End}_E(E_1)$. The decomposition $A(E_1) = A(E) \otimes_E \operatorname{End}_E(E_1)$ gives $\mathfrak{A}(E_1) = \mathfrak{A}(E) \otimes_{\mathfrak{o}_E} \mathfrak{C}$ and, further, $\mathfrak{B} = \mathfrak{C} \otimes_{\mathfrak{o}_{E_1}} \mathfrak{B}_1$.

Put $e_1 = e(\mathfrak{B}_1|\mathfrak{o}_{E_1})$. This gives us $\nu_{\mathfrak{A}(E_1)}(\beta) = -n/e_1$, $\nu_{\mathfrak{A}(E_1)}(c) = -r/e_1$. The strata $[\mathfrak{A}(E_1), n/e_1, r/e_1, \beta]$, $[\mathfrak{C}, r/e_1, r/e_1 - 1, c]$ are simple and $k_0(\beta, \mathfrak{A}(E_1)) = k_0(\beta, \mathfrak{A})/e_1$, by (1.4.13). Choose $b_1 \in A(E_1)$ with $\nu_{\mathfrak{A}(E_1)}(b_1) = -r/e_1$ such that $s'(b_1) = c$. By (2.2.3), the stratum $[\mathfrak{A}(E_1), n/e_1, r/e_1 - 1, \beta + b_1]$ is simple with the necessary properties. Consider the stratum $[\mathfrak{A}, n, r-1, (\beta + b_1) \otimes 1]$. We have $F[\beta + b_1]^\times \subset \mathfrak{K}(\mathfrak{A}(E_1)) \subset \mathfrak{K}(\mathfrak{A})$, so this stratum is pure, and it refines $[\mathfrak{A}, n, r, \beta]$. Moreover,

$$[\mathfrak{B}, r, r-1, s(b_1 \otimes 1)] = [\mathfrak{B}, r, r-1, s'(b_1) \otimes 1] = [\mathfrak{B}, r, r-1, c].$$

(We can identify c with $c \otimes 1$ in (2.2.9).) We can use (2.2.1) to find a conjugate of $[\mathfrak{A}, n, r-1, (\beta + b_1) \otimes 1]$ equivalent to $[\mathfrak{A}, n, r-1, \beta + b]$. To prove the theorem, therefore, we have only to compute $k_0((\beta + b_1) \otimes 1, \mathfrak{A})$. Set $K = F[\beta + b_1]$. By (2.2.3), we have $e(K|F) = e(E_1|F)$, $[K :$

$F] = [E_1 : F]$, and K normalises $\mathfrak{A}(E_1)$. Thus we have $e(\mathfrak{A}(E_1)|\mathfrak{o}_F) = e(E_1|F) = e(K|F) = e(\mathfrak{A}(K)|\mathfrak{o}_F)$ whence, by (1.4.13), $k_0(\beta+b, \mathfrak{A}(E_1)) = k_0(\beta + b, \mathfrak{A}(K))$. However, (1.4.13) further gives us $k_0(\beta + b_1, \mathfrak{A}) = e_K k_0(\beta + b_1, \mathfrak{A}(K))$, where e_K is the ramification over $\mathfrak{o}_K$ of the $\mathfrak{A}$-centraliser of K. This is just $e(\mathfrak{A}|\mathfrak{o}_F)/e(K|F) = e_1$, by (2.2.3). The theorem follows. ■

We shall see below that all simple strata in A can be constructed by iterating the process of (2.2.8), and that this process has strong uniqueness properties.

(2.3) Split refinements

We now need to consider a rather different sort of refinement of a simple stratum. This is also a convenient place to set down, in our present language, stronger versions of some standard results. The justifications for some of these are postponed to **(2.5)**.

As before, $\mathfrak{A}$ is a hereditary $\mathfrak{o}_F$-order in A, with radical $\mathfrak{P}$. Recall **[Bu1]** that a stratum of the form $[\mathfrak{A}, n, n-1, b]$ is called *fundamental* if $b + \mathfrak{P}^{1-n}$ does not contain a nilpotent element of A. Of course, this property only depends on the equivalence class of the stratum. The main result concerning these is:

(2.3.1) *Suppose that $[\mathfrak{A}, n, n-1, b]$ is not fundamental. There exists a hereditary order $\mathfrak{A}_1$ in A, with radical $\mathfrak{P}_1$, and an integer n_1, such that*

$$b + \mathfrak{P}^{1-n} \subset \mathfrak{P}_1^{-n_1} \quad \text{and} \quad n/e > n_1/e_1,$$

where $e = e(\mathfrak{A}|\mathfrak{o}_F)$, $e_1 = e(\mathfrak{A}_1|\mathfrak{o}_F)$.

For a proof, see **[Bu1]** Th. 1. One can further choose the order $\mathfrak{A}_1$ in (2.3.1) so that the pair $(\mathfrak{A}, \mathfrak{A}_1)$ satisfies (1.1.9):— see **[KM2]** Prop. 3.5.

It is also necessary to observe

(2.3.2) *Let $[\mathfrak{A}, n, r, \beta]$ be a pure stratum in A. Then $[\mathfrak{A}, n, n-1, \beta]$ is fundamental.*

For, since $\beta \in \mathfrak{K}(\mathfrak{A})$ and $\nu_{\mathfrak{A}}(\beta) = -n$, we have $\beta + \mathfrak{P}^{1-n} = \beta \boldsymbol{U}^1(\mathfrak{A}) \subset \mathfrak{K}(\mathfrak{A}) \subset G$, which surely contains no nilpotent element of A.

One can give a rough classification of fundamental strata as in **[K4]**. Let $[\mathfrak{A}, n, n-1, b]$ be fundamental. Choose a prime element π_F of F, and set

$$y = y_b = b^{e/g}\pi_F^{n/g} + \mathfrak{P},$$

where $e = e(\mathfrak{A})$ and $g = \gcd(e,n)$. As an element of $\mathfrak{A}/\mathfrak{P}$, this depends only on the equivalence class of the stratum. If $\mathfrak{A} = \mathrm{End}^0_{\mathfrak{o}_F}(\mathcal{L}), \mathcal{L} = \{L_i\}$, we can identify

$$\mathfrak{A}/\mathfrak{P} = \coprod_{i=0}^{e-1} \mathrm{End}_{\mathrm{k}_F}(L_i/L_{i+1}) \subset \mathrm{End}_{\mathrm{k}_F}(L_0/L_e).$$

We define $\phi_b(X) \in \mathrm{k}_F[X]$ by

$$\phi_b(X) = \text{ the characteristic polynomial of } y \in \mathrm{End}_{\mathrm{k}_F}(L_0/L_e).$$

Of course, this property only depends on the *equivalence class* of the original stratum $[\mathfrak{A}, n, n-1, b]$. Alternatively, we can form the element $Y = b^{e/g}\pi_F^{n/g} \in A$, in the notation above, and let $\Phi(X) \in F[X]$ be the characteristic polynomial of Y as an F-endomorphism of V. Then, since $Y \in \mathfrak{A}$, we have $\Phi(X) \in \mathfrak{o}_F[X]$, and $\phi_b(X)$ is just the reduction of $\Phi(X)$ modulo $\mathfrak{p}_F$.

(2.3.3) Definition: *A fundamental stratum* $[\mathfrak{A}, n, n-1, b]$ *is called* split *if the polynomial* $\phi_b(X)$ *has at least two distinct irreducible factors in* $\mathrm{k}_F[X]$.

Otherwise, if $\phi_b(X)$ is a power of some irreducible polynomial over k_F, we say that the stratum $[\mathfrak{A}, n, n-1, b]$ is *nonsplit.*

(2.3.4) *Let* $[\mathfrak{A}, n, n-1, b]$ *be a nonsplit fundamental stratum. There is a simple stratum* $[\mathfrak{A}', n', n'-1, \alpha]$ *such that*

$$b + \mathfrak{P}^{1-n} \subset \alpha + \mathfrak{P}'^{1-n'},$$

where $\mathfrak{P}' = \mathrm{rad}(\mathfrak{A}')$, $\mathfrak{P} = \mathrm{rad}(\mathfrak{A})$. *Moreover,* $n/e(\mathfrak{A}) = n'/e(\mathfrak{A}')$, *and the lattice chain defining* $\mathfrak{A}'$ *contains that which defines* $\mathfrak{A}$. *In particular,* $[\mathfrak{A}', n', n'-1, b]$ *is a stratum and it is equivalent to* $[\mathfrak{A}', n', n'-1, \alpha]$.

This is a slightly more general version of some ideas in **[K4]** (see especially the proof of Th. 3.2). However, **[K4]** imposes the hypothesis $\gcd(n, e(\mathfrak{A})) = 1$, which we wish to avoid here. The necessary justification is given in **(2.5)** below.

We now turn to the split case and recall some more of the ideas of **[K4]**. First, an *(F-)splitting* of the vector space V is an ordered pair (V^1, V^2) of proper (F-) subspaces of V such that

$$V = V^1 \oplus V^2.$$

Given such a splitting of V, we define $A_{ij} = \mathrm{Hom}_F(V^j, V^i)$, $i, j \in \{1, 2\}$, so that

$$A = \coprod_{i,j} A_{ij},$$

which we view as a block matrix decomposition of A in the customary manner. If $x \in A$, we write x_{ij} for its component in A_{ij}. We say (V^1, V^2) *splits* x if $x_{ij} = 0$ when $i \neq j$, i.e.

$$x = \begin{pmatrix} x_{11} & 0 \\ 0 & x_{22} \end{pmatrix}.$$

(2.3.5) Lemma: *Let $a \in A$, and let (V^1, V^2) be a splitting of V which splits a. Define*

$$\partial_a : A_{12} \longrightarrow A_{12},$$
$$\partial_a(x) = a_{11}x - xa_{22}, \quad x \in A_{12}.$$

Then ∂_a is an isomorphism if and only if the characteristic polynomials of a_{11}, a_{22} are relatively prime.

Proof: The eigenvalues of the operator ∂_a are differences of the eigenvalues of a_{11}, a_{22}. ∎

Next, let $\mathcal{L} = \{L_i : i \in \mathbb{Z}\}$ be an $\mathfrak{o}_F$-lattice chain in V and (V^1, V^2) an F-splitting of V. We say that (V^1, V^2) *splits* $\mathcal{L}$ if

(2.3.6) *(i) $L_i = L_i^1 \oplus L_i^2$, $i \in \mathbb{Z}$, where $L_i^j = L_i \cap V^j$;*
(ii) $\mathcal{L}^1 = \{L_i^1 : i \in \mathbb{Z}\}$ is a uniform $\mathfrak{o}_F$-lattice chain in V^1 with period $e(\mathcal{L}^1) = e(\mathcal{L})$.

In this situation, the set $\{L_i^2\}$ (after a renumbering) is a lattice chain in V^2 of period at most $e(\mathcal{L})$. Also, if $\mathfrak{A} = \mathrm{End}^0_{\mathfrak{o}_F}(\mathcal{L})$, and $\mathfrak{P} = \mathrm{rad}(\mathfrak{A})$, it is easy to check that

(2.3.7) $$\mathfrak{P}^n = \coprod_{i,j} \mathfrak{P}^n \cap A_{ij}, \quad n \in \mathbb{Z},$$

since the projections $V \to V^i$ lie in $\mathfrak{A}$. In particular, the set $\mathcal{M} = \{M_j : j \in \mathbb{Z}\}$ given by

$$M_j = \mathfrak{P}^j \cap A_{12} = \{x \in A_{12} : xL_i^2 \subset L_{i+j}^1, \quad i \in \mathbb{Z}\}$$

is a uniform $\mathfrak{o}_F$-lattice chain in A_{12} with period $e(\mathcal{M}) = e(\mathcal{L})$, (see **[K4]** (2.2)).

Now let $[\mathfrak{A}, n, n-1, b]$ be a stratum in A and (V^1, V^2) a splitting of V. Write $\mathfrak{A} = \mathrm{End}^0_{\mathfrak{o}_F}(\mathcal{L})$ for a lattice chain $\mathcal{L}$. We say that (V^1, V^2) *splits* $[\mathfrak{A}, n, n-1, b]$ if

(2.3.8) *(i)* (V^1, V^2) *splits* $\mathcal{L}$ *and* b;
(ii) $b_{11} L^1_i = L^1_{i-n}$ *for all* i;
(iii) $\partial_b(M_i) = M_{i-n}$ *for all* i.

It is not hard to show that if $[\mathfrak{A}, n, n-1, b]$ possesses a splitting, then it is split fundamental. Further, the condition *(iii)* implies that ∂_b acts as an automorphism of A_{12}. (2.3.5) now implies that, if the stratum $[\mathfrak{A}, n, n-1, b]$ has a splitting, then the algebra $F[b]$ cannot be a field.

In the opposite direction, let $[\mathfrak{A}, n, n-1, b]$ be split fundamental, and assume that

(2.3.9) $$\gcd(n, e(\mathfrak{A})) = 1.$$

We recall the construction of a splitting of $[\mathfrak{A}, n, n-1, b]$. Fix a prime element π_F of F and put $Y = \pi_F^n b^e$, $e = e(\mathfrak{A})$. Let $\Phi(X)$ be the characteristic polynomial of Y. We write "bar" for the map $\mathfrak{o}_F[X] \to \mathrm{k}_F[X]$ given by reduction mod $\mathfrak{p}_F$, so that $\bar{\Phi}(X) = \phi_b(X)$ in our earlier notation. By Hensel's Lemma we have

$$\Phi(X) = f(X)g(X)$$

for some monic nonconstant polynomials $f(X), g(X) \in \mathfrak{o}_F[X]$ satisfying

$$\gcd(\bar{f}(X), \bar{g}(X)) = 1.$$

In particular, f, g are relatively prime in $F[X]$. We further arrange matters so that

$$\bar{\Phi}(0) = 0 \Rightarrow \bar{g}(0) = 0 \quad \textit{whence } \bar{f}(0) \neq 0.$$

Set $V^1 = \mathrm{Ker}(f(Y))$, $V^2 = \mathrm{Ker}(g(Y))$. Then by **[K4]** (3.4),

(2.3.10) (V^1, V^2) *splits* $[\mathfrak{A}, n, n-1, b]$.

Combining this with (2.3.2) and the remark following (2.3.8), we have

(2.3.11) *Let* $[\mathfrak{A}, n, r, \beta]$ *be a pure stratum in* A. *Then* $[\mathfrak{A}, n, n-1, \beta]$ *is nonsplit fundamental.*

We can now give the one genuinely new result of this section. Observe that this is another result, like (2.2.2) and (2.2.3), which derives precise conclusions from approximate data.

(2.3.12) Proposition: *Let $[\mathfrak{A}, n, r, \beta]$ be a simple stratum in A, and use the associated notations of* **(2.2)**. *Let $b \in \mathfrak{P}^{-r}$ and suppose that the stratum $[\mathfrak{B}, r, r-1, s(b)]$ is split fundamental. Then the algebra $F[\beta+b]$ is not a field.*

Remark: The hypothesis here only depends on $b \bmod \mathfrak{P}^{1-r}$. The proposition therefore asserts that $F[\beta']$ is not a field, for any $\beta' \in \beta + b + \mathfrak{P}^{1-r}$.

Proof: Write $e_\beta = e(\mathfrak{B}|\mathfrak{o}_E) = e(\mathfrak{A}|\mathfrak{o}_F)/e(E|F)$. We first reduce to the case where $\gcd(r, e_\beta) = 1$. Put $g = \gcd(r, e_\beta)$, $\mathfrak{A} = \mathrm{End}^0_{\mathfrak{o}_F}(\mathcal{L})$, $\mathcal{L} = \{L_i\}$. Define an $\mathfrak{o}_E$-lattice chain $\mathcal{L}' = \{L'_i\}$ in V by setting $L'_i = L_{ig}$, $i \in \mathbb{Z}$. Then $\mathcal{L}'$ has $\mathfrak{o}_E$-period e_β/g. If we write $\mathfrak{A}' = \mathrm{End}^0_{\mathfrak{o}_F}(\mathcal{L}')$, $\mathfrak{P}' = \mathrm{rad}(\mathfrak{A}')$, we have

$$\mathfrak{P}'^m \supset \mathfrak{P}^{mg} \supset \mathfrak{P}^{mg+1} \supset \mathfrak{P}'^{m+1}, \quad m \in \mathbb{Z}.$$

Intersecting with B we get

$$\mathfrak{Q}'^m \supset \mathfrak{Q}^{mg} \supset \mathfrak{Q}^{mg+1} \supset \mathfrak{Q}'^{m+1},$$

where $\mathfrak{Q}' = \mathfrak{P}' \cap B = \mathrm{End}^1_{\mathfrak{o}_E}(\mathcal{L}')$. The coset $\beta + b + \mathfrak{P}^{1-r}$ is therefore a union of cosets $\beta + b + c + \mathfrak{P}'^{1-r/g}$ with $c \in \mathfrak{P}^{1-r}$. The stratum $[\mathfrak{A}', n/g, r/g, \beta]$ is simple by (1.4.13), and we show that $[\mathfrak{B}', r/g, r/g - 1, s(b+c)]$ is split fundamental, where $\mathfrak{B}' = \mathfrak{A}' \cap B$. Since $\gcd(r/g, e_\beta/g)$ is 1, this gives our desired reduction. We compute the charactersitic polynomial $\phi_{s(b+c)}(X)$ relative to a prime element π_E of E. This is the reduction modulo $\mathfrak{p}_E$ of the characteristic polynomial of $\pi_E^{r/g}.s(b+c)^{e_\beta/g}$ and is therefore the same whether we calculate it relative to $\mathfrak{B}$ or $\mathfrak{B}'$. Thus, since $s(c) \in \mathfrak{Q}^{1-r}$, we have $\phi_{s(b+c)} = \phi_{s(b)}$, so $[\mathfrak{B}', r/g, r/g - 1, s(b+c)]$ is split fundamental, as required.

Therefore we now assume that $(r, e_\beta) = 1$ and write $\gamma = \beta + b$. Let (V^1, V^2) be a splitting of $[\mathfrak{B}, r, r-1, s(b)]$ as in (2.3.10) (but observe that we now work relative to the field $E = F[\beta]$). Since, in particular, the V^i are E-spaces, (V^1, V^2) splits β. We define

$$b' = \begin{pmatrix} b_{11} & 0 \\ 0 & b_{22} \end{pmatrix}.$$

Then $b' \in \mathfrak{P}^{-r}$ and $s(b') = s(b)$.

We observe that, in this situation, we can use our (a_β, s)-calculus "block by block". More precisely, write e_i for the projection $V \to V^i$ with kernel V^j, $j \neq i$. Then $\mathrm{e}_i \in B$, $i = 1, 2$, and $A_{ij} = \mathrm{Hom}_F(V^j, V^i) = \mathrm{e}_i A \mathrm{e}_j$. Indeed, we have $\mathrm{e}_i \in \mathfrak{B}$. Thus, if M is any $(\mathfrak{B}, \mathfrak{B})$-bimodule in

A, we get

$$M \cap A_{ij} = e_i M e_j \quad \text{and}$$
$$M = \coprod_{i,j \in \{1,2\}} e_i M e_j .$$

Moreover, for any $(\mathfrak{B}, \mathfrak{B})$-bimodule homomorphism $\phi : M \to M'$, for example $\phi = a_\beta$ or s, we have $\phi(e_i M e_j) \subset e_i M' e_j$. In particular, a sequence

$$M \xrightarrow{s} M' \xrightarrow{a_\beta} M''$$

is exact if and only if all the sequences

$$e_i M e_j \xrightarrow{s} e_i M' e_j \xrightarrow{a_\beta} e_i M'' e_j$$

are exact, for $i, j \in \{1, 2\}$. Likewise with the roles of s and a_β reversed.

(2.3.13) Lemma: *Write $M_i = \mathfrak{P}^i \cap A_{12}$ as before. Then*

$$M_{i-r} \subset \partial_{\beta+b'}(M_i) + M_{1+i-r}, \quad i \in \mathbb{Z}.$$

Proof: Since, as we have just noted, the map a_β respects the decomposition $A = \coprod A_{ij}$, (1.4.10) and (2.3.7) together imply

$$M_i \cap a_\beta(A) \subset a_\beta(M_{i-k}), \quad i \in \mathbb{Z},$$

where $k = k_0(\beta, \mathfrak{A})$. In particular,

$$M_{i-r} \cap a_\beta(A) \subset a_\beta(M_{i-r-k}).$$

For $x \in M_{i-r-k}$, we have

$$\partial_{\beta+b'}(x) \equiv a_\beta(x) \pmod{M_{1+i-r}}$$

since $b' \in \mathfrak{P}^{-r}$ and $-(r+k) \geq 1$. Therefore

$$M_{i-r} \cap a_\beta(A) \subset \partial_{\beta+b'}(M_{i+1}) + M_{1+i-r} \tag{2.3.14}$$

which, of course, implies the weaker statement

$$M_{i-r} \cap a_\beta(A) \subset \partial_{\beta+b'}(M_i) + M_{1+i-r}.$$

So, to prove the lemma, we need only check that

$$s(M_{i-r}) \subset s(\partial_{\beta+b'}(M_i)) + s(M_{1+i-r}).$$

Since s respects the blocks A_{ij}, we have $s(M_j) = \mathfrak{Q}^j \cap A_{12}$, $j \in \mathbb{Z}$. Also, if $y \in \mathfrak{Q}^i \cap A_{12}$ (which is contained in M_i), then

$$s(\partial_{\beta+b'}(y)) = \partial_{s(b)}(y)$$

since $s(b') = s(b)$. From the definition of splitting, we have $s(M_{i-r}) = \mathfrak{Q}^{i-r} \cap A_{12} \subset \partial_{s(b)}(\mathfrak{Q}^i \cap A_{12})$, and hence the result. ■

(2.3.15) Corollary: *In the situation of* (2.3.13), *we have* $M_{i-r} \subset \partial_{\beta+b'}(M_i)$, $i \in \mathbb{Z}$, *and so* $\partial_{\beta+b'} : A_{12} \to A_{12}$ *is an isomorphism.*

Proof: The first assertion follows from iteration and a standard completeness argument. It implies, in particular, that $\partial_{\beta+b'}(A_{12}) \supset M_{i-r}$, whence the linear map $\partial_{\beta+b'}$ induces a surjection $A_{12} \to A_{12}$. This surjection must be an isomorphism. ■

In particular, we now know from (2.3.5) that $F[\beta + b']$ cannot be a field. Our original element $\gamma = \beta + b$ now has the form

$$\gamma = \begin{pmatrix} \beta + b_{11} & y \\ z & \beta + b_{22} \end{pmatrix}$$

with $y, z \in \mathfrak{P}^{-r} \cap a_\beta(A)$. We now construct a conjugate

$$\tilde{\gamma} = \begin{pmatrix} 1 & t \\ 0 & 1 \end{pmatrix} \gamma \begin{pmatrix} 1 & -t \\ 0 & 1 \end{pmatrix} \tag{2.3.16}$$

of γ, where $t \in M_1$, of the form

$$\tilde{\gamma} = \begin{pmatrix} \beta + \tilde{b}_{11} & 0 \\ z & \beta + \tilde{b}_{22} \end{pmatrix}, \quad \tilde{b} = \begin{pmatrix} \tilde{b}_{11} & 0 \\ 0 & \tilde{b}_{22} \end{pmatrix}$$

such that $\partial_{\beta+\tilde{b}}$ is an automorphism of A_{12}. Then $F[\gamma] \cong F[\tilde{\gamma}]$ as F-algebras, and $F[\tilde{\gamma}]$ is not a field by (2.3.5).

We construct the element t by successive approximation. The first step is to use (2.3.14) to find $t_0 \in M_1$ such that

$$y \equiv \partial_{\beta+b'}(t_0) \pmod{M_{1-r}}.$$

Then

$$\begin{pmatrix} 1 & t_0 \\ 0 & 1 \end{pmatrix} \gamma \begin{pmatrix} 1 & -t_0 \\ 0 & 1 \end{pmatrix} = \begin{pmatrix} \beta + b_{11} + t_0 z & y_1 \\ z & \beta + b_{22} - z t_0 \end{pmatrix}$$

where $y_1 \in M_{1-r}$. The terms zt_0, $t_0 z$ lie in $\mathfrak{P}^{1-r}$. Put

$$b^{(1)} = \begin{pmatrix} b_{11} + t_0 z & 0 \\ 0 & b_{22} - z t_0 \end{pmatrix}.$$

Then, for $x \in M_i$, we have

$$\partial_{\beta+b^{(1)}}(x) \equiv \partial_{\beta+b'}(x) \pmod{M_{1+i-r}}.$$

Therefore, by (2.3.13),

$$M_{i-r} \subset \partial_{\beta+b^{(1)}}(M_i) + M_{1+i-r}, \quad i \in \mathbb{Z}.$$

Inductively, take $m \geq 1$, and suppose we have found $t_{m-1} \in M_1$ such that

$$\gamma_{m-1} = \begin{pmatrix} 1 & t_{m-1} \\ 0 & 1 \end{pmatrix} \gamma \begin{pmatrix} 1 & -t_{m-1} \\ 0 & 1 \end{pmatrix} = \begin{pmatrix} \beta + b_{11}^{(m)} & y_m \\ z & \beta + b_{22}^{(m)} \end{pmatrix}$$

with $y_m \in M_{m-r}$ and

$$b^{(m)} = \begin{pmatrix} b_{11}^{(m)} & 0 \\ 0 & b_{22}^{(m)} \end{pmatrix}$$

such that

$$\partial_{\beta+b^{(m)}}(x) \equiv \partial_{\beta+b'}(x) \pmod{M_{1+i-r}}, \quad x \in M_i.$$

In particular, we have

$$M_{i-r} \subset \partial_{\beta+b^{(m)}}(M_i) + M_{1+i-r}, \quad i \in \mathbb{Z}.$$

Take $v \in M_m$ such that $\partial_{\beta+b^{(m)}}(v) \equiv y_m \pmod{M_{1+m-r}}$ and consider

$$\gamma_m = \begin{pmatrix} 1 & v \\ 0 & 1 \end{pmatrix} \gamma_{m-1} \begin{pmatrix} 1 & -v \\ 0 & 1 \end{pmatrix} = \begin{pmatrix} \beta + b_{11}^{(m)} + vz & y_{m+1} \\ z & \beta + b_{22}^{(m)} - zv \end{pmatrix},$$

so that $y_{m+1} \in M_{m+1-r}$. Here, $vz, zv \in \mathfrak{P}^{m-r}$. We put

$$b^{(m+1)} = \begin{pmatrix} b_{11}^{(m)} + vz & 0 \\ 0 & b_{22}^{(m)} - zv \end{pmatrix},$$
$$t_m = t_{m-1} + v.$$

The map $\partial_{\beta+b^{(m+1)}}$ has the desired property

$$\partial_{\beta+b^{(m+1)}}(x) \equiv \partial_{\beta+b'}(x) \pmod{M_{1+i-r}}$$

for $x \in M_i$ and $i \in \mathbb{Z}$. Our induction therefore continues. We have $t_m \equiv t_{m-1} \pmod{M_{m-r}}$, so the sequence $\{t_m\}$ converges to a limit $t \in M_1$, while $y_m \to 0$. This limit t is the element we require. ■

(2.4) Approximation of simple strata

In **(2.2)**, we showed how to construct simple strata by a refinement process. This section considers the converse of this, and the main result demonstrates that every simple stratum can be obtained by iterating the procedure of **(2.2)**. The construction is by no means unique, but it does have very useful uniqueness properties.

(2.4.1) Theorem: *(i) Let $[\mathfrak{A}, n, r, \beta]$ be a pure stratum in A. There exists a simple stratum $[\mathfrak{A}, n, r, \gamma]$ in A such that*

$$[\mathfrak{A}, n, r, \gamma] \sim [\mathfrak{A}, n, r, \beta].$$

For any simple stratum $[\mathfrak{A}, n, r, \gamma]$ satisfying this condition, $e(F[\gamma]|F)$ divides $e(F[\beta]|F)$ and $f(F[\gamma]|F)$ divides $f(F[\beta]|F)$.

In particular, among all the pure strata $[\mathfrak{A}, n, r, \beta']$ equivalent to the given $[\mathfrak{A}, n, r, \beta]$, the simple ones are precisely those for which the field extension $F[\beta']/F$ has minimal degree.

(ii) Let $[\mathfrak{A}, n, r, \gamma_1]$, $[\mathfrak{A}, n, r, \gamma_2]$ be simple strata in A which are equivalent to each other. Then

(a) $k_0(\gamma_1, \mathfrak{A}) = k_0(\gamma_2, \mathfrak{A})$;

(b) $e(F[\gamma_1]|F) = e(F[\gamma_2]|F)$ and $f(F[\gamma_1]|F) = f(F[\gamma_2]|F)$;

(c) Let s_1 be a tame corestriction on A relative to $F[\gamma_1]/F$. Then there exists $\delta \in F[\gamma_1]$ such that

$$s_1(\gamma_1 - \gamma_2) \equiv \delta \pmod{\mathfrak{P}^{1-r}},$$

where $\mathfrak{P} = \mathrm{rad}(\mathfrak{A})$.

(iii) Let $[\mathfrak{A}, n, r, \beta]$ be a pure stratum in A with $r = -k_0(\beta, \mathfrak{A})$. Let $[\mathfrak{A}, n, r, \gamma]$ be a simple stratum in A which is equivalent to $[\mathfrak{A}, n, r, \beta]$, let s_γ be a tame corestriction on A relative to $F[\gamma]/F$, let B_γ be the A-centraliser of γ, and $\mathfrak{B}_\gamma = \mathfrak{A} \cap B_\gamma$. Then $[\mathfrak{B}_\gamma, r, r-1, s_\gamma(\beta - \gamma)]$ is equivalent to a simple stratum in B_γ.

Before embarking on the (lengthy and elaborate) proof of this result, we make a few remarks. First, parts *(ii)(a)* and *(b)* have already been established in (2.1.4). This shows that in *(i)*, the quantities $e(F[\gamma]|F)$, $f(F[\gamma]|F)$ depend only on the equivalence class of $[\mathfrak{A}, n, r, \beta]$.

It follows from *(i)* that we can construct a "defining sequence" for a given *pure* $[\mathfrak{A}, n, r, \beta]$. This is a family $[\mathfrak{A}, n, r_i, \gamma_i]$, $0 \leq i \leq s$, of simple strata such that

(2.4.2) *(i) $[\mathfrak{A}, n, r_i, \gamma_i]$ is simple, $0 \leq i \leq s$;*

(ii) $[\mathfrak{A}, n, r_0, \gamma_0] \sim [\mathfrak{A}, n, r, \beta]$;

(iii) $r = r_0 < r_1 < \ldots < r_s < n$;

(iv) $r_{i+1} = -k_0(\gamma_i, \mathfrak{A})$, *and* $[\mathfrak{A}, n, r_{i+1}, \gamma_{i+1}]$ *is equivalent to* $[\mathfrak{A}, n, r_{i+1}, \gamma_i]$, $0 \le i \le s-1$;

(v) $k_0(\gamma_s, \mathfrak{A}) = -n$ *or* $-\infty$.

(vi) Let $\mathfrak{B}_i$ *be the* $\mathfrak{A}$*-centraliser of* γ_i *and* s_i *a tame corestriction on* A *relative to* $F[\gamma_i]/F$. *The derived stratum* $[\mathfrak{B}_i, r_i, r_i - 1, s_i(\gamma_{i-1} - \gamma_i)]$ *is equivalent to a simple stratum, for* $1 \le i \le s$.

This is a clumsy device, but it can be a useful expedient in certain circumstances.

Now let us prove *(i)* of (2.4.1). We can exclude the case in which $[\mathfrak{A}, n, r, \beta]$ is already simple:— in the first statement there is nothing to prove, and the assertions concerning ramification indices and residue class degrees follow from (2.1.4).

Therefore we assume that $r \ge -k_0(\beta, \mathfrak{A})$. It is then enough to treat the case $r = -k_0(\beta, \mathfrak{A})$. For, if $r > -k_0(\beta, \mathfrak{A})$, we use this special case to find a simple stratum $[\mathfrak{A}, n, -k_0(\beta, \mathfrak{A}), \gamma]$ equivalent to $[\mathfrak{A}, n, -k_0(\beta, \mathfrak{A}), \beta]$. Then $[\mathfrak{A}, n, r, \gamma] \sim [\mathfrak{A}, n, r, \beta]$. Of course, the stratum $[\mathfrak{A}, n, r, \gamma]$ need not be simple, but $n - r \le n + k_0(\gamma, \mathfrak{A}) < n + k_0(\beta, \mathfrak{A})$, and we can iterate the procedure.

We next reduce to the case in which $E = F[\beta]$ is a *maximal* subfield of A. In general, we write $A(E) = \mathrm{End}_F(E)$, $\mathfrak{A}(E) = \mathrm{End}^0_{\mathfrak{o}_F}(\{\mathfrak{p}_E^i : i \in \mathbb{Z}\})$, as in §**1**. Thus $\mathfrak{A}(E)$ is the unique hereditary $\mathfrak{o}_F$-order in $A(E)$ which is normalised by E. Let B be the A-centraliser of β, $\mathfrak{B} = \mathfrak{A} \cap B$, $e = e(\mathfrak{B}|\mathfrak{o}_E)$. Then, by (1.4.13), $k_0(\beta, \mathfrak{A}) = ek_0(\beta, \mathfrak{A}(E))$. Take $r = -k_0(\beta, \mathfrak{A})$, $r(E) = -k_0(\beta, \mathfrak{A}(E))$. Suppose we have found a simple stratum $[\mathfrak{A}(E), n(E), r(E), \gamma]$ in $A(E)$ equivalent to $[\mathfrak{A}(E), n(E), r(E), \beta]$, where $n(E) = n/e = -\nu_{\mathfrak{A}(E)}(\beta)$. Take a (W, E)-decomposition (1.2.9) of $\mathfrak{A}$, and consider the stratum $[\mathfrak{A}, n, r, \gamma \otimes 1]$. We have $F[\gamma \otimes 1]^\times = F[\gamma]^\times \otimes 1$, which is contained in $\mathfrak{K}(\mathfrak{A})$ by (1.2.11). (1.4.13) gives us

$$\begin{aligned} k_0(\gamma \otimes 1, \mathfrak{A}) &= k_0(\gamma, \mathfrak{A}(F[\gamma]))e(\mathfrak{A}|\mathfrak{o}_F)/e(F[\gamma]|F) \\ &= e.e(\mathfrak{A}(E)|\mathfrak{o}_F)/e(F[\gamma]|F) = ek_0(\gamma, \mathfrak{A}(E)), \end{aligned}$$

so $[\mathfrak{A}, n, r, \gamma \otimes 1]$ is simple. Moreover,

$$\beta - \gamma \otimes 1 \in \mathfrak{P}(E)^{-r(E)} \otimes 1 \subset \mathfrak{P}^{-r},$$

where $\mathfrak{P}(E) = \mathrm{rad}(\mathfrak{A}(E))$. Altogether, $[\mathfrak{A}, n, r, \gamma \otimes 1]$ is simple and equivalent to $[\mathfrak{A}, n, r, \beta]$ as required.

So, we now proceed under the twin assumption that $r = -k_0(\beta, \mathfrak{A})$ and E/F is a maximal subfield of A. Consider first the stratum $[\mathfrak{A}, n, n-1, \beta]$. This is pure, and so nonsplit fundamental by (2.3.11). We use (2.3.4) to find a simple stratum $[\mathfrak{A}', n', n'-1, \alpha]$ equivalent to $[\mathfrak{A}', n', n'-$

$1, \beta]$. We assert that $F[\alpha]$ is not a maximal subfield of A. If it is, the equivalence $[\mathfrak{A}', n', n'-1, \beta] \sim [\mathfrak{A}', n', n'-1, \alpha]$ implies that $[\mathfrak{A}', n', n'-1, \beta]$ is simple by (2.2.2). In particular, $F[\beta]$ normalises $\mathfrak{A}'$ and hence, since $F[\beta]$ is maximal, $\mathfrak{A}' = \mathfrak{A}$. This further implies $n' = n$, and $k_0(\beta, \mathfrak{A}) \leq -n < -r$, which is nonsense. We deduce

(2.4.3) *There exists a simple stratum* $[\mathfrak{A}_0, n_0, r_0, \gamma_0]$ *in* A *with* $\beta \equiv \gamma_0 \pmod{\mathfrak{P}_0^{-r_0}}$, *where* $\mathfrak{P}_0 = \mathrm{rad}(\mathfrak{A}_0)$, *and such that* $F[\gamma_0]$ *is not a maximal subfield of* A.

Observe that the congruence connecting γ_0 and β in (2.4.3) forces $\beta \in \mathfrak{K}(\mathfrak{A}_0)$ and $n_0 = -\nu_{\mathfrak{A}_0}(\beta) = ne(\mathfrak{A}_0)/e(\mathfrak{A})$. In particular, $[\mathfrak{A}_0, n_0, r_0, \beta]$ is a stratum in A equivalent to $[\mathfrak{A}_0, n_0, r_0, \gamma_0]$.

To save some writing we adhere to the following convention throughout the proof of (2.4.1)*(i)*:

(2.4.4) Notation: *(i) If* $\mathfrak{A}_i$ *(or* $\mathfrak{A}'$*) is a hereditary* $\mathfrak{o}_F$*-order in* A, *then* $\mathrm{rad}(\mathfrak{A}_i) = \mathfrak{P}_i$ *(or* $\mathrm{rad}(\mathfrak{A}') = \mathfrak{P}'$*).*

(ii) If $F[\gamma]$ *is a subfield of* A, *then* s_γ *is a tame corestriction on* A *relative to* $F[\gamma]/F$.

Before proceeding, we need another definition. A stratum $[\mathfrak{A}, n, n-1, b]$ in A is called *scalar fundamental* if it is fundamental and there exists $x \in F \cap \mathfrak{P}^{-n}$ such that $[\mathfrak{A}, n, n-1, b-x]$ is not fundamental (possibly null). Equivalently, $e(\mathfrak{A})$ divides n and $\phi_b(X) = (X-\alpha)^N$ for some $\alpha \in \mathrm{k}_F^\times$. If we apply (2.3.4) to a scalar fundamental stratum $[\mathfrak{A}, n, n-1, b]$, we get a simple $[\mathfrak{A}', n', n'-1, x]$ equivalent to $[\mathfrak{A}', n', n'-1, b]$, where $x \in F$ (and this x may be taken the same as the one in the definition).

We now choose a stratum $[\mathfrak{A}_0, n_0, r_0, \gamma_0]$ satisfying (2.4.3) and such that the degree $[F[\gamma_0] : F]$ is maximal, and then take r_0 minimal for the condition $\gamma_0 \equiv \beta \pmod{\mathfrak{P}_0^{-r_0}}$. Fix an $\mathfrak{A}_0$-lattice L_0 in V.

(2.4.5) Lemma: *Let* $[\mathfrak{A}', n', r', \gamma']$ *be a simple stratum in* A, L' *an* $\mathfrak{A}'$*-lattice in* V, *and* β' *a* G*-conjugate of* β. *Suppose that* $\beta' \equiv \gamma' \pmod{\mathfrak{P}'^{-r'}}$. *Let* B' *be the* A*-centraliser of* γ', $\mathfrak{B}' = \mathfrak{A}' \cap B'$. *Suppose that the stratum* $[\mathfrak{B}', r', r'-1, s_{\gamma'}(\beta'-\gamma')]$ *is either non-fundamental or scalar fundamental. Then there exists a simple stratum* $[\mathfrak{A}'', n'', r'', \gamma'']$ *and a* G*-conjugate* β'' *of* β *such that*

(i) $\beta'' \equiv \gamma'' \pmod{\mathfrak{P}''^{-r''}}$;

(ii) L' *is an* $\mathfrak{A}''$*-lattice;*

(iii) $r''/e(\mathfrak{A}'') < r'/e(\mathfrak{A}')$;

(iv) $e(F[\gamma'']|F) = e(F[\gamma']|F)$ *and* $f(F[\gamma'']|F) = f(F[\gamma']|F)$;

(v) $k_0(\gamma'', \mathfrak{A}'')/e(\mathfrak{A}'') = k_0(\gamma', \mathfrak{A}')/e(\mathfrak{A}')$.

Proof: Take first the case in which $[\mathfrak{B}', r', r'-1, s_{\gamma'}(b')]$ is *not fundamental*, where $b' = \beta' - \gamma'$. We apply (2.3.1) to find a hereditary order

$\mathfrak{B}''$ in B' and an integer r'' such that

$$s_{\gamma'}(b') + \mathfrak{Q}'^{1-r'} \subset \mathfrak{Q}''^{-r''} \quad \textit{and} \quad r''/e(\mathfrak{B}'') < r'/e(\mathfrak{B}').$$

(Here we put $\mathfrak{Q}' = \mathrm{rad}(\mathfrak{B}')$, $\mathfrak{Q}'' = \mathrm{rad}(\mathfrak{B}'')$.) We also ensure that $\mathfrak{B}'$, $\mathfrak{B}''$ satisfy (1.1.9). Let $\mathfrak{A}''$ be the unique hereditary $\mathfrak{o}_F$-order in A such that $F[\gamma']^\times \subset \mathfrak{K}(\mathfrak{A}'')$ and $\mathfrak{A}'' \cap B' = \mathfrak{B}''$. We have

$$r'' < r'e(\mathfrak{B}'')/e(\mathfrak{B}') = r'e(\mathfrak{A}'')/e(\mathfrak{A}') < -k_0(\gamma', \mathfrak{A}''),$$

by (1.4.13), so $[\mathfrak{A}'', n'', r'', \gamma']$ is simple, with $n'' = n'e(\mathfrak{A}'')/e(\mathfrak{A}')$. Take a $(W, F[\gamma'])$-decomposition of A where W is the F-span of a common $\mathfrak{o}_{F[\gamma']}$-basis of the lattice chains defining $\mathfrak{A}'$ and $\mathfrak{A}''$. Since $\mathfrak{Q}'^{1-r'} \subset \mathfrak{Q}''^{-r''}$, it follows from (1.2.10) that $\mathfrak{P}'^{1-r'} \subset \mathfrak{P}''^{-r''}$.

The conditions $s_{\gamma'}(b') \in \mathfrak{Q}''^{-r''}$, $b' \in \mathfrak{P}'^{-r'}$ imply that

$$b' \in (\mathfrak{P}''^{-r''} + a_{\gamma'}(A)) \cap \mathfrak{P}'^{-r'}.$$

We assert

$$(\mathfrak{P}''^{-r''} + a_{\gamma'}(A)) \cap \mathfrak{P}'^{-r'} = \mathfrak{P}''^{-r''} \cap \mathfrak{P}'^{-r'} + a_{\gamma'}(A) \cap \mathfrak{P}'^{-r'}.$$

We certainly have the containment $\supset$. The two sides have the same intersection with $a_{\gamma'}(A)$. Moreover, applying $s_{\gamma'}$ to both sides we get

$$s_{\gamma'}((\mathfrak{P}''^{-r''} + a_{\gamma'}(A)) \cap \mathfrak{P}'^{-r'}) \subset \mathfrak{Q}''^{-r''} \cap \mathfrak{Q}'^{-r'} = s_{\gamma'}(\mathfrak{P}''^{-r''} \cap \mathfrak{P}'^{-r'})$$

by (1.3.12). The desired equality follows, and by (1.4.10) there exists $x \in \mathfrak{P}'$ such that

$$b' \equiv a_{\gamma'}(x) \pmod{\mathfrak{P}''^{-r''}}.$$

(We could have taken the modulus $\mathfrak{P}''^{-r''} \cap \mathfrak{P}'^{r'}$ here.) Now we assert that

$$(1+x)(\gamma' + b')(1+x)^{-1} \equiv \gamma' \pmod{\mathfrak{P}''^{-r''}}.$$

Remark *(ii)* following (1.1.9) shows that $\mathfrak{P}' \subset \mathfrak{A}''$, so $1+x$ is a unit of $\mathfrak{A}''$. Therefore, we have to show

$$(1+x)(\gamma' + b') \equiv \gamma'(1+x) \pmod{\mathfrak{P}''^{-r''}}.$$

The difference of the two sides here is $xb' \in \mathfrak{P}'^{1-r'} \subset \mathfrak{P}''^{-r''}$, and we have the desired congruence.

Now we write $\beta_1' = (1+x)\beta'(1+x)^{-1}$, so that $\beta_1' \equiv \gamma' \pmod{\mathfrak{P}''^{-r''}}$. The pair $([\mathfrak{A}'', n'', r'', \gamma'], \beta_1')$ now satisfies all the required conditions except possibly *(ii)*. The lattice L' is an $\mathfrak{o}_{F[\gamma']}$-lattice, and so are all the lattices in the chain defining $\mathfrak{A}''$. The group $B'^{\times}$ acts transitively on the set of $\mathfrak{o}_{F[\gamma']}$-lattices in V, so there exists $y \in B'^{\times}$ such that L' is a $y^{-1}\mathfrak{A}''y$-lattice. Then the pair $([y^{-1}\mathfrak{A}''y, n'', r'', \gamma'], y^{-1}\beta_1'y)$ satisfies the conditions of the lemma.

Now suppose $[\mathfrak{B}', r', r'-1, s_{\gamma'}(b')]$ is *scalar fundamental.* We apply (2.3.4) to get a simple $[\mathfrak{B}'', r'', r''-1, \alpha]$ equivalent to $[\mathfrak{B}'', r'', r''-1, s_{\gamma'}(b')]$ with $\mathfrak{B}'' \subset \mathfrak{B}'$ and $\alpha \in F[\gamma']$. Let $\mathfrak{A}''$ be the unique hereditary $\mathfrak{o}_F$-order in A with $F[\gamma']^{\times} \subset \mathfrak{K}(A'')$ and $\mathfrak{A}'' \cap B' = \mathfrak{B}''$. Then $r''/e(\mathfrak{A}'') = r'/e(\mathfrak{A}')$ and $[\mathfrak{A}'', n'', r'', \gamma']$ is a simple stratum with $n'' = n'e(\mathfrak{A}'')/e(\mathfrak{A}')$. Now we apply (2.2.8) to get a simple $[\mathfrak{A}'', n'', r''-1, \gamma'']$ such that $\gamma'' \equiv \gamma' \pmod{\mathfrak{P}''^{-r''}}$ and $[\mathfrak{B}'', r'', r''-1, s_{\gamma'}(\gamma''-\gamma')] \sim [\mathfrak{B}'', r'', r''-1, \alpha]$. (2.2.8) further gives us $e(F[\gamma'']|F) = e(F[\gamma']|F)$, $f(F[\gamma'']|F) = f(F[\gamma']|F)$ and $k_0(\gamma'', \mathfrak{A}'') = k_0(\gamma', \mathfrak{A}'') = k_0(\gamma', \mathfrak{A}')e(\mathfrak{A}'')/e(\mathfrak{A}')$.

Our application of (2.3.4) above gives

$$s_{\gamma'}(b') + \mathfrak{Q}'^{1-r'} \subset \alpha + \mathfrak{Q}''^{1-r''}$$

and so $\mathfrak{Q}'^{1-r'} \subset \mathfrak{Q}''^{1-r''}$. We therefore have

$$s_{\gamma'}(b') \equiv s_{\gamma'}(\gamma''-\gamma') \pmod{\mathfrak{Q}''^{1-r''}}.$$

Now we use (2.2.1) to find $x \in \mathfrak{P}''$ such that

$$(1+x)\beta'(1+x)^{-1} \equiv \gamma'' \pmod{\mathfrak{P}''^{1-r''}}.$$

Put $\beta'' = (1+x)\beta'(1+x)^{-1}$, and then the pair $([\mathfrak{A}'', n'', r''-1, \gamma''], \beta'')$ satisfies all the conditions of (2.4.5) except possibly *(ii)*, and we arrange this as in the first case. ∎

Now we return to the stratum $[\mathfrak{A}_0, n_0, r_0, \gamma_0]$ chosen above just before (2.4.5)), and the fixed $\mathfrak{A}_0$-lattice L_0. If the pair $([\mathfrak{A}_0, n_0, r_0, \gamma_0], \beta)$ satisfies the hypothesis of (2.4.5), we apply this lemma to get a pair $([\mathfrak{A}_1, n_1, r_1, \gamma_1], \beta_1)$. If this also satisfies the hypothesis of (2.4.5), we iterate the procedure. We repeat as many times as we can. The next step is to show that the process terminates after finitely many steps. Suppose, for a contradiction, that it does not. We therefore obtain a sequence of pairs $([\mathfrak{A}_m, n_m, r_m, \gamma_m], \beta_m)$, $m \geq 0$, consisting of a simple stratum $[\mathfrak{A}_m, n_m, r_m, \gamma_m]$ and a G-conjugate β_m of β, satisfying

(2.4.6) *(i)* $\beta_m \equiv \gamma_m \pmod{\mathfrak{P}_m^{-r_m}}$;
(ii) L_0 *is an* $\mathfrak{A}_m$*-lattice;*
(iii) $r_m/e(\mathfrak{A}_m) < r_{m-1}/e(\mathfrak{A}_{m-1})$, $m \geq 1$;
(iv) $e(F[\gamma_m]|F) = e(F[\gamma_0]|F)$ *and* $f(F[\gamma_m]|F) = f(F[\gamma_0]|F)$;
(v) $k_0(\gamma_m, \mathfrak{A}_m)/e(\mathfrak{A}_m) = k_0(\gamma_0, \mathfrak{A}_0)/e(\mathfrak{A}_0)$.

(We put $\beta_0 = \beta$ here.) By *(ii)*, all of the orders $\mathfrak{A}_m$ are contained in the maximal order $\mathfrak{M} = \mathrm{End}_{\mathfrak{o}_F}(L_0)$. The congruence *(i)* implies that $\beta_m \in \mathfrak{K}(\mathfrak{A}_m)$ and $-n_m = \nu_{\mathfrak{A}_m}(\gamma_m) = \nu_{\mathfrak{A}_m}(\beta_m) = -n_0 e(\mathfrak{A}_m)/e(\mathfrak{A}_0)$. Therefore $\gamma_m \in \mathfrak{p}_F^g \mathfrak{M}$ for all m, where $g = [-n_0/e(\mathfrak{A}_0)]$. The sequence $\{\gamma_m\}$ therefore has a convergent subsequence $\{\gamma_{m_j} : j \geq 0\}$ with limit γ, say. The denominators $e(\mathfrak{A}_m)$ in *(iii)* are bounded above, so $r_m/e(\mathfrak{A}_m) \to -\infty$ as $m \to \infty$. Putting $g_m = [-r_m/e(\mathfrak{A}_m)]$, we have $g_m \to \infty$, and $\beta_m - \gamma_m \in \mathfrak{p}_F^{g_m}\mathfrak{M}$ for all m. It follows that $\beta_{m_j} \to \gamma$ as $j \to \infty$.

Let $f_\beta(X)$ denote the characteristic polynomial of β as an F-endomorphism of V. Since $F[\beta]$ is a maximal subfield of A, the polynomial $f_\beta(X)$ is irreducible over F. Each β_{m_j} has characteristic polynomial $f_\beta(X)$, hence so does the element γ. On the other hand, put

$$d = [F[\beta] : F]/[F[\gamma_m] : F] = [F[\beta] : F]/[F[\gamma_0] : F].$$

We have $d > 1$ by (2.4.3). The characteristic polynomial of γ_m is therefore the d-th power of an irreducible polynomial over F. The same therefore applies to γ, and we have our desired contradiction.

So, iterated applications of (2.4.5) give us finally a pair

$$([\mathfrak{A}_m, n_m, r_m, \gamma_m], \beta_m)$$

satisfying (2.4.6) *but such that* $[\mathfrak{B}_m, r_m, r_m - 1, s_{\gamma_m}(b_m)]$ *is fundamental, and not scalar fundamental.* Here, B_m is the A-centraliser of γ_m, $\mathfrak{B}_m = \mathfrak{A}_m \cap B_m$, $b_m = \beta_m - \gamma_m$. The coset $\beta_m + \mathfrak{P}_m^{1-r_m} = \gamma_m + b_m + \mathfrak{P}_m^{1-r_m}$ contains the field element β_m, so the fundamental stratum $[\mathfrak{B}_m, r_m, r_m - 1, s_{\gamma_m}(b_m)]$ is nonsplit by (2.3.12). We now go through the same procedure as in (2.4.5). By (2.3.4), there is a simple stratum $[\mathfrak{B}', r', r' - 1, \alpha] \sim [\mathfrak{B}', r', r' - 1, s_{\gamma_m}(b_m)]$. Let $\mathfrak{A}'$ be the hereditary $\mathfrak{o}_F$-order in A with $F[\gamma_m]^\times \subset \mathfrak{K}(\mathfrak{A}')$ and $\mathfrak{A}' \cap B_m = \mathfrak{B}'$. We have $r'/e(\mathfrak{B}') = r_m/e(\mathfrak{B}_m)$ and $\alpha \notin F[\gamma_m] + \mathfrak{P}'^{1-r'}$ since the original stratum was not scalar fundamental. We use (2.2.8) to find a simple stratum $[\mathfrak{A}', n', r' - 1, \gamma]$ such that

(a) $[\mathfrak{A}', n', r', \gamma] \sim [\mathfrak{A}', n', r', \gamma_m]$;
(b) $[\mathfrak{B}', r', r' - 1, s_{\gamma_m}(\gamma - \gamma_m)] \sim [\mathfrak{B}', r', r' - 1, \alpha]$;
(c) $k_0(\gamma, \mathfrak{A}') = -r'$;
(d) $n' = n_m e(\mathfrak{A}')/e(\mathfrak{A}_m)$.

We also find (using (2.2.1)) a $\boldsymbol{U}^1(\mathfrak{A}')$-conjugate β' of β_m with $[\mathfrak{A}', n', r'-1, \beta'] \sim [\mathfrak{A}', n', r'-1, \gamma]$.

At this point, we recall that β_m is G-conjugate to β. Therefore $\beta = g\beta' g^{-1}$, for some $g \in G$. Replacing all of the objects $\mathfrak{A}'$, γ, γ_m etc. by their g-conjugates, we can assume that $\beta' = \beta$. We therefore have a simple stratum $[\mathfrak{A}', n', r'-1, \gamma]$ such that $\beta \equiv \gamma \pmod{\mathfrak{P}'^{1-r'}}$. However, $[F[\gamma] : F] = [F[\gamma_m, \alpha] : F] > [F[\gamma_m] : F] = [F[\gamma_0] : F]$, and γ_0 was chosen to maximise the field degree $[F[\gamma_0] : F]$, subject to (2.4.3). Therefore $F[\gamma]$ is a maximal subfield of A. (2.2.8) now implies that $F[\gamma_m, \alpha]$ is a maximal subfield of B_m. (2.2.2) allows us to apply (2.2.3) to deduce that $[\mathfrak{A}', n', r'-1, \beta]$ is simple. In particular, $F[\beta]^\times \subset \mathfrak{K}(\mathfrak{A}')$. Since $F[\beta]$ is a maximal subfield of A, this means $\mathfrak{A}' = \mathfrak{A}$, whence $n' = n$. (2.2.3) further tells us that $-k_0(\beta, \mathfrak{A}) = r'$, from which we deduce $r' = r$. Thus we have

$$[\mathfrak{A}, n, r, \beta] \sim [\mathfrak{A}, n, r, \gamma] \sim [\mathfrak{A}, n, r, \gamma_m],$$

and the last of these strata is simple. Moreover, we observe that $e(F[\beta]|F) = e(F[\gamma]|F) = e(F[\gamma_m]|F)e(F[\gamma_m, \alpha]|F[\gamma_m])$ by (2.2.3), and likewise for residue class degrees. Therefore $[\mathfrak{A}, n, r, \gamma_m]$ satisfies all the requirements of (2.4.1)*(i)*.

Remark: This argument has produced a simple stratum $[\mathfrak{A}, n, r, \gamma_m]$ equivalent to $[\mathfrak{A}, n, r, \beta]$ such that the derived stratum $[\mathfrak{B}_m, r, r-1, s_{\gamma_m}(\beta - \gamma_m)]$ is simple. It operated under the hypotheses that $F[\beta]$ is a maximal subfield of A, and that $r = -k_0(\beta, \mathfrak{A})$. We can remove the first of these, using a (W, E)-decomposition, just as in the preliminary reduction stages of this last proof. We deduce:

(2.4.7) *Let $[\mathfrak{A}, n, r, \beta]$ be a pure stratum in A with $r = -k_0(\beta, \mathfrak{A})$. There exists a simple stratum $[\mathfrak{A}, n, r, \gamma]$ in A with the following properties:*

(i) $[\mathfrak{A}, n, r, \gamma] \sim [\mathfrak{A}, n, r, \beta]$;

(ii) the stratum $[\mathfrak{B}_\gamma, r, r-1, s_\gamma(\beta - \gamma)]$ is equivalent to a simple stratum, where $\mathfrak{B}_\gamma$ denotes the $\mathfrak{A}$-centraliser of γ and s_γ is a tame corestriction on A relative to $F[\gamma]/F$.

We now turn to (2.4.1)*(ii)*, recalling that parts *(a)* and *(b)* have already been done. Write $\gamma_2 = \gamma_1 + d$, $d \in \mathfrak{P}^{-r}$, $B =$ the A-centraliser of γ_1, $\mathfrak{B} = \mathfrak{A} \cap B$, $\mathfrak{Q} = \mathfrak{P} \cap B$, and abbreviate $s_1 = s$. We work out the $B^\times$-intertwining of the coset $s(d) + \mathfrak{Q}^{1-r}$ (i.e. the formal intertwining in $B^\times$ of the stratum $[\mathfrak{B}, r, r-1, s(d)]$, in the terminology of **(1.5)**). Take $x \in B^\times$. Then x intertwines $s(d) + \mathfrak{Q}^{1-r}$ if and only if

$$xs(d)x^{-1} \equiv s(d) \pmod{\mathfrak{Q}^{1-r} + x\mathfrak{Q}^{1-r}x^{-1}}.$$

We apply (1.4.16) (with $-i$ and n both very large). This condition on x is equivalent to

$$\text{(2.4.8)} \qquad xdx^{-1} - d \equiv a_{\gamma_1}(y) \pmod{\mathfrak{P}^{1-r} + x\mathfrak{P}^{1-r}x^{-1}}$$

for some $y \in \mathfrak{Q}^{-r-k}\mathfrak{N} + x\mathfrak{Q}^{-r-k}\mathfrak{N}x^{-1}$, where we write $k = k_0(\gamma_1, \mathfrak{A})$ and $\mathfrak{N} = \mathfrak{N}_k(\gamma_1, \mathfrak{A})$. Write $y = t + xzx^{-1}$, $t, z \in \mathfrak{Q}^{-r-k}\mathfrak{N}$. We have

$$(1+z)(\gamma_1 + d)(1+z)^{-1} \equiv \gamma_1 + d - a_{\gamma_1}(z) \pmod{\mathfrak{P}^{1-r}}.$$

Conjugating this by x, using (2.4.8) and the definition of y, we get

$$\begin{aligned} &x(1+z)(\gamma_1 + d)(1+z)^{-1}x^{-1} \\ &\qquad \equiv \gamma_1 + d + a_{\gamma_1}(t) \pmod{\mathfrak{P}^{1-r} + x\mathfrak{P}^{1-r}x^{-1}} \\ &\qquad \equiv (1+t)^{-1}(\gamma_1 + d)(1+t) \pmod{\mathfrak{P}^{1-r} + x\mathfrak{P}^{1-r}x^{-1}}. \end{aligned}$$

The ideal $\mathfrak{P}^{1-r}$ is invariant under conjugation by $1+t$ and $1+z$, so this congruence is equivalent to

$$w(\gamma_1 + d)w^{-1} \equiv \gamma_1 + d \pmod{\mathfrak{P}^{1-r} + w\mathfrak{P}^{1-r}w^{-1}},$$

where $w = (1+t)x(1+z)$. We have proved

(2.4.9) Lemma: *An element $x \in B^\times$ intertwines $s(d) + \mathfrak{Q}^{1-r}$ if and only if there exist $t, z \in \mathfrak{Q}^{-r-k}\mathfrak{N}$ such that $(1+t)x(1+z)$ intertwines $\gamma_1 + d + \mathfrak{P}^{1-r} = \gamma_2 + \mathfrak{P}^{1-r}$.*

Observe that, in proving this result, we only used the simplicity of $[\mathfrak{A}, n, r, \gamma_1]$ and the fact $d \in \mathfrak{P}^{-r}$.

Now let B_2 be the A-centraliser of γ_2, $\mathfrak{Q}_2 = \mathfrak{P} \cap B_2$. Put $k = k_0(\gamma_1, \mathfrak{A})$ ($= k_0(\gamma_2, \mathfrak{A})$ by (2.4.1)*(ii)*(a)) and $\mathfrak{N} = \mathfrak{N}_k(\gamma_1, \mathfrak{A})$ ($= \mathfrak{N}_k(\gamma_2, \mathfrak{A})$ by (2.1.3)). (2.1.3) also gives $\mathfrak{Q}_2^{-r-k}\mathfrak{N} = \mathfrak{Q}^{-r-k}\mathfrak{N}$, so (1.5.8) applied to $[\mathfrak{A}, n, r, \gamma_1] \sim [\mathfrak{A}, n, r, \gamma_2]$ yields

$$(1 + \mathfrak{Q}^{-r-k}\mathfrak{N})B^\times(1 + \mathfrak{Q}^{-r-k}\mathfrak{N}) = (1 + \mathfrak{Q}^{-r-k}\mathfrak{N})B_2^\times(1 + \mathfrak{Q}^{-r-k}\mathfrak{N}).$$

That is, given $x \in B^\times$, there exist $t, z \in \mathfrak{Q}^{-r-k}\mathfrak{N}$ such that $(1+t)x(1+z) \in B_2^\times$, and $B_2^\times$ surely intertwines $\gamma_2 + \mathfrak{P}^{1-r}$. Therefore

(2.4.10) *The coset $s(d) + \mathfrak{Q}^{1-r}$ is intertwined by every $x \in B^\times$.*

The assertion (2.4.1)*(ii)(c)* is now implied by the following general result.

(2.4.11) Lemma: *Let $\mathfrak{A}$ be a hereditary $\mathfrak{o}_F$-order in A, $n \in \mathbb{Z}$, $b \in \mathfrak{P}^n$. Suppose that the coset $b + \mathfrak{P}^{1+n}$ is intertwined by every element of $G = A^\times$. Then*

$$(b + \mathfrak{P}^{1+n}) \cap F \neq \emptyset.$$

Further, let χ be a character of $U^{m+1}(\mathfrak{A})$, where $m \geq 0$, which is intertwined by every element of G. Then χ factors through the determinant mapping $\det : G \to F^\times$.

This result is best treated via the explicit block matrix description of hereditary orders. We therefore postpone it till **(2.5)**.

We have now completed the proof of (2.4.1)*(ii)*.

We now come to the proof of part *(iii)*, which we start with a sequence of lemmas.

(2.4.12) Lemma: *Let $[\mathfrak{A}, n, r, \gamma_i]$, $i = 1, 2$, be simple strata in A which are equivalent to each other. Put $E_i = F[\gamma_i]$. Then*

(i) $\mathfrak{o}_{E_1}$, $\mathfrak{o}_{E_2}$ have the same images in $\mathfrak{A}/\mathfrak{P}$;

(ii) for $i = 1, 2$, there exists a tame corestriction s_i on A relative to $F[\gamma_i]/F$ such that

$$s_1(x) \equiv s_2(x) \pmod{\mathfrak{P}^{j+1}}, \quad j \in \mathbb{Z},\ x \in \mathfrak{P}^j.$$

Proof: We start by treating the special case in which $j = 0$ and $\mathfrak{A}$ is maximal among hereditary $\mathfrak{o}_F$-orders in A normalised by E_1. The latter is equivalent to the condition $e(\mathfrak{A}|\mathfrak{o}_F) = e(E_1|F)$. By (2.4.1)*(ii)*, we have $e(E_2|F) = e(E_1|F)$, so $\mathfrak{A}$ is also maximal among orders normalised by E_2. Write B_i for the A-centraliser of E_i, $\mathfrak{B}_i = \mathfrak{A} \cap B_i$, $\mathfrak{Q}_i = \mathrm{rad}(\mathfrak{B}_i)$. We have $\mathfrak{B}_1 + \mathfrak{P} = \mathfrak{B}_2 + \mathfrak{P}$ by (2.1.4), so that $\mathfrak{B}_1/\mathfrak{Q}_1$, $\mathfrak{B}_2/\mathfrak{Q}_2$ coincide when viewed as subalgebras of $\mathfrak{A}/\mathfrak{B}$ in the natural way. However, since $\mathfrak{B}_i$ is a maximal order, the centre of $\mathfrak{B}_i/\mathfrak{Q}_i$ is $\mathfrak{o}_{E_i}/\mathfrak{p}_{E_i}$. This proves (2.4.12)*(i)* in the present case.

To prove *(ii)*, we start by choosing the tame corestrictions s_i at random. The map $s_2 : \mathfrak{A}/\mathfrak{P} \to \mathfrak{A}/\mathfrak{P}$ is a bimodule homomorphism for the ring $\mathfrak{B}_1/\mathfrak{Q}_1$, by what we have just proved. It has the same image, namely $\mathfrak{B}_1 + \mathfrak{P}/\mathfrak{P}$ as the corresponding map induced by s_1. We next show that the s_i have the same kernel on $\mathfrak{A}/\mathfrak{P}$. If we write $k = k_0(\gamma_i, \mathfrak{A})$, the kernel of s_i on $\mathfrak{A}/\mathfrak{P}$ is just $\mathfrak{P} + a_{\gamma_i}(\mathfrak{Q}_i^{-k}\mathfrak{N}_k(\gamma_i, \mathfrak{A}))$, by (1.4.10). However, (2.1.3) shows that $\mathfrak{Q}_1^{-k}\mathfrak{N}_k(\gamma_1, \mathfrak{A}) = \mathfrak{Q}_2^{-k}\mathfrak{N}_k(\gamma_2, \mathfrak{A})$. Further, for $z \in \mathfrak{P}^{-k}$, we have $a_{\gamma_1}(z) \equiv a_{\gamma_2}(z) \pmod{\mathfrak{P}}$, so indeed the s_i have the same kernel on $\mathfrak{A}/\mathfrak{P}$. The s_i therefore differ by a $\mathfrak{B}_1/\mathfrak{Q}_1$-bimodule automorphism of $\mathfrak{B}_1/\mathfrak{Q}_1$, i.e. $s_2 \equiv \xi s_1 \pmod{\mathfrak{P}}$ for some $\xi \in \mathfrak{o}_{E_1}^\times$. However, ξs_1 is again a tame corestriction on A relative to E_1/F, and we have the result in this case.

Now we treat the general case. Let $\mathfrak{A} = \mathrm{End}^0_{\mathfrak{o}_F}(\mathcal{L})$, for a lattice chain $\mathcal{L}$ in V, and fix a lattice $L \in \mathcal{L}$. Let π_i be a prime element of E_i, $i = 1, 2$. Then $(L : \pi_1 L) = (L : \pi_2 L)$ by (2.4.1)*(ii)(b)*, whence $\pi_1^j L = \pi_2^j L$ for all j. Let $\mathfrak{A}_0 = \mathrm{End}^0_{\mathfrak{o}_F}(\{\pi_i^j L\})$, and let $e_0 = e(\mathfrak{A})/e(E_i|F)$, $r_0 = [r/e_0]$. The order $\mathfrak{A}_0$ is normalised by both E_1 and E_2 (by (1.2.1)), and is maximal for this property.

We assert that $[\mathfrak{A}, n, e_0r_0, \gamma_1] \sim [\mathfrak{A}, n, e_0r_0, x\gamma_2x^{-1}]$ for some $x \in U^1(\mathfrak{A})$ satisfying $[\mathfrak{A}, n, r, x^{-1}\gamma_ix] \sim [\mathfrak{A}, n, r, \gamma_i]$. (The set of x with this second property is given explicitly by (1.5.8) and is independent of i.) To prove this, let t be the least integer, $r \geq t \geq e_0r_0$, such that $[\mathfrak{A}, n, t, \gamma_1] \sim [\mathfrak{A}, n, t, x\gamma_2x^{-1}]$, for some x as above. Assume that $t > e_0r_0$, and we may as well take $x = 1$. Choose a tame corestriction s_1 on A relative to E_1/F. By part *(ii)(c)* of the present theorem, the element $\delta = s_1(\gamma_2 - \gamma_1)$ lies in the set

$$(F[\gamma_1] + \mathfrak{P}^{1-t}) \cap \mathfrak{P}^{-t}$$

which, since e_0 does not divide t, is precisely $\mathfrak{P}^{1-t}$. Thus we may take $\delta = 0$ and apply (2.2.1) to get the assertion.

Conjugating γ_2 by an element x of this type has no effect on the conclusions of the lemma, so we can assume that $[\mathfrak{A}, n, e_0r_0, \gamma_2]$ is equivalent to $[\mathfrak{A}, n, e_0r_0, \gamma_1]$. Taking $\mathfrak{A}_0 = \mathrm{End}_{\mathfrak{o}_F}(L)$ as above and $\mathfrak{P}_0 = \mathrm{rad}(\mathfrak{A}_0)$, we have $\mathfrak{P}_0^{-r_0} \supset \mathfrak{P}^{-e_0r_0}$. Setting $n_0 = n/e_0$, the strata $[\mathfrak{A}_0, n_0, r_0, \gamma_1]$, $[\mathfrak{A}_0, n_0, r_0, \gamma_2]$ are simple and equivalent to each other. We know that (2.4.12) holds for these strata, in the case $j = 0$. The natural embedding $\mathfrak{o}_{E_i} \to \mathfrak{A}_0$ factors through the inclusion $\mathfrak{A} \to \mathfrak{A}_0$ and we are assuming $\mathfrak{o}_{E_1} + \mathfrak{P}_0 = \mathfrak{o}_{E_2} + \mathfrak{P}_0$. Adding $\mathfrak{P}$ (which contains $\mathfrak{P}_0$) to this equation, we get the first assertion of (2.4.12) in general. Now take s_1, s_2 to satisfy (2.4.12)*(ii)* relative to the order $\mathfrak{A}_0$ and $j = 0$. Then s_1, s_2 surely agree on $\mathfrak{A}/\mathfrak{P}$. This proves the lemma for all orders $\mathfrak{A}$ in the case $j = 0$.

Now we remove the restriction on j. Putting $k = k_0(\gamma_i, \mathfrak{A})$ as before, (2.1.3) gives us

$$\mathfrak{Q}_1^j \mathfrak{N}_{k+1}(\gamma_1, \mathfrak{A}) = \mathfrak{Q}_2^j \mathfrak{N}_{k+1}(\gamma_2, \mathfrak{A}).$$

If we add $\mathfrak{P}^{j+1}$ to each side here, and use the definition (1.4.5) of k, we get

$$\mathfrak{Q}_1^j + \mathfrak{P}^{j+1} = \mathfrak{Q}_2^j + \mathfrak{P}^{j+1}.$$

Now take $y \in \mathfrak{Q}_1^j$, $z \in \mathfrak{A}$, and choose $y' \in \mathfrak{Q}_2^j$ such that $y' \equiv y \pmod{\mathfrak{P}^{j+1}}$. Thus $yz = y'z + w$, for some $w \in \mathfrak{P}^{j+1}$. We have

$$s_1(yz) + \mathfrak{P}^{j+1} = ys_1(z) + \mathfrak{P}^{j+1} = ys_2(z) + \mathfrak{P}^{j+1}$$

by the case $j = 0$. By definition

$$\begin{aligned} ys_2(z) + \mathfrak{P}^{j+1} &= y's_2(z) + \mathfrak{P}^{j+1} \\ &= s_2(y'z) + \mathfrak{P}^{j+1} \\ &= s_2(yz) + \mathfrak{P}^{j+1}. \end{aligned}$$

We have $\mathfrak{P}^j = \mathfrak{Q}_i^j\mathfrak{A}$, so any element of $\mathfrak{P}^j$ is a sum of elements of the form yz. Therefore the s_i agree on $\mathfrak{P}^j/\mathfrak{P}^{j+1}$, as required. This completes the proof of the lemma. ■

(2.4.13) Lemma: *Let $[\mathfrak{A}, n, n-1, \alpha]$ be a nonsplit fundamental stratum in A. Define $\mathcal{R} = \{x \in \mathfrak{A}/\mathfrak{P} : x\alpha \equiv \alpha x \pmod{\mathfrak{P}^{1-n}}\}$. Then $[\mathfrak{A}, n, n-1, \alpha]$ is equivalent to a simple stratum if and only if the ring $\mathcal{R}$ is a semisimple k_F-algebra.*

The proof of this lemma requires rather different techniques from those we are currently using. We therefore defer it until **(2.5)** below, and now proceed with the proof of (2.4.1)*(iii)*. We are given a pure stratum $[\mathfrak{A}, n, r, \beta]$ with $r = -k_0(\beta, \mathfrak{A})$, and a simple stratum $[\mathfrak{A}, n, r, \gamma_1]$ equivalent to it. (2.4.7) shows that there exists a simple stratum $[\mathfrak{A}, n, r, \gamma_2]$ which is equivalent to $[\mathfrak{A}, n, r, \beta]$ with the desired property:— if B_2 is the A-centraliser of γ_2, $\mathfrak{B}_2 = B_2 \cap \mathfrak{A}$ and s_2 is a tame corestriction on A relative to $F[\gamma_2]/F$, then $[\mathfrak{B}_2, r, r-1, s_2(\beta - \gamma_2)]$ is equivalent to a simple stratum. This property is, of course, independent of the choice of tame corestriction s_2.

We now choose tame corestrictions s_i relative to $F[\gamma_i]/F$, $i = 1, 2$, to satisfy condition *(ii)* of (2.4.12). Put $\beta = \gamma_1 + b_1 = \gamma_2 + b_2$ so that $s_1(b_2) = s_1(\gamma_1 - \gamma_2 + b_1) \equiv \delta_1 + s_1(b_1) \pmod{\mathfrak{P}^{1-r}}$, for some $\delta_1 \in F[\gamma_1]$. We can adjust γ_1, without affecting the relevant properties of either $[\mathfrak{A}, n, r, \gamma_1]$ or s_1, to ensure that δ_1 is zero mod $\mathfrak{P}^{1-r}$. Put $\alpha_i = s_i(b_i)$, $i = 1, 2$, so that, in particular, $\alpha_2 \in \mathfrak{K}(\mathfrak{B}_2) \subset \mathfrak{K}(\mathfrak{A})$, by the choice of γ_2. Since $\alpha_1 \equiv \alpha_2 \pmod{\mathfrak{P}^{1-r}}$, it follows that $\alpha_1 \in \mathfrak{K}(\mathfrak{A})$, and further $\alpha_1\mathfrak{A} = \mathfrak{P}^{-r}$. Thus, if $\mathfrak{B}_1$ denotes the $\mathfrak{A}$-centraliser of γ_1, the stratum $[\mathfrak{B}_1, r, r-1, \alpha_1]$ is fundamental, and it is nonsplit by (2.3.12). The cosets $\alpha_i + \mathfrak{P}^{1-r}$, for $i = 1, 2$, are equal and certainly have the same centralisers in $\mathfrak{A}/\mathfrak{P}$, and therefore also in the subalgebra $\mathfrak{B}_1/\mathfrak{B}_1 \cap \mathfrak{P} = \mathfrak{B}_2/\mathfrak{B}_2 \cap \mathfrak{P}$. The result now follows from (2.4.13), and this finally completes the proof of (2.4.1). ■

Remark: Let $[\mathfrak{A}, n, m, \beta]$ be a simple stratum in A. We can use (1.4.9) and the notion of defining sequence (2.4.2) to give an inductive description of $\mathfrak{N}_t(\beta, \mathfrak{A})$. The case where β is minimal over F is given by the Remark following the proof of (1.4.15). In general, if $k = k_0(\beta, \mathfrak{A})$, (1.4.9) gives

$$\mathfrak{N}_{k+r}\beta, \mathfrak{A}) = \mathfrak{B} + \mathfrak{Q}^r\mathfrak{N}_k(\beta, \mathfrak{A}), \quad r \geq 0,$$

where $\mathfrak{B}$ is the $\mathfrak{A}$-centraliser of β and $\mathfrak{Q} = \mathrm{rad}(\mathfrak{B})$. If we choose a simple stratum $[\mathfrak{A}, n, -k, \gamma]$ equivalent to $[\mathfrak{A}, n, -k, \beta]$, we have

$$\mathfrak{N}_k(\beta, \mathfrak{A}) = \mathfrak{N}_k(\gamma, \mathfrak{A}),$$

by, for example, (2.1.3). Setting $k' = k_0(\gamma, \mathfrak{A})$, we have $k' < k$, and

$$\mathfrak{N}_k(\gamma, \mathfrak{A}) = \mathfrak{B}_\gamma + \mathfrak{Q}_\gamma^{k-k'} \mathfrak{N}_{k'}(\gamma, \mathfrak{A}),$$

and so on.

(2.5) Nonsplit fundamental strata

The aim of this section is to produce a "standard form" (virtually a Jordan canonical form) for equivalence classes of nonsplit fundamental strata. We will need this in the next section, but we use it here to dispose of the outstanding proofs of (2.3.4) and (2.4.13). We also prove (2.4.11), which is somewhat easier. However, these results have in common the requirement for explicit block matrix calculations. It therefore seems a good idea to put these together, and deal with all arguments of this type in one place.

Let us start with some arcane terminology. Given a vector of positive integers $\boldsymbol{n} = (n_0, n_1, \dots n_{e-1})$ and a commutative ring R, write $\mathbb{M}(\boldsymbol{n}, R)$ for the collection of $e \times e$ block matrices over R in which the (i,j)-block has dimensions $n_i \times n_j$. We refer to the elements of $\mathbb{M}(\boldsymbol{n}, R)$ as "$\boldsymbol{n}$-block matrices over R".

We take a hereditary $\mathfrak{o}_F$-order $\mathfrak{A} = \mathrm{End}^0_{\mathfrak{o}_F}(\mathcal{L})$, for some lattice chain $\mathcal{L} = \{L_i : i \in \mathbb{Z}\}$ in V. We abbreviate $e = e(\mathfrak{A}|\mathfrak{o}_F)$, and put $n_i = \dim_{\mathrm{k}_F}(L_i/L_{i+1})$, $i \in \mathbb{Z}$. If we choose a basis of the lattice chain $\mathcal{L}$ as in §**1**, which is also a basis of the lattice $L_0 \in \mathcal{L}$, and use this to identify A with $\mathbb{M}(N, F)$, then $\mathfrak{A}$ becomes identified with the ring of all $\boldsymbol{n}$-block matrices over $\mathfrak{o}_F$ of the form

$$\text{(2.5.1)} \qquad \mathfrak{A} = \begin{pmatrix} \mathfrak{o}_F & \mathfrak{o}_F & \mathfrak{o}_F & \cdots & \cdots & \mathfrak{o}_F \\ \mathfrak{p}_F & \mathfrak{o}_F & \mathfrak{o}_F & \cdots & \cdots & \mathfrak{o}_F \\ \mathfrak{p}_F & \mathfrak{p}_F & \mathfrak{o}_F & \cdots & \cdots & \mathfrak{o}_F \\ \vdots & & \ddots & \ddots & & \vdots \\ \vdots & & & \ddots & \ddots & \vdots \\ \mathfrak{p}_F & \cdots & \cdots & \cdots & \mathfrak{p}_F & \mathfrak{o}_F \end{pmatrix},$$

where $\boldsymbol{n} = (n_0, n_1, \dots, n_{e-1})$. The radical $\mathfrak{P}$ of $\mathfrak{A}$ gets identified with the group of $\boldsymbol{n}$-block matrices over F with entries thus:

$$\mathfrak{P} = \begin{pmatrix} \mathfrak{p}_F & \mathfrak{o}_F & \mathfrak{o}_F & \cdots & \cdots & \mathfrak{o}_F \\ \mathfrak{p}_F & \mathfrak{p}_F & \mathfrak{o}_F & \cdots & \cdots & \mathfrak{o}_F \\ \mathfrak{p}_F & \mathfrak{p}_F & \mathfrak{p}_F & \cdots & \cdots & \mathfrak{o}_F \\ \vdots & & \ddots & \ddots & & \vdots \\ \vdots & & & \ddots & \mathfrak{p}_F & \mathfrak{o}_F \\ \mathfrak{p}_F & \cdots & \cdots & \cdots & \mathfrak{p}_F & \mathfrak{p}_F \end{pmatrix}.$$

We can indeed describe any power $\mathfrak{P}^n$ of $\mathfrak{P}$ in these terms. Since $\pi_F^t\mathfrak{P}^n = \mathfrak{P}^{n+et}$, where π_F is a prime element of F, we need only treat the case $0 \le n \le e-1$. Then:

(2.5.2) *Let $\mathfrak{A}$, $\mathfrak{P}$ be as above, and let $0 \le n \le e-1$. Then $\mathfrak{P}^n$ is the set of all $\boldsymbol{n}$-block matrices over F, $\boldsymbol{n} = (n_0, n_1, \dots, n_{e-1})$, with the (i,j)-block having entries as follows:*

$$\left.\begin{array}{l} \mathfrak{o}_F \text{ if } j-i \ge n, \\ \mathfrak{p}_F^2 \text{ if } j-i < n-e, \\ \mathfrak{p}_F \text{ otherwise,} \end{array}\right\} \quad 0 \le i,j \le e-1.$$

Now we prove (2.4.11). We are given a hereditary $\mathfrak{o}_F$-order $\mathfrak{A}$ in A with radical $\mathfrak{P}$, and an element $b \in \mathfrak{P}^n$ such that the coset $b + \mathfrak{P}^{n+1}$ is intertwined by every $x \in G$, in other words

$$x(b+\mathfrak{P}^{n+1}) \cap (b+\mathfrak{P}^{n+1})x \ne \emptyset, \quad x \in G.$$

We have to prove

$$(b+\mathfrak{P}^{n+1}) \cap F \ne \emptyset.$$

If $b \in \mathfrak{P}^{n+1}$, then $b+\mathfrak{P}^{n+1} = \mathfrak{P}^{n+1}$, and surely $0 \in F \cap \mathfrak{P}^{n+1}$. We can therefore assume $b \notin \mathfrak{P}^{n+1}$.

We next observe that scaling the coset $b+\mathfrak{P}^{n+1}$ by an element of $F^\times$ has no effect on either the hypothesis or conclusion, so we may as well assume that

$$0 \le n \le e-1.$$

We show first that, under this hypothesis, we must have $n = 0$. Suppose otherwise. Let $\mathcal{L} = \{L_i\}$ denote the lattice chain defining $\mathfrak{A}$, so that we may identify

$$\mathfrak{P}^n/\mathfrak{P}^{n+1} = \coprod_{i=0}^{e-1} \mathrm{Hom}_{\mathrm{k}_F}(L_i/L_{i+1}, L_{n+i}/L_{n+i+1}),$$

where $e = e(\mathfrak{A})$. Shifting the index in $\mathcal{L}$, we can assume that b defines a nonzero homomorphism $L_0/L_1 \to L_n/L_{n+1}$. We choose a basis of the lattice chain $\mathcal{L}$ whose $\mathfrak{o}_F$-span is L_0, and use it to identify $\mathfrak{A}$ with an order of block matrices as in (2.5.1). We retain the other associated notations. The $(0,n)$-block of b is then nonzero modulo $\mathfrak{p}_F$. Take for x an element of the form

$$x \in (\mathrm{diag}(1,0,\dots,0) + \mathfrak{p}_F^M.\mathbb{M}(N,\mathfrak{o}_F)) \cap G,$$

where M is some large positive integer (and 1 denotes the $n_0 \times n_0$ identity matrix). Consider the sets $x(b+\mathfrak{P}^{n+1})$, $(b+\mathfrak{P}^{n+1})x$. The first consists of matrices with a block in the $(0,n)$-place which is nonzero (modulo $\mathfrak{p}_F$), while any element of the second has all of its columns (of blocks), except possibly the first, divisible by $\mathfrak{p}_F^M$. We conclude that $x(b+\mathfrak{P}^{n+1}) \cap (b+\mathfrak{P}^{n+1})x = \emptyset$ in this situation, so we must have $n = 0$ here.

Therefore we have $b \in \mathfrak{A}$, $b \notin \mathfrak{P}$. We may now assume that b is of the form

$$b = \operatorname{diag}(b_0, b_1, \dots, b_{e-1}),$$

where $b_i \in \mathbb{M}(n_i, \mathfrak{o}_F)$, and some b_i is nonzero modulo $\mathfrak{p}_F$. Our hypothesis implies that $x^{-1}(b+\mathfrak{P})x = (b+\mathfrak{P})$ for all $x \in \boldsymbol{U}(\mathfrak{A})$, which implies that there exists $a_i \in \mathfrak{o}_F$ such that

$$b_i \equiv a_i I_i \pmod{\mathfrak{p}_F \mathbb{M}(n_i, \mathfrak{o}_F)}, \quad 0 \le i \le e-1,$$

where I_i denotes the $n_i \times n_i$ identity matrix. Further, some $a_i \notin \mathfrak{p}_F$ since $b \notin \mathfrak{P}$. Finally, let x range over the permutation matrices in $GL(N, \mathfrak{o}_F)$. For any $y \in \mathfrak{P}$ the diagonal entries of $x(b+y)x^{-1}$, taken mod $\mathfrak{p}_F$, are just some permutation (in fact the one corresponding to x) of the a_i (mod $\mathfrak{p}_F$). The intertwining condition for such x therefore implies that there exists $a \in \mathfrak{o}_F$ such that $a_i \equiv a \pmod{\mathfrak{p}_F}$ for all i. The coset $b+\mathfrak{P}$ therefore contains the element $a \in F$. This proves the first assertion of (2.4.11).

To prove the second, let $\chi \in \boldsymbol{U}^{m+1}(\mathfrak{A})^{\wedge}$ be a character which is intertwined by the whole of G, $m \ge 0$. Let $n \ge m$ be the least integer such that χ is trivial on $\boldsymbol{U}^{n+1}(\mathfrak{A})$. If $n = m$, there is nothing to prove, so assume otherwise. According to the first part, the restriction $\chi \mid \boldsymbol{U}^n(\mathfrak{A})$ is of the form ψ_b, for some $b \in F \cap \mathfrak{P}^{-n}$. By definition, $b \notin \mathfrak{P}^{1-n}$, so n is divisible by $e = e(\mathfrak{A}|\mathfrak{o}_F)$, $n = n_0 e$ say. An elementary argument then gives

$$\psi_b \mid \boldsymbol{U}^n(\mathfrak{A}) = (\psi_{F,b} \mid \boldsymbol{U}^{n_0}(\mathfrak{o}_F)) \circ \det,$$

where $\psi_{F,b}$ is the function $x \mapsto \psi_F(b(x-1))$, $x \in F$. Therefore $\chi = \chi_1 . \chi_0 \circ \det$, for some character χ_0 of $F^\times$ (restricted to $\det(\boldsymbol{U}^{m+1}(\mathfrak{A}))$) and a character χ_1 of $\boldsymbol{U}^{m+1}(\mathfrak{A})$ which is trivial on $\boldsymbol{U}^n(\mathfrak{A})$. This character χ_1 is also intertwined by every element of G, so the result follows by induction.

This completes the proof of (2.4.11).

We continue with our hereditary order $\mathfrak{A}$ as in (2.5.1). The notations e, n_i are as before, and we put $\boldsymbol{n} = (n_0, n_1, \dots, n_{e-1})$. It will now be convenient to have an explicit set of representatives for the cosets $\mathfrak{P}^m/\mathfrak{P}^{m+1}$, for any integer m. We start by choosing (and fixing) a prime element π_F of F. We define integers ℓ, k by

(2.5.3) $$0 \le \ell = m - ek \le e-1.$$

We consider the set $\boldsymbol{B}_\ell = \boldsymbol{B}_\ell(\mathfrak{A})$ of e-tuples of matrices

$$\boldsymbol{b} = (b_0, b_1, \ldots, b_{e-1}),$$

where b_i has entries in $\mathfrak{o}_F$ and dimensions $n_i \times n_{i+\ell}$. The set $\boldsymbol{B}_\ell$ admits the obvious equivalence relation of congruence (entry by entry) modulo $\mathfrak{p}_F$. Given m, k as in (2.5.3), and an element $\boldsymbol{b} \in \boldsymbol{B}_\ell$, we define an $\boldsymbol{n}$-block matrix $\boldsymbol{r}_m(\boldsymbol{b})$ over F by

(2.5.4)
$$\boldsymbol{r}_m(\boldsymbol{b})_{ij} = \begin{cases} \pi_F^k b_i \text{ if } j \leq e-1 \text{ and } j-i=\ell, \\ \pi_F^{k+1} b_i \text{ if } j \geq e-1 \text{ and } j-i=\ell, \\ 0 \text{ otherwise.} \end{cases}$$

Then (2.5.2) gives

(2.5.5) *Given an integer m, define integers k, ℓ by* (2.5.3). *Then the map $\boldsymbol{b} \mapsto \boldsymbol{r}_m(\boldsymbol{b})$ induces a bijection between $\boldsymbol{B}_\ell(\mathfrak{A})$ (modulo $\mathfrak{p}_F$) and $\mathfrak{P}^m/\mathfrak{P}^{m+1}$.*

Of course, the sizes of the matrices b_i, for $\boldsymbol{b} = (b_0, b_1, \ldots, b_{e-1}) \in \boldsymbol{B}_\ell$, do indeed depend on both $\mathfrak{A}$ (or, more precisely, the sequence $\boldsymbol{n} = (n_0, n_1, \ldots, n_{e-1})$) and the "level" ℓ. (2.5.5) can be very useful for analysing the cosets $b + \mathfrak{P}^{m+1}$, $b \in \mathfrak{P}^m$. For example,

(2.5.6) *In the notation of* (2.5.5), *we have $\mathfrak{P}^m \cap \mathfrak{K}(\mathfrak{A}) \subset \mathfrak{P}^{m+1}$ unless $n_i = n_{i+\ell}$ for all i. If this condition holds, and $\boldsymbol{b} = (b_0, b_1, \ldots, b_{e-1}) \in \boldsymbol{B}_\ell$, then $\boldsymbol{r}_m(\boldsymbol{b}) \in \mathfrak{K}(\mathfrak{A})$ if and only if $b_i \in GL(n_i, \mathfrak{o}_F)$ for all i.*

The proof is immediate. Suppose now that we are given a monic polynomial $f(X) \in \mathfrak{o}_F[X]$ such that $\overline{f}(X) \in \mathrm{k}_F[X]$ is irreducible, where we write "bar" for the obvious map $\mathfrak{o}_F[X] \to \mathrm{k}_F[X]$ of reduction mod $\mathfrak{p}_F$. *Throughout, we exclude the possibility $\overline{f}(X) = X$.* Let $d = \deg(f(X))$, and choose a matrix $\gamma \in \mathbb{M}(d, \mathfrak{o}_F)$ whose minimal polynomial is $f(X)$. Of course, this matrix γ is determined up to conjugation by an element of $GL(d, \mathfrak{o}_F)$. The algebra $F[\gamma]$ is a field, indeed an unramified extension of F of degree d, and, moreover, $\mathfrak{o}_{F[\gamma]} = \mathfrak{o}_F[\gamma]$.

We also write $\overline{\gamma}$ for the image of γ in $GL(d, \mathrm{k}_F)$.

We can now use this data to define a family of "standard" strata. Fix a family $\mathcal{C}$ of coset representatives in $\mathbb{Z}/e\mathbb{Z}$ of the subgroup generated by $m + e\mathbb{Z}$ (or $\ell + e\mathbb{Z}$). Of course, this subgroup is just $g\mathbb{Z}$, where $g = \gcd(e, m)$.

(2.5.7) Definition: *(i) Given γ as above, an element $\boldsymbol{b} \in \boldsymbol{B}_\ell(\mathfrak{A})$ is called a γ-element if*

(a) *b_i is an identity matrix whenever $i \notin \mathcal{C}$;*

(b) for $i \in \mathcal{C}$, write $\bar{b}_i = b_i \bmod \mathfrak{p}_F$. Then

$$\bar{b}_i = \begin{pmatrix} \bar{\gamma} & & & & \\ 0 & \bar{\gamma} & & * & \\ 0 & 0 & \bar{\gamma} & & \\ \vdots & & & \ddots & \\ 0 & \dots & \dots & 0 & \bar{\gamma} \end{pmatrix},$$

and this matrix is in Jordan canonical form over k_F.

*(ii) A stratum $[\mathfrak{A}, m, m-1, b]$ is in γ-*standard form *if $b = \boldsymbol{r}_{-m}(\boldsymbol{b})$, for some γ-element $\boldsymbol{b} \in \boldsymbol{B}_\ell(\mathfrak{A})$, where $\ell \equiv -m \pmod{e}$, $0 \le \ell \le e-1$.*

In *(i)(b)*, what we sometimes require is that the upper triangular matrix $\bar{b}_i - \mathrm{diag}(\bar{\gamma}, \bar{\gamma}, \dots, \bar{\gamma})$ is nilpotent and can be expressed as a polynomial in $\bar{b}_i$ with coefficients in k_F. This, of course, is exactly what the Jordan canonical form provides.

Remarks: *(i)* The existence of a stratum $[\mathfrak{A}, m, m-1, b]$ in γ-standard form imposes quite stringent conditions on $\mathfrak{A}$ and m. In particular, we must have $n_i = n_{i+\ell}$ for all i. Further, if $[\mathfrak{A}, m, m-1, b]$ is in γ-standard form, then $b \in \mathfrak{K}(\mathfrak{A})$ and $b\mathfrak{A} = \mathfrak{P}^{-m}$.

(ii) The choice of coset representatives $\mathcal{C}$ is essentially irrelevant here. Conjugating $\boldsymbol{r}_m(\boldsymbol{b})$ by an element of $\boldsymbol{U}(\mathfrak{A})$, we can move the nontrivial block b_j to the i-th place, for any i in the same coset as j, without affecting b_k, for any k in a different coset.

(2.5.8) Proposition: *Let the stratum $[\mathfrak{A}, m, m-1, b]$ be in γ-standard form. Then $[\mathfrak{A}, m, m-1, b]$ is nonsplit fundamental and, in the notation of* **(2.3)**, *the characteristic polynomial $\phi_b(X)$ of $[\mathfrak{A}, m, m-1, b]$ is a power of $\bar{f}(X)$.*

Moreover, the stratum $[\mathfrak{A}, m, m-1, b]$ is equivalent to a simple stratum if and only if the matrices $\bar{b}_i$ in (2.5.7)(i)(b) are all semisimple (i.e. of the form $\mathrm{diag}(\bar{\gamma}, \dots, \bar{\gamma})$).

Before proving this, let us note a few consequences. First, if we have a stratum $[\mathfrak{A}, m, m-1, b]$ in γ-standard form, and $d = \deg(f(X))$, where $f(X)$ is the minimal polynomial of γ, then the block sizes n_i in $\mathfrak{A}$ are all divisible by d. Thus $\mathfrak{A}$ contains the principal order $\mathfrak{A}'$, which is upper triangular (in blocks) mod $\mathfrak{p}_F$, and all of whose blocks are of dimension $d \times d$. In particular, $e(\mathfrak{A}'|\mathfrak{o}_F) = N/d$. Writing $\mathfrak{P}'$ for the radical of $\mathfrak{A}'$, the block picture shows that we have $b \in \mathfrak{P}'^{-m'}$, where

$$\begin{aligned} d\ell' &= n_0 + n_1 + \dots + n_{\ell-1}, \\ -m' &= \ell' + Nk/d, \\ -m &= \ell + ek. \end{aligned}$$

Indeed, $b \in \mathfrak{K}(\mathfrak{A}')$ and $b\mathfrak{A}' = \mathfrak{P}'^{-m'}$. Write $\boldsymbol{b}$ for a γ-element of $\boldsymbol{B}_\ell(\mathfrak{A})$ such that $b = \boldsymbol{r}_m(\boldsymbol{b})$, and define $\boldsymbol{b}' \in \boldsymbol{B}_\ell$ by $\boldsymbol{b}' = (b'_0, \dots, b'_{e-1})$, $b'_i = b_i$ if $i \notin \mathcal{C}$, and $b'_i = \operatorname{diag}(\gamma, \dots, \gamma)$ otherwise. Put $b' = \boldsymbol{r}_m(\boldsymbol{b}')$. Then $b' \equiv b \pmod{\mathfrak{P}'^{1-m'}}$ and:

(2.5.9) Corollary: *The stratum* $[\mathfrak{A}', m', m'-1, b']$ *is simple. In particular,* (2.3.4) *holds for a stratum in* γ*-standard form.*

Proof: Put $g' = \gcd(m', e')$, $e' = e(\mathfrak{A}'|\mathfrak{o}_F)$, and form the element

$$y' = \pi_F^{m'/g'} b'^{e'/g'}.$$

We have $b \in \mathfrak{K}(\mathfrak{A})$ and $b \in \mathfrak{K}(\mathfrak{A}')$. Further, $b\mathfrak{A} = \mathfrak{P}^{-m}$, and $b\mathfrak{A}' = \mathfrak{P}'^{-m'}$. It follows that $m/e = m'/e'$, and thus $m'/g' = m/g$ and $e'/g' = e/g$. This observation makes it easy to compute the element y'. It is the diagonal $(d, d, \dots, d)$-block matrix $\operatorname{diag}(\gamma, \gamma, \dots, \gamma)$. Thus b' is a root of the polynomial $f((\pi_F^{-m'/g'} X)^{e'/g'})$. This polynomial is irreducible over F, so $F[b']$ is a field. The ramification index $e(F[b']|F)$ is exactly e'/g', and y' generates the residue field of $F[b']$ over k_F. Since $b' \in \mathfrak{K}(\mathfrak{A}')$, (1.5.6) shows that $[\mathfrak{A}', m', m'-1, b']$ is simple. The containment $b + \mathfrak{P}^{1-m} \subset b' + \mathfrak{P}'^{1-m'}$ demanded by (2.3.4) follows from inspection of the block pictures. ■

Returning to the stratum $[\mathfrak{A}, m, m-1, b]$ of (2.5.8), we define

$$\mathcal{R} = \{x \in \mathfrak{A}/\mathfrak{P} : xb \equiv bx \pmod{\mathfrak{P}^{1-m}}\},$$

just as in (2.4.13). We can calculate this explicitly. It is the product, over all $i \in \mathcal{C}$, of the $\mathbb{M}(n_i, \mathrm{k}_F)$-centraliser of the block $\overline{b}_i$. Thus $\mathcal{R}$ is semisimple if and only if all the matrices $\overline{b}_i$, $i \in \mathcal{C}$, are semisimple. Therefore:

(2.5.10) Corollary: (2.4.13) *holds for strata in* γ*-standard form.*

Now let us prove (2.5.8). We form $y = \pi_F^{m/g} b^{e/g}$, $g = \gcd(e, m)$. This is a diagonal block matrix in $\mathfrak{A}$, in which the j-th diagonal block is b_i, where $i \in \mathcal{C}$ is the chosen representative of the coset $i + g\mathbb{Z}$ in $\mathbb{Z}/e\mathbb{Z}$. The characteristic polynomial of y (taken modulo $\mathfrak{p}_F$) is then clearly a power of the minimal polynomial of $\overline{\gamma}$, i. e. a power of $\overline{f}(X)$. The remark preceding (2.3.3) now shows that the characteristic polynomial of $[\mathfrak{A}, m, m-1, b]$ is a power of $\overline{f}$. In particular, it is not a power of X, so the stratum is fundamental, and indeed nonsplit fundamental.

For the second assertion, if all of the $\overline{b}_i$ are semisimple, which amounts to saying that they are of the form $\operatorname{diag}(\overline{\gamma}, \overline{\gamma}, \dots, \overline{\gamma})$, we can assume that $b_i = \operatorname{diag}(\gamma, \dots, \gamma)$, $i \in \mathcal{C}$, and work out explicitly the structure of the algebra $F[b]$. Just as above, it is a field, b is minimal over F, and the stratum $[\mathfrak{A}, m, m-1, b]$ is simple.

Conversely, suppose that $[\mathfrak{A}, m, m-1, b]$ is equivalent to a simple stratum $[\mathfrak{A}, m, m-1, \beta]$, but that $\bar{b}_i$ is not semisimple, for some $i \in C$. Form the element y as above. The $\mathfrak{A}/\mathfrak{P}$-centraliser of $y + \mathfrak{P}$ is therefore not semisimple, and indeed contains a nontrivial central nilpotent element $\mathfrak{n}$ which we may take to be a polynomial in $y + \mathfrak{P}$ (over k_F). Write $\mathfrak{B}$ for the $\mathfrak{A}$-centraliser of β (so that $\mathfrak{B}$ is a hereditary $\mathfrak{o}_{F[\beta]}$-order). We have $y \equiv \pi_F^{m/g} \beta^{e/g} \pmod{\mathfrak{P}}$, so y commutes with the image of $\mathfrak{B}$ in $\mathfrak{A}/\mathfrak{P}$. The same therefore applies to $\mathfrak{n}$. Likewise, $\beta y \beta^{-1} \equiv y \pmod{\mathfrak{P}}$, so $\beta \mathfrak{n} \beta^{-1} \equiv \mathfrak{n} \pmod{\mathfrak{P}}$. However, since $[\mathfrak{A}, m, m-1, \beta]$ is simple, the kernel of the endomorphism $x \mapsto \beta x \beta^{-1} - x$ of $\mathfrak{A}/\mathfrak{P}$ is just $\mathfrak{N}_{1-m}(\beta, \mathfrak{A}) = \mathfrak{B} + \mathfrak{P}$ (taken modulo $\mathfrak{P}$). It follows that $\mathfrak{n}$ is a central nilpotent element of the semisimple k_F-algebra $\mathfrak{B} + \mathfrak{P}/\mathfrak{P}$. This is nonsense, and the contradiction proves (2.5.8). ■

Therefore, all that remains is to prove:

(2.5.11) Proposition: *Let $[\mathfrak{A}, m, m-1, b]$ be a nonsplit fundamental stratum in $\mathbb{M}(N, F)$ with $\mathfrak{A}$ as in (2.5.1), and let $\phi(X) \in \mathsf{k}_F[X]$ be the unique (monic) irreducible factor of the characteristic polynomial of $[\mathfrak{A}, m, m-1, b]$. Choose a monic polynomial $f(X) \in \mathfrak{o}_F[X]$ with $\bar{f} = \phi$ and $\deg(f) = \deg(\phi) = d$, say. Choose a matrix $\gamma \in \mathbb{M}(d, \mathfrak{o}_F)$ with minimal polynomial $f(X)$. Then there exists $x \in \boldsymbol{U}(\mathfrak{A})$ such that $[\mathfrak{A}, m, m-1, xbx^{-1}]$ is equivalent to a stratum in γ-standard form.*

Proof: Scaling by an element of $F^\times$ changes nothing, so we may as well assume that $0 \leq -m \leq e-1$. Likewise, moving to an equivalent stratum changes nothing, so we can assume that the (i, j)-block of b is zero unless $j - i \equiv -m \pmod{e}$. For $j - i \equiv -m \pmod{e}$, define an $(n_i \times n_j)$ matrix over $\mathfrak{o}_F$ by

$$b_i = \begin{cases} b_{ij} & \text{if } i \leq j, \\ \pi_F^{-1} b_{ij} & \text{otherwise.} \end{cases}$$

It will be easier now if we think of the index i on b_i as being an integer mod e. We have $(b_0, b_1, \dots, b_{e-1}) = \boldsymbol{b} \in \boldsymbol{B}_{-m}(\mathfrak{A})$ and $b = \boldsymbol{r}_{-m}(\boldsymbol{b})$. Form the element $y = \pi_F^{m/g} b^{e/g}$ as usual, with $g = \gcd(m, e)$. This is a diagonal block matrix $y = \operatorname{diag}(y_0, y_1, \dots, y_{e-1})$, in which

$$y_i = b_i b_{i-m} \dots b_{i-(e/g-1)m}.$$

Each y_i is a square matrix over $\mathfrak{o}_F$ whose characteristic polynomial (taken mod $\mathfrak{p}_F$) is a power of $\phi(X)$. In particular, y_i is invertible over $\mathfrak{o}_F$. It follows that all of the b_i are square and invertible over $\mathfrak{o}_F$. Indeed, if we allow i to range over one coset in $\mathbb{Z}/e\mathbb{Z}$ of the subgroup generated by $-m$, all the b_i have the same size $n_i \times n_i$.

If we take a diagonal block matrix $a = \mathrm{diag}(a_0, a_1, \ldots, a_{e-1}) \in \boldsymbol{U}(\mathfrak{A})$, and replace b by aba^{-1}, then b_i gets replaced by $a_i b_i a_{i-m}^{-1}$ (where the indices are again taken mod e). Therefore, using conjugations of this sort, we can arrange for all of the b_i to be identity matrices with the exception of one in each coset of $\mathbb{Z}/e\mathbb{Z}$ modulo $e\mathbb{Z} + m\mathbb{Z}$. (We can further arrange for the exceptions to appear at any preordained place in their cosets.) Now we can conjugate to get these exceptions to get them in Jordan canonical form (mod $\mathfrak{p}_F$) as in (2.5.7). ■

(2.3.4), (2.4.13) now follow from (2.5.11) and (2.5.9), (2.5.10) respectively.

(2.6) Intertwining and conjugacy

We conclude this section with a quite remarkable "rigidity" or "anti-intertwining" property of simple strata. It is possible to prove much more general results than the one we give here, but we do not need them for our applications, and the extra effort involved is substantial, so we simply give a statement at the end.

For $i = 1, 2$, let $[\mathfrak{A}_i, n_i, r_i, b_i]$ be a stratum in A, and put $\mathfrak{P}_i = \mathrm{rad}(\mathfrak{A}_i)$. We say that these strata *intertwine in* G if there exists $x \in G$ such that

$$x^{-1}(b_2 + \mathfrak{P}_2^{-r_2})x \cap (b_1 + \mathfrak{P}_1^{-r_1}) \neq \emptyset.$$

Of course, this only depends on the equivalence classes of the strata involved. Also, an element x which satisfies this condition is said to *intertwine* the strata.

(2.6.1) Theorem: *For $i = 1, 2$, let $[\mathfrak{A}, n, r, \beta_i]$ be simple strata in A, and suppose that they intertwine in G. Then there exists $x \in \boldsymbol{U}(\mathfrak{A})$ such that*

$$[\mathfrak{A}, n, r, x\beta_2 x^{-1}] \sim [\mathfrak{A}, n, r, \beta_1].$$

Remark: In the statement of (2.6.1), we only really need the hypothesis that the $[\mathfrak{A}, n, r, \beta_i]$ are *equivalent to* simple strata, since intertwining is a property of equivalence classes of strata, and conjugation by $\boldsymbol{U}(\mathfrak{A})$ also preserves equivalence.

Proof: We show by induction that, for each integer t, $r \leq t \leq n-1$, there exists $x_t \in \boldsymbol{U}(\mathfrak{A})$ such that the (pure) strata $[\mathfrak{A}, n, t, \beta_1]$, $[\mathfrak{A}, n, t, x_t\beta_2 x_t^{-1}]$ are equivalent. We start with the case $t = n-1$. For each i, we use (2.4.1) to choose a simple stratum $[\mathfrak{A}, n, n-1, \alpha_i] \sim [\mathfrak{A}, n, n-1, \beta_i]$. The strata $[\mathfrak{A}, n, n-1, \alpha_1]$, $[\mathfrak{A}, n, n-1, \alpha_2]$ then intertwine. Therefore $\alpha_1 + \mathfrak{P}^{1-n}$ contains a conjugate of some element of $\alpha_2 + \mathfrak{P}^{1-n}$. The polynomial $\phi_{\alpha_i}(X)$ only depends on the coset $\alpha_i + \mathfrak{P}^{1-n}$, and so it follows that $\phi_{\alpha_1} = \phi_{\alpha_2}$. By (2.3.2), the strata $[\mathfrak{A}, n, n-1, \alpha_i]$ are nonsplit fundamental, so we can apply (2.5.11). We choose a basis of V to identify A with $\mathbb{M}(N, F)$ so that $\mathfrak{A}$ has the form (2.5.1). We can replace α_i by a $\boldsymbol{U}(\mathfrak{A})$-conjugate and assume that

$$[\mathfrak{A}, n, n-1, \alpha_i] \sim [\mathfrak{A}, n, n-1, \alpha_i'], \quad i = 1, 2,$$

where the $[\mathfrak{A}, n, n-1, \alpha_i']$ are both in $\boldsymbol{\gamma}$-standard form for the same matrix $\boldsymbol{\gamma}$. However, all of these strata are equivalent to simple strata, so (2.5.8) (together with Remark *(ii)* preceding it) shows that $[\mathfrak{A}, n, n-1, \alpha_1']$ is equivalent to a $\boldsymbol{U}(\mathfrak{A})$-conjugate of $[\mathfrak{A}, n, n-1, \alpha_2']$. Altogether, we have shown that there exists $x \in \boldsymbol{U}(\mathfrak{A})$ such that

$$[\mathfrak{A}, n, n-1, x^{-1}\alpha_1 x] \sim [\mathfrak{A}, n, n-1, \alpha_2],$$

as required.

So, we now take an integer t, $r < t \leq n-1$ and an $x_t \in \boldsymbol{U}(\mathfrak{A})$ such that $[\mathfrak{A}, n, t, x_t\beta_2 x_t^{-1}] \sim [\mathfrak{A}, n, t, \beta_1]$. We can always replace $[\mathfrak{A}, n, r, \beta_2]$ by a $\boldsymbol{U}(\mathfrak{A})$-conjugate without changing anything, so we assume further that $[\mathfrak{A}, n, t, \beta_2] \sim [\mathfrak{A}, n, t, \beta_1]$. We use (2.4.1) again to choose a simple stratum $[\mathfrak{A}, n, t, \gamma]$ equivalent to $[\mathfrak{A}, n, t, \beta_i]$. Let B denote the A-centraliser of γ, $\mathfrak{B} = B \cap \mathfrak{A}$, and let s be a tame corestriction on A relative to $F[\gamma]/F$. Write $b_i = \beta_i - \gamma$, $i = 1, 2$, and consider the strata $[\mathfrak{B}, t, t-1, s(b_i)]$. For each i, choose a simple stratum $[\mathfrak{A}, n, t-1, \gamma_i]$ equivalent to $[\mathfrak{A}, n, t-1, \beta_i]$ (as we may, by (2.4.1)*(i)*). We then have $[\mathfrak{B}, t, t-1, s(b_i)] \sim [\mathfrak{B}, t, t-1, s(\gamma_i - \gamma)]$. (2.4.1)*(ii), (iii)* now show that the stratum $[\mathfrak{B}, t, t-1, s(\gamma_i - \gamma)]$ is equivalent to either a simple stratum or the null stratum $[\mathfrak{B}, t, t-1, 0]$. The same therefore applies to $[\mathfrak{B}, t, t-1, s(b_i)]$.

However, the first step is to show that these strata $[\mathfrak{B}, t, t-1, s(b_1)]$, $[\mathfrak{B}, t, t-1, s(b_2)]$ intertwine in $B^\times$. By hypothesis, the strata $[\mathfrak{A}, n, t-1, \beta_i]$ are intertwined by some element $x \in G$. This element certainly intertwines $[\mathfrak{A}, n, t, \gamma]$ with itself and therefore, by (1.5.8),

$$x \in (1 + \mathfrak{Q}^{-t-k}\mathfrak{N})B^\times(1 + \mathfrak{Q}^{-t-k}\mathfrak{N}),$$

where $\mathfrak{Q} = \mathrm{rad}(\mathfrak{B})$, $k = k_0(\gamma, \mathfrak{A})$ and $\mathfrak{N} = \mathfrak{N}_k(\gamma, \mathfrak{A})$. We write $x = (1+y)\delta(1+z)$ according to this decomposition and substitute in the intertwining relation

$$x^{-1}(\gamma + b_2)x \equiv \gamma + b_1 \pmod{x^{-1}\mathfrak{P}^{1-t}x + \mathfrak{P}^{1-t}}.$$

We can conjugate by $(1+z)$ (which normalises the modulus $\mathfrak{P}^{1-t}$, since $(1+z) \in \boldsymbol{U}(\mathfrak{A})$) to get

$$\delta^{-1}(1+y)^{-1}(\gamma + b_2)(1+y)\delta \equiv (1+z)(\gamma + b_1)(1+z)^{-1} \pmod{\delta^{-1}\mathfrak{P}^{1-t}\delta + \mathfrak{P}^{1-t}}.$$

The next step is to show

$$(1+z)(\gamma+b_1)(1+z)^{-1} \equiv \gamma - a_\gamma(z) + b_1 \pmod{\mathfrak{P}^{1-t}}.$$

This sort of straightforward computation will recur frequently, so we write it out fully this time. Since $(1+z)$ is a unit of $\mathfrak{A}$, the desired congruence is equivalent to

$$(1+z)(\gamma+b_1) \equiv (\gamma - a_\gamma(z) + b_1)(1+z) \pmod{\mathfrak{P}^{1-t}}.$$

Expanding, this is equivalent to

$$zb_1 \equiv b_1 z - a_\gamma(z)z \pmod{\mathfrak{P}^{1-t}}.$$

However, $b_1 \in \mathfrak{P}^{-t}$ and $z \in \mathfrak{Q}^{-t-k}\mathfrak{N} \subset \mathfrak{P}$. The terms zb_1, b_1z are therefore null. Further, $a_\gamma(z) \in \mathfrak{P}^{-t}$, so $a_\gamma(z)z \in \mathfrak{P}^{1-t}$ as required.

Exactly the same computation on the other side gives

$$(1+y)^{-1}(\gamma+b_2)(1+y) \equiv \gamma + a_\gamma(y) + b_2 \pmod{\mathfrak{P}^{1-t}},$$

and conjugating by δ (which commutes with γ), we get

$$\delta^{-1}(1+y)^{-1}(\gamma+b_2)(1+y)\delta \equiv \gamma + a_\gamma(\delta^{-1}y\delta) + \delta^{-1}b_2\delta \pmod{\delta^{-1}\mathfrak{P}^{1-t}\delta}.$$

In all, this gives us

$$\gamma + a_\gamma(\delta^{-1}y\delta) + \delta^{-1}b_2\delta \equiv \gamma - a_\gamma(z) + b_1 \pmod{\delta^{-1}\mathfrak{P}^{1-t}\delta + \mathfrak{P}^{1-t}}.$$

Applying the map s, we get

$$\delta^{-1}s(b_2)\delta \equiv s(b_1) \pmod{\delta^{-1}\mathfrak{Q}^{1-t}\delta + \mathfrak{Q}^{1-t}}.$$

This relation simply says that the element $\delta \in B^\times$ intertwines $[\mathfrak{B}, t, t-1, s(b_2)]$ with $[\mathfrak{B}, t, t-1, s(b_1)]$, as asserted.

Suppose to start with that these strata are both *equivalent to simple strata*. Since intertwining of strata is a property of equivalence classes, there is no harm in proceeding as if they are actually simple. By the first case above, there exists $x \in \boldsymbol{U}(\mathfrak{B})$ such that $[\mathfrak{B}, t, t-1, x^{-1}s(b_2)x]$ is equivalent to $[\mathfrak{B}, t, t-1, s(b_1)]$. (2.2.1) now implies that there exists $y \in \boldsymbol{U}(\mathfrak{A})$ such that $[\mathfrak{A}, n, t-1, y^{-1}\beta_2 y] \sim [\mathfrak{A}, n, t-1, \beta_1]$, as required.

For the next case, suppose that the strata $[\mathfrak{B}, t, t-1, s(b_i)]$ both null, i.e. equivalent to $[\mathfrak{B}, t, t-1, 0]$. Thus, for example, $b_1 \in \mathfrak{P}^{-t}$ while $s(b_1) \in \mathfrak{Q}^{1-t}$, so (1.4.10) gives $z \in \mathfrak{Q}^{-(t+k)}\mathfrak{N}$ (with k, $\mathfrak{N}$ as above) such that $b_1 \equiv a_\gamma(z) \pmod{\mathfrak{P}^{1-t}}$. Since $-(t+k) \geq 1$, we have $(1+z) \in U^1(\mathfrak{A})$, and a simple commutator computation gives

$$[\mathfrak{A}, n, t-1, (1+z)\beta_1(1+z)^{-1}] \sim [\mathfrak{A}, n, t-1, \gamma].$$

The same argument applies to the other stratum, so there exists $x \in U(\mathfrak{A})$ with $[\mathfrak{A}, n, t-1, x^{-1}\beta_2 x] \sim [\mathfrak{A}, n, t-1, \beta_1]$.

For the final case, suppose, for example, that $[\mathfrak{B}, t, t-1, s(b_1)]$ is equivalent to a simple stratum (and hence fundamental), while $[\mathfrak{B}, t, t-1, s(b_2)]$ is null. We know that these strata intertwine in $B^\times$, so this case cannot arise because of the following general result.

(2.6.2) Lemma: *Let $[\mathfrak{A}, m, m-1, x]$ be a stratum in A, and suppose that it intertwines with a null stratum $[\mathfrak{A}', m', m'-1, 0]$, such that $(m'-1)/e' < m/e$, where $e = e(\mathfrak{A})$, $e' = e(\mathfrak{A}')$. Then the stratum $[\mathfrak{A}, m, m-1, x]$ is not fundamental.*

Proof: Write $\mathfrak{P} = \mathrm{rad}(\mathfrak{A})$ and $\mathfrak{P}' = \mathrm{rad}(\mathfrak{A}')$. Replacing $[\mathfrak{A}', m', m'-1, 0]$ by an appropriate conjugate, we may assume that $(x + \mathfrak{P}^{1-m}) \cap \mathfrak{P}'^{1-m'}$ is not empty. Indeed, we may as well assume that the element x lies in this intersection, i.e. $x \in \mathfrak{P}^{-m} \cap \mathfrak{P}'^{1-m'}$. Now we argue as in the proof of **[Bu1]** (2.9) to show that $x + \mathfrak{P}^{1-m}$ contains a nilpotent, as required. ■

This completes the proof of (2.6.1). ■

(2.6.3) Remark: The converse of the first case in this proof also holds:— two simple strata $[\mathfrak{A}_i, n_i, n_i - 1, \alpha_i]$ intertwine if and only if $n_1/e(\mathfrak{A}_1) = n_2/e(\mathfrak{A}_2)$ and $\phi_{\alpha_1}(X) = \phi_{\alpha_2}(X)$. One can also easily reverse the other steps in the argument. If $[\mathfrak{A}, n, r, \gamma]$ is simple, and $b_1, b_2 \in \mathfrak{P}^{-r}$ are such that the strata $[\mathfrak{B}, r, r-1, s(b_i)]$ intertwine and are equivalent to simple ones, then the (equivalent to simple) strata $[\mathfrak{A}, n, r-1, \gamma + b_i]$ intertwine, and are therefore conjugate. Thus, in principal, it is not difficult to count conjugacy classes of simple strata (up to equivalence).

(2.6.4) Remark: As we mentioned earlier, it is possible to prove substantially more general results of this kind. For $i = 1, 2$, let $[\mathfrak{A}_i, n_i, r_i, \beta_i]$ be simple strata in A, and suppose that they intertwine. By symmetry, we can assume that

$$k_0(\beta_1, \mathfrak{A}_1)/e_1 \leq k_0(\beta_2, \mathfrak{A}_2)/e_2,$$

where $e_i = e(\mathfrak{A}_i)$, so that

$$-r_2 > k_0(\beta_1, \mathfrak{A}_1)e_2/e_1.$$

Using methods of the above type, and the more general version (1.5.12) of the intertwining theorem (1.5.8), one can show that there exists $x \in G$

such that $[\mathfrak{A}_2, n_2, t, x\beta_1 x^{-1}]$ is pure and equivalent to $[\mathfrak{A}_2, n_2, t, \beta_2]$ for all integers t satisfying

$$n_2 - 1 \geq t \geq r_2, \quad \textit{and} \quad t > r_1 e_2/e_1 - 1 \ \ (\text{i.e. } t \geq [r_1 e_2/e_1]).$$

(2.6.1) is the special case $\mathfrak{A}_1 = \mathfrak{A}_2 = \mathfrak{A}$, $r_1 = r_2 = r$.

3. THE SIMPLE CHARACTERS OF A SIMPLE STRATUM

We now take a simple stratum in our algebra $A = \mathrm{End}_F(V)$ of the form $[\mathfrak{A}, n, 0, \beta]$. Since the integer n is determined by β and $\mathfrak{A}$, in fact $n = -\nu_{\mathfrak{A}}(\beta)$, we often omit it from the notation. In this section, we attach to the stratum a pair of $\mathfrak{o}_F$-orders

$$\mathfrak{H}(\beta, \mathfrak{A}) \subset \mathfrak{J}(\beta, \mathfrak{A}) \subset \mathfrak{A}.$$

This gives us two compact open subgroups $H(\beta, \mathfrak{A}) = \mathfrak{H}(\beta, \mathfrak{A})^\times$ and $J(\beta, \mathfrak{A}) = \mathfrak{J}(\beta, \mathfrak{A})^\times$ of G, filtered by $H^i(\beta, \mathfrak{A}) = H(\beta, \mathfrak{A}) \cap \boldsymbol{U}^i(\mathfrak{A})$, $J^i(\beta, \mathfrak{A}) = J(\beta, \mathfrak{A}) \cap \boldsymbol{U}^i(\mathfrak{A})$, $i \geq 0$. For such i, we define a set $\mathcal{C}(\mathfrak{A}, i, \beta)$ of abelian characters of the group $H^{i+1}(\beta, \mathfrak{A})$, which we call "simple characters". These will be of fundamental importance in our treatment below of the representations of $G = \mathrm{Aut}_F(V)$. Here we confine ourselves to defining them and establishing their basic properties.

(3.1) The rings of a simple stratum

We start by establishing a system of notation and recalling a few approximation lemmas. Suppose we are given a pure stratum $[\mathfrak{A}, n, r, \beta]$. We write B_β for the A-centraliser of β, $\mathfrak{B}_\beta = \mathfrak{A} \cap B_\beta$, $\mathfrak{Q}_\beta = \mathrm{rad}(\mathfrak{B}_\beta)$. We abbreviate

(3.1.1) $$\mathfrak{N}(\beta, \mathfrak{A}) = \mathfrak{N}_{k_0(\beta,\mathfrak{A})}(\beta, \mathfrak{A})$$

and, since the order $\mathfrak{A}$ will be fixed for a long period, we often omit it from these notations. Thus $\mathfrak{N}_m(\beta)$ means $\mathfrak{N}_m(\beta, \mathfrak{A})$, $m \in \mathbb{Z}$.

If we have another pure stratum $[\mathfrak{A}, n, r, \beta']$ equivalent to $[\mathfrak{A}, n, r, \beta]$, i.e., in our earlier notation, $[\mathfrak{A}, n, r, \beta'] \sim [\mathfrak{A}, n, r, \beta]$, we recall from (2.1.1) and (2.1.3) that

(3.1.2) $$\mathfrak{Q}_\beta^t \mathfrak{N}_m(\beta) = \{x \in \mathfrak{P}^t : a_\beta(x) \in \mathfrak{P}^{t+m}\};$$

(3.1.3) $$\mathfrak{Q}_{\beta'}^t \mathfrak{N}_m(\beta') = \mathfrak{Q}_\beta^t \mathfrak{N}_m(\beta), \quad m \leq -r \textit{ and } t \in \mathbb{Z}.$$

In particular,

(3.1.4) $$\mathfrak{N}_m(\beta') = \mathfrak{N}_m(\beta), \quad m \leq -r.$$

(3.1.5) Lemma: *Let* $[\mathfrak{A}, n, r, \beta]$ *be a pure stratum with* $k_0(\beta, \mathfrak{A}) = -r$. *Let* $[\mathfrak{A}, n, r, \gamma]$ *be a simple stratum equivalent to* $[\mathfrak{A}, n, r, \beta]$ *and set* $s = -k_0(\gamma, \mathfrak{A})$. *Then*

$$\mathfrak{Q}_\beta^t \subset \mathfrak{Q}_\gamma^t + \mathfrak{Q}_\gamma^{t+s-r}\mathfrak{N}(\gamma), \quad t \in \mathbb{Z},$$
$$\mathfrak{K}(\mathfrak{B}_\beta) \subset \mathfrak{K}(\mathfrak{B}_\gamma).(1 + \mathfrak{Q}_\gamma^{s-r}\mathfrak{N}(\gamma)).$$

Proof: We have $\mathfrak{Q}_\beta^t \subset \mathfrak{Q}_\beta^t \mathfrak{N}(\beta) = \mathfrak{Q}_\gamma^t \mathfrak{N}_{-r}(\gamma) = \mathfrak{Q}_\gamma^t(\mathfrak{B}_\gamma + \mathfrak{Q}_\gamma^{s-r}\mathfrak{N}(\gamma))$ (see (1.4.9), and the first assertion follows. By (1.5.8), the $\mathfrak{K}(\mathfrak{A})$-normaliser of the coset $\gamma + \mathfrak{P}^{-r} = \beta + \mathfrak{P}^{-r}$ is $\mathfrak{K}(\mathfrak{B}_\gamma).(1 + \mathfrak{Q}_\gamma^{s-r}\mathfrak{N}(\gamma))$, and this normaliser surely contains $\mathfrak{K}(\mathfrak{B}_\beta)$. ∎

We now fix a simple stratum $[\mathfrak{A}, n, 0, \beta]$, and set

(3.1.6) $$r = -k_0(\beta, \mathfrak{A}).$$

(3.1.7) Definition: *(i) Suppose that* β *is minimal over* F *(so that* $r = n$ *or* ∞*). Put*

$$\mathfrak{H}(\beta) = \mathfrak{H}(\beta, \mathfrak{A}) = \mathfrak{B}_\beta + \mathfrak{P}^{[\frac{n}{2}]+1}.$$

(ii) Suppose that $r < n$, *and let* $[\mathfrak{A}, n, r, \gamma]$ *be a simple stratum equivalent to* $[\mathfrak{A}, n, r, \beta]$. *Put*

$$\mathfrak{H}(\beta, \mathfrak{A}) = \mathfrak{H}(\beta) = \mathfrak{B}_\beta + \mathfrak{H}(\gamma) \cap \mathfrak{P}^{[\frac{r}{2}]+1}.$$

Remark: In case *(i)* above when $\beta \in F$, we get $\mathfrak{H}(\beta) = \mathfrak{A}$. In case *(ii)*, it is not immediately apparent that the definition of $\mathfrak{H}(\beta)$ is independent of the choice of the simple stratum $[\mathfrak{A}, n, r, \gamma]$, although we will later see that it is. For the time being, we have to proceed as if $\mathfrak{H}(\beta)$ is only defined relative to a fixed choice of γ. By the same token, $\mathfrak{H}(\gamma)$ is only defined relative to the choice of a simple $[\mathfrak{A}, n, s, \delta]$ equivalent to $[\mathfrak{A}, n, s, \gamma]$ where $s = -k_0(\gamma, \mathfrak{A})$, and so on. In other words, $\mathfrak{H}(\beta)$ is at present only defined relative to a defining sequence for $[\mathfrak{A}, n, 0, \beta]$ in the sense of (2.4.2).

There is a parallel object $\mathfrak{J}(\beta, \mathfrak{A}) = \mathfrak{J}(\beta)$ defined as follows.

(3.1.8) Definition: *(i) Suppose that* β *is minimal over* F. *Put*

$$\mathfrak{J}(\beta) = \mathfrak{B}_\beta + \mathfrak{P}^{[\frac{n+1}{2}]}.$$

(ii) Suppose that β *is not minimal over* F, *and let* $[\mathfrak{A}, n, r, \gamma]$ *be a simple stratum equivalent to* $[\mathfrak{A}, n, r, \beta]$. *Put*

$$\mathfrak{J}(\beta) = \mathfrak{B}_\beta + \mathfrak{J}(\gamma) \cap \mathfrak{P}^{[\frac{r+1}{2}]}.$$

The same caveats apply. However, we always have

$$\mathfrak{A} \supset \mathfrak{H}(\beta) \supset \mathfrak{P}^{[\frac{n}{2}]+1},$$

so $\mathfrak{H}(\beta)$ is an $\mathfrak{o}_F$-lattice in A, and a similar comment applies to $\mathfrak{J}$. We set

$$\mathfrak{H}^k(\beta) = \mathfrak{H}(\beta) \cap \mathfrak{P}^k, \quad \mathfrak{J}^k(\beta) = \mathfrak{J}(\beta) \cap \mathfrak{P}^k, \quad k \geq 0.$$

In particular, if β is *minimal* over F, we get

$$\mathfrak{H}^k(\beta) = \begin{cases} \mathfrak{Q}_\beta^k + \mathfrak{P}^{[\frac{n}{2}]+1}, \ 0 \leq k \leq \left[\frac{n}{2}\right], \\ \mathfrak{P}^k, \ k \geq \left[\frac{n}{2}\right] + 1. \end{cases}$$

In the general case (using the notation of (3.1.7)), we shall see in (3.1.9) below that

$$\mathfrak{H}^k(\beta) = \begin{cases} \mathfrak{Q}_\beta^k + \mathfrak{H}^{[\frac{r}{2}]+1}(\gamma), \ 0 \leq k \leq \left[\frac{r}{2}\right], \\ \mathfrak{H}^k(\gamma), \ k \geq \left[\frac{r}{2}\right] + 1. \end{cases}$$

Similar remarks apply to $\mathfrak{J}$.

(3.1.9) Proposition: *In the above notation, we have:*

(i) For $-1 \leq t \leq r$, the lattice $\mathfrak{H}^{[\frac{t}{2}]+1}(\beta)$ is a bimodule over the ring $\mathfrak{N}_{-t}(\beta)$.

(ii) If $r < n$, we have $\mathfrak{H}^k(\beta) = \mathfrak{H}^k(\gamma)$ for $k \geq \left[\frac{r}{2}\right] + 1$.

(iii) For $k \geq 0$, $\mathfrak{H}^k(\beta)$ is a $\mathfrak{B}_\beta$-bimodule satisfying $\mathfrak{Q}_\beta \mathfrak{H}^k(\beta) = \mathfrak{H}^k(\beta)\mathfrak{Q}_\beta$.

(iv) $\mathfrak{H}(\beta)$ is a ring, in particular an $\mathfrak{o}_F$-order in A, and $\mathfrak{H}^k(\beta)$ is a two-sided ideal of $\mathfrak{H}(\beta)$, $k \geq 0$.

(v) Let $t \leq r - 1$, and let $[\mathfrak{A}, n, t, \beta']$ be a simple stratum equivalent to $[\mathfrak{A}, n, t, \beta]$. Then

$$\mathfrak{H}^k(\beta') = \mathfrak{H}^k(\beta), \quad \textit{for } k \geq \max\left\{0, t + 1 - \left[\tfrac{r+1}{2}\right]\right\}.$$

Remark: In part *(v)*, we must provisionally regard $\mathfrak{H}(\beta')$ as having been defined relative to the same simple $[\mathfrak{A}, n, r, \gamma]$ equivalent to both $[\mathfrak{A}, n, r, \beta]$, $[\mathfrak{A}, n, r, \beta']$. However, once established on these terms, *(v)* shows that the definition of $\mathfrak{H}(\beta)$ is indeed independent of the choice of the simple stratum $[\mathfrak{A}, n, r, \gamma]$ equivalent to $[\mathfrak{A}, n, r, \beta]$. For, suppose we have another simple $[\mathfrak{A}, n, r, \gamma'] \sim [\mathfrak{A}, n, r, \beta]$. Then $[\mathfrak{A}, n, r, \gamma] \sim [\mathfrak{A}, n, r, \gamma']$. If $s = -k_0(\gamma, \mathfrak{A}) = -k_0(\gamma', \mathfrak{A})$, we have

$$\left[\tfrac{r}{2}\right] + 1 \geq r + 1 - \left[\tfrac{s+1}{2}\right],$$

so $\mathfrak{H}^{[\frac{r}{2}]+1}(\gamma) = \mathfrak{H}^{[\frac{r}{2}]+1}(\gamma')$, whence the two definitions $\mathfrak{B}_\beta + \mathfrak{H}^{[\frac{r}{2}]+1}(\gamma)$, $\mathfrak{B}_\beta + \mathfrak{H}^{[\frac{r}{2}]+1}(\gamma')$ of $\mathfrak{H}(\beta)$ coincide.

Proof: We proceed by "induction along β", starting with the case in which β is minimal over F. There is nothing to prove if $\beta \in F$, so we assume that $r = n$. In *(i)*, we have $\mathfrak{H}^{[\frac{t}{2}]+1}(\beta) = \mathfrak{Q}_\beta^{[\frac{t}{2}]+1} + \mathfrak{P}^{[\frac{n}{2}]+1}$, while $\mathfrak{N}_{-t}(\beta) = \mathfrak{B}_\beta + \mathfrak{P}^{n-t}$. The observation

$$n \geq t \quad \Rightarrow \quad n - t + \left[\tfrac{t}{2}\right] \geq \left[\tfrac{n}{2}\right]$$

and a simple computation yields *(i)* in this case. Assertion *(ii)* is empty here, while *(iii)* is immediate. Also, *(iv)* is clear. In *(v)*, we apply (3.1.3), namely

$$\mathfrak{Q}_\beta^{[\frac{t}{2}]+1}\mathfrak{N}_m(\beta) = \mathfrak{Q}_{\beta'}^{[\frac{t}{2}]+1}\mathfrak{N}_m(\beta'),$$

with $m = -\left[\frac{t}{2}\right] - n + \left[\frac{n}{2}\right]$. Since $m > -n$, we have $\mathfrak{N}_m(\beta) = \mathfrak{B}_\beta + \mathfrak{P}^{m+n}$ by (1.4.9), and so $\mathfrak{Q}_\beta^{[\frac{t}{2}]+1}\mathfrak{N}_m(\beta) = \mathfrak{Q}_\beta^{[\frac{t}{2}]+1} + \mathfrak{P}^{[\frac{n}{2}]+1} = \mathfrak{H}^{[\frac{t}{2}]+1}(\beta)$. Likewise, the right hand side of the equation is $\mathfrak{H}^{[\frac{t}{2}]+1}(\beta')$, and the result follows.

We now turn to the general case $r < n$. As above, the stratum $[\mathfrak{A}, n, r, \gamma]$ is chosen simple and equivalent to $[\mathfrak{A}, n, r, \beta]$. By inductive hypothesis, $\mathfrak{H}^{[\frac{r}{2}]+1}(\gamma)$ is independent of the choice of γ subject to this condition. Put $s = -k_0(\gamma, \mathfrak{A})$.

In *(i)* we have to check that

$$\mathfrak{N}_{-t}(\beta)(\mathfrak{Q}_\beta^{[\frac{t}{2}]+1} + \mathfrak{H}^{[\frac{r}{2}]+1}(\gamma)) \subset \mathfrak{Q}_\beta^{[\frac{t}{2}]+1} + \mathfrak{H}^{[\frac{r}{2}]+1}(\gamma),$$

and symmetrically. Now, since $t \leq r$, we have $\mathfrak{N}_{-t}(\beta) \subset \mathfrak{N}(\beta) = \mathfrak{N}_{-r}(\gamma)$, and $\mathfrak{N}_{-r}(\gamma)\mathfrak{H}^{[\frac{r}{2}]+1}(\gamma) \subset \mathfrak{H}^{[\frac{r}{2}]+1}(\gamma)$ by induction, since $r \leq s$. Thus we need only check that

$$\mathfrak{N}_{-t}(\beta)\mathfrak{Q}_\beta^{[\frac{t}{2}]+1} \subset \mathfrak{Q}_\beta^{[\frac{t}{2}]+1} + \mathfrak{H}^{[\frac{r}{2}]+1}(\gamma).$$

Since $\mathfrak{N}_{-t}(\beta) = \mathfrak{B}_\beta + \mathfrak{Q}_\beta^{r-t}\mathfrak{N}(\beta)$ by (1.4.9), we are done if we can show that

$$\mathfrak{Q}_\beta^{r-t}\mathfrak{N}(\beta)\mathfrak{Q}_\beta^{[\frac{t}{2}]+1} \subset \mathfrak{H}^{[\frac{r}{2}]+1}(\gamma).$$

However, (1.4.11) gives $\mathfrak{Q}_\beta^{r-t}\mathfrak{N}(\beta)\mathfrak{Q}_\beta^{[\frac{t}{2}]+1} = \mathfrak{Q}_\beta^{r-t+[\frac{t}{2}]+1}\mathfrak{N}(\beta)$, and we have $r - t + \left[\frac{t}{2}\right] + 1 \geq \left[\frac{r}{2}\right] + 1$. Therefore

$$\mathfrak{Q}_\beta^{r-t}\mathfrak{N}(\beta)\mathfrak{Q}_\beta^{[\frac{t}{2}]+1} \subset \mathfrak{Q}_\beta^{[\frac{r}{2}]+1}\mathfrak{N}(\beta) = \mathfrak{Q}_\gamma^{[\frac{r}{2}]+1}\mathfrak{N}_{-r}(\gamma)$$

by (3.1.3). Further, $\mathfrak{Q}_\gamma^{[\frac{r}{2}]+1} \subset \mathfrak{H}^{[\frac{r}{2}]+1}(\gamma)$, so

$$\mathfrak{Q}_\gamma^{[\frac{r}{2}]+1}\mathfrak{N}_{-r}(\gamma) \subset \mathfrak{H}^{[\frac{r}{2}]+1}(\gamma)$$

by induction, and we have proved *(i)*.

We only need prove *(ii)* for the case $k = [\frac{r}{2}] + 1$, when $\mathfrak{H}^k(\beta)$ is $\mathfrak{Q}_\beta^{[\frac{r}{2}]+1} + \mathfrak{H}^{[\frac{r}{2}]+1}(\gamma)$. However, $\mathfrak{Q}_\beta^{[\frac{r}{2}]+1} \subset \mathfrak{Q}_\beta^{[\frac{r}{2}]+1}\mathfrak{N}(\beta)$ which equals $\mathfrak{Q}_\gamma^{[\frac{r}{2}]+1}\mathfrak{N}_{-r}(\gamma)$ by (3.1.3), and this is contained in $\mathfrak{H}^{[\frac{r}{2}]+1}(\gamma)$, as required.

In *(iii)*, the first statement follows from *(i)*, since $\mathfrak{B}_\beta \subset \mathfrak{N}_t(\beta)$ for all t. For the second statement, we first take $k \geq [\frac{r}{2}] + 1$. In this case, $\mathfrak{H}^k(\beta) = \mathfrak{H}^k(\gamma)$ by *(ii)*, and this is an $\mathfrak{N}(\beta) = \mathfrak{N}_{-r}(\gamma)$-module. We have $\mathfrak{Q}_\beta\mathfrak{H}^k(\beta) = \mathfrak{Q}_\beta\mathfrak{N}(\beta)\mathfrak{H}^k(\beta) = \mathfrak{Q}_\gamma\mathfrak{N}_{-r}(\gamma)\mathfrak{H}^k(\gamma)$ by (3.1.3) and part *(ii)*. However, this equals $\mathfrak{Q}_\gamma\mathfrak{H}^k(\gamma)$, by *(i)*, which, by induction, equals $\mathfrak{H}^k(\gamma)\mathfrak{Q}_\gamma$. Repeating the argument from the other side, $\mathfrak{H}^k(\gamma)\mathfrak{Q}_\gamma = \mathfrak{H}^k(\beta)\mathfrak{Q}_\beta$, as required. The case $k \leq [\frac{r}{2}]$ now follows from this one and the definition.

We have just proved that $\mathfrak{H}^{[\frac{r}{2}]+1}(\gamma)$ is a $\mathfrak{B}_\beta$-bimodule. It is certainly closed under multiplication, being an ideal of the ring $\mathfrak{H}(\gamma)$. Therefore $\mathfrak{H}(\beta) = \mathfrak{B}_\beta + \mathfrak{H}^{[\frac{r}{2}]+1}(\gamma)$ is a ring, of which $\mathfrak{H}^k(\beta) = \mathfrak{H}(\beta) \cap \mathfrak{P}^k$ is surely an ideal. This proves *(iv)*.

In *(v)*, we observe that the assertion follows from *(ii)* in the case $k \geq [\frac{r}{2}] + 1$. Therefore it is enough to show

$$\mathfrak{Q}_\beta^k + \mathfrak{H}^{[\frac{r}{2}]+1}(\gamma) = \mathfrak{Q}_{\beta'}^k + \mathfrak{H}^{[\frac{r}{2}]+1}(\gamma)$$

under the assumption $[\frac{r}{2}] + 1 \geq k \geq \max\{0, t + 1 - [\frac{r+1}{2}]\}$. Now, by *(i)* we have $\mathfrak{H}^{[\frac{r}{2}]+1}(\gamma) \supset \mathfrak{Q}_\gamma^{[\frac{r}{2}]+1}\mathfrak{N}_{-r}(\gamma)$, which equals $\mathfrak{Q}_\beta^{[\frac{r}{2}]+1}\mathfrak{N}(\beta)$ by (3.1.3). However, by (2.1.3) and (1.4.9), we have

$$\mathfrak{Q}_\beta^{t+1-[\frac{r+1}{2}]} + \mathfrak{Q}_\beta^{[\frac{r}{2}]+1}\mathfrak{N}(\beta) = \mathfrak{Q}_{\beta'}^{t+1-[\frac{r+1}{2}]} + \mathfrak{Q}_{\beta'}^{[\frac{r}{2}]+1}\mathfrak{N}(\beta').$$

Intersecting with $\mathfrak{P}^k$, the assertion follows. ∎

There are parallel results for the rings $\mathfrak{J}$. The proofs are virtually identical to those for $\mathfrak{H}$, so we just give statements.

(3.1.10) Proposition: *(i)* $\mathfrak{J}^{[\frac{t+1}{2}]}(\beta)$ *is an* $(\mathfrak{N}_{-t}(\beta), \mathfrak{N}_{-t}(\beta))$*-bimodule, for* $-1 \leq t \leq r$.

(ii) If $r < n$, *we have* $\mathfrak{J}^k(\beta) = \mathfrak{J}^k(\gamma)$ *for* $k \geq [\frac{r+1}{2}]$.

(iii) For $k \geq 0$, $\mathfrak{J}^k(\beta)$ *is a* $\mathfrak{B}_\beta$*-bimodule satisfying* $\mathfrak{Q}_\beta\mathfrak{J}^k(\beta) = \mathfrak{J}^k(\beta)\mathfrak{Q}_\beta$.

(iv) $\mathfrak{J}(\beta)$ *is a ring, and* $\mathfrak{J}^k(\beta)$ *is a two-sided ideal of* $\mathfrak{J}(\beta)$*, for* $k \geq 0$.

(v) Let $t \leq r-1$*, and let* $[\mathfrak{A}, n, t, \beta']$ *be simple and equivalent to* $[\mathfrak{A}, n, t, \beta]$*. Then*

$$\mathfrak{J}^k(\beta) = \mathfrak{J}^k(\beta'), \quad k \geq \max\{0, t - [\tfrac{r}{2}]\}.$$

Again we conclude that $\mathfrak{J}(\beta)$ is well-defined, independent of choices. It is also worth noting the following consequences.

(3.1.11) Corollary: *For* $m \in \mathbb{Z}$, $k \geq 0$*, we have* $\mathfrak{Q}_\beta^m \mathfrak{H}^k(\beta) = \mathfrak{H}^k(\beta)\mathfrak{Q}_\beta^m$ *and* $\mathfrak{Q}_\beta^m \mathfrak{J}^k(\beta) = \mathfrak{J}^k(\beta)\mathfrak{Q}_\beta^m$*. Moreover,* $\mathfrak{H}^k(\beta)$ *and* $\mathfrak{J}^k(\beta)$ *are invariant under conjugation by the group* $\mathfrak{K}(\mathfrak{B}_\beta)$*.*

Proof: For $m \geq 0$, the first statement follows from (3.1.9)*(iii)* by induction. To get the case $m < 0$, we take the equation $\mathfrak{Q}_\beta \mathfrak{H}^k = \mathfrak{H}^k \mathfrak{Q}_\beta$ given by (3.1.9)*(iii)* and multiply on either side by $\mathfrak{Q}_\beta^{-1}$ to get $\mathfrak{B}_\beta \mathfrak{H}^k \mathfrak{Q}_\beta^{-1} = \mathfrak{Q}_\beta^{-1} \mathfrak{H}^k \mathfrak{B}_\beta$. (3.1.9)*(iii)* now allows us to absorb the factors $\mathfrak{B}_\beta$, to get $\mathfrak{Q}_\beta^{-1} \mathfrak{H}^k = \mathfrak{H}^k \mathfrak{Q}_\beta^{-1}$. The assertion for $\mathfrak{H}^k$ now follows by induction, and likewise for $\mathfrak{J}^k$.

For the second assertion, we note that if $x \in \mathfrak{K}(\mathfrak{B}_\beta)$, then $x\mathfrak{B}_\beta = \mathfrak{B}_\beta x = \mathfrak{Q}_\beta^m$, for some $m \in \mathbb{Z}$. The assertion now follows from the first one. ■

Since $[\frac{t}{2}] + 1 \geq [\frac{t+1}{2}]$ for all t, we have

(3.1.12)
$$\mathfrak{H}^k(\beta) \subset \mathfrak{J}^k(\beta), \quad k \geq 0.$$

(3.1.13) Proposition: *(i)* $\mathfrak{Q}_\beta \mathfrak{J}^k(\beta) \subset \mathfrak{H}^{k+1}(\beta)$*, for all* $k \geq 0$.

(ii) For $k, \ell \geq 1$*, we have* $\mathfrak{J}^k(\beta)\mathfrak{J}^\ell(\beta) \subset \mathfrak{H}^{k+\ell}(\beta)$.

(iii) For $k \geq 1$, $\mathfrak{H}^k(\beta)$ *is a two-sided ideal of* $\mathfrak{J}(\beta)$.

Proof: When β is minimal over F, all of these assertions follow from trivial computations. In the general case, we take γ as before. In *(i)*, we just have to check that $\mathfrak{Q}_\beta \mathfrak{J}(\beta)$ is contained in $\mathfrak{H}(\beta)$, for which it is enough to show that $\mathfrak{Q}_\beta \mathfrak{J}^m(\gamma) \subset \mathfrak{H}(\gamma)$ for $m \geq [\frac{r+1}{2}]$. However, $\mathfrak{Q}_\beta \mathfrak{J}^m(\gamma) = \mathfrak{Q}_\beta(\mathfrak{N}_{-r}(\gamma)\mathfrak{J}^m(\gamma)) = \mathfrak{Q}_\beta \mathfrak{N}(\beta)\mathfrak{J}^m(\gamma) = \mathfrak{Q}_\gamma \mathfrak{N}_{-r}(\gamma)\mathfrak{J}^m(\gamma)$ which is contained in $\mathfrak{H}(\gamma)$ by induction. This proves *(i)*.

For *(ii)*, it is enough to check that $\mathfrak{J}^1(\beta)\mathfrak{J}^1(\beta)$ is contained in $\mathfrak{H}(\beta)$. For this, we use *(i)* and then we just have to show that $\mathfrak{J}^{[\frac{r+1}{2}]}(\gamma)\mathfrak{J}^{[\frac{r+1}{2}]}(\gamma)$ is contained in $\mathfrak{H}^{[\frac{r}{2}]+1}(\gamma)$, which is immediate from induction and the fact that $r \geq 1$.

In *(iii)*, we have to show that $\mathfrak{J}(\beta)\mathfrak{H}^1(\beta) \subset \mathfrak{H}(\beta)$ and symmetrically. We write $\mathfrak{J}(\beta) = \mathfrak{B}_\beta + \mathfrak{J}^{[\frac{r+1}{2}]}(\beta)$. Expanding the product, the required statement follows from part *(ii)*, (3.1.12) and (3.1.9)*(ii)*, *(iii)*. ■

We now turn to multiplicative structures and define two families of compact open subgroups $H^m(\beta, \mathfrak{A})$, $J^m(\beta, \mathfrak{A})$ of G by

(3.1.14) $$\left\{ \begin{array}{l} H^m(\beta, \mathfrak{A}) = \mathfrak{H}(\beta, \mathfrak{A}) \cap \boldsymbol{U}^m(\mathfrak{A}) \\ J^m(\beta, \mathfrak{A}) = \mathfrak{J}(\beta, \mathfrak{A}) \cap \boldsymbol{U}^m(\mathfrak{A}) \end{array} \right\} \quad \text{for } m \geq 0.$$

As above, we frequently omit $\mathfrak{A}$ from this notation. We also prefer to write $J(\beta, \mathfrak{A})$ rather that $J^0(\beta, \mathfrak{A})$. The additive properties of $\mathfrak{H}$, $\mathfrak{J}$ above translate to give multiplicative properties of the groups H^m, J^m. For example:

(3.1.15) Proposition: *(i) For $0 \leq m \leq [\frac{r}{2}] + 1$, we have*

$$H^m(\beta) = \boldsymbol{U}^m(\mathfrak{B}_\beta) H^{[\frac{r}{2}]+1}(\beta),$$

and, for $0 \leq m \leq [\frac{r+1}{2}]$,

$$J^m(\beta) = \boldsymbol{U}^m(\mathfrak{B}_\beta) J^{[\frac{r+1}{2}]}(\beta).$$

(ii) For $m \geq 0$, the groups $H^m(\beta)$, $J^m(\beta)$ are normalised by $\mathfrak{K}(\mathfrak{B}_\beta)$.

(iii) We have $J^m(\beta) \supset H^m(\beta)$ and $H^{m+1}(\beta)$ is a normal subgroup of $J^0(\beta)$, for all $m \geq 0$.

(iv) For $k, \ell \geq 1$, the commutator group $[J^k(\beta), J^\ell(\beta)]$ is contained in $H^{k+\ell}(\beta)$.

We now develop some "exactness" properties of $\mathfrak{H}$ and $\mathfrak{J}$, analogous to those given in §**1**. We fix an additive character $\psi = \psi_A$ of A as in **(1.1)**, and recall the "star" operation of (1.1.4). We choose a tame corestriction map s_β on A relative to the subfield $F[\beta]/F$.

(3.1.16) Proposition: *For $r = -k_0(\beta, \mathfrak{A})$ as above, and $-1 \leq m \leq r - 1$, we have an exact sequence*

$$\mathfrak{Q}_\beta^{r-m}\mathfrak{N}(\beta) + \mathfrak{J}^{[\frac{r+1}{2}]}(\beta) \xrightarrow{a_\beta} (\mathfrak{H}^{m+1}(\beta))^* \xrightarrow{s_\beta} \mathfrak{Q}_\beta^{-m} \to 0.$$

Proof: We first show that

(3.1.17) $$a_\beta(\mathfrak{J}^{[\frac{r+1}{2}]}) \subset (\mathfrak{H}^{m+1})^*, \quad m \geq -1,$$

abbreviating $\mathfrak{H} = \mathfrak{H}(\beta)$, $\mathfrak{J} = \mathfrak{J}(\beta)$, when there is no fear of confusion. Suppose first that $m \leq [\frac{r}{2}]$, so that we have $\mathfrak{H}^{m+1} = \mathfrak{Q}_\beta^{m+1} + \mathfrak{H}^{[\frac{r}{2}]+1}$. Then

$$(\mathfrak{H}^{m+1})^* = (\mathfrak{P}^{-m} + a_\beta(A)) \cap (\mathfrak{H}^{[\frac{r}{2}]+1})^*,$$

whence

$$\textbf{(3.1.18)} \qquad (\mathfrak{H}^{m+1})^* \cap a_\beta(A) = (\mathfrak{H}^{[\frac{r}{2}]+1})^* \cap a_\beta(A), \quad m \le \left[\tfrac{r}{2}\right].$$

This means that we may as well take $m \ge \left[\frac{r}{2}\right]$ for this argument. Indeed, since $(\mathfrak{H}^{m+1})^*$ increases with m, we can assume here that $m = \left[\frac{r}{2}\right]$. If β is minimal over F, the assertion (3.1.17) is given by a simple computation. Otherwise, we choose a simple $[\mathfrak{A}, n, r, \gamma]$ equivalent to $[\mathfrak{A}, n, r, \beta]$ and write $\beta = \gamma + c$, $c \in \mathfrak{P}^{-r}$. Then $a_c(\mathfrak{J}^{[\frac{r+1}{2}]}(\beta)) \subset \mathfrak{P}^{-[\frac{r}{2}]} \subset (\mathfrak{H}^{[\frac{r}{2}]+1}(\beta))^*$. However, by (3.1.10)*(ii)*,

$$\mathfrak{J}^{[\frac{r+1}{2}]}(\beta) = \mathfrak{J}^{[\frac{r+1}{2}]}(\gamma) = \mathfrak{Q}_\gamma^{[\frac{r+1}{2}]} + \mathfrak{J}^{[\frac{s+1}{2}]}(\gamma),$$

where $s = -k_0(\gamma, \mathfrak{A})$, and by induction we have

$$a_\gamma(\mathfrak{J}^{[\frac{r+1}{2}]}(\gamma)) = a_\gamma(\mathfrak{J}^{[\frac{s+1}{2}]}(\gamma)) \subset (\mathfrak{H}^{[\frac{s}{2}]+1}(\gamma))^*.$$

This shows that $a_\gamma(\mathfrak{J}^{[\frac{r+1}{2}]}) \subset (\mathfrak{H}^{[\frac{r}{2}]+1}(\gamma))^*$ by (3.1.18) applied to γ. Thus indeed

$$a_\beta(\mathfrak{J}^{[\frac{r+1}{2}]}(\beta)) \subset a_\gamma(\mathfrak{J}^{[\frac{r+1}{2}]}(\beta)) + a_c(\mathfrak{J}^{[\frac{r+1}{2}]}(\beta)) \subset (\mathfrak{H}^{[\frac{r}{2}]+1}(\beta))^*,$$

and we have proved (3.1.17). The next step is to show:

(3.1.19) Lemma: *For $m \ge -1$, we have*

$$(\mathfrak{H}^{m+1}(\beta))^* = a_\beta(\mathfrak{J}^{[\frac{r+1}{2}]}(\beta)) + \mathfrak{P}^{-m}.$$

Proof: Take first the case $m = \left[\frac{r}{2}\right]$. Then $(\mathfrak{H}^{m+1}(\beta))^*$ surely contains the group $a_\beta(\mathfrak{J}^{[\frac{r+1}{2}]}(\beta)) + \mathfrak{P}^{-m}$. The opposite containment is trivial if β is minimal over F, while in general, since $a_c(\mathfrak{J}^{[\frac{r+1}{2}]}) \subset \mathfrak{P}^{[\frac{r+1}{2}]-r} = \mathfrak{P}^{-m}$, we have

$$\begin{aligned} a_\beta(\mathfrak{J}^{[\frac{r+1}{2}]}(\beta)) + \mathfrak{P}^{-m} &= a_\gamma(\mathfrak{J}^{[\frac{r+1}{2}]}(\gamma)) + \mathfrak{P}^{-m} \\ &= a_\gamma(\mathfrak{J}^{[\frac{s+1}{2}]}(\gamma)) + \mathfrak{P}^{-m} \\ &= (\mathfrak{H}^{m+1}(\gamma))^* \end{aligned}$$

by induction. We note ((3.1.9)) that $\mathfrak{H}^{m+1}(\beta) = \mathfrak{H}^{m+1}(\gamma)$, and this gives the lemma in this case.

If $m > \left[\frac{r}{2}\right]$, we have $\mathfrak{H}^{m+1} = \mathfrak{H}^{[\frac{r}{2}]+1} \cap \mathfrak{P}^{m+1}$, so $(\mathfrak{H}^{m+1})^* = (\mathfrak{H}^{[\frac{r}{2}]+1})^* + \mathfrak{P}^{-m}$, and the assertion follows immediately.

This leaves us with the case $m < \left[\frac{r}{2}\right]$, where we have

$$\begin{aligned}(\mathfrak{H}^{m+1}(\beta))^* &= (\mathfrak{P}^{-m} + a_\beta(A)) \cap (\mathfrak{H}^{[\frac{r}{2}]+1}(\beta))^* \\ &= (\mathfrak{P}^{-m} + a_\beta(A)) \cap (a_\beta(\mathfrak{J}^{[\frac{r+1}{2}]}(\beta)) + \mathfrak{P}^{-[\frac{r}{2}]})\end{aligned}$$

by the first case. This last group surely contains $\mathfrak{P}^{-m} + a_\beta(\mathfrak{J}^{[\frac{r+1}{2}]}(\beta))$. We have to show that this containment is equality. Now, from the last equation above,

$$\begin{aligned}s_\beta((\mathfrak{H}^{m+1}(\beta))^*) &\subset s_\beta(\mathfrak{P}^{-m} + a_\beta(A)) \cap s_\beta(a_\beta(\mathfrak{J}^{[\frac{r+1}{2}]}(\beta)) + \mathfrak{P}^{-[\frac{r}{2}]}) \\ &= \mathfrak{Q}_\beta^{-m} = s_\beta(\mathfrak{P}^{-m} + a_\beta(\mathfrak{J}^{[\frac{r+1}{2}]}(\beta))).\end{aligned}$$

It follows that

(3.1.20) $$s_\beta((\mathfrak{H}^{m+1}(\beta))^*) = \mathfrak{Q}_\beta^{-m}, \quad -1 \leq m \leq \left[\tfrac{r}{2}\right].$$

Moreover, to get our desired equality, we just have to show

$$(\mathfrak{H}^{m+1}(\beta))^* \cap \operatorname{Ker}(s_\beta) \subset (\mathfrak{P}^{-m} + a_\beta(\mathfrak{J}^{[\frac{r+1}{2}]}(\beta)) \cap \operatorname{Ker}(s_\beta).$$

Of course, $\operatorname{Ker}(s_\beta) = a_\beta(A)$, so these intersections are respectively

$$a_\beta(A) \cap \mathfrak{P}^{-[\frac{r}{2}]} + a_\beta(\mathfrak{J}^{[\frac{r+1}{2}]}(\beta)), \quad a_\beta(A) \cap \mathfrak{P}^{-m} + a_\beta(\mathfrak{J}^{[\frac{r+1}{2}]}(\beta)),$$

by (3.1.18) and the case $m = \left[\frac{r}{2}\right]$ above. However, for $m \leq \left[\frac{r}{2}\right]$, we have $\mathfrak{P}^{-m} \cap a_\beta(A) = a_\beta(\mathfrak{Q}_\beta^{r-m}\mathfrak{N}(\beta))$ (by (1.4.10)), while $\mathfrak{Q}_\beta^{r-m}\mathfrak{N}(\beta) \subset \mathfrak{J}^{[\frac{r+1}{2}]}(\beta)$ by (3.1.10)*(i)* (with $t = r$). This containment gives our desired equality, and proves the lemma. ■

Now we can prove the proposition. First, the equality $\mathfrak{Q}_\beta^{-m} = s_\beta((\mathfrak{H}^{m+1})^*)$ follows from (3.1.19). By the same lemma, $(\mathfrak{H}^{m+1})^* \cap \operatorname{Ker}(s_\beta) = a_\beta(\mathfrak{J}^{[\frac{r+1}{2}]}) + \mathfrak{P}^{-m} \cap a_\beta(A) = a_\beta(\mathfrak{J}^{[\frac{r+1}{2}]} + \mathfrak{Q}_\beta^{r-m}\mathfrak{N}(\beta))$ for $-1 \leq m \leq r-1$, and this completes the proof. ■

There are parallel results, proved in exactly the same fashion, interchanging the roles of $\mathfrak{H}$ and $\mathfrak{J}$, for example:—

(3.1.21) $$(\mathfrak{J}^m(\beta))^* = a_\beta(\mathfrak{H}^{[\frac{r}{2}]+1}(\beta)) + \mathfrak{P}^{1-m}, \quad m \geq 0.$$

(3.1.22) *For $0 \leq m \leq r$, we have an exact sequence*

$$\mathfrak{Q}_\beta^{1+r-m}\mathfrak{N}(\beta) + \mathfrak{H}^{[\frac{r}{2}]+1}(\beta) \xrightarrow{a_\beta} (\mathfrak{J}^m(\beta))^* \xrightarrow{s_\beta} \mathfrak{Q}_\beta^{1-m} \to 0,$$

and, in particular, for $0 \leq m \leq \left[\frac{r+1}{2}\right]$, an exact sequence

$$0 \to \mathfrak{Q}_\beta^{[\frac{r}{2}]+1} \to \mathfrak{H}^{[\frac{r}{2}]+1}(\beta) \xrightarrow{a_\beta} (\mathfrak{J}^m(\beta))^* \xrightarrow{s_\beta} \mathfrak{Q}_\beta^{1-m} \to 0.$$

The second assertion here follows from the first and (3.1.9)*(i)*.

(3.2) Characters and commutators

We now fix a simple stratum $[\mathfrak{A}, n, 0, \beta]$ with $r = -k_0(\beta, \mathfrak{A})$, and use the associated notations introduced in **(3.1)**. We shall define a family $\mathcal{C}(\mathfrak{A}, m, \beta)$ of very special characters θ of the group $H^{m+1}(\beta) = H^{m+1}(\beta, \mathfrak{A})$, $m \geq 0$. This family will depend only on the equivalence class of $[\mathfrak{A}, n, m, \beta]$, but it will take some time for this fact to emerge. We therefore have to provisionally define $\mathcal{C}(\mathfrak{A}, m, \beta)$ relative to a defining sequence for β in the sense of (2.4.2). Further, it is certainly not immediately clear from the definition that the set $\mathcal{C}(\mathfrak{A}, m, \beta)$ is nonempty, except in the case where β is minimal over F. We proceed by first deriving numerous consequences of the definition of the characters $\theta \in \mathcal{C}(\mathfrak{A}, m, \beta)$, and compute their intertwining. Finally, in (3.3.18), we are able to show that $\mathcal{C}(\mathfrak{A}, m, \beta)$ is nonempty and describe its elements.

Throughout, we write $\det_A$ for the determinant homomorphism $A^\times \to F^\times$ and use the analogous notation for centralisers of subfields of A. We also recall the additive character $\psi = \psi_A$ used in §**1**. Throughout this section, we use the notation

$$\psi_x(y) = \psi(x(y-1)), \quad x, y \in A.$$

We note one particular case here. Take $b \in \mathfrak{P}^{-n} \cap F$, $n \geq 1$, so that ψ_b defines a character of $\boldsymbol{U}^n(\mathfrak{A})$ (or $\boldsymbol{U}^n(\mathfrak{A})/\boldsymbol{U}^{n+1}(\mathfrak{A})$). Then

$$\psi_b(x) = \psi_{F,b}(\det(x)), \quad x \in \boldsymbol{U}^n(\mathfrak{A}).$$

(3.2.1) Definition: *Suppose that β is minimal over F. For $0 \leq m \leq n-1$, let $\mathcal{C}(\mathfrak{A}, m, \beta)$ denote the set of characters θ of $H^{m+1}(\beta)$ such that*

(a) $\theta \mid H^{m+1}(\beta) \cap \boldsymbol{U}^{[\frac{n}{2}]+1}(\mathfrak{A}) = \psi_\beta$;

(b) $\theta \mid H^{m+1}(\beta) \cap B_\beta^\times$ factors through $\det_{B_\beta} : B_\beta^\times \to F[\beta]^\times$.

Observe that the function ψ_β defines a character of $\boldsymbol{U}^{[\frac{n}{2}]+1}(\mathfrak{A})/\boldsymbol{U}^{n+1}(\mathfrak{A})$ in the situation of (3.2.1). Moreover, $\psi_\beta \mid \boldsymbol{U}^{[\frac{n}{2}]+1}(\mathfrak{A}) \cap B_\beta$ factors through $\det_{B_\beta}$ (by (2.4.11), for example), and it follows that $\mathcal{C}(\mathfrak{A}, m, \beta)$ is nonempty for all $m \geq 0$.

It is also worth noting, at this stage, a simple identity. Let s be some tame corestriction on A relative to the field E/F, where $E = F[\beta]$. Let $\psi_{B_\beta} = \psi_E \circ \mathrm{tr}$ be the character of B_β connecting ψ_A and s as in (1.3.5). Then $s(\beta) \in E$ by (1.3.2), and

$$\psi_\beta \mid \boldsymbol{U}^{[\frac{n}{2}]+1}(\mathfrak{B}_\beta) = \psi_{E,s(\beta)} \circ \det_{B_\beta}.$$

Immediately from the definition and (3.1.15) we get:

(3.2.2) Proposition: *In the situation of* (3.2.1), *we have*
(i) $\mathcal{C}(\mathfrak{A}, m, \beta) = \{\psi_\beta\}$ *for* $\left[\frac{n}{2}\right] \le m \le n-1$.
(ii) Every $\theta \in \mathcal{C}(\mathfrak{A}, m, \beta)$ *is normalised by* $\mathfrak{K}(\mathfrak{B}_\beta)$.

In the general case, our definition has to be "inductive along β". We take a simple stratum $[\mathfrak{A}, n, r, \gamma]$ equivalent to $[\mathfrak{A}, n, r, \beta]$ (and until we have established independence of choices, it has to be the first term in our fixed defining sequence for β).

(3.2.3) Definition: *Suppose that* $r < n$. *Then, for* $0 \le m \le r-1$, *let* $\mathcal{C}(\mathfrak{A}, m, \beta)$ *be the set of characters* θ *of* $H^{m+1}(\beta)$ *such that*
(a) $\theta \mid H^{m+1}(\beta) \cap B_\beta^\times$ *factors through* $\det_{B_\beta}$;
(b) θ *is normalised by* $\mathfrak{K}(\mathfrak{B}_\beta)$;
(c) if $m' = \max\{m, \left[\frac{r}{2}\right]\}$, *the restriction* $\theta \mid H^{m'+1}(\beta)$ *is of the form* $\theta_0\psi_c$, *for some* $\theta_0 \in \mathcal{C}(\mathfrak{A}, m', \gamma)$, *where* $c = \beta - \gamma$.

Note that in (3.2.3)*(c)* we have $H^{m'+1}(\beta) = H^{m'+1}(\gamma)$. Observe also that ψ_c is null on $H^{r+1}(\beta)$, since $2(m'+1) \ge r+1$. It therefore defines a character of $H^{m'+1}(\beta)/H^{r+1}(\beta)$. It is useful to extend our notation and, in the above situation, set

$$\mathcal{C}(\mathfrak{A}, m, \beta) = \mathcal{C}(\mathfrak{A}, m, \gamma), \quad m \ge r.$$

We start with a couple of simple properties.

(3.2.4) Proposition: *For* $m \ge \left[\frac{n}{2}\right]$, *we have* $\mathcal{C}(\mathfrak{A}, m, \beta) = \{\psi_\beta\}$.

Proof: This follows easily from the definitions and induction along β. ■

(3.2.5) Proposition: *For* $0 \le m \le \left[\frac{r}{2}\right]$, *restriction induces a surjective map*

$$\mathcal{C}(\mathfrak{A}, m, \beta) \to \mathcal{C}(\mathfrak{A}, \left[\tfrac{r}{2}\right], \beta).$$

The fibres of this map (if nonempty) are of the form $\theta.X$, *where* $\theta \in \mathcal{C}(\mathfrak{A}, m, \beta)$ *and* X *is the group of characters of* $U^{m+1}(\mathfrak{B}_\beta)/U^{[\frac{r}{2}]+1}(\mathfrak{B}_\beta)$ *which factor through the determinant* $\det_{B_\beta}$.

Proof: Take $\theta \in \mathcal{C}(\mathfrak{A}, \left[\frac{r}{2}\right], \beta)$, so that $\theta \mid H^{[\frac{r}{2}]+1}(\beta) \cap B_\beta^\times = \theta \mid U^{[\frac{r}{2}]+1}(\mathfrak{B}_\beta) = \chi \circ \det_{B_\beta}$, for some character χ of the closed subgroup $\det_{B_\beta}(U^{[\frac{r}{2}]+1}(\mathfrak{B}_\beta))$ of $F[\beta]^\times$. We have

$$H^{m+1}(\beta) = 1 + \mathfrak{Q}_\beta^{m+1} + \mathfrak{H}^{[\frac{r}{2}]+1}(\beta) = U^{m+1}(\mathfrak{B}_\beta).H^{[\frac{r}{2}]+1}(\beta),$$

with the first factor normalising the second. Take any extension χ' of χ to a character of $F[\beta]^\times$ and define θ' by

$$\theta'(uh) = \chi'(\det_{B_\beta}(u))\theta(h), \quad u \in U^{m+1}(\mathfrak{B}_\beta),\ h \in H^{[\frac{r}{2}]+1}(\beta).$$

Since $\mathbf{U}^{m+1}(\mathfrak{B}_\beta)$ normalises θ on $H^{[\frac{r}{2}]+1}(\beta)$, this defines a character of $H^{m+1}(\beta)$, and indeed $\theta' \in \mathcal{C}(\mathfrak{A}, m, \beta)$. This proves the first assertion, and the second one follows similarly. ∎

We now embark on a lengthy sequence of commutator calculations, of a progressively more general nature.

(3.2.6) Proposition: *Let $\theta \in \mathcal{C}(\mathfrak{A}, m, \beta)$, $[\frac{r}{2}] \le m \le r-1$. Let $k, \ell \ge 1$ and suppose*

$$k + \ell \ge m+1, \quad k + 2\ell \ge r+1.$$

Let $x \in \mathfrak{Q}_\beta^k \mathfrak{N}(\beta)$, $y \in \mathfrak{Q}_\beta^\ell \mathfrak{N}(\beta)$. Then the commutator $[1+x, 1+y]$ lies in $H^{m+1}(\beta)$ and

$$\theta[1+x, 1+y] = \psi_{(1+x)^{-1}\beta(1+x)-\beta}(1+y).$$

Proof: The commutator $[1+x, 1+y] = (1+x)(1+y)(1+x)^{-1}(1+y)^{-1}$ lies in $1 + \mathfrak{Q}_\beta^{m+1}\mathfrak{N}(\beta)$, which is contained in $1 + \mathfrak{H}^{m+1}(\beta)$ by (3.1.9)*(i)*.

We proceed by induction along β, starting with the case in which β is minimal over F. The case $\beta \in F$ is trivial so we may as well assume that $\beta \notin F$. We write $(1+x)^{-1} = 1 + \bar{x}$, $(1+y)^{-1} = 1 + \bar{y}$, so that, for example,

$$\textbf{(3.2.7)} \qquad \bar{x} \in \mathfrak{Q}_\beta^k \mathfrak{N}(\beta) \quad \text{and} \quad x + \bar{x} + x\bar{x} = x + \bar{x} + \bar{x}x = 0.$$

In this case, we have $[1+x, 1+y] - 1 \in \mathfrak{P}^{k+\ell} \subset \mathfrak{P}^{[\frac{n}{2}]+1}$, so $\theta[1+x, 1+y] = \psi_\beta[1+x, 1+y]$ by (3.2.2). We expand the commutator, and use relations like (3.2.7) to get

$$[1+x, 1+y] \equiv 1 + xy + y\bar{x} + xy\bar{x} \pmod{\mathfrak{P}^{n+1}},$$

since all terms involving two y's and an x lie in $\mathfrak{P}^{k+2\ell} \subset \mathfrak{P}^{n+1}$. Therefore

$$[1+x, 1+y] \equiv (1+x)(1+y)(1+x)^{-1} - y \pmod{\mathfrak{P}^{n+1}}.$$

Evaluating at $\theta = \psi_\beta$, we get

$$\begin{aligned} \theta[1+x, 1+y] &= \psi(\beta((1+x)(1+y)(1+x)^{-1} - y - 1)) \\ &= \psi_{(1+x)^{-1}\beta(1+x)-\beta}(1+y), \end{aligned}$$

as desired.

Now we treat the general case, writing $\theta = \theta_0 \psi_c$, where $\theta_0 \in \mathcal{C}(\mathfrak{A}, m, \gamma)$, $c = \beta - \gamma \in \mathfrak{P}^{-r}$, and $[\mathfrak{A}, n, r, \gamma]$ is as in (3.2.3). As before, set $s = -k_0(\gamma, \mathfrak{A})$. Then (3.1.3) and (1.4.9) give

$$\begin{aligned} \mathfrak{Q}_\beta^k \mathfrak{N}(\beta) &= \mathfrak{Q}_\gamma^k \mathfrak{N}_{-r}(\gamma) = \mathfrak{Q}_\gamma^k + \mathfrak{Q}_\gamma^{k+s-r}\mathfrak{N}(\gamma), \\ \mathfrak{Q}_\beta^\ell \mathfrak{N}(\beta) &= \mathfrak{Q}_\gamma^\ell + \mathfrak{Q}_\gamma^{\ell+s-r}\mathfrak{N}(\gamma). \end{aligned}$$

We accordingly write $(1+x) = (1+x_1)(1+x_2)$, with $x_1 \in \mathfrak{Q}_\gamma^k$, $x_2 \in \mathfrak{Q}_\gamma^{k+s-r}\mathfrak{N}(\gamma)$. With the usual notation ${}^a b = aba^{-1}$, we have

$$[1+x, 1+y] = {}^{(1+x_1)}[1+x_2, 1+y].[1+x_1, 1+y].$$

The factor $[1+x_2, 1+y]$ lies in $1+\mathfrak{Q}_\gamma^{k+\ell+s-r}\mathfrak{N}(\gamma)\mathfrak{N}_{-r}(\gamma)$, which equals $1+\mathfrak{Q}_\gamma^{k+\ell+s-r}\mathfrak{N}(\gamma)$ since $\mathfrak{N}(\gamma) \supset \mathfrak{N}_{-r}(\gamma)$ and $\mathfrak{N}(\gamma)$ is a ring. Certainly $k+\ell+s-r \geq m+1$ and $\left[\frac{s}{2}\right]+1$, so this factor lies in $H^{m+1}(\gamma)$, the domain of θ_0. By definition, $(1+x_1)$ normalises θ_0, so we can ignore the conjugation to get

$$\theta_0[1+x, 1+y] = \theta_0[1+x_2, 1+y].\theta_0[1+x_1, 1+y].$$

We now show that $\theta_0[1+x_1, 1+y] = 1$. To do this, we write $(1+y) = (1+y_1)(1+y_2)$, with $y_1 \in \mathfrak{Q}_\gamma^\ell$, $y_2 \in \mathfrak{Q}_\gamma^{\ell+s-r}\mathfrak{N}(\gamma)$, and expand the commutator. As before, we get

$$\theta_0[1+x_1, 1+y] = \theta_0[1+x_1, 1+y_1].\theta_0[1+x_1, 1+y_2].$$

The first factor is null since $\theta_0|U^{m+1}(\mathfrak{B}_\gamma)$ factors through the determinant $\det_{B_\gamma}$. In the second, we have $x_1 \in \mathfrak{Q}_\gamma^k \subset \mathfrak{Q}_\gamma^k\mathfrak{N}(\gamma)$ and $y_2 \in \mathfrak{Q}_\gamma^{\ell+s-r}\mathfrak{N}(\gamma)$. Further,

$$\begin{gathered} k+\ell+s-r \geq 1+\max\left\{m, \left[\tfrac{s}{2}\right]\right\}, \\ k+2\ell+2s-2r \geq s+1, \end{gathered}$$

so we may apply our inductive hypothesis. This gives

$$\theta_0[1+x_1, 1+y_2] = \psi_{(1+x_1)^{-1}\gamma(1+x_1)-\gamma}(1+y_2) = 1$$

since x_1 commutes with γ.

We are therefore left with

$$\theta_0[1+x, 1+y] = \theta_0[1+x_2, 1+y],$$

where we have $x_2 \in \mathfrak{Q}_\gamma^{k+s-r}\mathfrak{N}(\gamma)$, $y \in \mathfrak{Q}_\gamma^\ell\mathfrak{N}_{-r}(\gamma) \subset \mathfrak{Q}_\gamma^\ell\mathfrak{N}(\gamma)$. Also,

$$\begin{gathered} k+s-r+\ell \geq 1+\max\left\{m, \left[\tfrac{s}{2}\right]\right\}, \\ k+s-r+2\ell \geq s+1, \end{gathered}$$

so our inductive hypothesis applies to give

$$\begin{aligned} \theta_0[1+x, 1+y] = \theta_0[1+x_2, 1+y] &= \psi_{(1+x_2)^{-1}\gamma(1+x_2)-\gamma}(1+y) \\ &= \psi_{(1+x)^{-1}\gamma(1+x)-\gamma}(1+y), \end{aligned}$$

since x_1 commutes with γ.

We now compute the other factor $\psi_c[1+x, 1+y]$. Proceeding exactly as in the first part of the proof (i.e. where β was minimal), and recalling that $c \in \mathfrak{P}^{-r}$, we find

$$\psi_c[1+x, 1+y] = \psi_{(1+x)^{-1}c(1+x)-c}(1+y),$$

and this gives us

$$\begin{aligned}\theta[1+x, 1+y] &= \theta_0[1+x, 1+y].\psi_c[1+x, 1+y] \\ &= \psi_{(1+x)^{-1}\beta(1+x)-\beta}(1+y),\end{aligned}$$

as desired. ∎

(3.2.8) Proposition: *Let m, θ, k, ℓ be as in* (3.2.6). *Let $x \in \mathfrak{Q}_\beta^k \mathfrak{N}(\beta)$, $y \in \mathfrak{J}^\ell(\beta)$. Then $[1+x, 1+y] \in H^{m+1}(\beta)$ and*

$$\theta[1+x, 1+y] = \psi_{(1+x)^{-1}\beta(1+x)-\beta}(1+y).$$

Proof: The commutator $[1+x, 1+y]$ lies in $1+\mathfrak{Q}_\beta^k\mathfrak{J}^\ell(\beta)\mathfrak{N}(\beta)$. We have $\mathfrak{Q}_\beta^k\mathfrak{J}^\ell \subset \mathfrak{H}^{k+\ell}$ by (3.1.13), and this is contained in $\mathfrak{H}^{[\frac{r}{2}]+1}$. Therefore, $\mathfrak{Q}_\beta^k\mathfrak{J}^\ell\mathfrak{N}(\beta) \subset \mathfrak{H}^{[\frac{r}{2}]+1}\mathfrak{N}(\beta) = \mathfrak{H}^{[\frac{r}{2}]+1}$ by (3.1.9). It is also contained in $\mathfrak{P}^{k+\ell}$, so altogether

$$[1+x, 1+y] \in H^{k+\ell}(\beta) \subset H^{m+1}(\beta),$$

as required.

To prove the main statement, we work by induction along β again. If β is minimal over F, we have $\mathfrak{N}(\beta)$ and $\mathfrak{J}^\ell(\beta) \subset \mathfrak{P}^\ell = \mathfrak{Q}_\beta^\ell\mathfrak{N}(\beta)$, so the assertion is weaker than that of (3.2.6) in this case. So we assume that β is not minimal over F, and write $\theta = \theta_0\psi_c$ as in the proof of (3.2.6). The same direct computation as there yields

$$\psi_c[1+x, 1+y] = \psi_{(1+x)^{-1}c(1+x)-c}(1+y), \tag{3.2.9}$$

so we just have to show

$$\theta_0[1+x, 1+y] = \psi_{(1+x)^{-1}\gamma(1+x)-\gamma}(1+y). \tag{3.2.10}$$

We show first that we may assume $\ell \geq [\frac{r+1}{2}]$. For, assuming the result in this case and taking $\ell < [\frac{r+1}{2}]$, we have $\mathfrak{J}^\ell(\beta) = \mathfrak{Q}_\beta^\ell + \mathfrak{J}^{[\frac{r+1}{2}]}(\beta)$. We therefore write $(1+y) = (1+y_1)(1+y_2)$ with $y_1 \in \mathfrak{Q}_\beta^\ell$ and $y_2 \in \mathfrak{J}^{[\frac{r+1}{2}]}$. Expanding the commutator we get

$$\theta[1+x, 1+y] = \theta[1+x, 1+y_1].\theta[1+x, 1+y_2],$$

since $(1+y_1)$ normalises θ. Applying (3.2.6) to the first factor and the case $\ell \geq \left[\frac{r+1}{2}\right]$ to the second, we get

$$\theta[1+x, 1+y] = \psi_{(1+x)^{-1}\beta(1+x)-\beta}(1+y_1+y_2).$$

This differs from the desired result by a factor $\psi(((1+x)^{-1}\beta(1+x) - \beta)y_1y_2)$. However, $(1+x)^{-1}\beta(1+x) \equiv \beta \pmod{\mathfrak{P}^{k-r}}$ since $x \in \mathfrak{Q}_\beta^k\mathfrak{N}(\beta)$. Also, $y_1y_2 \in \mathfrak{P}^{\ell+\left[\frac{r+1}{2}\right]}$, and $\ell+\left[\frac{r+1}{2}\right]+k-r \geq m+1-\left[\frac{r}{2}\right] \geq 1$. Therefore

$$((1+x)^{-1}\beta(1+x) - \beta)y_1y_2 \in \mathfrak{P} \subset \mathrm{Ker}(\psi),$$

the extra factor is null, and we have the desired result.

So we now assume that $\ell \geq \left[\frac{r+1}{2}\right]$, and prove (3.2.10). We have $x \in \mathfrak{Q}_\beta^k\mathfrak{N}(\beta) = \mathfrak{Q}_\gamma^k + \mathfrak{Q}_\gamma^{k+s-r}\mathfrak{N}(\gamma)$, $y \in \mathfrak{J}^\ell(\beta) = \mathfrak{J}^\ell(\gamma)$. We write $(1+x) = (1+x_1)(1+x_2)$ with $x_1 \in \mathfrak{Q}_\gamma^k$, $x_2 \in \mathfrak{Q}_\gamma^{k+s-r}\mathfrak{N}(\gamma)$, and this gives as usual

$$\theta_0[1+x, 1+y] = \theta_0[1+x_2, 1+y].\theta_0[1+x_1, 1+y].$$

By induction, the first factor here is $\psi_{(1+x_2)^{-1}\gamma(1+x_2)-\gamma}(1+y)$. To evaluate the second, we write $(1+y) = (1+y_1)(1+y_2)$, with $y_1 \in \mathfrak{Q}_\gamma^\ell$, $y_2 \in \mathfrak{J}^{\ell'}(\gamma)$, where $\ell' = \max\{\ell, \left[\frac{s+1}{2}\right]\}$, and decompose the commutator. The factor $\theta_0[1+x_1, 1+y_1] = 1$ since θ_0 factors through $\det_{B_\gamma}$ on $U^{m+1}(\mathfrak{B}_\gamma)$. Our inductive hypothesis applies to the other factor to give

$$\theta_0[1+x_1, 1+y_2] = \psi_{(1+x_1)^{-1}\gamma(1+x_1)-\gamma}(1+y_2) = 1$$

since $(1+x_1)$ commutes with γ. For the same reason, when we assemble all these factors, we get the desired equality (3.2.10). ■

(3.2.11) Corollary: *Let $\theta \in \mathcal{C}(\mathfrak{A}, m, \beta)$, with $\left[\frac{r}{2}\right] \leq m \leq r-1$. Let $k, \ell \geq 1$, $k+\ell \geq m+1$, $k+2\ell \geq r+1$. Let $x \in \mathfrak{Q}_\beta^k\mathfrak{N}(\beta)$, $y \in \mathfrak{H}^\ell(\beta)$. Then $[1+x, 1+y] \in H^{m+1}(\beta)$ and*

$$\theta[1+x, 1+y] = \psi_{(1+x)^{-1}\beta(1+x)-\beta}(1+y).$$

The assertion of the corollary is, of course, weaker than that of the proposition. The final result in this sequence is:—

(3.2.12) Proposition: *Let $\theta \in \mathcal{C}(\mathfrak{A}, m, \beta)$, with $\left[\frac{r}{2}\right] \leq m \leq r-1$. Let $k, \ell \geq 1$, $k+\ell \geq m+1$, $k+2\ell \geq r+1$. Let $x \in \mathfrak{J}^k(\beta)$, $y \in \mathfrak{J}^\ell(\beta)$. Then $[1+x, 1+y] \in H^{m+1}(\beta)$ and*

$$\theta[1+x, 1+y] = \psi_{(1+x)^{-1}\beta(1+x)-\beta}(1+y).$$

Proof: When β is minimal over F, the assertion here is weaker than previous ones. By (3.2.8), we may assume in the general case that $k, \ell \geq \left[\frac{r+1}{2}\right]$. The proof then goes by induction, exactly as before. ■

(3.3) Intertwining

We continue with the same notation. In particular, $[\mathfrak{A}, n, 0, \beta]$ is simple, $r = -k_0(\beta, \mathfrak{A})$, and $[\mathfrak{A}, n, r, \gamma]$ is simple and equivalent to $[\mathfrak{A}, n, r, \beta]$ in the case $r < n$. (Provisionally, it still has to be the first term in a chosen defining sequence for β.) In this section, we compute the intertwining of the "simple characters" $\theta \in \mathcal{C}(\mathfrak{A}, m, \beta)$, $0 \le m \le r-1$. It will follow that the set $\mathcal{C}(\mathfrak{A}, m, \beta)$ only depends on the equivalence class of $[\mathfrak{A}, n, m, \beta]$ and not on any of the choices made in its definition. We will also be able to list its elements:— in particular it is non-empty.

(3.3.1) Proposition: *Let* $\theta \in \mathcal{C}(\mathfrak{A}, m, \beta)$, $0 \le m \le r-1$, *and let* $j \in J(\beta)$. *Then* $\theta(j(1+y)j^{-1}) = \theta(1+y)$, $y \in \mathfrak{H}^{m+1}(\beta)$.

Proof: We have $J(\beta) = \boldsymbol{U}(\mathfrak{B}_\beta) J^{[\frac{r+1}{2}]}(\beta)$, and $\boldsymbol{U}(\mathfrak{B}_\beta)$ normalises θ by definition. Therefore it is enough to take $j = 1+x \in J^{[\frac{r+1}{2}]}(\beta)$. Suppose first that $m \ge \left[\frac{r}{2}\right]$. The hypotheses of (3.2.12) then apply to give

$$\theta((1+x)(1+y)(1+x)^{-1}) = \theta(1+y)\psi_{(1+x)^{-1}\beta(1+x)-\beta}(1+y).$$

We have $(1+x)^{-1}\beta(1+x) - \beta = (1+x)^{-1}a_\beta(x)$. (3.1.19) implies $a_\beta(x) \in (\mathfrak{H}^{m+1})^*$ and (3.1.13) that $y(1+x)^{-1} \in \mathfrak{H}^{m+1}$, so

$$\psi_{(1+x)^{-1}\beta(1+x)-\beta}(1+y) = \psi(y(1+x)^{-1}a_\beta(x)) = 1.$$

The result follows in this case.

Now take the case $m < \left[\frac{r}{2}\right]$. We have

$$H^{m+1}(\beta) = \boldsymbol{U}^{m+1}(\mathfrak{B}_\beta) H^{[\frac{r}{2}]+1}(\beta),$$

so by the first part we need only consider the case $y \in \mathfrak{Q}_\beta^{m+1}$. Now we can apply (3.2.8) to the character $\theta \mid H^{[\frac{r}{2}]+1}(\beta) \in \mathcal{C}(\mathfrak{A}, \left[\frac{r}{2}\right], \beta)$ and get

$$\theta[1+y, 1+x] = \psi_{(1+y)^{-1}\beta(1+y)-\beta}(1+x) = 1,$$

since y commutes with β. The result follows. ■

We now recall a standard definition. Temporarily, let H_1, H_2 be subgroups of G, and ϕ_i an abelian character of H_i, $i = 1, 2$. Define

$$\begin{aligned} I_G(\phi_1, \phi_2) &= I_G(\phi_1 \mid H_1, \phi_2 \mid H_2) \\ &= \{x \in G : \phi_1^x(h) = \phi_2(h),\ h \in x^{-1}H_1x \cap H_2\}. \end{aligned}$$

Here, ϕ_1^x denotes the character $h \mapsto \phi_1(xhx^{-1})$ of the group $x^{-1}H_1x$. We say that an element $g \in G$ *intertwines* ϕ_1, ϕ_2 if $g \in I_G(\phi_1, \phi_2)$. If $H_1 = H_2 = H$ and $\phi_1 = \phi_2 = \phi$, we often abbreviate

$$I_G(\phi, \phi) = I_G(\phi) = I_G(\phi \mid H).$$

(3.3.2) Theorem: *Let $[\mathfrak{A}, n, 0, \beta]$ be a simple stratum in A, and put $r = -k_0(\beta, \mathfrak{A})$. Let $0 \le m \le r-1$, and $\theta \in \mathcal{C}(\mathfrak{A}, m, \beta)$. Then*

$$I_G(\theta \mid H^{m+1}(\beta)) \\ = (1 + \mathfrak{Q}_\beta^{r-m}\mathfrak{N}(\beta) + \mathfrak{J}^{[\frac{r+1}{2}]}(\beta))B_\beta^\times(1 + \mathfrak{Q}_\beta^{r-m}\mathfrak{N}(\beta) + \mathfrak{J}^{[\frac{r+1}{2}]}(\beta)).$$

Remark: We can rewrite

$$(1 + \mathfrak{Q}_\beta^{r-m}\mathfrak{N}(\beta) + \mathfrak{J}^{[\frac{r+1}{2}]}(\beta)) = (1 + \mathfrak{Q}_\beta^{r-m}\mathfrak{N}(\beta))J^{[\frac{r+1}{2}]}(\beta),$$

noting that, since $(1+\mathfrak{Q}_\beta^{r-m}\mathfrak{N}(\beta))$ is contained in $\mathfrak{N}(\beta)^\times$, the first factor normalises the second. Moreover, when $m \le \left[\frac{r}{2}\right]$, we have $\mathfrak{Q}_\beta^{r-m}\mathfrak{N}(\beta) \subset \mathfrak{J}^{[\frac{r+1}{2}]}(\beta)$, so the assertion in this case amounts to

$$I_G(\theta \mid H^{m+1}) = J^{[\frac{r+1}{2}]}(\beta)B_\beta^\times J^{[\frac{r+1}{2}]}(\beta), \quad m \le \left[\tfrac{r}{2}\right].$$

Proof: As usual, we proceed by induction along β, starting with the case $r = n$, the case $r = \infty$ being trivial. Suppose first that $m \ge \left[\frac{n}{2}\right]$. Then $\theta = \psi_\beta$ and by (1.5.8) (or **[KM1]** Th.2.4), the intertwining set $I_G(\theta)$ is just $(1 + \mathfrak{P}^{n-m})B_\beta^\times(1 + \mathfrak{P}^{n-m})$, while the asserted value is

$$(1 + \mathfrak{Q}_\beta^{n-m}\mathfrak{N}(\beta) + \mathfrak{J}^{[n+1/2]}(\beta))B_\beta^\times(1 + \mathfrak{Q}_\beta^{n-m}\mathfrak{N}(\beta) + \mathfrak{J}^{[n+1/2]}(\beta)).$$

However, $\mathfrak{J}^{[\frac{n+1}{2}]} = \mathfrak{P}^{[\frac{n+1}{2}]}$, and $\mathfrak{Q}_\beta^{n-m}\mathfrak{N}(\beta) = \mathfrak{P}^{n-m}$, so

$$(1 + \mathfrak{Q}_\beta^{n-m}\mathfrak{N}(\beta) + \mathfrak{J}^{[\frac{n+1}{2}]}(\beta)) = 1 + \mathfrak{P}^{n-m},$$

as $m \ge \left[\frac{n}{2}\right]$. The assertion of the theorem therefore holds in this case.

Now suppose that $m < \left[\frac{n}{2}\right]$. Then certainly $I_G(\theta)$ is contained in

$$I_G(\theta \mid U^{[\frac{n}{2}]+1}(\mathfrak{A})) = (1 + \mathfrak{P}^{[\frac{n+1}{2}]})B_\beta^\times(1 + \mathfrak{P}^{[\frac{n+1}{2}]}) = J^{[\frac{n+1}{2}]}B_\beta^\times J^{[\frac{n+1}{2}]}.$$

However,

$$\mathfrak{P}^{n-m} = \mathfrak{Q}_\beta^{n-m}\mathfrak{N}(\beta) \subset \mathfrak{J}^{[\frac{n+1}{2}]}(\beta), \quad m < \left[\tfrac{n}{2}\right].$$

Therefore we need only check that θ is intertwined by the whole of $J^{[\frac{n+1}{2}]}B_\beta^\times J^{[\frac{n+1}{2}]}$ on $H^{m+1}(\beta)$. The factor $J^{[\frac{n+1}{2}]}(\beta)$ normalises θ by (3.3.1), so we only have to show

(3.3.3) *$\theta \mid H^{m+1}(\beta)$ is intertwined by every $x \in B_\beta^\times$.*

To prove this, we first take $x \in B_\beta^\times$ and compute $x^{-1}H^{m+1}x \cap H^{m+1}$. This is just $1 + x^{-1}\mathfrak{H}^{m+1}x \cap \mathfrak{H}^{m+1}$, and

$$x^{-1}\mathfrak{H}^{m+1}x \cap \mathfrak{H}^{m+1} = x^{-1}(\mathfrak{Q}_\beta^{m+1} + \mathfrak{P}^{[\frac{n}{2}]+1})x \cap (\mathfrak{Q}_\beta^{m+1} + \mathfrak{P}^{[\frac{n}{2}]+1}).$$

We assert

$$\textbf{(3.3.4)} \quad \begin{aligned} &x^{-1}(\mathfrak{Q}_\beta^{m+1} + \mathfrak{P}^{[\frac{n}{2}]+1})x \cap (\mathfrak{Q}_\beta^{m+1} + \mathfrak{P}^{[\frac{n}{2}]+1}) \\ &\qquad = x^{-1}\mathfrak{Q}_\beta^{m+1}x \cap \mathfrak{Q}_\beta^{m+1} + x^{-1}\mathfrak{P}^{[\frac{n}{2}]+1}x \cap \mathfrak{P}^{[\frac{n}{2}]+1}. \end{aligned}$$

We surely have containment in the direction $\supset$. On the other hand,

$$x^{-1}(\mathfrak{Q}_\beta^{m+1} + \mathfrak{P}^{[\frac{n}{2}]+1})x \cap (\mathfrak{Q}_\beta^{m+1} + \mathfrak{P}^{[\frac{n}{2}]+1}) \cap B_\beta = x^{-1}\mathfrak{Q}_\beta^{m+1}x \cap \mathfrak{Q}_\beta^{m+1}.$$

This is contained in $(x^{-1}\mathfrak{Q}_\beta^{m+1}x \cap \mathfrak{Q}_\beta^{m+1} + x^{-1}\mathfrak{P}^{[\frac{n}{2}]+1}x \cap \mathfrak{P}^{[\frac{n}{2}]+1}) \bigcap B_\beta$. The two sides of (3.3.4) thus have the same intersection with $B_\beta = \mathrm{Ker}(a_\beta)$. On the other hand,

$$\begin{aligned} &a_\beta(x^{-1}\mathfrak{Q}_\beta^{m+1}x \cap \mathfrak{Q}_\beta^{m+1} + x^{-1}\mathfrak{P}^{[\frac{n}{2}]+1}x \cap \mathfrak{P}^{[\frac{n}{2}]+1}) \\ &\qquad = a_\beta(x^{-1}\mathfrak{P}^{[\frac{n}{2}]+1}x \cap \mathfrak{P}^{[\frac{n}{2}]+1}) \\ &\qquad\qquad = x^{-1}a_\beta(\mathfrak{P}^{[\frac{n}{2}]+1})x \cap a_\beta(\mathfrak{P}^{[\frac{n}{2}]+1}) \end{aligned}$$

by (1.4.16). This last set contains the image under a_β of the left hand side of (3.3.4). Thus both sides have the same image under a_β, and the same intersection with $\mathrm{Ker}(a_\beta)$. They are therefore the same.

Now consider the characters θ, θ^x of $x^{-1}H^{m+1}x \cap H^{m+1}$. (3.3.4) shows that

$$\begin{aligned} &x^{-1}H^{m+1}x \cap H^{m+1} \\ &\quad = (x^{-1}\boldsymbol{U}^{m+1}(\mathfrak{B}_\beta)x \cap \boldsymbol{U}^{m+1}(\mathfrak{B}_\beta)).(x^{-1}\boldsymbol{U}^{[\frac{n}{2}]+1}(\mathfrak{A})x \cap \boldsymbol{U}^{[\frac{n}{2}]+1}(\mathfrak{A})) \end{aligned}$$

The characters agree on both factors, therefore they agree, as required.

Now we turn to the case $r < n$. We let $[\mathfrak{A}, n, r, \gamma]$ be a simple stratum equivalent to $[\mathfrak{A}, n, r, \beta]$ (provisionally, it is the one used to define $\mathcal{C}(\mathfrak{A}, m, \beta)$). Put $s = -k_0(\gamma, \mathfrak{A})$. We deal first with the case $m \geq [\frac{r}{2}]$. Thus $H^{m+1}(\beta) = H^{m+1}(\gamma)$ and, for $\theta \in \mathcal{C}(\mathfrak{A}, m, \beta)$, we certainly have $I_G(\theta) \subset I_G(\theta \mid H^{r+1}(\beta))$. Of course, $H^{r+1}(\beta) = H^{r+1}(\gamma)$. Also, if we write $\beta = \gamma + c$, then ψ_c is null on H^{r+1}, so the restriction of θ to H^{r+1} lies in $\mathcal{C}(\mathfrak{A}, r, \gamma)$. Put $I^r(\gamma) = I_G(\theta \mid H^{r+1})$ so that, by induction, we have

$$I^r(\gamma) = (1 + \mathfrak{Q}_\gamma^{s-r}\mathfrak{N}(\gamma) + \mathfrak{J}^{[\frac{s+1}{2}]}(\gamma))B_\gamma^\times(1 + \mathfrak{Q}_\gamma^{s-r}\mathfrak{N}(\gamma) + \mathfrak{J}^{[\frac{s+1}{2}]}(\gamma)),$$

and this contains $I_G(\theta)$. We now define another set

$$I_m^+(\beta) = \{x \in G : x^{-1}(\beta + \mathfrak{H}^{m+1}(\beta)^*)x \cap (\beta + \mathfrak{H}^{m+1}(\beta)^*) \neq \emptyset\},$$

i.e., $I_m^+(\beta)$ is the formal intertwining of the coset $\beta + \mathfrak{H}^{m+1}(\beta)^*$. The main step in the proof is:

(3.3.5) Lemma: $I_G(\theta) = I_m^+(\beta) \cap I^r(\gamma)$.

Proof: We first show that

$$\text{(3.3.6)} \qquad J^{[\frac{r+1}{2}]}(\beta).I_m^+(\beta).J^{[\frac{r+1}{2}]}(\beta) = I_m^+(\beta).$$

This will follow if we can show

$$\text{(3.3.7)} \qquad j^{-1}\beta j \equiv \beta \pmod{\mathfrak{H}^{m+1}(\beta))^*}, \quad j \in J^{[\frac{r+1}{2}]}(\beta),$$

since $J^{[\frac{r+1}{2}]}(\beta)$ normalises $\mathfrak{H}^{m+1}(\beta)$ by (3.1.15) and hence also its dual. Indeed, by (3.1.13) and duality,

$$J^{[\frac{r+1}{2}]}(\beta).(\mathfrak{H}^{m+1}(\beta))^* \subset (\mathfrak{H}^{m+1}(\beta))^*,$$

so (3.3.7) comes down to showing that $a_\beta(x) \in (\mathfrak{H}^{m+1}(\beta))^*$ for $x \in \mathfrak{J}^{[\frac{r+1}{2}]}(\beta)$. This is given by (3.1.16), so we have proved (3.3.6).

We now take $x \in I^r(\gamma)$ and show that $x \in I_m^+(\beta)$ if and only if $x \in I_G(\theta)$. Since all the sets in question are bi-invariant under $J^{[\frac{r+1}{2}]}(\beta) = J^{[\frac{r+1}{2}]}(\gamma)$, we may as well take $x \in (1 + \mathfrak{Q}_\gamma^{s-r}\mathfrak{N}(\gamma))B_\gamma^\times(1 + \mathfrak{Q}_\gamma^{s-r}\mathfrak{N}(\gamma))$. We therefore write $x = (1+y)t(1+z)$, $t \in B_\gamma^\times$, $y, z \in \mathfrak{Q}_\gamma^{s-r}\mathfrak{N}(\gamma)$. Take $h \in x^{-1}H^{m+1}(\beta)x \cap H^{m+1}(\beta)$, and consider $\theta^x(h)\theta(h)^{-1}$. We put $\beta = \gamma + c$, $c \in \mathfrak{P}^{-r}$, write $\theta = \theta_0\psi_c$, $\theta_0 \in \mathcal{C}(\mathfrak{A}, m, \gamma)$, and treat the factors separately. First we have

$$\theta_0^x(h) = \theta_0((1+y)t(1+z)h(1+z)^{-1}t^{-1}(1+y)^{-1}).$$

Let us now abbreviate $H^{m+1} = H^{m+1}(\beta)$. The element $t(1+z)h(1+z)^{-1}t^{-1}$ lies in H^{m+1}, since this group is normalised by $(1+y)$. We have

$$s - r + m + 1 \geq 1 + \max\left\{\left[\tfrac{s}{2}\right], m\right\},$$
$$s - r + 2(m+1) \geq s + 1,$$

so we can apply (3.2.8) to get

$$\theta_0^x(h) = \theta_0(t(1+z)h(1+z)^{-1}t^{-1})\psi_{(1+y)^{-1}\gamma(1+y)-\gamma}(t(1+z)h(1+z)^{-1}t^{-1}).$$

We have $(1+z)h(1+z)^{-1} \in t^{-1}H^{m+1}t \cap H^{m+1}$, and $t \in B_\gamma^\times$ intertwines θ_0, so this comes down to

$$\begin{aligned}&\theta_0^x(h)\\ &= \theta_0((1+z)h(1+z)^{-1})\psi_{(1+y)^{-1}\gamma(1+y)-\gamma}(t(1+z)h(1+z)^{-1}t^{-1})\\ &= \theta_0(h)\psi_{(1+z)^{-1}\gamma(1+z)-\gamma}(h)\psi_{(1+y)^{-1}\gamma(1+y)-\gamma}(t(1+z)h(1+z)^{-1}t^{-1})\end{aligned}$$

using (3.2.8) again. Since t commutes with γ, a simple calculation further reduces this to

$$\theta_0^x(h) = \theta_0(h)\psi_{x^{-1}\gamma x-\gamma}(h), \quad h \in x^{-1}H^{m+1}x \cap H^{m+1},\ x \in I^r(\gamma).$$

We now compute $(\psi_c)^x(h)$ directly to get

$$(\psi_c)^x(h) = \psi_{x^{-1}cx-c}(h)\psi_c(h),$$

and altogether

$$\theta^x(h) = \theta(h)\psi_{x^{-1}\beta x-\beta}(h),$$

for $h \in x^{-1}H^{m+1}x \cap H^{m+1}$, $x \in I^r(\gamma)$. Thus x intertwines θ if and only if $\psi_{x^{-1}\beta x-\beta}(h) = 1$ for all $h \in x^{-1}H^{m+1}x \cap H^{m+1}$, i.e. if and only if $x^{-1}\beta x - \beta \in (x^{-1}\mathfrak{H}^{m+1}x \cap \mathfrak{H}^{m+1})^* = x^{-1}(\mathfrak{H}^{m+1})^*x + (\mathfrak{H}^{m+1})^*$. This is equivalent to $x \in I_m^+(\beta)$, and we have proved (3.3.5). ■

The next step of the proof is to compute $I_m^+(\beta)$. This will imply

(3.3.8) $$I_m^+(\beta) \subset I^r(\gamma),$$

whence $I_m^+(\beta) = I_G(\theta)$, and we will have the result. First, however, it is worth recording, for future reference, the conclusion of the computation in the proof of (3.3.5):—

(3.3.9) Proposition: *Let $[\mathfrak{A}, n, 0, \beta]$ be simple with $r = -k_0(\beta, \mathfrak{A})$. Let $[\mathfrak{A}, n, r, \gamma]$ be simple and equivalent to $[\mathfrak{A}, n, r, \beta]$, with $s = -k_0(\gamma, \mathfrak{A})$. Let $\left[\frac{r}{2}\right] \le m \le r-1$, $\theta \in \mathcal{C}(\mathfrak{A}, m, \beta)$, $\theta_0 \in \mathcal{C}(\mathfrak{A}, m, \gamma)$. Let x be an element of*

$$(1 + \mathfrak{Q}_\gamma^{s-r}\mathfrak{N}(\gamma) + \mathfrak{J}^{[\frac{s+1}{2}]}(\gamma))B_\gamma^\times(1 + \mathfrak{Q}_\gamma^{s-r}\mathfrak{N}(\gamma) + \mathfrak{J}^{[\frac{s+1}{2}]}(\gamma)).$$

Then for $h \in x^{-1}H^{m+1}(\beta)x \cap H^{m+1}(\beta)$, we have

$$\begin{aligned}\theta^x(h) &= \theta(h)\psi_{x^{-1}\beta x-\beta}(h),\\ \theta_0^x(h) &= \theta_0(h)\psi_{x^{-1}\gamma x-\gamma}(h).\end{aligned}$$

So, we now have to investigate the set $I_m^+(\beta)$, which is the formal intertwining of the coset $\beta + (\mathfrak{H}^{m+1}(\beta))^*$. We have already observed that it is bi-invariant under multiplication by $J^{[\frac{r+1}{2}]}(\beta)$. Since $a_\beta(\mathfrak{Q}_\beta^{r-m}\mathfrak{N}(\beta)) \subset \mathfrak{P}^{-m} \subset (\mathfrak{H}^{m+1})^*$, the same proof shows that it is bi-invariant under $(1+\mathfrak{Q}_\beta^{r-m}\mathfrak{N}(\beta))$. Certainly $B_\beta^\times \subset I_m^+(\beta)$, so we conclude

(3.3.10)

$$I_m^+(\beta) \supset (1+\mathfrak{Q}_\beta^{r-m}\mathfrak{N}(\beta)) + \mathfrak{J}^{[\frac{r+1}{2}]}(\beta))B_\beta^\times(1+\mathfrak{Q}_\beta^{r-m}\mathfrak{N}(\beta)) + \mathfrak{J}^{[\frac{r+1}{2}]}(\beta)).$$

We need to show that we have equality here.

For $x \in I_m^+(\beta)$, we have by definition

$$x^{-1}(\beta + (\mathfrak{H}^{m+1})^*)x \cap (\beta + (\mathfrak{H}^{m+1})^*) \neq \emptyset.$$

Therefore there exist $\delta_1, \delta_2 \in (\mathfrak{H}^{m+1})^*$ such that

$$\textbf{(3.3.11)} \qquad x^{-1}(\beta + \delta_1 + \mathfrak{P}^{-m})x \cap (\beta + \delta_2 + \mathfrak{P}^{-m}) \neq \emptyset.$$

We prove later:—

(3.3.12) Lemma: *Let $\delta \in (\mathfrak{H}^{m+1}(\beta))^*$. There exists $y \in \mathfrak{J}^{[\frac{r+1}{2}]}(\beta)$ such that*

$$(1+y)^{-1}(\beta + \delta + \mathfrak{P}^{-m})(1+y) = \beta + \mathfrak{P}^{-m}.$$

Granting this, we can write in (3.3.11)

$$\beta + \delta_i + \mathfrak{P}^{-m} = (1+y_i)^{-1}(\beta + \mathfrak{P}^{-m})(1+y_i), \quad i = 1, 2,$$

for elements $y_1, y_2 \in \mathfrak{J}^{[\frac{r+1}{2}]}$. Thus, if $z = (1+y_1)x(1+y_2)^{-1}$, we have $z \in I_m^+(\beta)$ and

$$z^{-1}(\beta + \mathfrak{P}^{-m})z \cap (\beta + \mathfrak{P}^{-m}) \neq \emptyset.$$

Then, by (1.5.8), $z \in (1+\mathfrak{Q}_\beta^{r-m}\mathfrak{N}(\beta))B_\beta^\times(1+\mathfrak{Q}_\beta^{r-m}\mathfrak{N}(\beta))$, and x therefore lies in

$$(1+\mathfrak{Q}_\beta^{r-m}\mathfrak{N}(\beta) + \mathfrak{J}^{[\frac{r+1}{2}]}(\beta))B_\beta^\times(1+\mathfrak{Q}_\beta^{r-m}\mathfrak{N}(\beta) + \mathfrak{J}^{[\frac{r+1}{2}]}(\beta))$$

as desired, showing that we have equality in (3.3.10).

We can now prove (3.3.8). Since $I^r(\gamma)$ is bi-invariant under multiplication by $J^{[\frac{r+1}{2}]}(\beta) = J^{[\frac{r+1}{2}]}(\gamma)$, it is enough to show that

$$I^r(\gamma) \supset (1+\mathfrak{Q}_\beta^{r-m}\mathfrak{N}(\beta))B_\beta^\times(1+\mathfrak{Q}_\beta^{r-m}\mathfrak{N}(\beta)).$$

However, this last set is the formal intertwining of $\beta + \mathfrak{P}^{-m}$, and so is contained in the formal intertwining of $\beta + \mathfrak{P}^{-r} = \gamma + \mathfrak{P}^{-r}$. The formal intertwining of $\gamma + \mathfrak{P}^{-r}$ is $(1 + \mathfrak{Q}_\gamma^{s-r}\mathfrak{N}(\gamma))B_\gamma^\times(1 + \mathfrak{Q}_\gamma^{s-r}\mathfrak{N}(\gamma))$ (where $s = -k_0(\gamma, \mathfrak{A})$) and this is certainly contained in $I^r(\gamma)$. We deduce

$$\begin{aligned} I_G(\theta) &= I_m^+(\beta) \\ &= (1 + \mathfrak{Q}^{r-m}\mathfrak{N}(\beta) + \mathfrak{J}^{[\frac{r+1}{2}]}(\beta))B_\beta^\times(1 + \mathfrak{Q}^{r-m}\mathfrak{N}(\beta) + \mathfrak{J}^{[\frac{r+1}{2}]}(\beta)), \end{aligned}$$

as required.

Now we prove (3.3.12) to complete the proof of the theorem in the case $m \geq \left[\frac{r}{2}\right]$.

Proof of (3.3.12): We show

(3.3.13) *Let $k \in \mathbb{Z}$, $\delta \in (\mathfrak{H}^{m+1}(\beta)^* \cap \mathfrak{P}^k) + \mathfrak{P}^{-m}$. There exists $y \in \mathfrak{J}^{[\frac{r+1}{2}]}(\beta)$ such that*

$$(1+y)^{-1}(\beta+\delta)(1+y) \equiv \beta \pmod{(\mathfrak{H}^{m+1}(\beta)^* \cap \mathfrak{P}^{k+1}) + \mathfrak{P}^{-m}}.$$

(3.3.12) then follows by induction. We start by taking $\delta \in (\mathfrak{H}^{m+1})^* \cap \mathfrak{P}^k$, $y \in \mathfrak{J}^{[\frac{r+1}{2}]}$, and expanding $(1+y)^{-1}(\beta+\delta)(1+y)$. First we note that

$$\begin{aligned} &(1+y)^{-1}\delta(1+y) \equiv \delta \\ &\qquad\qquad \pmod{((\mathfrak{H}^{m+1})^* \cap \mathfrak{P}^k)\mathfrak{J}^{[\frac{r+1}{2}]} + \mathfrak{J}^{[\frac{r+1}{2}]}((\mathfrak{H}^{m+1})^* \cap \mathfrak{P}^k)}. \end{aligned}$$

We have $\mathfrak{J}^{[\frac{r+1}{2}]}\mathfrak{H}^{m+1} \subset \mathfrak{H}^{m+1}$, so $(\mathfrak{H}^{m+1})^*\mathfrak{J}^{[\frac{r+1}{2}]} \subset (\mathfrak{H}^{m+1})^*$, and symmetrically. Also, $\mathfrak{J}^{[\frac{r+1}{2}]}\mathfrak{P}^k \subset \mathfrak{P}^{k+1}$, so

$$(1+y)^{-1}\delta(1+y) \equiv \delta \pmod{(\mathfrak{H}^{m+1})^* \cap \mathfrak{P}^{k+1}}.$$

Further, $(1+y)^{-1}\beta(1+y) = \beta + (1+y)^{-1}a_\beta(y)$. If we write $(1+y)^{-1} = 1+\bar{y}$, for some $\bar{y} \in \mathfrak{J}^{[\frac{r+1}{2}]}$, then $\bar{y}a_\beta(y) \in \mathfrak{J}^{[\frac{r+1}{2}]}(\mathfrak{H}^{m+1})^*$ by the same reasoning, so

$$(1+y)^{-1}a_\beta(y) \equiv a_\beta(y) \pmod{(\mathfrak{H}^{m+1})^* \cap \mathfrak{P}^{k+1}}.$$

Altogether, we have shown

$$(1+y)^{-1}(\beta+\delta)(1+y) \equiv \beta + \delta + a_\beta(y) \pmod{(\mathfrak{H}^{m+1})^* \cap \mathfrak{P}^{k+1}},$$

for $\delta \in (\mathfrak{H}^{m+1})^* \cap \mathfrak{P}^k$, $y \in \mathfrak{J}^{[\frac{r+1}{2}]}$. Since the group $J(\beta)$ normalises $\mathfrak{P}^{-m}$, this holds for $\delta \in ((\mathfrak{H}^{m+1})^* \cap \mathfrak{P}^k) + \mathfrak{P}^{-m}$. By (3.1.18), we can choose $y \in \mathfrak{J}^{[\frac{r+1}{2}]}$ so that $\delta + a_\beta(y) \in \mathfrak{P}^{-m}$. Thus, for this y, we have

$$(1+y)^{-1}(\beta+\delta)(1+y) \equiv \beta \pmod{((\mathfrak{H}^{m+1})^* \cap \mathfrak{P}^{k+1}) + \mathfrak{P}^{-m}},$$

as required. ∎

This completes the proof of the theorem in the case $m \geq \left[\frac{r}{2}\right]$. Now we assume that $m < \left[\frac{r}{2}\right]$. We note that

$$\mathfrak{Q}_\beta^{r-[\frac{r}{2}]}\mathfrak{N}(\beta) = \mathfrak{Q}_\beta^{[\frac{r+1}{2}]}\mathfrak{N}(\beta) \subset \mathfrak{J}^{[\frac{r+1}{2}]}(\beta)$$

by (3.1.10). Therefore the assertion of the theorem amounts to

$$I_G(\theta) = J^{[\frac{r+1}{2}]}(\beta)B_\beta^\times J^{[\frac{r+1}{2}]}(\beta), \quad m < \left[\tfrac{r}{2}\right].$$

We certainly have

$$I_G(\theta) \subset I_G(\theta \mid H^{[\frac{r}{2}]+1}) = J^{[\frac{r+1}{2}]}(\beta)B_\beta^\times J^{[\frac{r+1}{2}]}(\beta),$$

by the first part. By (3.3.1), the group $J^{[\frac{r+1}{2}]}(\beta)$ normalises θ, so we just have to show

(3.3.14) $$B_\beta^\times \subset I_G(\theta).$$

We take $x \in B_\beta^\times$ and start by proving

(3.3.15)

$$x^{-1}\mathfrak{H}^{m+1}x \cap \mathfrak{H}^{m+1} = x^{-1}\mathfrak{Q}_\beta^{m+1}x \cap \mathfrak{Q}_\beta^{m+1} + x^{-1}\mathfrak{H}^{[\frac{r}{2}]+1}x \cap \mathfrak{H}^{[\frac{r}{2}]+1},$$

when $0 \leq m \leq \left[\frac{r}{2}\right]$. Then we can argue exactly as in the first case (where β was minimal over F) to get (3.3.14) and the theorem.

In (3.3.15), we certainly have containment in the direction $\supset$. The left hand side intersects B_β in $x^{-1}\mathfrak{Q}_\beta^{m+1}x \cap \mathfrak{Q}_\beta^{m+1}$, which is contained in the intersection of the right hand side with B_β. Therefore both sides intersect B_β in $x^{-1}\mathfrak{Q}_\beta^{m+1}x \cap \mathfrak{Q}_\beta^{m+1}$.

(3.3.16) Lemma: *For $0 \leq m \leq \left[\frac{r}{2}\right]$ and $x \in B_\beta^\times$, we have*

$$a_\beta(x^{-1}\mathfrak{H}^{m+1}x \cap \mathfrak{H}^{m+1}) = x^{-1}a_\beta(\mathfrak{H}^{m+1})x \cap a_\beta(\mathfrak{H}^{m+1}).$$

Proof: We certainly have the containment $\subset$. For the opposite one, take $h \in \mathfrak{H}^{m+1}$, $k \in x^{-1}\mathfrak{H}^{m+1}x$ with $a_\beta(h) = a_\beta(k)$. There exists $u \in B_\beta$ with $u + h = k$. Thus $u = k - h \in (\mathfrak{H}^{m+1} + x^{-1}\mathfrak{H}^{m+1}x) \cap B_\beta$. However,

$$\begin{aligned}\mathfrak{Q}_\beta^{m+1} + x^{-1}\mathfrak{Q}_\beta^{m+1}x &\subset (\mathfrak{H}^{m+1} + x^{-1}\mathfrak{H}^{m+1}x) \cap B_\beta \\ &\subset (\mathfrak{P}^{m+1} + x^{-1}\mathfrak{P}^{m+1}x) \cap B_\beta = \mathfrak{Q}_\beta^{m+1} + x^{-1}\mathfrak{Q}_\beta^{m+1}x,\end{aligned}$$

the last equality coming from (1.3.16). Thus $u = u' - v'$ where $u' \in \mathfrak{Q}_\beta^{m+1}$, $v' \in x^{-1}\mathfrak{Q}_\beta^{m+1}x$. Therefore $u' + h = v' + k \in \mathfrak{H}^{m+1} \cap x^{-1}\mathfrak{H}^{m+1}x$ and so $a_\beta(h)$ lies in $a_\beta(\mathfrak{H}^{m+1} \cap x^{-1}\mathfrak{H}^{m+1}x)$, which proves the lemma. ■

Now we apply a_β to both sides of (3.3.15), noting that $a_\beta(\mathfrak{H}^{m+1}) = a_\beta(\mathfrak{H}^{[\frac{r}{2}]+1})$. Both sides have the same image. Therefore (3.3.15) holds, and we have finished the proof of the theorem. ■

(3.3.17) Corollary: *Let $\theta \in \mathcal{C}(\mathfrak{A}, m, \beta)$, $0 \le m \le r-1$. The G-normaliser of θ is*

$$(1 + \mathfrak{Q}_\beta^{r-m}\mathfrak{N}(\beta) + \mathfrak{J}^{[\frac{r+1}{2}]}(\beta))\mathfrak{K}(\mathfrak{B}_\beta).$$

Proof: We have already noted in (3.3.1) that the group $J^{[\frac{r+1}{2}]}(\beta)$ normalises θ. The group $1+\mathfrak{Q}_\beta^{r-m}\mathfrak{N}(\beta)$ normalises $H^{m+1}(\beta)$ by (3.1.9), and it intertwines the character θ, by the theorem. Therefore it normalises θ. The group $\mathfrak{K}(\mathfrak{B}_\beta)$ normalises θ by definition. The normaliser of θ therefore contains the group

$$(1 + \mathfrak{Q}_\beta^{r-m}\mathfrak{N}(\beta))J^{[\frac{r+1}{2}]}(\beta)\mathfrak{K}(\mathfrak{B}_\beta) = (1 + \mathfrak{Q}_\beta^{r-m}\mathfrak{N}(\beta) + \mathfrak{J}^{[\frac{r+1}{2}]}(\beta))\mathfrak{K}(\mathfrak{B}_\beta).$$

Now let $x \in G$ normalise θ. Then x certainly intertwines θ so, by the theorem,

$$x \in (1 + \mathfrak{Q}_\beta^{r-m}\mathfrak{N}(\beta) + \mathfrak{J}^{[\frac{r+1}{2}]}(\beta))B_\beta^\times(1 + \mathfrak{Q}_\beta^{r-m}\mathfrak{N}(\beta) + \mathfrak{J}^{[\frac{r+1}{2}]}(\beta)).$$

We may ignore the outer factors, since they normalise θ, and assume that $x \in B_\beta^\times$. This element x normalises θ, and the group $H^{m+1}(\beta)$. It must also normalise $H^{m+1}(\beta) \cap B_\beta^\times = \boldsymbol{U}^{m+1}(\mathfrak{B}_\beta)$. However, the $B_\beta^\times$-normaliser of $\boldsymbol{U}^{m+1}(\mathfrak{B}_\beta)$ is $\mathfrak{K}(\mathfrak{B}_\beta)$ (*cf.* **(1.1)**), so $x \in \mathfrak{K}(\mathfrak{B}_\beta)$, as required. ■

(3.3.18) Corollary: *Let $\left[\frac{r}{2}\right] \le m \le r-1$, and assume that $r < n$. Let $[\mathfrak{A}, n, r, \gamma]$ be simple and equivalent to $[\mathfrak{A}, n, r, \beta]$, and put $c = \beta - \gamma$. The map*

$$\mathcal{C}(\mathfrak{A}, m, \beta) \to \mathcal{C}(\mathfrak{A}, m, \gamma),$$
$$\theta \mapsto \theta\psi_c^{-1}$$

is bijective.

Remark: For the time being, we must continue to assume that this γ is the one chosen to define $\mathcal{C}(\mathfrak{A}, m, \beta)$.

Proof: From the definition (3.2.3), $\theta \mapsto \theta\psi_c^{-1}$ gives an injective map $\mathcal{C}(\mathfrak{A}, m, \beta) \to \mathcal{C}(\mathfrak{A}, m, \gamma)$. We just have to show that if $\theta_0 \in \mathcal{C}(\mathfrak{A}, m, \gamma)$, then $\theta = \theta_0\psi_c$ lies in $\mathcal{C}(\mathfrak{A}, m, \beta)$. We start by demonstrating

(3.3.19) *θ is intertwined by every $x \in B_\beta^\times$.*

For this, we take $x \in B_\beta^\times$, and put $s = -k_0(\gamma, \mathfrak{A})$. Then $B_\beta^\times$ is contained in $(1 + \mathfrak{Q}_\gamma^{s-r}\mathfrak{N}(\gamma))B_\gamma^\times(1 + \mathfrak{Q}_\gamma^{s-r}\mathfrak{N}(\gamma))$, since this set is the formal intertwining of the coset $\gamma + \mathfrak{P}^{-r} = \beta + \mathfrak{P}^{-r}$ by (1.5.8). Now we use (3.3.9), applied to θ_0, to get

$$\theta^x(h) = \theta(h)\psi_{x^{-1}\beta x - \beta}(h), \quad h \in x^{-1}H^{m+1}x \cap H^{m+1}.$$

Since our element x commutes with β, we get $\theta^x(h) = \theta(h)$, whence x intertwines θ, as required for (3.3.19).

It follows that $\mathfrak{K}(\mathfrak{B}_\beta)$ normalises θ. Moreover, every $x \in B_\beta^\times$ intertwines $\theta \mid \boldsymbol{U}^{m+1}(\mathfrak{B}_\beta)$ so, by (2.4.11), $\theta \mid (H^{m+1} \cap B_\beta^\times)$ factors through $\det_{B_\beta}$. Therefore $\theta \in \mathcal{C}(\mathfrak{A}, m, \beta)$, as required. ■

(3.3.20) Corollary: *(i) The definition* (3.2.3) *of the set $\mathcal{C}(\mathfrak{A}, m, \beta)$ is independent of the choice of simple stratum $[\mathfrak{A}, n, r, \gamma]$ equivalent to $[\mathfrak{A}, n, r, \beta]$.*

(ii) Let $0 \leq \ell \leq r - 1$, and let $[\mathfrak{A}, n, \ell, \beta']$ be simple and equivalent to $[\mathfrak{A}, n, \ell, \beta]$. Then

$$\mathcal{C}(\mathfrak{A}, m, \beta') = \mathcal{C}(\mathfrak{A}, m, \beta).\psi_{\beta'-\beta}, \quad \left[\tfrac{\ell}{2}\right] \leq m \leq r - 1.$$

Proof: We show first that *(ii)* implies *(i)*. Let $[\mathfrak{A}, n, r, \gamma]$ be the simple stratum equivalent to $[\mathfrak{A}, n, r, \beta]$ originally chosen to define $\mathcal{C}(\mathfrak{A}, m, \beta)$. Let $[\mathfrak{A}, n, r, \gamma']$ be some other simple stratum equivalent to $[\mathfrak{A}, n, r, \beta]$, so that $[\mathfrak{A}, n, r, \gamma'] \sim [\mathfrak{A}, n, r, \gamma]$. Temporarily write $\mathcal{C}'(\mathfrak{A}, m, \beta)$ for the set of characters of $H^{m+1}(\beta)$ defined by (3.2.3) using γ' in place of γ. Set $\beta = \gamma + c = \gamma' + c'$. By (3.3.18), we have

$$\mathcal{C}'(\mathfrak{A}, m, \beta) = \mathcal{C}(\mathfrak{A}, m, \gamma')\psi_{c'} = \mathcal{C}(\mathfrak{A}, m, \gamma)\psi_{\gamma'-\gamma}\psi_{c'} = \mathcal{C}(\mathfrak{A}, m, \beta),$$

for $\left[\frac{r}{2}\right] \leq m \leq r - 1$. For $m < \left[\frac{r}{2}\right]$, we have

$$H^{m+1}(\beta) = \boldsymbol{U}^{m+1}(\mathfrak{B}_\beta)H^{[\frac{r}{2}]+1}(\beta),$$

and the elements of $\mathcal{C}(\mathfrak{A}, m, \beta)$ are the characters of H^{m+1} of the form

$$uh \mapsto \phi(u)\theta(h), \quad u \in \boldsymbol{U}^{m+1}(\mathfrak{B}_\beta), \ h \in H^{[\frac{r}{2}]+1}(\beta),$$

where θ ranges over $\mathcal{C}(\mathfrak{A}, \left[\frac{r}{2}\right], \beta)$ and ϕ ranges over those characters of the group $\boldsymbol{U}^{m+1}(\mathfrak{B}_\beta)$ which factor through $\det_{B_\beta}$ and agree with θ on

$U^{[\frac{r}{2}]+1}(\mathfrak{B}_\beta)$. The same comment applies to $\mathcal{C}'(\mathfrak{A}, m, \beta)$, so these sets are equal, as desired.

We now prove *(ii)*. We recall from (3.1.9) that the groups $H^{m+1}(\beta)$, $H^{m+1}(\beta')$ coincide in the relevant range. We take $\theta \in \mathcal{C}(\mathfrak{A}, m, \beta)$, and show that $\theta\psi_{\beta'-\beta} \in \mathcal{C}(\mathfrak{A}, m, \beta')$. The assertion then follows by symmetry. For the purposes of this argument, we assume that the character sets for β and β' are defined relative to the same simple $[\mathfrak{A}, n, r, \gamma]$ equivalent to $[\mathfrak{A}, n, r, \beta]$ and $[\mathfrak{A}, n, r, \beta']$. Let $m' = \max\{m, [\frac{r}{2}]\}$. Then, on $H^{m'+1}$, we have $\theta = \theta_0\psi_c$, where $\theta_0 \in \mathcal{C}(\mathfrak{A}, m', \gamma)$, $c = \beta - \gamma$, and $\theta\psi_{\beta'-\beta} = \theta_0\psi_{c'}$, where $c' = \beta' - \gamma$, as required by (3.2.3)*(c)*. We show that the character $\theta' = \theta\psi_{\beta'-\beta}$ is intertwined on H^{m+1} by all of $B^\times_{\beta'}$. It will then follow from (2.4.11) (as in the proof of (3.3.18)) that $\theta' \in \mathcal{C}(\mathfrak{A}, m, \beta')$, as required. The relation $\beta + \mathfrak{P}^{-\ell} = \beta' + \mathfrak{P}^{-\ell}$ and (1.5.8) give

$$B^\times_{\beta'} \subset (1 + \mathfrak{Q}_\beta^{r-\ell}\mathfrak{N}(\beta))B^\times_\beta(1 + \mathfrak{Q}_\beta^{r-\ell}\mathfrak{N}(\beta)).$$

We take $x \in B^\times_{\beta'}$ and compute $\theta^x(h)$ using the *second* assertion of (3.3.9), applied to β in place of γ: the conditions on ℓ and m allow us to do this. We find

$$\theta^x(h) = \theta(h)\psi_{x^{-1}\beta x - \beta}(h), \quad h \in x^{-1}H^{m+1}x \cap H^{m+1}.$$

For such h, we also have

$$(\psi_{\beta'-\beta})^x(h) = \psi_{x^{-1}(\beta'-\beta)x}(h),$$

so that

$$\theta'^x(h) = \theta'(h).\psi_{x^{-1}\beta' x - \beta'}(h) = \theta'(h),$$

since x commutes with β'. Therefore x intertwines θ' as desired. ■

(3.3.21) Corollary: *Let* $0 \leq m \leq r-1$. *Restriction induces a surjective map* $\mathcal{C}(\mathfrak{A}, m, \beta) \to \mathcal{C}(\mathfrak{A}, m+1, \beta)$. *If* $m < [\frac{r}{2}]$, *the fibres of this map are given by* (3.2.5). *Otherwise, they are in bijection with the fibres of the restriction map* $\mathcal{C}(\mathfrak{A}, m, \gamma) \to \mathcal{C}(\mathfrak{A}, m+1, \gamma)$.

Proof: If β is minimal over F, surjectivity of restriction is given by (3.2.4), (3.2.5). The general case follows from (3.3.18) by induction. ■

This corollary gives an easy inductive method of computing the fibres of these restriction maps.

(3.4) A nondegeneracy property

Again, $[\mathfrak{A}, n, 0, \beta]$ is a simple stratum in A, with $r = -k_0(\beta, \mathfrak{A})$. We use all the other notations of the preceding subsections as well. The object of this section is to prove the following nondegeneracy result which will be of importance in the representation theory of the groups J^m.

(3.4.1) Theorem: *Let $1 \le m \le r$, and let $\theta \in \mathcal{C}(\mathfrak{A}, m-1, \beta)$. The pairing*

$$\boldsymbol{k}_\theta : (1+x, 1+y) \mapsto \theta[1+x, 1+y], \qquad x, y \in \mathfrak{J}^m(\beta),$$

induces a nondegenerate alternating bilinear form

$$J^m(\beta)/H^m(\beta) \times J^m(\beta)/H^m(\beta) \to \mathbb{C}^\times.$$

Proof: The commutator group $[J^m(\beta), J^m(\beta)]$ lies in $H^m(\beta)$. Moreover, since $J^m(\beta)$ normalises θ, the subgroup $\mathrm{Ker}(\theta)$ is a normal subgroup of $J^m(\beta)$, and $H^m(\beta)/\mathrm{Ker}(\theta)$ is central in $J^m(\beta)/\mathrm{Ker}(\theta)$. It follows that $\boldsymbol{k}_\theta$ defines an alternating bilinear form on $J^m(\beta)/H^m(\beta)$. To establish the nondegeneracy, we just have to show that for $x \in \mathfrak{J}^m(\beta)$,

$$\theta[1+x, 1+y] = 1 \quad \forall y \in \mathfrak{J}^m(\beta) \quad \Leftrightarrow \quad x \in \mathfrak{H}^m(\beta).$$

We observe that the implication $\Leftarrow$ is immediate.

We deal first with the critical case $m = \left[\frac{r+1}{2}\right]$. We use (3.2.12) to get

$$\theta[1+x, 1+y] = \psi_{(1+x)^{-1}\beta(1+x)-\beta}(1+y), \qquad x, y \in \mathfrak{J}^{\left[\frac{r+1}{2}\right]}(\beta).$$

We have $a_\beta(\mathfrak{J}^{\left[\frac{r+1}{2}\right]})\mathfrak{J}^{\left[\frac{r+1}{2}\right]} \subset (\mathfrak{H}^{\left[\frac{r}{2}\right]+1})^*\mathfrak{J}^{\left[\frac{r+1}{2}\right]}$ by (3.1.17) and this is contained in $(\mathfrak{J}^{\left[\frac{r+1}{2}\right]})^*$. Therefore above we have $(1+x)^{-1}\beta(1+x) \equiv \beta + a_\beta(x) \pmod{(\mathfrak{J}^{\left[\frac{r+1}{2}\right]})^*}$. Hence $\theta[1+x, 1+y]$ is null for all y if and only if $a_\beta(x) \in (\mathfrak{J}^{\left[\frac{r+1}{2}\right]})^*$. By (3.1.22), this is the same as $x \in (B_\beta + \mathfrak{H}^{\left[\frac{r}{2}\right]+1}) \cap \mathfrak{J}^{\left[\frac{r+1}{2}\right]} = \mathfrak{H}^{\left[\frac{r+1}{2}\right]}$.

Now suppose $m < \left[\frac{r+1}{2}\right]$, so that $J^m(\beta) = \boldsymbol{U}^m(\mathfrak{B}_\beta).J^{\left[\frac{r+1}{2}\right]}(\beta)$. Since $\boldsymbol{U}^m(\mathfrak{B}_\beta)$ normalises θ, the commutator groups

$$[\boldsymbol{U}^m(\mathfrak{B}_\beta), \boldsymbol{U}^m(\mathfrak{B}_\beta)], \ [\boldsymbol{U}^m(\mathfrak{B}_\beta), J^{\left[\frac{r+1}{2}\right]}]$$

are both contained in $\mathrm{Ker}(\theta)$. Take a typical element $j = u.j'$ of J^m, $u \in \boldsymbol{U}^m(\mathfrak{B}_\beta)$, $j' \in J^{\left[\frac{r+1}{2}\right]}$. Then $\theta[j, J^m] = 1$ if and only if $\theta[j', J^m] = 1$. By the first part, this implies $j' \in H^{\left[\frac{r+1}{2}\right]}$. We deduce that $\theta[j, J^m] = 1$ implies $j \in \boldsymbol{U}^m(\mathfrak{B}_\beta)H^{\left[\frac{r+1}{2}\right]} = H^m$. We know $\theta[H^m, J^m] = 1$, so we have the result in this case.

Now suppose that $m > \left[\frac{r+1}{2}\right]$. If β is minimal over F, this means $m \geq \left[\frac{r}{2}\right] + 1$ and $J^m = H^m = \boldsymbol{U}^m(\mathfrak{A})$. The assertion is therefore trivial. Otherwise, we choose a simple stratum $[\mathfrak{A}, n, r, \gamma]$ equivalent to $[\mathfrak{A}, n, r, \beta]$. We have

$$[J^m(\beta), J^m(\beta)] \subset H^{2m}(\beta)$$

by (3.1.15), and $H^{2m}(\beta) = H^{2m}(\gamma)$. Moreover, we have $2m \geq r+1$, so $\theta \mid H^{2m}(\beta) \in \mathcal{C}(\mathfrak{A}, 2m-1, \gamma)$. The result now follows by induction along β, observing that $H^m(\gamma) = H^m(\beta)$. ■

(3.5) Intertwining and conjugacy

The main object of this section is to prove a "rigidity" or "intertwining implies conjugacy" theorem for simple characters. We take two simple strata of the form $[\mathfrak{A}, n, m, \beta_i]$, $m \geq 0$, and simple characters $\theta_i \in \mathcal{C}(\mathfrak{A}, m, \beta_i)$, $i = 1, 2$. We show that, if the θ_i intertwine in G, then there exists $x \in \boldsymbol{U}(\mathfrak{A})$ such that $H^{m+1}(\beta_2, \mathfrak{A}) = x^{-1}H^{m+1}(\beta_1, \mathfrak{A})x$ and $\theta_2 = \theta_1^x$. In the case $m \geq \left[\frac{n}{2}\right]$, we have $H^{m+1}(\beta_i, \mathfrak{A}) = \boldsymbol{U}^{m+1}(\mathfrak{A})$ and $\theta_i = \psi_{\beta_i}$, so the assertion here is equivalent to (2.6.1). We therefore only have to deal with the case $m < \left[\frac{n}{2}\right]$.

However, the situation here is considerably more complicated than for intertwining of simple strata. There are certain new phenomena which arise as a consequence of the fact that the set $\mathcal{C}(\mathfrak{A}, m, \beta)$ does not determine the equivalence class of the simple stratum $[\mathfrak{A}, n, m, \beta]$, although it does determine the hereditary order $\mathfrak{A}$.

To get an easy example of this, take two simple strata $[\mathfrak{A}, n, n-1, \alpha_i]$, with $n \geq 1$ and the α_i both minimal over F. Assume further that $[\mathfrak{A}, n, \left[\frac{n}{2}\right], \alpha_1] \sim [\mathfrak{A}, n, \left[\frac{n}{2}\right], \alpha_2]$. (3.1.9) then gives $H^1(\alpha_1, \mathfrak{A}) = H^1(\alpha_2, \mathfrak{A})$, $J^1(\alpha_1, \mathfrak{A}) = J^1(\alpha_2, \mathfrak{A})$. (1.5.8) applied to the equivalent strata $[\mathfrak{A}, n, \left[\frac{n}{2}\right], \alpha_i]$ and (2.4.11) then imply readily that $\mathcal{C}(\mathfrak{A}, 0, \alpha_1) = \mathcal{C}(\mathfrak{A}, 0, \alpha_2)$, while it is perfectly easy to arrange (using, e.g., (2.2.8)) that $[\mathfrak{A}, n, 0, \alpha_1] \not\sim [\mathfrak{A}, n, 0, \alpha_2]$.

To deal with this problem, we have first to establish a couple of subsidiary results, namely a *coherence principle* (3.5.8), which shows that the characters θ in a given set $\mathcal{C}(\mathfrak{A}, 0, \beta)$ behave uniformly in this regard, and also an *extension property* (3.5.9).

We do not pursue the matter here, but one could put an equivalence relation $\approx$ on the set of simple strata in A by setting

$$[\mathfrak{A}, n, m, \beta_1] \approx [\mathfrak{A}, n, m, \beta_2]$$

when $H^{m+1}(\beta_1, \mathfrak{A}) = H^{m+1}(\beta_2, \mathfrak{A})$ and $\mathcal{C}(\mathfrak{A}, m, \beta_1) = \mathcal{C}(\mathfrak{A}, m, \beta_2)$. (We could in fact replace the last condition by $\mathcal{C}(\mathfrak{A}, m, \beta_1) \cap \mathcal{C}(\mathfrak{A}, m, \beta_2) \neq \emptyset$, without changing anything.) The results given here could then be used to find a set of representatives for the equivalence classes of simple strata relative to $\approx$.

(3.5.1) Proposition: *For $i = 1, 2$, and some $m \geq 0$, let $[\mathfrak{A}, n, m, \beta_i]$ be simple strata such that $\mathcal{C}(\mathfrak{A}, m, \beta_1) \cap \mathcal{C}(\mathfrak{A}, m, \beta_2) \neq \emptyset$ (in particular, $\mathfrak{H}^{m+1}(\beta_1) = \mathfrak{H}^{m+1}(\beta_2)$). Then*

$$k_0(\beta_1, \mathfrak{A}) = k_0(\beta_2, \mathfrak{A}),$$
$$e(F[\beta_1]|F) = e(F[\beta_2]|F),$$
$$f(F[\beta_1]|F) = f(F[\beta_2]|F).$$

Proof: Computing the intertwining of a character $\theta \in \mathcal{C}(\mathfrak{A}, m, \beta_1) \cap \mathcal{C}(\mathfrak{A}, m, \beta_2)$, by (3.3.2), we get $\mathcal{S}(\beta_1)B_{\beta_1}^{\times}\mathcal{S}(\beta_1) = \mathcal{S}(\beta_2)B_{\beta_2}^{\times}\mathcal{S}(\beta_2)$, where

$$\mathcal{S}(\beta_i) = 1 + \mathfrak{Q}_{\beta_i}^{r_i - m}\mathfrak{N}(\beta_i) + \mathfrak{J}^{[\frac{r_i+1}{2}]}(\beta_i), \quad r_i = -k_0(\beta_i, \mathfrak{A})$$

and the other notations are as usual (see the beginning of **(3.1)**). Let $\mathcal{G}_i$ denote the additive group generated by $\mathcal{S}(\beta_i)B_{\beta_i}^{\times}\mathcal{S}(\beta_i) \cap \mathfrak{A}_i$. We assert that

$$\mathcal{G}_i = \mathfrak{B}_{\beta_i} + \mathfrak{Q}_{\beta_i}^{r_i - m}\mathfrak{N}(\beta_i) + \mathfrak{J}^{[\frac{r_i+1}{2}]}(\beta_i).$$

We surely have containment in the direction $\subset$, and equality if we replace the term $\mathfrak{B}_{\beta_i}$ by the additive closure of $B_{\beta_i}^{\times} \cap \mathfrak{B}_{\beta_i}$. However, a standard density argument with matrices shows that this additive closure is indeed $\mathfrak{B}_{\beta_i}$, as required. We conclude

$$\mathfrak{B}_{\beta_1} + \mathfrak{Q}_{\beta_1}^{r_1 - m}\mathfrak{N}(\beta_1) + \mathfrak{J}^{[\frac{r_1+1}{2}]}(\beta_1) = \mathfrak{B}_{\beta_2} + \mathfrak{Q}_{\beta_2}^{r_2 - m}\mathfrak{N}(\beta_2) + \mathfrak{J}^{[\frac{r_2+1}{2}]}(\beta_2).$$

Adding $\mathfrak{P}$ to each side, we get $\mathfrak{B}_{\beta_1} + \mathfrak{P} = \mathfrak{B}_{\beta_2} + \mathfrak{P}$ and we have an isomorphism of algebras

$$\mathfrak{B}_{\beta_1}/\mathfrak{Q}_{\beta_1} \cong \mathfrak{B}_{\beta_2}/\mathfrak{Q}_{\beta_2}.$$

We can recover the quantities $e(F[\beta_i]|F)$, $f(F[\beta_i]|F)$ from the Wedderburn structure constants of $\mathfrak{B}_{\beta_i}/\mathfrak{Q}_{\beta_i}$ and those of $\mathfrak{A}/\mathfrak{P}$ (as in the proof of (2.1.4)), so we get the second and third assertions.

To get the other, suppose that, say, $r_1 < r_2$. Consider the restriction of θ to $H^{r_1+1} = H^{m+1} \cap U^{r_1+1}(\mathfrak{A})$. Going through the same procedure, we would get an algebra isomorphism between $\mathfrak{B}_{\beta_2}/\mathfrak{Q}_{\beta_2}$ and $\mathfrak{B}_{\gamma}/\mathfrak{Q}_{\gamma}$, where $[\mathfrak{A}, n, r_1, \gamma]$ is a simple stratum equivalent to $[\mathfrak{A}, n, r_1, \beta_1]$. This would imply $[F[\gamma] : F] = [F[\beta_2] : F]$. However, by (2.4.1) and (2.2.8), the field degree $[F[\gamma] : F]$ is strictly less than $[F[\beta_1] : F] = [F[\beta_2] : F]$, which gives a contradiction. Therefore, $r_1 = r_2$. ■

We now have a sequence of "extrapolation" results, concerned with deducing properties of simple characters $\theta \in \mathcal{C}(\mathfrak{A}, m, \beta)$ from information about their restrictions in $\mathcal{C}(\mathfrak{A}, m+1, \beta)$. These will form the basis of subsequent inductive arguments.

(3.5.2) Proposition: *Let $[\mathfrak{A}, n, m, \beta]$, $m \geq 0$, be a simple stratum, let $b \in \mathfrak{P}^{-(1+m)}$ and view ψ_b as a character of $H^{m+1}(\beta)$ which is null on $H^{m+2}(\beta)$. Suppose there exists $\theta \in \mathcal{C}(\mathfrak{A}, m, \beta)$ such that $\theta\psi_b \in \mathcal{C}(\mathfrak{A}, m, \beta)$. Then $\theta\psi_b \in \mathcal{C}(\mathfrak{A}, m, \beta)$ for all $\theta \in \mathcal{C}(\mathfrak{A}, m, \beta)$.*

Proof: We proceed by induction along β. If β is minimal over F and $m \geq \left[\frac{n}{2}\right]$, then the set $\mathcal{C}(\mathfrak{A}, m, \beta)$ has only the one element θ, and there is nothing to prove. If $m < \left[\frac{n}{2}\right]$, then $\psi_b \mid H^{m+1}$, which is effectively a character of $U^{m+1}(\mathfrak{B}_\beta)/U^{m+2}(\mathfrak{B}_\beta)$, is intertwined by the whole of $B_\beta^\times$, since this is so for both θ and $\theta\psi_b$. It therefore factors through the determinant $\det_{B_\beta}$ by (2.4.11). The definition (3.2.1) then shows that $\theta'\psi_b \in \mathcal{C}(\mathfrak{A}, m, \beta)$ for all $\theta' \in \mathcal{C}(\mathfrak{A}, m, \beta)$.

So, suppose that β is not minimal over F, and take a simple stratum $[\mathfrak{A}, n, r, \gamma]$ equivalent to $[\mathfrak{A}, n, r, \beta]$, where $r = -k_0(\beta, \mathfrak{A})$. Assume first $m \geq \left[\frac{r}{2}\right]$. Put $\beta = \gamma + c$, and $\theta = \theta_0\psi_c$, for some $\theta_0 \in \mathcal{C}(\mathfrak{A}, m, \gamma)$. Then $\theta_0\psi_b \in \mathcal{C}(\mathfrak{A}, m, \gamma)$ by (3.3.18), and the result follows by induction. The case $m < \left[\frac{r}{2}\right]$ follows exactly as in the minimal case. ■

Now we need some new terminology. We take our simple stratum $[\mathfrak{A}, n, m, \beta]$, $m \geq 0$, and define a finite increasing sequence of integers (and maybe ∞) $\{r_j\} = \{r_j(\beta)\}$ which we call the *jumps* of β (relative to $[\mathfrak{A}, n, m, \beta]$). Define

(3.5.3) *(i) $r_0(\beta) = m$;*

(ii) $r_1(\beta) = r = -k_0(\beta, \mathfrak{A})$;

(iii) if $r < n$ and $j \geq 2$, put $r_j(\beta) = r_{j-1}(\gamma)$, where $[\mathfrak{A}, n, r, \gamma]$ is simple and equivalent to $[\mathfrak{A}, n, r, \beta]$.

In effect, for $j \geq 1$, $r_j(\beta) = -k_0(\gamma_{j-1}, \mathfrak{A})$, where $\{\gamma_k\}$ is a defining sequence for $[\mathfrak{A}, n, m, \beta]$, in the sense of (2.4.2).

(3.5.4) Lemma: *For $i = 1, 2$, let $[\mathfrak{A}_i, n, m, \beta_i]$ be a simple stratum. Suppose that $H^{m+1}(\beta_1, \mathfrak{A}) = H^{m+1}(\beta_2, \mathfrak{A})$ and $\mathcal{C}(\mathfrak{A}, m, \beta_1) \cap \mathcal{C}(\mathfrak{A}, m, \beta_2) \neq \emptyset$. Then β_1, β_2 have the same jump sequences.*

Proof: We recall (remark following (3.2.3)) that the symbol $\mathcal{C}(\mathfrak{A}, \ell, \beta_i)$ conventionally means $\mathcal{C}(\mathfrak{A}, \ell, \gamma)$, for any simple stratum $[\mathfrak{A}, n, \ell, \gamma]$ equivalent to $[\mathfrak{A}, n, \ell, \beta_i]$. (3.3.21) and (3.3.18) together show that the restriction maps

$$\mathcal{C}(\mathfrak{A}, m, \beta_i) \to \mathcal{C}(\mathfrak{A}, \ell, \beta_i)$$

are surjective, for $m \leq \ell \leq n$. In particular, $\mathcal{C}(\mathfrak{A}, \ell, \beta_1) \cap \mathcal{C}(\mathfrak{A}, \ell, \beta_2) \neq \emptyset$ for $\ell \geq m$. The lemma now follows from (3.5.1) by induction. ■

Given $[\mathfrak{A}, n, m, \beta]$ as above, with $m < \left[\frac{n}{2}\right]$, a *core approximation to* $[\mathfrak{A}, n, m, \beta]$ is a simple stratum $[\mathfrak{A}, n, r_j(\beta), \delta]$ equivalent to $[\mathfrak{A}, n, r_j(\beta), \beta]$ where $r_{j+1}(\beta)$ is the least jump of β such that $j \geq 0$ and $m < \left[\frac{r_{j+1}}{2}\right]$. Then, for such a δ, we have $r_{j+1}(\beta) = -k_0(\delta, \mathfrak{A})$. Also, $m \geq \left[\frac{r_j}{2}\right]$, and the definition of $\mathfrak{H}$ shows that

$$\mathfrak{H}^{m+1}(\beta) = \mathfrak{H}^{m+1}(\delta) = \mathfrak{H}^{m+2}(\delta) + \mathfrak{Q}_\delta^{m+1}.$$

Moreover, δ is the "simplest" element (i.e. the one with the fewest jumps) which has this property. Also, (3.3.18) applied repeatedly gives

$$\textbf{(3.5.5)} \qquad \mathcal{C}(\mathfrak{A}, m, \beta) = \mathcal{C}(\mathfrak{A}, m, \delta)\psi_{\beta-\delta}.$$

Observe that if $m < \left[\frac{r}{2}\right]$, then $[\mathfrak{A}, n, m, \beta]$ is a core approximation to itself.

(3.5.6) Proposition: *In the situation of* (3.5.2), *let* $[\mathfrak{A}, n, r_j, \delta]$ *be a core approximation to* $[\mathfrak{A}, n, m, \beta]$. *Then* $\mathcal{C}(\mathfrak{A}, m, \beta)\psi_b$ *meets* $\mathcal{C}(\mathfrak{A}, m, \beta)$ *if and only if* $s_\delta(b) \in F[\delta] + \mathfrak{Q}_\delta^{-m}$, *for a tame corestriction* s_δ *on* A *relative to* $F[\delta]/F$.

Proof: The sets $\mathcal{C}(\mathfrak{A}, m, \beta)$, $\mathcal{C}(\mathfrak{A}, m, \beta)\psi_b$ meet nontrivially if and only if $\mathcal{C}(\mathfrak{A}, m, \delta) \cap \mathcal{C}(\mathfrak{A}, m, \delta)\psi_b \neq \emptyset$ (by (3.5.5)). From (3.2.5), this holds if and only if $\psi_b \mid U^{m+1}(\mathfrak{B}_\delta)$ factors through $\det_{B_\delta}$. This holds if and only if $s_\delta(b) \in F[\delta] + \mathfrak{Q}_\delta^{-m}$, as required (*cf.* the remarks at the beginning of **(3.2)**). ■

(3.5.7) Proposition: *Let* $[\mathfrak{A}, n, m, \beta]$, $0 \leq m < \left[\frac{n}{2}\right]$, *be a simple stratum, and let* $[\mathfrak{A}, n, r_j, \delta]$ *be a core approximation to it. Let* θ_1, $\theta_2 \in \mathcal{C}(\mathfrak{A}, m, \beta)$. *Then the character* θ_1/θ_2 *of* $H^{m+1}(\beta) = H^{m+1}(\delta)$ *is intertwined by the set* $\mathcal{S}(\delta)B_\delta^\times \mathcal{S}(\delta)$, *where*

$$\mathcal{S}(\delta) = 1 + \mathfrak{Q}_\delta^{r_{j+1}-r_j}\mathfrak{N}(\delta) + \mathfrak{J}^{\left[\frac{r_{j+1}+1}{2}\right]}(\delta).$$

Proof: If $r_{j+1} = r_1$, there is nothing to prove because we can take $\delta = \beta$, so we assume $j \geq 1$. Let $[\mathfrak{A}, n, r_{j-1}, \epsilon]$ be a simple stratum equivalent to $[\mathfrak{A}, n, r_{j-1}, \beta]$. Then $\mathcal{C}(\mathfrak{A}, m, \epsilon) = \mathcal{C}(\mathfrak{A}, m, \delta)\psi_{\epsilon-\delta}$. We therefore have $\theta_1/\theta_2 = \eta_1/\eta_2$ for characters $\eta_i \in \mathcal{C}(\mathfrak{A}, m, \epsilon)$. For $x \in \mathcal{S}(\delta)B_\delta^\times \mathcal{S}(\delta)$, we compute the character η_i^x of $x^{-1}H^{m+1}x \cap H^{m+1}$ from (3.3.9), with $(\delta, \epsilon, r_j, r_{j+1})$ in place of (γ, β, r, s). In this situation, we get

$$\eta_i^x(h) = \eta_i(h)\psi_{x^{-1}\epsilon x-\epsilon}(h), \quad h \in x^{-1}H^{m+1}x \cap H^{m+1}.$$

The assertion follows immediately. ■

(3.5.8) Theorem: *For $i = 1, 2$, let $[\mathfrak{A}, n, m, \beta_i]$ be a simple stratum. Suppose that $\mathcal{C}(\mathfrak{A}, m, \beta_1) \cap \mathcal{C}(\mathfrak{A}, m, \beta_2) \neq \emptyset$ (in particular, $H^{m+1}(\beta_1, \mathfrak{A}) = H^{m+1}(\beta_2, \mathfrak{A})$). Then*

$$\mathcal{C}(\mathfrak{A}, m, \beta_1) = \mathcal{C}(\mathfrak{A}, m, \beta_2).$$

Proof: By (3.5.4), the β_i have the same jump sequences, so we may set $r = -k_0(\beta_1, \mathfrak{A}) = -k_0(\beta_2, \mathfrak{A})$. We work by induction along the β's as usual. Take first the case where the β_i are minimal over F. If $m \geq \left[\frac{n}{2}\right]$, the assertion is trivial, since the sets $\mathcal{C}(\mathfrak{A}, m, \beta_i)$ are singletons. We therefore assume that $m < \left[\frac{n}{2}\right]$. The hypothesis, together with surjectivity of restriction, implies that the $\mathcal{C}(\mathfrak{A}, m+1, \beta_i)$ meet nontrivially, so we assume inductively that $\mathcal{C}(\mathfrak{A}, m+1, \beta_1) = \mathcal{C}(\mathfrak{A}, m+1, \beta_2)$. Computing (via (3.3.2)) the intertwining of any character $\theta \in \mathcal{C}(\mathfrak{A}, m, \beta_1) \cap \mathcal{C}(\mathfrak{A}, m, \beta_2)$, we get

$$(1+\mathfrak{J}^{[\frac{n+1}{2}]}(\beta_1))B_{\beta_1}^{\times}(1+\mathfrak{J}^{[\frac{n+1}{2}]}(\beta_1)) = (1+\mathfrak{J}^{[\frac{n+1}{2}]}(\beta_2))B_{\beta_2}^{\times}(1+\mathfrak{J}^{[\frac{n+1}{2}]}(\beta_2)).$$

However, by (3.3.2), the right hand side here is the intertwining of any character $\theta' \in \mathcal{C}(\mathfrak{A}, m, \beta_2)$. Thus any $\theta' \in \mathcal{C}(\mathfrak{A}, m, \beta_2)$ is intertwined by the whole of $B_{\beta_1}^{\times}$. Any such character θ' therefore factors through the determinant $\det_{B_{\beta_1}}$ on $U^{m+1}(\mathfrak{B}_{\beta_1})$ by (2.4.11), and agrees with $\psi_{\beta_2} = \psi_{\beta_1}$ on $H^{[\frac{n}{2}]+1}$. Therefore $\theta' \in \mathcal{C}(\mathfrak{A}, m, \beta_1)$, and the result follows by symmetry.

We assume therefore that $r < n$. As in the first case, we can also assume inductively that $\mathcal{C}(\mathfrak{A}, m+1, \beta_1) = \mathcal{C}(\mathfrak{A}, m+1, \beta_2)$. There is again nothing to prove if $m \geq \left[\frac{n}{2}\right]$, so we assume the contrary. Let $\theta \in \mathcal{C}(\mathfrak{A}, m, \beta_1) \cap \mathcal{C}(\mathfrak{A}, m, \beta_2)$. For $\phi' \in \mathcal{C}(\mathfrak{A}, m, \beta_2)$, there exists $\phi \in \mathcal{C}(\mathfrak{A}, m, \beta_1)$, and $b(\phi) \in \mathfrak{P}^{-(m+1)}$ such that $\phi' = \phi\psi_{b(\phi)}$. We investigate these elements $b(\phi)$. We apply (3.5.7) to the quotients ϕ/θ, $\phi\psi_{b(\phi)}/\theta$. The β_i have the same jump sequences $\{r_j\}$ by (3.5.4), and we take a core approximation $[\mathfrak{A}, n, r_j, \delta_i]$ to $[\mathfrak{A}, n, m, \beta_i]$, for $i = 1, 2$.

The restriction of θ to $H^{r_j+1}(\beta_i) = H^{r_j+1}(\delta_i)$ lies in $\mathcal{C}(\mathfrak{A}, r_j, \delta_i)$, so the intertwining of θ on $H^{r_j+1}(\delta_i)$ is $\mathcal{S}(\delta_i)B_{\delta_i}^{\times}\mathcal{S}(\delta_i)$, where

$$\mathcal{S}(\delta_i) = 1 + \mathfrak{Q}_{\delta_i}^{r_{j+1}-r_j}\mathfrak{N}(\delta_i) + \mathfrak{J}^{\left[\frac{r_{j+1}+1}{2}\right]}(\delta_i),$$

and $r_{j+1} = r_{j+1}(\beta_i) = -k_0(\delta_i, \mathfrak{A})$. We conclude that

$$\mathcal{S}(\delta_1)B_{\delta_1}^{\times}\mathcal{S}(\delta_1) = \mathcal{S}(\delta_2)B_{\delta_2}^{\times}\mathcal{S}(\delta_2).$$

This set therefore intertwines the quotients ϕ/θ, $\phi\psi_{b(\phi)}/\theta$ by (3.5.7). In particular, the character $\psi_{b(\phi)}$ is intertwined by the whole of $B_{\delta_1}^{\times}$ on

$H^{m+1} \cap B_{\delta_1} = U^{m+1}(\mathfrak{B}_{\delta_1})$. Thus it factors through the determinant $\det_{B_{\delta_1}}$ (by (2.4.11)) and we have $\phi' = \phi\psi_{b(\phi)} \in \mathcal{C}(\mathfrak{A}, m, \beta_1)$ by (3.5.6). However, ϕ' was an arbitrary element of $\mathcal{C}(\mathfrak{A}, m, \beta_2)$, so $\mathcal{C}(\mathfrak{A}, m, \beta_1) \supset \mathcal{C}(\mathfrak{A}, m, \beta_2)$, and the assertion follows by symmetry. ■

(3.5.9) Theorem: *Let $[\mathfrak{A}, n, m, \beta_i]$ be simple strata, for $i = 1, 2$, with $m \geq 1$. Suppose that $\mathcal{C}(\mathfrak{A}, m, \beta_1) = \mathcal{C}(\mathfrak{A}, m, \beta_2)$. Then $H^m(\beta_1) = H^m(\beta_2)$ and there exists a simple stratum $[\mathfrak{A}, n, m, \beta_1']$ equivalent to $[\mathfrak{A}, n, m, \beta_1]$ such that $\mathcal{C}(\mathfrak{A}, m-1, \beta_1') = \mathcal{C}(\mathfrak{A}, m-1, \beta_2)$.*

Proof: We start by proving that $H^m(\beta_1) = H^m(\beta_2)$, first reducing to the case $m \geq \left[\frac{r}{2}\right] + 1$. So, suppose that the result holds here, and let $m \leq \left[\frac{r}{2}\right]$. Comparing intertwining in the usual way, we get $\mathfrak{J}(\beta_1) = \mathfrak{J}(\beta_2)$, whence $\mathfrak{J}^m(\beta_1) = \mathfrak{J}^m(\beta_2)$. The commutator group $[J^m(\beta_i), J^m(\beta_i)]$ is contained in $H^{m+1}(\beta_1) = H^{m+1}(\beta_2)$. By (3.4.1), $H^m(\beta_i)$ is the inverse image of the centre of the group $J^m(\beta_i)/\mathcal{K}_i$, where $\mathcal{K}_i$ is the kernel of $\theta_i \mid H^{m+1}(\beta_i)$, for any $\theta_i \in \mathcal{C}(\mathfrak{A}, m, \beta_i)$. It is therefore independent of i and $H^m(\beta_1) = H^m(\beta_2)$, as required.

We therefore assume that $m \geq \left[\frac{r}{2}\right] + 1$. If β is minimal over F, there is nothing to prove, so we assume that $r < n$. There is again nothing to prove if $m \geq \left[\frac{n}{2}\right] + 1$, since then $H^m(\beta_i) = U^m(\mathfrak{A})$, $\mathcal{C}(\mathfrak{A}, m-1, \beta_i) = \{\psi_{\beta_i}\}$, and the strata $[\mathfrak{A}, n, m, \beta_i]$ are equivalent.

So, we have reduced to the case $m \leq \left[\frac{n}{2}\right]$, $r < n$. We take a core approximation $[\mathfrak{A}, n, r_j, \delta_i]$ to $[\mathfrak{A}, n, m-1, \beta_i]$. Here, $r_j = r_j(\beta_1) = r_j(\beta_2)$ (*cf.* (3.5.4)). Then $r_j > m$, and $\mathcal{C}(\mathfrak{A}, r_j, \delta_i)$ consists of the restrictions to $H^{r_j+1}(\delta_i) = H^{r_j+1}(\beta_i)$ of the characters in $\mathcal{C}(\mathfrak{A}, m, \beta_i)$. In particular, we have $\mathcal{C}(\mathfrak{A}, r_j, \delta_1) = \mathcal{C}(\mathfrak{A}, r_j, \delta_2)$. Consider the intertwining of some character $\phi \in \mathcal{C}(\mathfrak{A}, r_j, \delta_1) = \mathcal{C}(\mathfrak{A}, r_j, \delta_2)$. Intersecting this intertwining set with $\mathfrak{P}^m$ and taking additive closures, as in the proof of (3.5.1), we find that the set

$$\mathfrak{Q}_{\delta_i}^m + \mathfrak{Q}_{\delta_i}^{m+r_{j+1}-r_j}\mathfrak{N}(\delta_i) + \mathfrak{Q}_{\delta_i}^m \mathfrak{J}^{\left[\frac{r_{j+1}+1}{2}\right]}(\delta_i)$$

is independent of i, where $r_{j+1} = r_{j+1}(\beta_i) = -k_0(\delta_i, \mathfrak{A})$. It is, moreover, contained in $\mathfrak{H}^m(\delta_i) = \mathfrak{H}^m(\beta_i)$. Adding $\mathfrak{H}^{m+1}(\delta_1) = \mathfrak{H}^{m+1}(\delta_2)$, we get

$$\mathfrak{H}^m(\beta_1) = \mathfrak{H}^m(\delta_1) = \mathfrak{Q}_{\delta_1}^m + \mathfrak{H}^{m+1}(\delta_1) = \mathfrak{H}^m(\beta_2),$$

as required.

Take $\theta \in \mathcal{C}(\mathfrak{A}, m-1, \beta_1)$ and an element $b \in \mathfrak{P}^{-m}$ such that $\theta\psi_b \in \mathcal{C}(\mathfrak{A}, m-1, \beta_2)$. If $r = -k_0(\beta_i, \mathfrak{A})$, we have $\mathcal{S}(\beta_1)B_{\beta_1}^\times\mathcal{S}(\beta_1) = \mathcal{S}(\beta_2)B_{\beta_2}^\times\mathcal{S}(\beta_2)$ where $\mathcal{S}(\beta_i) = 1 + \mathfrak{Q}_{\beta_i}^{r-m}\mathfrak{N}(\beta_i) + \mathfrak{J}^{\left[\frac{r+1}{2}\right]}(\beta_i)$, by computing the intertwining of any character in $\mathcal{C}(\mathfrak{A}, m, \beta_1) = \mathcal{C}(\mathfrak{A}, m, \beta_2)$. Any $x \in \mathcal{S}(\beta_1)$ normalises H^m and fixes ψ_b. Moreover, for such x,

θ^x agrees with θ on $U^m(\mathfrak{B}_{\beta_1})$ (see (3.5.10) below). Take any element $t \in B_{\beta_1}^\times$. There exist $x, y \in \mathcal{S}(\beta_1)$ such that $xty \in B_{\beta_2}^\times$. Thus xty intertwines the character $\theta\psi_b \in \mathcal{C}(\mathfrak{A}, m-1, \beta_2)$. This says that t intertwines $(\theta\psi_b)^x = \theta^x\psi_b$ with $(\theta\psi_b)^{y^{-1}} = \theta^{y^{-1}}\psi_b$. By (3.5.10) again, the characters θ, θ^x, $\theta^{y^{-1}}$ all agree on $U^m(\mathfrak{B}_{\beta_1})$, so it follows that $\psi_b \mid U^m(\mathfrak{B}_{\beta_1})$ is intertwined by every element $t \in B_{\beta_1}^\times$. Therefore, by (2.4.11), if s denotes a tame corestriction on A relative to $F[\beta_1]/F$, then $s(b) \in F[\beta_1] + \mathfrak{Q}_{\beta_1}^{1-m}$. By (2.2.8), there is a simple $[\mathfrak{A}, n, m-1, \beta_1']$ equivalent to $[\mathfrak{A}, n, m-1, \beta_1 + b]$. We have $\theta\psi_b \in \mathcal{C}(\mathfrak{A}, m-1, \beta_1')$ by (3.3.20), and the result follows from (3.5.7).

It remains only to prove:

(3.5.10) Lemma: *Let $[\mathfrak{A}, n, m, \beta]$ be a simple stratum, with $m \geq 1$. Let $\theta \in \mathcal{C}(\mathfrak{A}, m-1, \beta)$, and let x normalise the character $\theta \mid H^{m+1}(\beta, \mathfrak{A})$. Then*

$$\theta^x \mid U^m(\mathfrak{B}_\beta) = \theta \mid U^m(\mathfrak{B}_\beta).$$

Proof: Set $r = -k_0(\beta, \mathfrak{A})$. If $m \leq \left[\frac{r}{2}\right]$, we have $\theta^x = \theta$ by (3.3.2) (see also the Remark following the statement of that theorem). Therefore we assume $m > \left[\frac{r}{2}\right]$. By (3.3.1), (3.2.3) and (3.3.17), we may as well take $x \in 1 + \mathfrak{Q}_\beta^{r-m}\mathfrak{N}(\beta)$. We use the second assertion of (3.3.9) (with β playing the role of γ) to get

$$\theta^x(h) = \theta(h)\psi_{x^{-1}\beta x - \beta}(h), \quad h \in H^m(\beta).$$

Writing $x = 1 + y$, $y \in \mathfrak{Q}_\beta^{r-m}\mathfrak{N}(\beta)$, we find readily that $\psi_{x^{-1}\beta x-\beta} \mid U^m(\mathfrak{A}) = \psi_{a_\beta(y)} \mid U^m(\mathfrak{A})$. If $h = 1 + z$, $z \in \mathfrak{Q}_\beta^m$, we have

$$\psi_{x^{-1}\beta x-\beta}(h) = \psi(a_\beta(y)z) = \psi(a_\beta(yz)) = 1,$$

and the lemma follows. ■

This completes the proof of (3.5.9). ■

We now come to the main result of the section.

(3.5.11) Theorem: *For $i = 1, 2$, let $[\mathfrak{A}, n, m, \beta_i]$ be simple strata with $m \geq 0$. Suppose there exist $\theta_i \in \mathcal{C}(\mathfrak{A}, m, \beta_i)$ which intertwine in G. Then there exists $x \in U(\mathfrak{A})$ such that*

$$\mathcal{C}(\mathfrak{A}, m, \beta_2) = \mathcal{C}(\mathfrak{A}, m, x^{-1}\beta_1 x)$$

and conjugation by x carries θ_1 to θ_2, i.e. $\theta_2 = \theta_1^x$.

Proof: We start with a lemma.

(3.5.12) Lemma: *Let $[\mathfrak{A}, n, r-1, \beta]$ be a simple stratum in A, with $-k_0(\beta, \mathfrak{A}) = r$, $n > r \geq 1$. Let $[\mathfrak{A}, n, r, \gamma]$ be simple and equivalent* *to* $[\mathfrak{A}, n, r, \beta]$. *Let* $\theta \in \mathcal{C}(\mathfrak{A}, r-1, \beta)$, $\varphi \in \mathcal{C}(\mathfrak{A}, r-1, \gamma)$, *and suppose that* $\theta \mid H^{r+1} = \varphi \mid H^{r+1}$, *where* $H^{r+1} = H^{r+1}(\beta) = H^{r+1}(\gamma)$. *Then* $I_G(\theta, \varphi) = \emptyset$.

Proof: Write $\beta = \gamma + c$, so that $\theta = \varphi'\psi_c$, for some $\varphi' \in \mathcal{C}(\mathfrak{A}, r-1, \gamma)$. We also have $\varphi' = \varphi\psi_b$, for some $b \in \mathfrak{P}^{-r}$, $\mathfrak{P} = \text{rad}(\mathfrak{A})$. The group $B_\gamma^\times = \text{Aut}_{F[\gamma]}(V)$ intertwines each of the characters φ, φ', and it follows that $\psi_b \mid H^r$ is intertwined by all of $B_\gamma^\times$. The same therefore applies to the restriction $\psi_b \mid U^r(\mathfrak{B}_\gamma)$, in the obvious notation. However, if s is some tame corestriction on A relative to $F[\gamma]/F$, this restriction has the form $\psi_{B_\gamma, s(b)}$. (2.4.11) now implies that $(s(b) + \mathfrak{Q}_\gamma^{1-r}) \cap F[\gamma]$ is nonempty. In other words, we may assume $s(b) \in F[\gamma]$. On the other hand, the stratum $[\mathfrak{B}_\gamma, r, r-1, s(c)]$ is equivalent to a simple stratum $[\mathfrak{B}_\gamma, r, r-1, \delta]$, by (2.4.1). We have $F[\delta, \gamma] \neq F[\gamma]$ since otherwise $[\mathfrak{A}, n, r, \beta]$ would be simple. Thus the element $s(b) + \delta$ generates the field $F[\gamma, \delta]$ over $F[\gamma]$, and this field normalises $\mathfrak{A}$. In other words, the stratum $[\mathfrak{B}_\gamma, r, r-1, s(b+c)]$ in B_γ is equivalent to a pure, and hence to a simple, stratum.

It is worth exhibiting the result of this argument for future reference:

(3.5.13) Lemma: *Let* $[\mathfrak{A}, n, r-1, \beta]$ *be a simple stratum in* A *with* $r = -k_0(\beta, \mathfrak{A})$, $n > r \geq 1$. *Let* $[\mathfrak{A}, n, r, \gamma]$ *be a simple stratum equivalent to* $[\mathfrak{A}, n, r, \beta]$. *Let* $\theta \in \mathcal{C}(\mathfrak{A}, r-1, \beta)$, $\varphi \in \mathcal{C}(\mathfrak{A}, r-1, \gamma)$, *and suppose* $\theta \mid H^{r+1} = \varphi \mid H^{r+1}$, *where* $H^{r+1} = H^{r+1}(\beta, \mathfrak{A}) = H^{r+1}(\gamma, \mathfrak{A})$. *Choose* $b \in \mathfrak{P}^{-r}$ *such that* $\theta = \varphi\psi_b$. *Write* $B_\gamma = \text{End}_{F[\gamma]}(V)$, $\mathfrak{B}_\gamma = \mathfrak{A} \cap B_\gamma$, *and let* s *be a tame corestriction on* A *relative to* $F[\gamma]/F$. *Then* $[\mathfrak{B}_\gamma, r, r-1, s(b)]$ *is equivalent to a simple stratum.*

Returning to the proof of (3.5.12), suppose that $I_G(\theta, \varphi) \neq \emptyset$, and let $g \in I_G(\theta, \varphi)$. We have $g = xty$, with $t \in B_\gamma^\times$ and $x, y \in 1 + \mathfrak{Q}_\gamma^{s-r}\mathfrak{N}(\gamma) + \mathfrak{J}^{[\frac{s+1}{2}]}(\gamma)$, where we write $s = -k_0(\gamma, \mathfrak{A})$. Thus the element t intertwines θ^x with $\varphi^{y^{-1}}$. We have $\theta^x \mid H^{r+1}(\gamma) = \theta \mid H^{r+1}(\gamma)$, and likewise for $\varphi^{y^{-1}}$. Thus the characters θ^x, $\varphi^{y^{-1}}$ agree on $H^{m+1}(\gamma)$. By (3.5.10), $\varphi^{y^{-1}} \mid U^r(\mathfrak{B}_\gamma) = \varphi \mid U^r(\mathfrak{B}_\gamma)$. The characters ψ_{b+c}^x, ψ_{b+c} agree on $U^r(\mathfrak{A})$, so, likewise,

$$\begin{aligned}\theta^x \mid U^r(\mathfrak{B}_\gamma) &= (\varphi \mid U^r(\mathfrak{B}_\gamma))(\psi_{b+c} \mid U^r(\mathfrak{B}_\gamma)) \\ &= (\varphi \mid U^r(\mathfrak{B}_\gamma))\psi_{B_\gamma, s(b+c)}.\end{aligned}$$

However, the character $\varphi \mid U^r(\mathfrak{B}_\gamma)$ factors through $\det_{B_\gamma}$ and hence is intertwined by t. It follows that t intertwines $\psi_{B_\gamma, s(b+c)}$ with the trivial character of $U^r(\mathfrak{B}_\gamma)$ or, equivalently, it intertwines the (equivalent to) simple stratum $[\mathfrak{B}_\gamma, r, r-1, s(b+c)]$ with the null stratum $[\mathfrak{B}_\gamma, r, r-1, 0]$. This is impossible, by (2.6.2). ■

Now we prove the theorem. If $m \geq [\frac{n}{2}]$, the result follows from (2.6.1). We therefore assume that $m < [\frac{n}{2}]$. Inductively, we can assume that $\mathcal{C}(\mathfrak{A}, m+1, \beta_1) = \mathcal{C}(\mathfrak{A}, m+1, \beta_2)$, and $\theta_1 \mid H^{m+2} = \theta_2 \mid H^{m+2}$. By (3.5.9), we have $H^{m+1}(\beta_1) = H^{m+1}(\beta_2)$. We show first that $[\mathfrak{A}, n, m+1, \beta_1]$ is simple if and only if $[\mathfrak{A}, n, m+1, \beta_2]$ is simple. For, suppose that $[\mathfrak{A}, n, m+1, \beta_1]$ is simple, while $[\mathfrak{A}, n, m+1, \beta_2]$ is not. Choose $[\mathfrak{A}, n, m+1, \gamma_2]$ simple and equivalent to $[\mathfrak{A}, n, m+1, \beta_2]$. In particular, we get $\mathcal{C}(\mathfrak{A}, m+1, \beta_2) = \mathcal{C}(\mathfrak{A}, m+1, \gamma_2)$. By (3.5.9), we may further choose γ_2 so that $\mathcal{C}(\mathfrak{A}, m, \gamma_2) = \mathcal{C}(\mathfrak{A}, m, \beta_1)$. (3.5.12) then gives the contradiction.

So, let us first deal with the case in which the $[\mathfrak{A}, n, m+1, \beta_i]$ are both simple. We use (3.5.9) again:— there exists β_1' such that $[\mathfrak{A}, n, m+1, \beta_1']$ is simple and equivalent to $[\mathfrak{A}, n, m+1, \beta_2]$, while $\mathcal{C}(\mathfrak{A}, m, \beta_1') = \mathcal{C}(\mathfrak{A}, m, \beta_1)$. In other words, we may assume that $[\mathfrak{A}, n, m+1, \beta_1] \sim [\mathfrak{A}, n, m+1, \beta_2]$ and, in particular, $k_0(\beta_1, \mathfrak{A}) = k_0(\beta_2, \mathfrak{A})$. Write $\beta_2 = \beta_1 + b$, $b \in \mathfrak{P}^{-(m+1)}$, so that (by (3.3.20)) $\mathcal{C}(\mathfrak{A}, m, \beta_2) = \mathcal{C}(\mathfrak{A}, m, \beta_1)\psi_b$. Thus

$$\theta_2 = \theta_1 \psi_{b+c}, \text{ where } c \in \mathfrak{P}^{-(1+m)}, \ \theta_1\psi_c \in \mathcal{C}(\mathfrak{A}, m, \beta_1).$$

Suppose that $I_G(\theta_1, \theta_2) \neq \emptyset$, and let $g \in I_G(\theta_1, \theta_2)$. Since the θ_i agree on H^{m+2}, (3.3.2) implies $g = xty$, with $t \in B_{\beta_1}^{\times}$ and $x, y \in 1 + \mathfrak{Q}^{r-m-1}\mathfrak{N}(\beta_1) + \mathfrak{J}^{[\frac{r+1}{2}]}(\beta_1)$, where $r = -k_0(\beta_i, \mathfrak{A})$ and $\mathfrak{Q} = \mathfrak{Q}_{\beta_1}$, in our usual notation. Let s be a tame corestriction on A relative to $F[\beta_1]/F$. Restricting to $\boldsymbol{U}^{m+1}(\mathfrak{B}_{\beta_1})$, we find as above (in the proof of (3.5.12)) that the element t intertwines the stratum $[\mathfrak{B}_{\beta_1}, m+1, m, s(b+c)]$ with $[\mathfrak{B}_{\beta_1}, m+1, m, 0]$.

However, (2.4.1)*(i)* and (3.5.1) imply $s(b) \in F[\beta_1] + \mathfrak{Q}^{-m}$. On the other hand, the condition $\theta_1\psi_c \in \mathcal{C}(\mathfrak{A}, m, \beta_1)$ implies that the restrictions $\theta_1 \mid \boldsymbol{U}^{m+1}(\mathfrak{B}_{\beta_1})$, $\theta_1\psi_c \mid \boldsymbol{U}^{m+1}(\mathfrak{B}_{\beta_1})$ factor through the determinant, hence so does $\psi_c \mid \boldsymbol{U}^{m+1}(\mathfrak{B}_{\beta_1})$. Therefore $s(c) \in F[\beta_1] + \mathfrak{Q}^{-m}$. Altogether, we have $s(b+c) \in F[\beta_1] + \mathfrak{Q}^{-m}$, so either the stratum $[\mathfrak{B}_{\beta_1}, m+1, m, s(b+c)]$ is fundamental or else $s(b+c) \in \mathfrak{Q}^{-m}$. We just showed that $[\mathfrak{B}_{\beta_1}, m+1, m, s(b+c)]$ intertwines with $[\mathfrak{B}_{\beta_1}, m+1, m, 0]$, so (2.6.2) implies that $s(b+c) \in \mathfrak{Q}^{-m}$. It follows that θ_1, θ_2 agree on $\boldsymbol{U}^{m+1}(\mathfrak{B}_{\beta_1})$ also. If $m < \left[\frac{r}{2}\right]$, we are done. Assume therefore that $m \geq \left[\frac{r}{2}\right]$. We now know that $\theta_2 = \theta_1\psi_d$, for some $d \in (\mathfrak{H}^{m+2})^* \cap a_\beta(A)$. By (3.1.16), we can take $d = a_\beta(x)$, for some $x \in \mathfrak{Q}^{r-m-1}\mathfrak{N}(\beta_1) + \mathfrak{J}^{[\frac{r+1}{2}]}$. The second assertion of (3.3.9) then gives $\theta_2 = \theta_1^{(1+x)}$, and the theorem follows from (3.5.8) in this case.

We now assume that the strata $[\mathfrak{A}, n, m+1, \beta_i]$ are not simple. Choose $[\mathfrak{A}, n, m+1, \gamma_i]$ simple and equivalent to $[\mathfrak{A}, n, m+1, \beta_i]$, $i = 1, 2$. By inductive hypothesis, we have $\mathcal{C}(\mathfrak{A}, m+1, \gamma_1) = \mathcal{C}(\mathfrak{A}, m+1, \gamma_2)$ and $\theta_1 \mid H^{m+2} = \theta_2 \mid H^{m+2}$. We automatically have $H^{m+1}(\beta_1) = H^{m+1}(\beta_2)$

here. (3.5.9) allows us to further impose the condition $\mathcal{C}(\mathfrak{A}, m, \gamma_1) = \mathcal{C}(\mathfrak{A}, m, \gamma_2)$.

Choose $\varphi \in \mathcal{C}(\mathfrak{A}, m, \gamma_i)$ to agree with θ_1 and θ_2 on H^{m+2}. Thus $\theta_i = \varphi\psi_{b_i}$, where $b_i \in \mathfrak{P}^{-(m+1)}$, $i = 1, 2$. Also, if s_i is a tame corestriction on A relative to $F[\gamma_i]/F$, $[\mathfrak{B}_{\gamma_i}, m+1, m, s_i(b_i)]$ is equivalent to a simple stratum, by (3.5.13). Let $\mathcal{I}_i$ denote the G-intertwining of θ_i, and form $\mathcal{R}_i = (\mathcal{I}_i \cap \mathfrak{A}) + \mathfrak{P}/\mathfrak{P}$. The first step is to determine the set (in fact ring) $\mathcal{R}_i$.

(3.5.14) Lemma: *In the above situation, we have*

$$\mathcal{R}_i = \{x \in \mathfrak{B}_{\gamma_i}/\mathfrak{Q}_{\gamma_i} : xs_i(b_i) \equiv s_i(b_i)x \pmod{\mathfrak{Q}_{\gamma_i}^{-m}}\}, \quad i = 1, 2.$$

Further,

$$\mathcal{R}_2 = \{x \in \mathfrak{B}_{\gamma_1}/\mathfrak{Q}_{\gamma_1} : xs_1(b_2) \equiv s_1(b_2)x \pmod{\mathfrak{Q}_{\gamma_1}^{-m}}\}.$$

Proof: By (3.3.2) (see also the proof of (3.5.1)), $\mathcal{R}_i$ is just $\mathfrak{B}_{\beta_i}/\mathfrak{Q}_{\beta_i}$, viewed as a subring of $\mathfrak{A}/\mathfrak{P}$, for $i = 1, 2$. The first assertion is therefore implied by (2.4.9) and the remark following it.

To prove the second assertion, we compute the intertwining of θ_2 using the approach of (3.3.9). Write $\mathcal{I}_0$ for the G-intertwining of $\varphi \mid H^{m+2}$. For $x \in \mathcal{I}_0$, $h \in x^{-1}H^{m+1}x \cap H^{m+1}$, we compute $\theta_2^x(h)$. (3.3.9) itself gives

$$\varphi^x(h) = \varphi(h)\psi_{x^{-1}\gamma_1 x - \gamma_1}(h).$$

We compute the factor $\psi_{b_2}^x(h)$ directly, as in the proof of (3.3.9). Taken together, we get

$$\theta_2^x(h) = \theta_2(h)\psi_{x^{-1}(\gamma_1+b_2)x-(\gamma_1+b_2)}(h).$$

The G-intertwining of θ_2 is therefore equal to $\mathcal{I}_0 \cap \mathcal{I}^+$, where $\mathcal{I}^+$ denotes the formal intertwining of the coset $\gamma_1 + b_2 + (\mathfrak{H}^{m+1})^*$. Also, let $\mathcal{I}$ denote the formal intertwining of the coset $\gamma_1 + b_2 + \mathfrak{P}^{-m}$. Since $(\mathfrak{H}^{m+1})^* \supset \mathfrak{P}^{-m}$, we have $\mathcal{I}^+ \supset \mathcal{I}$. On the other hand, write $s = -k_0(\gamma_i, \mathfrak{A})$, and take $x \in J^{[\frac{s+1}{2}]}(\gamma_1)$, $\delta \in (\mathfrak{H}^{m+1})^*$. Then $x^{-1}(\gamma_1 + b_2 + \delta + \mathfrak{P}^{-m})x = b_2 + x^{-1}(\gamma_1 + \delta + \mathfrak{P}^{-m})x$. For given δ, there exists $x \in J^{[\frac{s+1}{2}]}(\gamma_1)$ such that $x^{-1}(\gamma_1 + \delta + \mathfrak{P}^{-m})x = \gamma_1 + \mathfrak{P}^{-m}$, by (3.3.12). It follows that $\mathcal{I}^+ \subset J^{[\frac{s+1}{2}]}(\gamma_1)\mathcal{I}J^{[\frac{s+1}{2}]}(\gamma_1)$. However, $\mathcal{I}^+$ is certainly bi-invariant under multiplication by $J^{[\frac{s+1}{2}]}(\gamma_1)$, so $\mathcal{I}^+ = J^{[\frac{s+1}{2}]}(\gamma_1)\mathcal{I}J^{[\frac{s+1}{2}]}(\gamma_1)$. This is contained in $\mathcal{I}_0$ by (3.3.2), so the G-intertwining of θ_2 is exactly $\mathcal{I}^+ = J^{[\frac{s+1}{2}]}(\gamma_1)\mathcal{I}J^{[\frac{s+1}{2}]}(\gamma_1)$. Finally, this relation shows

$$(\mathcal{I}^+ \cap \mathfrak{A}) + \mathfrak{P}/\mathfrak{P} = (\mathcal{I} \cap \mathfrak{A}) + \mathfrak{P}/\mathfrak{P} = \mathcal{R}_2,$$

and the result follows from (2.4.9). ∎

We deduce from (2.4.13) that the $\mathcal{R}_i$ are semisimple k_F-algebras, and moreover that $[\mathfrak{B}_{\gamma_1}, m+1, m, s_1(b_2)]$ is equivalent to a simple stratum. Now we proceed as in the first case. Let $g \in G$ intertwine θ_1 with θ_2. Write $g = xty$, with $t \in B_{\gamma_1}^{\times}$, $x, y \in 1 + \mathfrak{Q}_{\gamma_1}^{r-m-1}\mathfrak{N}(\gamma_1) + \mathfrak{J}^{[\frac{r+1}{2}]}(\gamma_1)$, $r = -k_0(\gamma_1, \mathfrak{A})$. Restricting to $\boldsymbol{U}^{m+1}(\mathfrak{B}_{\gamma_1})$, we find that t intertwines the strata $[\mathfrak{B}_{\gamma_1}, m+1, m, s_1(b_i)]$ with each other. These two strata are therefore conjugate by an element of $\boldsymbol{U}(\mathfrak{B}_{\gamma_1})$, by (2.6.1). Conjugating θ_1 by an element of $\boldsymbol{U}(\mathfrak{B}_\gamma)$ has no effect on $\theta_1 \mid H^{m+2}$. We can therefore assume further that the θ_i agree on $\boldsymbol{U}^{m+1}(\mathfrak{B}_{\gamma_1})$. (3.1.16) gives an element $1+x$, $x \in \mathfrak{Q}_{\gamma_1}^{r-m-1}\mathfrak{N}(\gamma_1) + \mathfrak{J}^{[\frac{r+1}{2}]}(\gamma_1)$, such that $\theta_2 = \theta_1^{1+x}$, and then $\mathcal{C}(\mathfrak{A}, m, \beta_1) = \mathcal{C}(\mathfrak{A}, m, \beta_2)^{1+x} = \mathcal{C}(\mathfrak{A}, m, (1+x)^{-1}\beta_2(1+x))$.

This completes the proof of the theorem. ∎

Remark: In particular, this result shows that if we have two characters $\theta_1, \theta_2 \in \mathcal{C}(\mathfrak{A}, m, \beta)$ which intertwine in G, then there exists $x \in G$ such that $\theta_2 = \theta_1^x$ and $\mathcal{C}(\mathfrak{A}, m, \beta)^x = \mathcal{C}(\mathfrak{A}, m, \beta)$. However, determining whether or not given $\theta_i \in \mathcal{C}(\mathfrak{A}, m, \beta)$ do intertwine with each other can be a nontrivial matter, unless the extension $F[\beta]/F$ is tamely ramified. In the tamely ramified case, two characters in $\mathcal{C}(\mathfrak{A}, m, \beta)$ intertwine if and only if they are equal. For a discussion of the simplest wildly ramified case, see **[K1]**.

(3.6) Change of rings

In this section, we suppose given a field $E = F[\beta]$, where $k_0(\beta, \mathfrak{A}(E))$ and $\nu_E(\beta)$ are both negative. Let V_1, V_2 be E-vector spaces (of finite dimension), and $\mathfrak{B}_i$ a hereditary $\mathfrak{o}_E$-order in $B_i = \mathrm{End}_E(V_i)$, $i = 1, 2$. Let $\mathfrak{A}_i$ be the hereditary $\mathfrak{o}_F$-order in $A_i = \mathrm{End}_F(V_i)$ defined by the same lattice chain as $\mathfrak{B}_i$. This gives us simple strata $[\mathfrak{A}_i, n_i, 0, \beta]$, $i = 1, 2$, where $n_i = -\nu_{\mathfrak{A}_i}(\beta)$. The object of this section is to construct a canonical bijection

$$\tau_\beta = \tau_{\mathfrak{A}_1, \mathfrak{A}_2, \beta} : \mathcal{C}(\mathfrak{A}_1, 0, \beta) \xrightarrow{\approx} \mathcal{C}(\mathfrak{A}_2, 0, \beta).$$

It is quite easy to write down an explicit bijection here, using the inductive construction of simple characters. It is somewhat less easy to show that this process is independent of the various choices made in its definition. We therefore proceed more circumspectly.

Example: In the case $E = F$, the correspondence is quite obvious in nature. We have $H^1(\beta, \mathfrak{A}_i) = \boldsymbol{U}^1(\mathfrak{A}_i)$. If we write $\nu = \nu_F(\beta)$, $n_i = -\nu_{\mathfrak{A}_i}(\beta) = -\nu e(\mathfrak{A}_i|\mathfrak{o}_F)$, we get $\det(\boldsymbol{U}^{[n_i/2]+1}(\mathfrak{A}_i)) = \boldsymbol{U}^{[-\nu/2]+1}(\mathfrak{o}_F)$.

The simple characters $\theta_i \in \mathcal{C}(\mathfrak{A}_i, 0, \beta)$ are then just the characters of $U^1(\mathfrak{A}_i)$ which factor through the determinant,

$$\theta_i = \chi_{\theta_i} \circ \det,$$

where χ_{θ_i} is a character of $U^1(\mathfrak{o}_F)$ which agrees with $\psi_{F,\beta}$ on the subgroup $U^{[-\nu/2]+1}(\mathfrak{o}_F)$. Then, given $\theta_1 \in \mathcal{C}(\mathfrak{A}_1, 0, \beta)$, we can define $\tau_\beta(\theta_1) = \theta_2$, where θ_2 is the unique element of $\mathcal{C}(\mathfrak{A}_2, 0, \beta)$ such that $\chi_{\theta_1} = \chi_{\theta_2}$.

We start by treating the case in which the two vector spaces underlying the given strata are identical.

(3.6.1) Theorem: *For $i = 1, 2$, let $[\mathfrak{A}_i, n_i, m_i, \beta]$ be simple strata in $A = \mathrm{End}_F(V)$, with $m_i \geq 0$. Write $E = F[\beta]$, $B = \mathrm{End}_E(V)$, $\mathfrak{B}_i = \mathfrak{A}_i \cap B$, $e_i = e(\mathfrak{B}_i|\mathfrak{o}_E)$, and assume that*

$$\left[\frac{m_1}{e_1}\right] = \left[\frac{m_2}{e_2}\right].$$

Then given $\theta_1 \in \mathcal{C}(\mathfrak{A}_1, m_1, \beta)$, there exists a unique $\theta_2 \in \mathcal{C}(\mathfrak{A}_2, m_2, \beta)$ such that

$$\theta_1(x) = \theta_2(x), \; x \in H^{m_1+1}(\beta, \mathfrak{A}_1) \cap H^{m_2+1}(\beta, \mathfrak{A}_2),$$

i.e. $1 \in I_G(\theta_1, \theta_2)$).
For $\theta_i \in \mathcal{C}(\mathfrak{A}_i, m_i, \beta)$, the following are equivalent:
(i) $1 \in I_G(\theta_1, \theta_2)$;
(ii) $B^\times \subset I_G(\theta_1, \theta_2)$;
(iii) $B^\times \cap I_G(\theta_1, \theta_2) \neq \emptyset$.

This result gives us a *canonical* bijection

$$\text{(3.6.2)} \qquad \tau_\beta = \tau_{\mathfrak{A}_1, \mathfrak{A}_2, \beta, m_0} : \mathcal{C}(\mathfrak{A}_1, m_1, \beta) \xrightarrow{\approx} \mathcal{C}(\mathfrak{A}_2, m_2, \beta),$$

where $m_0 = [m_i/e_i]$:— for $\theta_1 \in \mathcal{C}(\mathfrak{A}_1, m_1, \beta)$, we put $\tau_\beta(\theta_1) = \theta_2$, where θ_2 is the unique element of $\mathcal{C}(\mathfrak{A}_2, m_2, \beta)$ such that the pair (θ_1, θ_2) satisfies the equivalent conditions of (3.6.1).

(3.6.3) Corollary: *Suppose, in the situation of* (3.6.1), *we have another element β' such that the strata $[\mathfrak{A}_i, n_i, m_i, \beta']$ are simple and $\mathcal{C}(\mathfrak{A}_i, m_i, \beta) = \mathcal{C}(\mathfrak{A}_i, m_i, \beta')$ for $i = 1, 2$. Then*

$$\tau_{\mathfrak{A}_1, \mathfrak{A}_2, \beta, m_0} = \tau_{\mathfrak{A}_1, \mathfrak{A}_2, \beta', m_0}.$$

This is an immediate consequence of (3.6.1).

Proof of (3.6.1): The first step is to show that if the theorem holds for one pair $[\mathfrak{A}_i, n_i, m_i, \beta]$ of simple strata, $i = 1,2$, then it holds for the pair $([\mathfrak{A}_1, n_1, m_1, \beta], [y^{-1}\mathfrak{A}_2 y, n_2, m_2, \beta])$, for any $y \in B^\times$. Observe that, in this situation, we have $H^{m_2+1}(\beta, y^{-1}\mathfrak{A}_2 y) = y^{-1}H^{m_2+1}(\beta, \mathfrak{A}_2)y$, and $\theta_2 \mapsto \theta_2^y$ is a bijection

$$\mathcal{C}(\mathfrak{A}_2, m_2, \beta) \xrightarrow{\approx} \mathcal{C}(y^{-1}\mathfrak{A}_2 y, m_2, \beta).$$

Suppose we have the result for the pair $(\mathfrak{A}_1, \mathfrak{A}_2)$. Take $\theta_i \in \mathcal{C}(\mathfrak{A}_i, m_i, \beta)$, and assume these characters agree on $H^{m_1+1}(\beta, \mathfrak{A}_1) \cap H^{m_2+1}(\beta, \mathfrak{A}_2)$. Then, By part *(iii)*, the element $y \in B^\times$ intertwines the θ_i. This says that θ_1, θ_2^y agree on $H^{m_1+1}(\beta, \mathfrak{A}_1) \cap H^{m_2+1}(\beta, y^{-1}\mathfrak{A}_2 y)$. Conversely, if we have a character $\theta_3 \in \mathcal{C}(y^{-1}\mathfrak{A}_2 y, m_2, \beta)$ which agrees with θ_1 on $H^{m_1+1}(\beta, \mathfrak{A}_1) \cap H^{m_2+1}(\beta, y^{-1}\mathfrak{A}_2 y)$, we write $\theta_3 = \theta_2^y$, for some $\theta_2 \in \mathcal{C}(\mathfrak{A}_2, m_2, \beta)$, and then y intertwines θ_1 with θ_2. This defines θ_2, and hence also θ_3, uniquely, and the first part of the theorem holds for $(\mathfrak{A}_1, y^{-1}\mathfrak{A}_2 y)$. The remaining assertions are proved similarly.

When proving the theorem, we may therefore replace the order $\mathfrak{A}_2$ by a convenient $B^\times$-conjugate. Therefore, *we henceforward assume that the* $\mathfrak{o}_E$*-orders* $\mathfrak{B}_1$, $\mathfrak{B}_2$ *satisfy the conditions of* (1.1.9). In particular, the lattice chains defining the $\mathfrak{B}_i$ have a common $\mathfrak{o}_E$-basis, and the orders $\mathfrak{A}_1$, $\mathfrak{A}_2$ admit a common (W, E)-decomposition.

The proof proceeds by *constructing explicitly* a family of bijections $\boldsymbol{\tau}_{\mathfrak{A}_1,\mathfrak{A}_2,\beta,m_0}$ (usually abbreviated to $\boldsymbol{\tau}_\beta$) which commute with restriction (on the "level" m_0). We abbreviate

$$H_i^t = H^t(\beta, \mathfrak{A}_i), \quad t \geq 0, \ i = 1,2.$$

We then show

(3.6.4) *(i) For* $\theta_1 \in \mathcal{C}(\mathfrak{A}_1, m_1, \beta)$, *let* $\theta_2 = \boldsymbol{\tau}_\beta(\theta_1)$. *Then* θ_2 *is the unique element of* $\mathcal{C}(\mathfrak{A}_2, m_2, \beta)$ *which agrees with* θ_1 *on* $H_1^{m_1+1} \cap H_2^{m_2+1}$.

(ii) For θ_1, θ_2 *as in (i), we have* $B^\times \subset I_G(\theta_1, \theta_2)$.

(iii) For $i = 1,2$, *let* $\theta_i \in \mathcal{C}(\mathfrak{A}_i, m_i, \beta)$, *and suppose there exists* $y \in B^\times$ *which intertwines the* θ_i. *Then* $\theta_2 = \boldsymbol{\tau}_\beta(\theta_1)$.

This will certainly be enough to prove the theorem. This explicitly constructed correspondence $\boldsymbol{\tau}_\beta$ is then the same as the one given *ex post facto* by (3.6.2). In particular, it is canonical and independent of the many choices which will be made in its definition. The explicit construction has an incidental advantage in that, in this form, it is obviously transitive in the orders concerned.

We need an extensive list of preliminaries. We start by recalling some generalities and establishing some notation. For this, we simply need a hereditary $\mathfrak{o}_F$-order $\mathfrak{A}$ in A. Write $e = e(\mathfrak{A}|\mathfrak{o}_F)$. According to **[BF2]** (2.8.3), we have

$$\textbf{(3.6.5)} \qquad \begin{aligned} \det(\boldsymbol{U}^{et}(\mathfrak{A})) &= \boldsymbol{U}^{t}(\mathfrak{o}_F), \\ \det(\boldsymbol{U}^{et+1}(\mathfrak{A})) &\subset \boldsymbol{U}^{t+1}(\mathfrak{o}_F), \end{aligned}$$

for all $t \geq 0$. Here, det denotes the determinant map $A \to F$. We can rephrase this as

$$\det(\boldsymbol{U}^{m+1}(\mathfrak{A})) = \boldsymbol{U}^{m+1}(\mathfrak{A}) \cap \boldsymbol{U}(\mathfrak{o}_F) = \boldsymbol{U}^{[m/e]+1}(\mathfrak{o}_F),$$

for $m \geq 0$.

Now suppose we have a simple stratum $[\mathfrak{A}, n, m, \beta]$ in A, with $m \geq 0$. As usual, we put $E = F[\beta]$, $B = \mathrm{End}_E(V)$, $\mathfrak{B} = \mathfrak{A} \cap B$. We write $e = e(\mathfrak{B}|\mathfrak{o}_E)$. Put

$$\begin{gathered} m_0 = \left[\tfrac{m}{e}\right], \\ \overline{m} = (m_0 + 1)e - 1, \quad \widetilde{m} = m_0 e. \end{gathered}$$

Thus $\overline{m}$ (resp. $\widetilde{m}$) is the greatest (resp. least) integer k such that $\left[\frac{k}{e}\right] = m_0$. In particular, we get

$$\textbf{(3.6.6)} \quad \begin{aligned} \det_B(\boldsymbol{U}^{\widetilde{m}+1}(\mathfrak{B})) = \det_B(\boldsymbol{U}^{m+1}(\mathfrak{B})) &= \det_B(\boldsymbol{U}^{\overline{m}+1}(\mathfrak{B})) \\ &= \boldsymbol{U}^{m_0+1}(\mathfrak{o}_E). \end{aligned}$$

(3.6.7) Lemma: *In the situation above, the strata $[\mathfrak{A}, n, \widetilde{m}, \beta]$ and $[\mathfrak{A}, n, \overline{m}, \beta]$ are simple, and the restriction maps*

$$\begin{gathered} \mathcal{C}(\mathfrak{A}, \widetilde{m}, \beta) \to \mathcal{C}(\mathfrak{A}, m, \beta), \\ \mathcal{C}(\mathfrak{A}, m, \beta) \to \mathcal{C}(\mathfrak{A}, \overline{m}, \beta) \end{gathered}$$

are bijective.

Proof: For the first assertion, we recall (from (1.4.13) that e divides $k_0(\beta, \mathfrak{A})$. By construction, there is no integer between m and $\overline{m}$ which is divisible by e.

The second assertion follows immediately from the observation that the restriction map

$$\mathcal{C}(\mathfrak{A}, k, \beta) \to \mathcal{C}(\mathfrak{A}, k+1, \beta)$$

is always surjective (by (3.3.21)), while it can only fail to be injective if $k + 1$ is divisible by e. This is clear (from, e.g., (3.6.5)) if β is minimal over F. In general, we put $r = -k_0(\beta, \mathfrak{A})$ and choose a simple stratum

$[\mathfrak{A}, n, r, \gamma]$ equivalent to $[\mathfrak{A}, n, r, \beta]$. Set $\mathfrak{B}_\gamma = \mathfrak{A} \cap \text{End}_{F[\gamma]}(V)$, $e_\gamma = e(\mathfrak{B}_\gamma|\mathfrak{o}_{F[\gamma]})$. Then e_γ is divisible by e, by (2.4.1), and the assertion follows from (3.3.21) and induction. ■

Continue in the same situation, and set $r_0 = -k_0(\beta, \mathfrak{A}(E))$, $r = -k_0(\beta, \mathfrak{A}) = er_0$. Elementary calculations then yield

(3.6.8)
$$\begin{aligned} \widetilde{m} \le \left[\tfrac{r}{2}\right] &\Leftrightarrow m_0 \le \left[\tfrac{r_0}{2}\right], \\ \overline{m} \ge \left[\tfrac{r}{2}\right] &\Leftrightarrow m_0 + 1 > \tfrac{r_0}{2}. \end{aligned}$$

Now we pass to the notation of the theorem. In particular, we put

$$m_0 = \left[\frac{m_1}{e_1}\right] = \left[\frac{m_2}{e_2}\right].$$

For $\theta_i \in \mathcal{C}(\mathfrak{A}_i, m_i, \beta)$, write

$$\overline{\theta}_i = \theta_i \mid H_i^{\overline{m}_i+1} \in \mathcal{C}(\mathfrak{A}_i, \overline{m}_i, \beta),$$

and define $\widetilde{\theta}_i$ by

$$\widetilde{\theta}_i \in \mathcal{C}(\mathfrak{A}_i, \widetilde{m}_i, \beta), \quad \widetilde{\theta}_i \mid H_i^{m_i+1} = \theta_i.$$

According to (3.6.7), there is indeed a unique $\widetilde{\theta}_i$ with this property. Moreover, $\widetilde{\theta}_i \mid H^{\overline{m}_i+1} = \overline{\theta}_i$, and this condition also determines the character $\widetilde{\theta}_i$ uniquely.

Next, we need an intertwining lemma.

(3.6.9) Lemma: *In the situation above, let χ_i be a character of $U^{m_0+1}(\mathfrak{o}_E)$, and set $\phi_i = \chi_i \circ \det_B \mid U^{m_i+1}(\mathfrak{B}_i)$, $i = 1, 2$. Suppose that ϕ_1, ϕ_2 intertwine in $B^\times$. Then $\chi_1 = \chi_2$. Moreover, we have*

$$\begin{aligned} \det_B(U^{\widetilde{m}_i+1}(\mathfrak{B}_i) &= \det_B(U^{\overline{m}_1+1}(\mathfrak{B}_1) \cap y^{-1}U^{\overline{m}_2+1}(\mathfrak{B}_2)y) \\ &= U^{m_0+1}(\mathfrak{o}_E), \end{aligned}$$

for $i = 1, 2$ and any $y \in B^\times$.

Remark: We are operating under the hypothesis that the lattice chains defining the orders $\mathfrak{B}_1$, $\mathfrak{B}_2$ admit a common $\mathfrak{o}_E$-basis. This hypothesis is not necessary for (3.6.9):— the lemma holds for the pair $(\mathfrak{B}_1, \mathfrak{B}_2)$ if and only if it holds for $(\mathfrak{B}_1, z^{-1}\mathfrak{B}_2 z)$, $z \in B^\times$. Notice also that the result (and the proof below) holds in the case $m_1 = m_2 = -1$ (except that we have to extend the definition of $\widetilde{m}_i$ by setting $\widetilde{-1} = -1$).

Proof of (3.6.9): The first assertion of the lemma is implied by the second one and (3.6.6). To prove the second, we choose a minimal hereditary $\mathfrak{o}_E$-order $\mathfrak{B}_0 \subset \mathfrak{B}_1 \cap \mathfrak{B}_2$ (as we may, by (1.1.9)). Since e_i divides $\overline{m}_i + 1$, we have $\boldsymbol{U}^{\overline{m}_i+1}(\mathfrak{B}_i) \supset \boldsymbol{U}^m(\mathfrak{B}_0)$, where

$$m = \frac{\overline{m}_i + 1}{e_i} R = (m_0 + 1)R, \quad R = \dim_E(V).$$

We have $e(\mathfrak{B}_0|\mathfrak{o}_E) = R$, so $\det_B(\boldsymbol{U}^m(\mathfrak{B}_0)) = \boldsymbol{U}^{m_0+1}(\mathfrak{o}_E)$. It is enough to show that

$$\det_B(\boldsymbol{U}^m(\mathfrak{B}_0) \cap y^{-1}\boldsymbol{U}^m(\mathfrak{B}_0)y) = \boldsymbol{U}^{m_0+1}(\mathfrak{o}_E),$$

for any $y \in B^\times$.

We choose a basis of the lattice chain of $\mathfrak{B}_0$, and use this to identify B with $\mathbb{M}(R, E)$. Thus $\mathfrak{B}_0$ becomes the ring of all matrices in $\mathbb{M}(R, \mathfrak{o}_E)$ which are upper triangular mod $\mathfrak{p}_E$. Let $\widetilde{\boldsymbol{W}}$ be the affine Weyl group associated with this setup (see **(5.4)** below for a summary of this theory). Thus $\widetilde{\boldsymbol{W}}$ is the group of all monomial matrices (i.e. those with one nonzero entry in each row and column) whose nonzero entries are powers of some chosen prime element of E. Nothing is changed if we replace the element y above by uyv, $u, v \in \boldsymbol{U}(\mathfrak{B}_0)$. Also, we have $B^\times = \boldsymbol{U}(\mathfrak{B}_0)\widetilde{\boldsymbol{W}}\boldsymbol{U}(\mathfrak{B}_0)$, so we may as well take $y \in \widetilde{\boldsymbol{W}}$. Clearly, the group $\boldsymbol{U}^m(\mathfrak{B}_0) \cap y^{-1}\boldsymbol{U}^m(\mathfrak{B}_0)y$ contains the group of all diagonal matrices in $\boldsymbol{U}^m(\mathfrak{B}_0)$. However, using the block matrix descriptions of **(2.5)**, one sees that this diagonal subgroup of $\boldsymbol{U}^m(\mathfrak{B}_0)$ consists of all diagonal matrices with eigenvalues in $\boldsymbol{U}^{m_0+1}(\mathfrak{o}_E)$. The lemma follows immediately. ∎

We also record a technical lemma whose proof we defer to the end of the section.

(3.6.10) Lemma: *Let $x \in B^\times$. Use the notation above and assume $m_i \le \left[\frac{r_i}{2}\right]$, $i = 1, 2$. Assume also that the orders $\mathfrak{B}_1$, $\mathfrak{B}_2$ satisfy the conditions of* (1.1.9). *Then*

$$\begin{aligned} &x^{-1}\mathfrak{H}^{m_1+1}(\beta, \mathfrak{A}_1)x \cap \mathfrak{H}^{m_2+1}(\beta, \mathfrak{A}_2) \\ &= x^{-1}\mathfrak{Q}_1^{m_1+1}x \cap \mathfrak{Q}_2^{m_2+1} + x^{-1}\mathfrak{H}^{[r_1/2]+1}(\beta, \mathfrak{A}_1)x \cap \mathfrak{H}^{[r_2/2]+1}(\beta, \mathfrak{A}_2), \end{aligned}$$

where $\mathfrak{Q}_i = \mathrm{rad}(\mathfrak{B}_i)$.

Now we can start the proof of the theorem. Set $m_0 = [m_i/e_i]$ as before, and $r_0 = r_i/e_i = -k_0(\beta, \mathfrak{A}(E))$. We also put $n_0 = n_i/e_i$, and use the notations $\widetilde{m}_i$, $\overline{m}_i$ as above. We first construct the bijection $\boldsymbol{\tau}_\beta$, and prove (3.6.4), under the assumption that β is *minimal* over F.

We define a character χ_0 of $\boldsymbol{U}^{[\frac{n_0}{2}]+1}(\mathfrak{o}_E)$, as follows. Choose a character ψ_E of E with conductor $\mathfrak{p}_E$, and set $\psi_B = \psi_E \circ \mathrm{tr}_{B/E}$. Let s be the tame corestriction on A relative to E/F such that $\psi_A(ab) = \psi_B(s(a)b)$, $a \in A$, $b \in B$ (*cf.* (1.3.5)). We put

$$\chi_0 = \psi_{E,s(\beta)}.$$

We have $n_i = n_0 e_i$, so

$$\left[\frac{1}{e_i}\left[\frac{n_i}{2}\right]\right] = \left[\frac{n_0}{2}\right].$$

An elementary argument gives

$$\chi_0 \circ \det_B(x) = \psi_{B,s(\beta)}(x), \quad x \in \boldsymbol{U}^{[\frac{n_i}{2}]+1}(\mathfrak{B}_i),\ i = 1,2.$$

We therefore get a bijection between $\mathcal{C}(\mathfrak{A}_i, m_i, \beta)$ and the set of characters χ of $\boldsymbol{U}^{m_0+1}(\mathfrak{o}_E)$ which agree with χ_0 on $\boldsymbol{U}^{m_0+1}(\mathfrak{o}_E) \cap \boldsymbol{U}^{[\frac{n_0}{2}]+1}(\mathfrak{o}_E)$. This is given by $\theta \mapsto \chi_\theta$, where $\theta(x) = \chi(\det(x))$, $x \in \boldsymbol{U}^{m_i+1}(\mathfrak{B}_i)$. (3.6.6) then gives us the property

$$\chi_\theta = \chi_{\tilde{\theta}} = \chi_{\bar{\theta}}.$$

For $\theta_1 \in \mathcal{C}(\mathfrak{A}_1, m_1, \beta)$, we can then define $\boldsymbol{\tau}_\beta(\theta_1) = \theta_2$, where θ_2 is the unique element of $\mathcal{C}(\mathfrak{A}_2, m_2, \beta)$ which satisfies

$$\chi_{\theta_2} = \chi_{\theta_1}.$$

This family of bijections commutes with restriction. In particular, we get

$$\boldsymbol{\tau}_\beta(\widetilde{\theta}_1) = \widetilde{\boldsymbol{\tau}_\beta(\theta_1)}, \quad \boldsymbol{\tau}_\beta(\overline{\theta}_1) = \overline{\boldsymbol{\tau}_\beta(\theta_1)}.$$

We must now verify (3.6.4) in this situation.

In (3.6.4)*(i)*, if the characters $\theta_i \in \mathcal{C}(\mathfrak{A}_i, m_i, \beta)$ agree on $H_1^{m_1+1} \cap H_2^{m_2+1}$, then they surely agree on $\boldsymbol{U}^{m_1+1}(\mathfrak{B}_1) \cap \boldsymbol{U}^{m_2+1}(\mathfrak{B}_2)$. (3.6.9) implies $\chi_{\theta_1} = \chi_{\theta_2}$, whence $\theta_2 = \boldsymbol{\tau}_\beta(\theta_1)$, giving the required uniqueness property. We therefore just have to show that θ_1 agrees with $\theta_2 = \boldsymbol{\tau}_\beta(\theta_1)$ on $H_1^{m_1+1} \cap H_2^{m_2+1}$. Here, it is enough to treat the case where $m_i = \widetilde{m}_i$, for both values of i. By (3.6.8), we have

$$m_1 \leq \left[\frac{r_1}{2}\right] \quad \Leftrightarrow \quad m_2 \leq \left[\frac{r_2}{2}\right].$$

If $m_i \geq \left[\frac{r_i}{2}\right]$, there is nothing to prove, since the characters θ_i are both restrictions of the function ψ_β. We therefore assume that $m_i \leq \left[\frac{r_i}{2}\right]$, for both values of i. We apply (3.6.10) (with $x = 1$), and pass to unit groups to get

$$H_1^{m_1+1} \cap H_2^{m_2+1} = (U^{m_1+1}(\mathfrak{B}_1) \cap U^{m_2+1}(\mathfrak{B}_2)).(H_1^{[\frac{r_1}{2}]+1} \cap H_2^{[\frac{r_2}{2}]+1}).$$

The characters θ_i agree on the second factor here by the first case, and they agree on the first factor by construction. This proves (3.6.4)*(i)* in this case.

In (3.6.4)*(ii)*, we have to take $\theta_i \in \mathcal{C}(\mathfrak{A}_i, m_i, \beta)$, related by $\theta_2 = \tau_\beta(\theta_1)$, and show that every element of $B^\times$ intertwines the θ_i. Here, it is again enough to treat the case where $m_i = \widetilde{m}_i$ for both values of i. Suppose first that $m_i \geq [\frac{r_i}{2}]$, for $i = 1, 2$. Then $\theta_i = \psi_\beta \mid U^{m_i+1}(\mathfrak{A}_i)$, and the assertion is immediate. We therefore assume $m_i \leq [\frac{r_i}{2}]$, and take $y \in B^\times$. (3.6.10) gives us

$$\begin{aligned} &H_1^{m_1+1} \cap y^{-1} H_2^{m_2+1} y \\ &\qquad = (U^{m_1+1}(\mathfrak{B}_1) \cap y^{-1} U^{m_2+1}(\mathfrak{B}_2) y).(H_1^{[\frac{r_1}{2}]+1} \cap y^{-1} H_2^{[\frac{r_2}{2}]+1} y). \end{aligned}$$

The characters θ_1, θ_2^y agree on both factors here, and the assertion follows.

To prove (3.6.4)*(iii)*, we take $\theta_i \in \mathcal{C}(\mathfrak{A}_i, m_i, \beta)$, $i = 1, 2$, and assume that they are intertwined by some $y \in B^\times$. We have to show that $\theta_2 = \tau_\beta(\theta_1)$. By hypothesis, the characters θ_1, θ_2^y agree on $H_1^{m_1+1} \cap y^{-1} H_2^{m_2+1} y$, so they surely agree on $U^{m_1+1}(\mathfrak{B}_1) \cap y^{-1} U^{m_2+1}(\mathfrak{B}_2) y$. (3.6.9) now implies $\chi_{\theta_1} = \chi_{\theta_2}$, whence $\theta_2 = \tau_\beta(\theta_1)$, as required.

This completes the proof of (3.6.4) when β is minimal over F.

In the general case, the orders $\mathfrak{A}_i$ admit a common (W, E)-decomposition $\mathfrak{A}_i = \mathfrak{A}(E) \otimes_{\mathfrak{o}_E} \mathfrak{B}_i$. Thus, if we choose a simple stratum $[\mathfrak{A}(E), n_0, r_0, \gamma]$ in $A(E)$ equivalent to $[\mathfrak{A}(E), n_0, r_0, \beta]$, then the stratum $[\mathfrak{A}_i, n_i, r_i, \gamma]$ is simple and equivalent to $[\mathfrak{A}_i, n_i, r_i, \beta]$, for $i = 1, 2$. Write $B_\gamma = \mathrm{End}_{F[\gamma]}(V)$, $\mathfrak{C}_i = \mathfrak{A}_i \cap B_\gamma$, $e_{\gamma,i} = e(\mathfrak{C}_i | \mathfrak{o}_{F[\gamma]})$, $i = 1, 2$. Since e_i divides $e_{\gamma,i}$, we have

$$\left[\frac{m_1}{e_{\gamma,1}}\right] = \left[\frac{m_2}{e_{\gamma,2}}\right] = m_{0,\gamma},$$

say. Define integers $\widetilde{m}_{i,\gamma}$, $\overline{m}_{i,\gamma}$ by

$$\begin{aligned} \widetilde{m}_{i,\gamma} &= m_{0,\gamma} e_{\gamma,i}, \\ \overline{m}_{i,\gamma} &= (m_{0,\gamma} + 1) e_{\gamma,i} - 1. \end{aligned}$$

Then $\widetilde{m}_{i,\gamma} \leq \widetilde{m}_i \leq \overline{m}_i \leq \overline{m}_{i,\gamma}$, and we can assume inductively that we have defined a family of bijections

$$\tau_\gamma = \tau_{\mathfrak{A}_1, \mathfrak{A}_2, \gamma, m_{0,\gamma}} : \mathcal{C}(\mathfrak{A}_1, m_1, \gamma) \xrightarrow{\approx} \mathcal{C}(\mathfrak{A}_2, m_2, \gamma)$$

which satisfy (3.6.4) (relative to γ).

We now define the bijection $\boldsymbol{\tau}_\beta$. Write $\beta = \gamma + c$, and $m_i' = \max\{m_i, [\frac{r_i}{2}]\}$. (3.3.18) and (3.3.21) give us a bijection between the set $\mathcal{C}(\mathfrak{A}_i, m_i, \beta)$ and the set of pairs (χ, θ_0) of the following form

(i) $\theta_0 \in \mathcal{C}(\mathfrak{A}_i, , m_i', \gamma)$ *and* $\chi \in \boldsymbol{U}^{m_0+1}(\mathfrak{o}_E)\hat{}$;

(ii) $\theta_0 \psi_c \mid \boldsymbol{U}^{m_i'+1}(\mathfrak{B}_i) = \chi \circ \det_B \mid \boldsymbol{U}^{m_i'+1}(\mathfrak{B}_i)$.

For $\theta_i \in \mathcal{C}(\mathfrak{A}_i, m_i, \beta)$, this bijection is given by $\theta_i \mapsto (\Gamma(\theta_i), \chi_{\theta_i})$, where

$$\Gamma(\theta_i).\psi_c = \theta_i \mid H_i^{m_i'+1},$$
$$\theta_i \mid \boldsymbol{U}^{m_i+1}(\mathfrak{B}_i) = \chi_{\theta_i} \circ \det_B.$$

Then, for $\theta_1 \in \mathcal{C}(\mathfrak{A}_1, m_1, \beta)$, we define $\theta_2 = \boldsymbol{\tau}_\beta(\theta_1) \in \mathcal{C}(\mathfrak{A}_2, m_2, \beta)$ by

$$\Gamma(\theta_2) = \boldsymbol{\tau}_\gamma(\Gamma(\theta_1)), \quad \chi_{\theta_2} = \chi_{\theta_1}.$$

Observe here that, by inductive hypothesis, the characters $\Gamma(\theta_i)$ agree on $H_1^{m_1'+1} \cap H_2^{m_2'+1}$, and hence on $\boldsymbol{U}^{m_1'+1}(\mathfrak{B}_1) \cap \boldsymbol{U}^{m_2'+1}(\mathfrak{B}_2)$. This construction therefore does indeed define a character $\theta_2 \in \mathcal{C}(\mathfrak{A}_2, m_2, \beta)$. Note also that this map $\boldsymbol{\tau}_\beta$ commutes with the operations "bar" and "tilde" on simple characters.

Now let us prove (3.6.4)*(i)*. If $\theta_i \in \mathcal{C}(\mathfrak{A}_i, m_i, \beta)$, $i = 1, 2$, agree on $H_1^{m_1+1} \cap H_2^{m_2+1}$, then they surely agree on $H_1^{m_1'+1} \cap H_2^{m_2'+1}$, where $m_i' = \max\{m_i, [\frac{r_i}{2}]\}$, as above. Further, they agree on $\boldsymbol{U}^{m_1+1}(\mathfrak{B}_1) \cap \boldsymbol{U}^{m_2+1}(\mathfrak{B}_2)$. By (3.6.9) and induction (i.e. (3.6.4) applied to γ), this is enough to show that $\theta_2 = \boldsymbol{\tau}_\beta(\theta_1)$, and hence the uniqueness assertion.

This reduces us to showing that θ_1 and $\theta_2 = \boldsymbol{\tau}_\beta(\theta_1)$ agree on $H_1^{m_1+1} \cap H_2^{m_2+1}$. Here it is enough to treat the case where $m_i = \widetilde{m}_i$, $i = 1, 2$. Thus $m_1 \geq [\frac{r_1}{2}]$ if and only if $m_2 \geq [\frac{r_2}{2}]$, by (3.6.8). In the case $m_i \geq [\frac{r_i}{2}]$, the assertion follows immediately from the corresponding property for $\boldsymbol{\tau}_\gamma$. If $m_i < [\frac{r_i}{2}]$, we apply (3.6.10) to get

$$H_1^{m_1+1} \cap H_2^{m_2+1} = (\boldsymbol{U}^{m_1+1}(\mathfrak{B}_1) \cap \boldsymbol{U}^{m_2+1}(\mathfrak{B}_2)).(H_1^{[\frac{r_1}{2}]+1} \cap H_2^{[\frac{r_2}{2}]+1}).$$

By construction and the case above, the characters θ_i agree on both factors here.

To prove (3.6.4)*(ii)*, we take $\theta_i \in \mathcal{C}(\mathfrak{A}_i, m_i, \beta)$, related by $\theta_2 = \boldsymbol{\tau}_\beta(\theta_1)$, $y \in B^\times$, and show that y intertwines θ_1 with θ_2. It is enough to do this under the assumption that $m_i = \widetilde{m}_i$, for $i = 1, 2$. Thus $m_1 \geq [\frac{r_1}{2}]$ if and only if $m_2 \geq [\frac{r_2}{2}]$. Exactly as in the case above, where β was minimal over F, (3.6.10) reduces us to the case $m_i \geq [\frac{r_i}{2}]$ for both values of i. For this case, we need a lemma:

(3.6.11) Lemma: *In the situation above, write* $\mathfrak{R}_i = \mathrm{rad}(\mathfrak{C}_i)$, $s_i = -k_0(\gamma, \mathfrak{A}_i)$, $\mathfrak{N}(\gamma, \mathfrak{A}_i) = \mathfrak{N}_{-s_i}(\gamma, \mathfrak{A}_i)$. *Then*

$$B^\times \subset (1 + \mathfrak{R}_1^{s_1 - r_1}\mathfrak{N}(\gamma, \mathfrak{A}_1))B_\gamma^\times(1 + \mathfrak{R}_2^{s_2 - r_2}\mathfrak{N}(\gamma, \mathfrak{A}_2)).$$

Proof: Choose a minimal hereditary $\mathfrak{o}_E$-order $\mathfrak{B}_{\mathrm{m}}$ in B contained in $\mathfrak{B}_1 \cap \mathfrak{B}_2$: this is possible by (1.1.9) and the hypothesis on the pair $(\mathfrak{B}_1, \mathfrak{B}_2)$. Let $\mathfrak{A}_{\mathrm{m}}$ be the hereditary $\mathfrak{o}_F$-order in A defined by the same lattice chain as $\mathfrak{B}_{\mathrm{m}}$. Set $r_{\mathrm{m}} = -k_0(\beta, \mathfrak{A}_{\mathrm{m}})$, $s_{\mathrm{m}} = -k_0(\gamma, \mathfrak{A}_{\mathrm{m}})$, $n_{\mathrm{m}} = -\nu_{\mathfrak{A}_{\mathrm{m}}}(\beta)$. The stratum $[\mathfrak{A}_{\mathrm{m}}, n_{\mathrm{m}}, r_{\mathrm{m}}, \gamma]$ is simple. We adapt to $\mathfrak{A}_{\mathrm{m}}$ our other notations attached to the $\mathfrak{A}_i$ in the obvious way. (2.1.3) gives us

$$\mathfrak{R}_j^{s_j - r_j}\mathfrak{N}(\gamma, \mathfrak{A}_j) = \mathfrak{Q}_j^{s_j - r_j}\mathfrak{N}_{-s_j}(\beta, \mathfrak{A}_j), \tag{3.6.12}$$

where $\mathfrak{Q}_j = \mathrm{rad}(\mathfrak{B}_j)$, and $j \in \{1, 2, \mathrm{m}\}$. The integers s_j, r_j are divisible by e_j, and the quotients are independent of j. We deduce

$$\mathfrak{Q}_{\mathrm{m}}^{s_{\mathrm{m}} - r_{\mathrm{m}}} \subset \mathfrak{Q}_i^{s_i - r_i}, \quad i = 1, 2.$$

However, the orders $\mathfrak{A}_j$ admit a common (W, F)-decomposition. Therefore (1.4.13), (1.2.10) give us

$$\mathfrak{Q}_j^{s_j - r_j}\mathfrak{N}_{-s_j}(\beta, \mathfrak{A}_j) = \mathfrak{N}_{-s_j/e_j}(\beta, \mathfrak{A}(E)) \otimes \mathfrak{Q}_j^{s_i - r_j},$$

from which it now follows that

$$\mathfrak{R}_{\mathrm{m}}^{s_{\mathrm{m}} - r_{\mathrm{m}}}\mathfrak{N}(\gamma, \mathfrak{A}_{\mathrm{m}}) \subset \mathfrak{R}_i^{s_i - r_i}\mathfrak{N}(\gamma, \mathfrak{A}_i), \quad i = 1, 2.$$

In other words, we need only prove the lemma under the assumption that $\mathfrak{A}_1 = \mathfrak{A}_2 = \mathfrak{A}_{\mathrm{m}}$. Now it is immediate:— we apply (1.5.8) to the simple stratum $[\mathfrak{A}_{\mathrm{m}}, n_{\mathrm{m}}, r_{\mathrm{m}}, \gamma]$, which is equivalent to $[\mathfrak{A}_{\mathrm{m}}, n_{\mathrm{m}}, r_{\mathrm{m}}, \beta]$. ■

Returning to (3.6.4)*(ii)*, we put $\theta_i = \theta_{0i}\psi_c$, so that $\theta_{02} = \boldsymbol{\tau}_\gamma(\theta_{01})$. Inductively, the whole of $B_\gamma^\times$ intertwines the θ_{0i}, so the set

$$\mathcal{S} = (1 + \mathfrak{R}_1^{s_1 - r_1}\mathfrak{N}(\gamma, \mathfrak{A}_1))B_\gamma^\times(1 + \mathfrak{R}_2^{s_2 - r_2}\mathfrak{N}(\gamma, \mathfrak{A}_2))$$

intertwines θ_{01} with θ_{02}. We now take $x \in \mathcal{S}$ and work out the character θ_1^x on $x^{-1}H_1x \cap H_2$, where we abbreviate $H_i = H_i^{m_i+1}$. We simply imitate the corresponding calculation in the proof of (3.3.2) (i.e. the proof of (3.3.9)), to get

$$\theta_1^x(h) = \theta_2(h)\psi_{x^{-1}\beta x - \beta}(h), \quad h \in H_1^x \cap H_2.$$

Taking our element $y \in B^\times$, (3.6.11) gives $y \in \mathcal{S}$, and this last identity reduces to

$$\theta_1^y(h) = \theta_2(h), \quad h \in H_1^y \cap H_2,$$

which proves (3.6.4)*(ii)*.

Now we prove (3.6.4)*(iii)*: we take $\theta_i \in \mathcal{C}(\mathfrak{A}_i, m_i, \beta)$, assume there exists $x \in B^\times$ which intertwines θ_1 with θ_2, and show that $\theta_2 = \boldsymbol{\tau}_\beta(\theta_1)$. We already have this result in the case where β is minimal over F.

Here we can assume that $m_i = \overline{m}_i$, for both values of i. Further, (3.6.9) and (3.6.10) reduce us to the case where $m_i \geq \left[\frac{r_i}{2}\right]$ for both values of i. There exists $\phi \in \mathcal{C}(\mathfrak{A}_1, m_1, \beta)$ such that $\boldsymbol{\tau}_\beta(\phi) = \theta_2$. The element x then intertwines ϕ with θ_2 by (3.6.4)*(ii)*. It follows that x intertwines the character $\phi^{-1}\theta_1$ of $H_1^{m_1+1}$ with the trivial character of $H_2^{m_2+1}$. Write $\phi = \phi_0\psi_c$, $\theta_i = \theta_{0i}\psi_c$ as before. Therefore x intertwines $\phi_0^{-1}\theta_{01}$ with the trivial character of $H_2^{m_2+1}$. We now use (3.6.11) to write $x = u_1 y u_2$, with $y \in B_\gamma^\times$ and $u_i \in 1 + \mathfrak{P}_i^{s_i - r_i}\mathfrak{N}(\gamma, \mathfrak{A}_i)$. The factor u_2 is irrelevant, since it surely intertwines the trivial character of $H^{m_2+1}(\beta, \mathfrak{A}_2)$. By (3.3.9), the factor u_1 fixes the character $\phi_0^{-1}\theta_{01}$, so we conclude that y intertwines $\phi_0^{-1}\theta_{01}$ with the trivial character of $H_2^{m_2+1}$. By construction, y intertwines ϕ_0 with θ_{02}, and it therefore intertwines θ_{01} with θ_{02}. By inductive hypothesis (i.e. (3.6.4)*(iii)* applied to γ), we therefore have $\theta_{02} = \boldsymbol{\tau}_\gamma(\theta_{01})$, and hence $\theta_2 = \boldsymbol{\tau}_\beta(\theta_1)$.

This proves (3.6.4)*(iii)* and completes the proof of (3.6.1). ■

Now we give a more general version of this. We start, as before, with a field $E = F[\beta]$, and set $\nu = \nu_E(\beta)$, $r_0 = -k_0(\beta, \mathfrak{A}(E))$, (except in the trivial case $E = F$, when we put $r_0 = -\nu$). We assume throughout that $r_0 > 0$. We seek a family $\mathbf{T}$ of bijections

$$\boldsymbol{\tau}_{\mathfrak{A}_1, \mathfrak{A}_2, \beta, m_0} : \mathcal{C}(\mathfrak{A}_1, m_1, \beta) \xrightarrow{\approx} \mathcal{C}(\mathfrak{A}_2, m_2, \beta).$$

Here, for $i = 1, 2$, we are given a simple stratum $[\mathfrak{A}_i, n_i, m_i, \beta]$ in $A_i = \mathrm{End}_F(V_i)$. We put $B_i = \mathrm{End}_E(V_i)$, $\mathfrak{B}_i = \mathfrak{A}_i \cap B_i$, $e_i = e(\mathfrak{B}_i|\mathfrak{o}_E)$. Further, m_0 is a integer such that $0 \leq m_0 < r_0$, and

$$\left[\frac{m_1}{e_1}\right] = \left[\frac{m_2}{e_2}\right] = m_0.$$

This family is to satisfy the following conditions:

(3.6.13)*(i) The bijections $\boldsymbol{\tau}$ commute with restriction:— given $0 \leq m_0 \leq \ell_0 < r_0$, and integers ℓ_i such that $[\ell_i/e_i] = \ell_0$, the diagram*

$$\begin{array}{ccc}
\mathcal{C}(\mathfrak{A}_1, m_1, \beta) & \xrightarrow{\boldsymbol{\tau}_{\mathfrak{A}_1, \mathfrak{A}_2, \beta, m_0}} & \mathcal{C}(\mathfrak{A}_2, m_2, \beta) \\
\downarrow & & \downarrow \\
\mathcal{C}(\mathfrak{A}_1, \ell_1, \beta) & \xrightarrow[\boldsymbol{\tau}_{\mathfrak{A}_1, \mathfrak{A}_2, \beta, \ell_0}]{} & \mathcal{C}(\mathfrak{A}_2, \ell_2, \beta)
\end{array}$$

commutes, where the vertical maps are restriction.

(ii) The maps $\boldsymbol{\tau}_\beta$ preserve β-determinants, in the following sense. Let $\theta_i \in \mathcal{C}(\mathfrak{A}_i, m_i, \beta)$, $i = 1,2$, and suppose $\theta_2 = \boldsymbol{\tau}_\beta(\theta_1)$. Let χ_i denote the (uniquely determined) character of $\boldsymbol{U}^{m_0+1}(\mathfrak{o}_E)$ such that $\theta_i \mid \boldsymbol{U}^{m_i+1}(\mathfrak{B}_i) = \chi_i \circ \det_B$. Then $\chi_1 = \chi_2$.

(iii) The bijections $\boldsymbol{\tau}$ are inductive along β:— suppose β is not minimal over F, and suppose given a field $F[\gamma]$, together with embeddings $\mu_i : F[\gamma] \to A_i$ such that the stratum $[\mathfrak{A}_i, n_i, r_i, \mu_i(\gamma)]$ is simple and equivalent to $[\mathfrak{A}_i, n_i, r_i, \beta]$, $i = 1,2$. Suppose also that $m_i \geq [r_i/2]$. Set $c_i = \beta - \mu_i(\gamma)$. Take $\theta_{0i} \in \mathcal{C}(\mathfrak{A}_i, m_i, \mu_i(\gamma))$, and suppose

$$\theta_{02} = \boldsymbol{\tau}_{\mathfrak{A}_1,\mathfrak{A}_2,\gamma}(\theta_{01}).$$

Set $\theta_i = \theta_{0i}\psi_{c_i}$. Then

$$\theta_2 = \boldsymbol{\tau}_{\mathfrak{A}_1,\mathfrak{A}_2,\beta}(\theta_1).$$

These conditions surely determine the family $\mathbf{T}$ uniquely. It is rather more surprising that such a family actually manages to exist:

(3.6.14) Theorem: *There exists a unique family of bijections $\mathbf{T}$ satisfying the conditions* (3.6.13).

Proof: Observe first that the bijections of (3.6.2) (in the case $V_1 = V_2 = V$) satisfy the conditions (3.6.13) by construction.

In general, suppose we are given simple strata $[\mathfrak{A}_i, n_i, m_i, \beta]$ in A_i as above. For each i, choose a maximal $\mathfrak{o}_E$-order $\overline{\mathfrak{B}}_i$ in B_i which contains $\mathfrak{B}_i$. Let $\overline{\mathfrak{A}}_i$ be the corresponding $\mathfrak{o}_F$-order in A_i. (3.6.2) then gives us bijections

$$\boldsymbol{\tau}_{\mathfrak{A}_i,\overline{\mathfrak{A}}_i,\beta,m_0} : \mathcal{C}(\mathfrak{A}_i, m_i, \beta) \xrightarrow{\approx} \mathcal{C}(\overline{\mathfrak{A}}_i, m_0, \beta).$$

It is enough to produce a bijection $\boldsymbol{\tau}_{\overline{\mathfrak{A}}_1,\overline{\mathfrak{A}}_2,\beta,m_0}$, since we can then define $\boldsymbol{\tau}_{\mathfrak{A}_1,\mathfrak{A}_2,\beta,m_0}$ by composing

$$\boldsymbol{\tau}_{\mathfrak{A}_1,\mathfrak{A}_2,\beta,m_0} = \boldsymbol{\tau}_{\mathfrak{A}_2,\overline{\mathfrak{A}}_2,\beta,m_0}{}^{-1} \circ \boldsymbol{\tau}_{\overline{\mathfrak{A}}_1,\overline{\mathfrak{A}}_2,\beta,m_0} \circ \boldsymbol{\tau}_{\mathfrak{A}_1,\overline{\mathfrak{A}}_1,\beta,m_0}.$$

In other words, we may as well assume that the orders $\mathfrak{B}_i$ are both maximal, and so $m_1 = m_2 = m_0$. Let $\mathcal{L}^i = \{L^i_j : j \in \mathbb{Z}\}$ denote the $\mathfrak{o}_E$-lattice chain in V_i defining $\mathfrak{B}_i$. We define a lattice chain $\mathcal{M} = \{M_j : j \in \mathbb{Z}\}$ in $V = V_1 \oplus V_2$ by

$$M_j = L^1_j \oplus L^2_j, \quad j \in \mathbb{Z}.$$

Let $\mathfrak{B}$ (resp. $\mathfrak{A}$) be the hereditary $\mathfrak{o}_E$-order (resp. $\mathfrak{o}_F$-order) in $B = \mathrm{End}_E(V)$ (resp. $A = \mathrm{End}_F(V)$) defined by the lattice chain $\mathcal{M}$. Then $\mathfrak{B}$

is a maximal order in B, and the stratum $[\mathfrak{A}, -\nu, m_0, \beta]$ is simple. We produce a canonical bijection

$$\mathcal{C}(\mathfrak{A}, m_0, \beta) \xrightarrow{\approx} \mathcal{C}(\mathfrak{A}_i, m_0, \beta),$$

with the desired properties, for each i. This will give us the desired bijection between the $\mathcal{C}(\mathfrak{A}_i, m_0, \beta)$.

Write $\mathbf{e}_i$ for the canonical projection $V = V_1 \oplus V_2 \to V_i$.

(3.6.15) Lemma: *Let K/F be a subfield of A such that the V_i are K-subspaces of V, and $K^\times \subset \mathfrak{K}(\mathfrak{A})$. Then*

(i) $K^\times \cong \mathbf{e}_i K^\times \mathbf{e}_i \subset \mathfrak{K}(\mathfrak{A}_i)$, $i = 1, 2$.

Write $C = \mathrm{End}_K(V)$, $C_i = \mathrm{End}_K(V_i)$. Put $\mathfrak{C} = \mathfrak{A} \cap C$, $\mathfrak{C}_i = \mathfrak{A}_i \cap C_i$. Write $\mathfrak{R} = \mathrm{rad}(\mathfrak{C})$, $\mathfrak{R}_i = \mathrm{rad}(\mathfrak{C}_i)$. Let $\mathfrak{M}$ be some $(\mathfrak{C}, \mathfrak{C})$-bimodule in A. Then $\mathbf{e}_i \in \mathfrak{C}$ and:

(ii) $\mathbf{e}_i \mathfrak{M} \mathbf{e}_j = \mathfrak{M} \cap \mathbf{e}_i A \mathbf{e}_j$, $i, j \in \{1, 2\}$.

(iii) $\mathfrak{M} = \coprod_{i,j} \mathbf{e}_i \mathfrak{M} \mathbf{e}_j$.

In particular, we have

(iv) $\mathfrak{R}^m \cap A_i = \mathbf{e}_i \mathfrak{R}^m \mathbf{e}_i = \mathfrak{R}_i^m$, $m \in \mathbb{Z}$, $i = 1, 2$.

Proof: By hypothesis, K commutes with the idempotents $\mathbf{e}_i$, so, for $x \in K^\times$, $k \in \mathbb{Z}$, we have $xL_k^i = \mathbf{e}_i x M_k = \mathbf{e}_i M_{k+\nu_{\mathfrak{A}}(x)} = L^i_{k+\nu_{\mathfrak{A}}(x)} \in \mathcal{L}^i$. This proves *(i)*.

The condition $\mathbf{e}_i \in \mathfrak{C}$ is equivalent to $\mathbf{e}_i M_k \subset M_k$, for all $k \in \mathbb{Z}$. However, $\mathbf{e}_i M_k = L_k^i$, and the assertion follows.

In *(ii)*, we have $\mathfrak{M} \cap \mathbf{e}_i A \mathbf{e}_j \subset \mathbf{e}_i \mathfrak{M} \mathbf{e}_j$, while $\mathbf{e}_i \mathfrak{M} \mathbf{e}_j \subset \mathfrak{M}$, since the $\mathbf{e}_i \in \mathfrak{C}$ and $\mathfrak{M}$ is a $(\mathfrak{C}, \mathfrak{C})$-bimodule. This proves *(ii)*, and *(iii)* follows immediately.

In *(iv)*, we have $\mathfrak{R}^m \cap A_i \subset \mathfrak{R}_i^m$, simply by comparing actions on lattice chains. Likewise, if $x \in \mathfrak{R}_i^m$, we get $xM_k = xL_k^i \subset L^i_{m+k} \subset M_{m+k}$. This shows $x \in \mathfrak{R}^m$, and the result follows. ∎

In particular, the lemma applies to $K = E$. Since $\mathfrak{H}^k(\beta, \mathfrak{A})$ is a $(\mathfrak{B}, \mathfrak{B})$-bimodule, we get

$$\mathfrak{H}^k(\beta, \mathfrak{A}) \cap A_i = \mathbf{e}_i \mathfrak{H}^k(\beta, \mathfrak{A}) \mathbf{e}_i, \quad k \geq 0.$$

In fact, we have

$$\mathfrak{H}^k(\beta, \mathfrak{A}) \cap A_i = \mathfrak{H}^k(\beta, \mathfrak{A}_i). \tag{3.6.16}$$

It is enough to prove this in the case $k = 0$, since $\mathfrak{H}^k = \mathfrak{H} \cap \mathfrak{P}^k$, $\mathfrak{P} = \mathrm{rad}(\mathfrak{A})$. When β is minimal over F, we have $\mathfrak{H}(\beta, \mathfrak{A}) = \mathfrak{B} + \mathfrak{P}^{[-\nu/2]+1}$. Multiplying on either side by $\mathbf{e}_i$ and using (3.6.15), we get

$$\mathfrak{H}(\beta, \mathfrak{A}) \cap A_i = \mathbf{e}_i \mathfrak{B} \mathbf{e}_i + \mathbf{e}_i \mathfrak{P}^{[-\nu/2]+1} \mathbf{e}_i = \mathfrak{B}_i + \mathfrak{P}_i^{[-\nu/2]+1},$$

where $\mathfrak{P}_i = \mathrm{rad}(\mathfrak{A}_i)$, and this last expression is just $\mathfrak{H}(\beta, \mathfrak{A}_i)$.

In the general case, let $r_0 = -k_0(\beta, \mathfrak{A}_i) = -k_0(\beta, \mathfrak{A}(E))$. We choose a simple stratum $[\mathfrak{A}(E), -\nu, r_0, \gamma]$ in $A(E)$ which is equivalent to $[\mathfrak{A}(E), -\nu, r_0, \beta]$. We can then embed (for example, via a suitable (W, E)-decomposition) the field $F[\gamma]$ in A_i in such a way that $[\mathfrak{A}_i, -\nu, r_0, \gamma]$ is a simple stratum equivalent to $[\mathfrak{A}_i, -\nu, r_0, \beta]$. The direct sum of these embeddings gives an embedding $F[\gamma] \to A$. The stratum $[\mathfrak{A}, -\nu, r_0, \gamma]$ is then simple and equivalent to $[\mathfrak{A}, -\nu, r_0, \beta]$. Moreover, as a subfield of A, $F[\gamma]$ satisfies the hypotheses of (3.6.15). We then have

$$\mathfrak{H}(\beta, \mathfrak{A}) = \mathfrak{B} + \mathfrak{H}^{[r_0/2]+1}(\gamma, \mathfrak{A}),$$
$$\mathfrak{H}(\beta, \mathfrak{A}) \cap A_i = \mathfrak{B}_i + \mathfrak{H}^{[r_0/2]+1}(\gamma, \mathfrak{A}_i) = \mathfrak{H}(\beta, \mathfrak{A}_i),$$

by induction and (3.6.15). This proves (3.6.16) and implies

$$H^k(\beta, \mathfrak{A}_1) \times H^k(\beta, \mathfrak{A}_2) \subset H^k(\beta, \mathfrak{A}), \quad k \in \mathbb{Z}.$$

Indeed, $H^k(\beta, \mathfrak{A}_i) = \mathbf{e}_i H^k(\beta, \mathfrak{A})\mathbf{e}_i$. Given $\theta \in \mathcal{C}(\mathfrak{A}, m_0, \beta)$, we can then form $\theta_i = \theta \mid H^{m_0+1}(\beta, \mathfrak{A}_i)$. We have to show that $\theta_i \in \mathcal{C}(\mathfrak{A}_i, m_0, \beta)$, and that $\theta \mapsto \theta_i$ establishes a bijection of the desired sort between $\mathcal{C}(\mathfrak{A}, m_0, \beta)$ and $\mathcal{C}(\mathfrak{A}_i, m_0, \beta)$. We proceed by induction along β, using the explicit parametrisation of simple characters introduced in the proof of (3.6.1). First, however, we take our fixed additive character ψ_F of F and form

$$\psi = \psi_A = \psi_F \circ \mathrm{tr}_A, \quad \psi_i = \psi_{A_i} = \psi_F \circ \mathrm{tr}_{A_i}.$$

We then have $\psi \mid A_i = \psi_i$ and, if $a \in A$ is of the form $a = a_1 + a_2$, $a_i \in A_i$, we further get $\psi_a \mid A_i = \psi_{i,a_i}$. In particular, we have

$$\psi_{i,\beta} = \psi_\beta \mid A_i.$$

Suppose first that β is minimal over F, and let χ_0 be the unique character of $U^{[\frac{-\nu}{2}]+1}(\mathfrak{o}_E)$ such that $\psi_\beta \mid U^{[\frac{-\nu}{2}]+1}(\mathfrak{B}) = \chi_0 \circ \det_B$. Then we also have

$$\psi_{i,\beta} \mid U^{[\frac{-\nu}{2}]+1}(\mathfrak{B}_i) = \chi_0 \circ \det{}_B.$$

Take $\theta \in \mathcal{C}(\mathfrak{A}, m_0, \beta)$. If $m_0 \geq [\frac{-\nu}{2}]$, we have $\theta = \psi_\beta$, and $\theta_i = \psi_{i,\beta}$. This certainly lies in $\mathcal{C}(\mathfrak{A}_i, m_0, \beta)$, and $\theta \mapsto \theta_i$ is bijective, as required. Otherwise, let χ_θ be the character of $U^{m_0+1}(\mathfrak{o}_E)$ such that $\theta \mid U^{m_0+1}(\mathfrak{B}) = \chi_\theta \circ \det_B$. As in the proof of (3.6.1), the map $\theta \mapsto \chi_\theta$ gives a bijection between $\mathcal{C}(\mathfrak{A}, m_0, \beta)$ and the set of characters χ of $U^{m_0+1}(\mathfrak{o}_E)$ which agree with χ_0 on $U^{[\frac{-\nu}{2}]+1}(\mathfrak{o}_E)$. We have $\theta_i \mid U^{m_0+1}(\mathfrak{B}_i) = \chi_\theta \circ \det_{B_i}$. The assertions then follow.

In the general case, we take an element γ as above in the proof of (3.6.16). Put $c = \beta - \gamma$. Suppose first that $m_0 \geq [\frac{r_0}{2}]$, where $r_0 = -k_0(\beta, \mathfrak{A}) = -k_0(\beta, \mathfrak{A}_i)$. Take $\theta \in \mathcal{C}(\mathfrak{A}, m_0, \beta)$, and write it in the form $\theta = \theta_0\psi_c$, where $\theta_0 \in \mathcal{C}(\mathfrak{A}, m_0, \gamma)$. Then $\theta_i = \theta_0 \mid H^{m_0+1}(\beta, \mathfrak{A}_i).\psi_{i,c}$. In this case, the assertion follows by induction and (3.3.18). The case $m_0 < [\frac{r_0}{2}]$ now follows, as before, using the parametrisations of the proof of (3.6.1).

This bijection $\theta \mapsto \theta_i$ commutes with restriction, preserves β-determinants and is inductive, in the sense of (3.6.13). The composite map

$$\mathcal{C}(\mathfrak{A}_1, m_0, \beta) \xrightarrow{\approx} \mathcal{C}(\mathfrak{A}, m_0, \beta) \xrightarrow{\approx} \mathcal{C}(\mathfrak{A}_2, m_0, \beta)$$

therefore has all the desired properties. ■

We must now discharge the obligation of proving (3.6.10). Recall that we are given simple strata $[\mathfrak{A}_i, n_i, m_i, \beta]$ in $A = \mathrm{End}_F(V)$, $i = 1, 2$. We put $E = F[\beta]$, $B = \mathrm{End}_E(V)$, $\mathfrak{B}_i = \mathfrak{A}_i \cap B$. We assume that the orders $\mathfrak{B}_i$ satisfy the conditions of (1.1.9), so that their lattice chains admit a common basis and the orders $\mathfrak{A}_i$ admit a common (W, E)-decomposition. Abbreviate $\mathfrak{H}_i^k = \mathfrak{H}^k(\beta, \mathfrak{A}_i)$. We have to show that, for $x \in B^\times$, we have

$$x^{-1}\mathfrak{H}_1^{m_1+1}x \cap \mathfrak{H}_2^{m_2+1} = x^{-1}\mathfrak{Q}_1^{m_1+1}x \cap \mathfrak{Q}_2^{m_2+1} + x^{-1}\mathfrak{H}_1^{[r_1/2]+1}x \cap \mathfrak{H}_2^{[r_2/2]+1},$$

where $\mathfrak{Q}_i = \mathrm{rad}(\mathfrak{B}_i)$, $r_i = -k_0(\beta, \mathfrak{A}_i)$, and $m_i \leq [\frac{r_i}{2}]$, $i = 1, 2$.

We clearly have containment in the direction $\supset$. Moreover, both sides of the desired equation have the same intersection with B, namely $x^{-1}\mathfrak{Q}_1^{m_1+1}x \cap \mathfrak{Q}_2^{m_2+1}$. It is therefore enough to show that they have the same image under a_β. Applying a_β to each side, and remembering the obvious containment relation, we get

$$\begin{aligned}
a_\beta(x^{-1}&\mathfrak{H}_1^{[\frac{r_1}{2}]+1}x \cap \mathfrak{H}_2^{[\frac{r_2}{2}]+1}) \\
&= a_\beta(x^{-1}\mathfrak{Q}_1^{m_1+1}x \cap \mathfrak{Q}_2^{m_2+1} + x^{-1}\mathfrak{H}_1^{[\frac{r_1}{2}]+1}x \cap \mathfrak{H}_2^{[\frac{r_2}{2}]+1}) \\
&\subset a_\beta(x^{-1}\mathfrak{H}_1^{m_1+1}x \cap \mathfrak{H}_2^{m_2+1}) \\
&\subset a_\beta(x^{-1}\mathfrak{H}_1^{m_1+1}x) \cap a_\beta(\mathfrak{H}_2^{m_2+1}) \\
&= a_\beta(x^{-1}\mathfrak{H}_1^{[\frac{r_1}{2}]+1}x) \cap a_\beta(\mathfrak{H}_2^{[\frac{r_2}{2}]+1}).
\end{aligned}$$

We henceforward abbreviate $\mathfrak{H}_i^{[\frac{r_i}{2}]+1} = \mathfrak{h}_i$. We are reduced to proving

(3.6.17) $$a_\beta(x^{-1}\mathfrak{h}_1 x \cap \mathfrak{h}_2) = a_\beta(x^{-1}\mathfrak{h}_1 x) \cap a_\beta(\mathfrak{h}_2).$$

Now we note a simple general fact. Given an abelian group A, with an endomorphism f and subgroups L, M, there is a canonical isomorphism

$$\frac{C \cap (L+M)}{(C\cap L)+(C\cap M)} \xrightarrow{\approx} \frac{f(L)\cap f(M)}{f(L\cap M)},$$

where $C = \mathrm{Ker}(f)$. Explicitly, this isomorphism is given by $x+y \mapsto f(x)$, where $x \in L$, $y \in M$ and $x+y \in C\cap(L+M)$.

In our present situation, this says that (3.6.17) is implied by

$$(x^{-1}\mathfrak{h}_1 x + \mathfrak{h}_2)\cap B = x^{-1}\mathfrak{Q}_1^{[\frac{r_1}{2}]+1}x + \mathfrak{Q}_2^{[\frac{r_2}{2}]+1}. \tag{3.6.18}$$

Here we have

$$\begin{aligned} x^{-1}\mathfrak{Q}_1^{[\frac{r_1}{2}]+1}x + \mathfrak{Q}_2^{[\frac{r_2}{2}]+1} &\subset (x^{-1}\mathfrak{h}_1 x + \mathfrak{h}_2)\cap B \\ &\subset (x^{-1}\mathfrak{P}_1^{[\frac{r_1}{2}]+1}x + \mathfrak{P}_2^{[\frac{r_2}{2}]+1})\cap B. \end{aligned}$$

It is therefore enough to show

$$(x^{-1}\mathfrak{P}_1^{[\frac{r_1}{2}]+1}x + \mathfrak{P}_2^{[\frac{r_2}{2}]+1})\cap B = x^{-1}\mathfrak{Q}_1^{[\frac{r_1}{2}]+1}x + \mathfrak{Q}_2^{[\frac{r_2}{2}]+1},$$

or, equivalently,

$$(\mathfrak{P}_1^{[\frac{r_1}{2}]+1}x + x\mathfrak{P}_2^{[\frac{r_2}{2}]+1})\cap B = \mathfrak{Q}_1^{[\frac{r_1}{2}]+1}x + x\mathfrak{Q}_2^{[\frac{r_2}{2}]+1}. \tag{3.6.19}$$

We argue as in (1.3.16). Choose a minimal hereditary $\mathfrak{o}_E$-order $\mathfrak{B}_{\mathrm{m}}$ contained in $\mathfrak{B}_1 \cap \mathfrak{B}_2$ (as we may, by (1.1.9)). (3.6.19) holds for the element x if and only if it holds for $y = uxv$, where $u, v \in \boldsymbol{U}(\mathfrak{B}_{\mathrm{m}})$. As in (1.3.16), we can choose u and v to make the lattice $\mathfrak{P}_1^{[\frac{r_1}{2}]+1}y + y\mathfrak{P}_2^{[\frac{r_2}{2}]+1}$ E-exact, in the sense of (1.3.10). This implies (3.6.19) straightaway, and hence also (3.6.18) and (3.6.17).

This completes the proof of (3.6.10).

4. INTERLUDE WITH HECKE ALGEBRAS

By way of relief, we now recall, in convenient form with proofs, some essentially well-known general results on Hecke algebras. Throughout, we shall be concerned with a locally profinite (i.e. locally compact, totally disconnected) topological group G with a countable base of open sets, and a compact open subgroup K of G. To simplify matters, we shall also assume that G is *unimodular*.

(4.1) Induction and intertwining

We fix once for all a Haar measure dg on our group G, and use μ to denote the measure of a subset of G with respect to dg:

$$\mu(S) = \int_S dg.$$

Let ρ be a continuous representation of K on a finite-dimensional complex vector space W, and write $(\check{\rho}, \check{W})$ for its contragredient. We recall the definition of the Hecke algebra $\mathcal{H}(G,\rho)$ of ρ-spherical functions on G. The notation becomes simpler if we define instead $\mathcal{H}(G,\check{\rho})$. This is the $\mathbb{C}$-space of compactly supported functions $\Phi : G \to \mathrm{End}_{\mathbb{C}}(W)$ satisfying

$$\Phi(k_1gk_2) = \rho(k_1) \circ \Phi(g) \circ \rho(k_2), \quad k_i \in K, \ g \in G.$$

Note that such functions are automatically smooth. We make $\mathcal{H}(G,\check{\rho})$ into an associative $\mathbb{C}$-algebra under convolution:

$$\Phi_1 * \Phi_2(g) = \int_G \Phi_1(x) \circ \Phi_2(x^{-1}g)\, dx, \quad g \in G.$$

It also has the unit element

$$\mathrm{e}(x) = \begin{cases} \mu(K)^{-1}.\rho(x) & \textit{if } x \in K, \\ 0 & \textit{otherwise.} \end{cases}$$

Now write $\langle\, ,\, \rangle_W$ for the canonical K-invariant bilinear pairing between W and $\check{W}$. If $\alpha \in \mathrm{End}_{\mathbb{C}}(W)$, we can define $\check{\alpha} \in \mathrm{End}_{\mathbb{C}}(\check{W})$ by $\langle w, \check{\alpha}\check{w}\rangle = \langle \alpha w, \check{w}\rangle$ for $w \in W$, $\check{w} \in \check{W}$, whence in particular $\rho(k)\check{} = \check{\rho}(k^{-1})$. This operator "klick" gives an anti-isomorphism $\mathrm{End}_{\mathbb{C}}(W) \to \mathrm{End}_{\mathbb{C}}(\check{W})$ (i.e. a $\mathbb{C}$-linear isomorphism which reverses multiplication). We also denote its inverse by "klick". It further defines an anti-isomorphism $\mathcal{H}(G,\rho) \xrightarrow{\approx} \mathcal{H}(G,\check{\rho})$ by $\Phi \mapsto \Phi'$, where $\Phi'(g) = \Phi(g^{-1})\check{}$.

Now fix $g \in G$, and write ${}^gK$ for the group gKg^{-1}, and $({}^g\rho, {}^gW)$ for the representation $x \mapsto \rho(g^{-1}xg)$ of ${}^gK$. Recall that g is said to *intertwine* ρ if there is a nonzero ${}^gK \cap K$-homomorphism from ${}^gW$ to W.

(4.1.1) Proposition: *In the above situation, let $g \in G$. The following are equivalent:*

(i) g intertwines ρ;

(ii) there exists $\Phi \in \mathcal{H}(G, \check{\rho})$ with $\Phi(g) \neq 0$.

If the element g satisfies these conditions, then we have a canonical vector space isomorphism between $\mathrm{Hom}_{K \cap {}^gK}({}^gW, W)$ and the space of functions $\Phi \in \mathcal{H}(G, \check{\rho})$ which vanish outside the double coset KgK.

Proof: We identify the underlying vector spaces of W, ${}^gW$. If $\Phi \in \mathcal{H}(G, \check{\rho})$ has $\Phi(g) \neq 0$, then $\Phi(g)$ indeed defines a nonzero ${}^gK \cap K$-homomorphism from ${}^gW$ to W. Conversely, if $f \in \mathrm{Hom}_{{}^gK \cap K}({}^gW, W)$ is nonzero, we can define a function $\Phi \in \mathcal{H}(G, \check{\rho})$ supported on the double coset KgK by $\Phi(k_1gk_2) = \rho(k_1) \circ f \circ \rho(k_2)$. All assertions are now immediate. ∎

Remark: It is often convenient to assemble the various dualities behind (4.1.1) in a different way. In the same situation, take $g \in G$ and put $K^g = g^{-1}Kg$, and define (ρ^g, W^g) in the obvious manner. Then we have:

(4.1.2) *In the situation of* (4.1.1), *take $g \in G$. There is a canonical isomorphism of vector spaces between $\mathrm{Hom}_{K^g \cap K}(W, W^g)$ and the space of $\Phi \in \mathcal{H}(G, \rho)$ supported on the double coset KgK.*

Continuing in the same situation, let H be another compact open subgroup of G which contains K. We can therefore define the induced representation (ρ_*, W_*) of H. The underlying space W_* consists of all functions $\phi : H \to W$ which satisfy

$$\phi(kh) = \rho(k).\phi(h), \quad k \in K, \ h \in H.$$

(4.1.3) Proposition: *There is a canonical algebra isomorphism*

$$\mathcal{H}(G, \rho) \xrightarrow{\approx} \mathcal{H}(G, \rho_*).$$

Proof: We identify W with the subspace of W_* consisting of functions in W_* which are null outside K. We actually produce an isomorphism between $\mathcal{H}(G, \check{\rho})$ and $\mathcal{H}(G, \check{\rho}_*)$, which is enough, since contragredience commutes with induction.

We have a canonical projection $\varpi : W_* \to W$ given by restricting functions to K. Then $\mathrm{End}_{\mathbb{C}}(W)$ is the subspace of $\mathrm{End}_{\mathbb{C}}(W_*)$ consisting of those endomorphisms f such that $\varpi \circ f \circ \varpi = f$. Indeed, it is a subalgebra but has a different unit element. With this identification, we can view the elements of $\mathcal{H}(G, \check{\rho})$ as $\mathrm{End}_{\mathbb{C}}(W_*)$-valued functions, and convolve them with elements of $\mathcal{H}(G, \check{\rho}_*)$. We also note the identities

$$\varpi \circ \rho_*(k) = \rho_*(k) \circ \varpi = \rho(k), \quad k \in K,$$
$$\varpi \circ \rho_*(h) \circ \varpi = 0, \quad h \in H, \ h \notin K.$$

Write $\mathbf{E}$ for the unit element of $\mathcal{H}(G, \check{\rho}_*)$. We therefore have $\varpi \circ \mathbf{E} \circ \varpi = (H : K)^{-1}\mathbf{e}$. Further,

$$\mathbf{E} * \mathbf{e} * \mathbf{E} = (H : K)^{-1}\mathbf{E}.$$

To get this identity, note first that $\mathbf{E} * \mathbf{e} * \mathbf{E} \in \mathcal{H}(G, \check{\rho}_*)$ and has support contained in H. The value of this function at 1 is

$$\mu(H)^{-2} \int_H \rho_*(y) \circ \varpi \circ \rho_*(y^{-1})\, dy.$$

One evaluates this directly via its action on the functions in the space W_*, and gets $\mathbf{E} * \mathbf{e} * \mathbf{E}(1) = \mu(K).\mu(H)^{-2}.1_{W_*}$, as required. Using these identities, it is immediate that

$$\textbf{(4.1.4)} \qquad \begin{cases} \mathcal{H}(G, \check{\rho}_*) \to \mathcal{H}(G, \check{\rho}) & \text{by } \Phi \mapsto (H : K).\varpi \circ \Phi \circ \varpi, \\ \mathcal{H}(G, \check{\rho}) \to \mathcal{H}(G, \check{\rho}_*) & \text{by } \phi \mapsto (H : K).\mathbf{E} * \phi * \mathbf{E}, \end{cases}$$

are mutually inverse vector space isomorphisms which take $\mathbf{E}$ to $\mathbf{e}$ and vice-versa. To prove they are algebra isomorphisms, we consider the space $\mathcal{W}$ consisting of compactly supported functions

$$\Psi : G \to \mathrm{Hom}_{\mathbb{C}}(W_*, W),$$

satisfying

$$\Psi(kgh) = \rho(k) \circ \Psi(g) \circ \rho_*(h), \quad k \in K, \ h \in H, \ g \in G.$$

Using convolution of functions, $\mathcal{W}$ becomes a $(\mathcal{H}(G, \check{\rho}), \mathcal{H}(G, \check{\rho}_*))$-bimodule. However, we have a right $\mathcal{H}(G, \check{\rho}_*)$-isomorphism $\mathcal{H}(G, \check{\rho}_*) \to \mathcal{W}$ by $\Phi \mapsto \varpi \circ \Phi$, with inverse $\Psi \mapsto (H : K).\mathbf{E} * \Psi$. Thus $\mathcal{W}$ is free of rank 1 as right $\mathcal{H}(G, \check{\rho}_*)$-module. The left action of $\mathcal{H}(G, \check{\rho})$ on $\mathcal{W}$ implies an *algebra* homomorphism $\mathcal{H}(G, \check{\rho}) \to \mathcal{H}(G, \check{\rho}_*)$, which is easily seen to coincide with the vector space isomorphism in (4.1.4). ∎

The maps we have defined here respect the transitivity of induction.

For $g \in G$, we write $\mathcal{H}(G,\rho)_g$ for the subspace of $\mathcal{H}(G,\rho)$ consisting of functions which are null outside KgK. We use a similar notation for $\mathcal{H}(G,\rho_*)$.

(4.1.5) Corollary: *Let $g \in G$. The isomorphisms* (4.1.3) *restrict to give an isomorphism*

$$\mathcal{H}(G,\rho_*)_g \cong \coprod_{\substack{g' \in K\backslash G/K \\ Hg'H = HgH}} \mathcal{H}(G,\rho)_{g'}.$$

(4.2) Scalar Hecke algebras

Now we write $\mathcal{H}(G)$ for the algebra (under convolution of functions) consisting of smooth, compactly-supported functions $G \to \mathbb{C}$. Again we take a compact open subgroup K of G and a continuous *irreducible* representation (ρ, W) of K. We define a function $e_\rho \in \mathcal{H}(G)$ by

$$\textbf{(4.2.1)} \qquad e_\rho(x) = \begin{cases} \mu(K)^{-1}\dim(\rho)\mathrm{tr}(\rho(x^{-1})) & \textit{if } x \in K, \\ 0 & \textit{otherwise.} \end{cases}$$

This is an idempotent element of $\mathcal{H}(G)$, $e_\rho * e_\rho = e_\rho$. The space $e_\rho * \mathcal{H}(G) * e_\rho$ is therefore a subalgebra of $\mathcal{H}(G)$, and it has a unit element, namely e_ρ.

Let $(\pi, \mathcal{V})$ be a smooth representation of G (we refer to **[Cr]** and **[Cs]** for the elementary representation theory of groups like G). We write $(\check{\pi}, \check{\mathcal{V}})$ for the contragredient or smooth dual of $(\pi, \mathcal{V})$. As usual, we extend π to a representation of the algebra $\mathcal{H}(G)$ on $\mathcal{V}$ by

$$\langle \pi(\Phi)v, \check{v}\rangle = \int_G \Phi(g)\langle \pi(g)v, \check{v}\rangle \, dg,$$

where $\langle\, ,\, \rangle$ is the canonical evaluation pairing on $\mathcal{V} \times \check{\mathcal{V}}$ and $\Phi \in \mathcal{H}(G)$, $v \in \mathcal{V}$, $\check{v} \in \check{\mathcal{V}}$.

Viewed, by restriction, as a representation of K, we have a discrete direct sum decomposition

$$\mathcal{V} = \sum_{\sigma \in \widehat{K}} \mathcal{V}^\sigma,$$

where σ ranges over the set $\widehat{K}$ of equivalence classes of irreducible representations of K, and $\mathcal{V}^\sigma$ denotes the σ-isotypic component of $\mathcal{V}$.

(4.2.2) Proposition: *In the above situation, $\pi(e_\rho)$ is the K-projection of $\mathcal{V}$ onto the component $\mathcal{V}^\rho$, so that $\mathcal{V}^\rho$ is an $e_\rho * \mathcal{H}(G) * e_\rho$-module.*

This is, of course, completely elementary.

(4.2.3) Proposition: *The following sets are in natural bijection:*

(i) equivalence classes of irreducible smooth representations $(\pi, \mathcal{V})$ of G with $\mathcal{V}^\rho \neq 0$;

(ii) isomorphism classes of simple $e_\rho * \mathcal{H}(G) * e_\rho$-modules.

Proof: Let $(\pi, \mathcal{V})$ be an irreducible smooth representation of G with $\mathcal{V}^\rho$ nonzero. We assert that $\mathcal{V}^\rho$ is simple as $\mathcal{H}_\rho(G)$-module, where we temporarily abbreviate $e_\rho * \mathcal{H}(G) * e_\rho = \mathcal{H}_\rho(G)$. Suppose otherwise, and let $\mathcal{M}$ be a proper $\mathcal{H}_\rho(G)$-submodule of $\mathcal{V}^\rho$. Consider the $\mathcal{H}(G)$-submodule $\mathcal{W} = \mathcal{H}(G).\mathcal{M}$ of $\mathcal{V}$ generated by $\mathcal{M}$. Then $\mathcal{W}$ is a G-subspace of $\mathcal{V}$, which is nonzero since it contains $\mathcal{M}$, and therefore $\mathcal{W} = \mathcal{V}$. However, $\mathcal{W}^\rho = \pi(e_\rho)\mathcal{W} = e_\rho * \mathcal{H}(G).\mathcal{M} = (e_\rho * \mathcal{H}(G) * e_\rho).\mathcal{M}$, since $e_\rho.\mathcal{M} = \mathcal{M}$. Thus $\mathcal{W}^\rho = \mathcal{M} \neq \mathcal{V}^\rho$. This contradiction implies that $\mathcal{V}^\rho$ is a simple $\mathcal{H}_\rho(G)$-module. Thus we get a map $\mathcal{V} \mapsto \mathcal{V}^\rho$ from the irreducible smooth representations $\mathcal{V}$ of G which contain ρ to the simple $\mathcal{H}_\rho(G)$-modules, and this preserves equivalence.

Next, we take a simple $\mathcal{H}_\rho(G)$-module $\mathcal{M}$. We view $\mathcal{H}(G)$ as a right $\mathcal{H}_\rho(G)$-module, via the algebra inclusion $\mathcal{H}_\rho(G) \to \mathcal{H}(G)$, and form the left $\mathcal{H}(G)$-module $\mathcal{V}_\mathcal{M} = \mathcal{H}(G) \otimes_{\mathcal{H}_\rho(G)} \mathcal{M}$. This affords a smooth representation of G, where G acts via left translation on the factor $\mathcal{H}(G)$. We have $\mathcal{V}^\rho_\mathcal{M} = e_\rho * \mathcal{H}(G) \otimes \mathcal{M} = e_\rho * \mathcal{H}(G) * e_\rho \otimes \mathcal{M} = e_\rho \otimes \mathcal{M}$, which we may identify with $\mathcal{M}$. We show that $\mathcal{V}_\mathcal{M}$ has a canonical irreducible quotient $\mathcal{V}_\mathcal{M}/\mathcal{U}_\mathcal{M}$ with $(\mathcal{V}_\mathcal{M}/\mathcal{U}_\mathcal{M})^\rho = \mathcal{M}$. To do this, we choose a nonzero element $\mathfrak{m} \in \mathcal{M}$. This element generates $\mathcal{M}$ as $\mathcal{H}_\rho(G)$-module. Zorn's Lemma guarantees the existence of a G-subspace $\mathcal{U}_\mathcal{M}$ of $\mathcal{V}_\mathcal{M}$ which is maximal for the property $\mathfrak{m} \notin \mathcal{U}_\mathcal{M}$. We have $\mathcal{U}^\rho_\mathcal{M} = \mathcal{U}_\mathcal{M} \cap \mathcal{M} = \{0\}$, since otherwise, by the simplicity hypothesis on $\mathcal{M}$, we would have $\mathfrak{m} \in \mathcal{M} \subset \mathcal{U}$. Thus the space $\mathcal{U}_\mathcal{M}$ is uniquely determined, being the sum of all G-subspaces of $\mathcal{V}_\mathcal{M}$ contained in the kernel of the projection $\mathcal{V}_\mathcal{M} \to \mathcal{M}$ given by $v \mapsto e_\rho * v$. If we have a G-space $\mathcal{W}$, with $\mathcal{U}_\mathcal{M} \subsetneq \mathcal{W} \subset \mathcal{V}_\mathcal{M}$, then $\mathfrak{m} \in \mathcal{W}$ by the definition of $\mathcal{U}_\mathcal{M}$, so $\mathcal{W}$ contains $\mathcal{M}$ and hence is equal to $\mathcal{V}_\mathcal{M}$. It follows that the G-space $\mathcal{V}_\mathcal{M}/\mathcal{U}_\mathcal{M}$ is irreducible. Since $\mathcal{U}_\mathcal{M} \cap \mathcal{M} = \{0\}$, we have $(\mathcal{V}_\mathcal{M}/\mathcal{U}_\mathcal{M})^\rho \cong \mathcal{M}$.

Thus we get a pair of maps between equivalence classes of irreducible representations $\mathcal{V}$ of G which contain ρ and isomorphism classes of simple $\mathcal{H}_\rho(G)$-modules. We have to show that these are mutually inverse bijections. Starting with the simple $\mathcal{H}_\rho(G)$-module $\mathcal{M}$, we have just seen that $(\mathcal{V}_\mathcal{M}/\mathcal{U}_\mathcal{M})^\rho = \mathcal{M}$. In the other direction, starting with the representation $(\pi, \mathcal{V})$, we put $\mathcal{M} = \mathcal{V}^\rho$, and form $\mathcal{V}_\mathcal{M}$. There is a canonical G-surjection $\mathcal{V}_\mathcal{M} \to \mathcal{V}$ by $\phi \otimes v \mapsto \pi(\phi)v$, $\phi \in \mathcal{H}(G)$, $v \in \mathcal{V}$. This map induces an isomorphism $\mathcal{V}_{\mathcal{M}}^{\rho} \cong \mathcal{M}$. The kernel $\mathcal{U}$ of the map $\mathcal{V}_{\mathcal{M}} \to \mathcal{V}$ is thus a maximal G-subspace of $\mathcal{V}_{\mathcal{M}}$ which is contained in the kernel of the projection $\mathcal{V}_{\mathcal{M}} \to \mathcal{V}_{\mathcal{M}}^{\rho}$. Therefore $\mathcal{U} = \mathcal{U}_{\mathcal{M}}$, and $\mathcal{V} \cong \mathcal{V}_{\mathcal{M}}/\mathcal{U}_{\mathcal{M}}$.

This completes the proof of the Proposition. ∎

(4.2.4) Proposition: *Let (ρ, W) be an irreducible smooth representation of K. There is a canonical algebra isomorphism*

$$\Upsilon : \mathcal{H}(G,\rho) \otimes_{\mathbb{C}} \mathrm{End}_{\mathbb{C}}(W) \xrightarrow{\approx} e_\rho * \mathcal{H}(G) * e_\rho.$$

Proof: As before, we write $(\check{\rho}, \check{W})$ for the contragredient of (ρ, W), and $\langle\,,\rangle : W \times \check{W} \to \mathbb{C}$ for the canonical evaluation pairing. We recall to start with the Schur orthogonality relation:—

$$\int_K \langle w_1, \check{\rho}(k)\check{w}_1\rangle\langle w_2, \check{\rho}(k^{-1})\check{w}_2\rangle\, dk = \frac{\mu(K)}{\dim(\rho)}\langle w_1, \check{w}_2\rangle\langle w_2, \check{w}_1\rangle,$$

for any $w_i \in W$, $\check{w}_i \in \check{W}$.

We also identify the algebra $\mathrm{End}_{\mathbb{C}}(W)$ with $W \otimes_{\mathbb{C}} \check{W}$. Recall that this tensor acts on W by $(w \otimes \check{w}) : v \mapsto \langle v, \check{w}\rangle w$. The multiplication in $W \otimes \check{W}$ is given by $(w_1 \otimes \check{w}_1)(w_2 \otimes \check{w}_2) = \langle w_2, \check{w}_1\rangle w_1 \otimes \check{w}_2$. To get the unit element, we choose a basis $\{w_j\}$ of W and take the dual basis $\{\check{w}_j\}$ of $\check{W}$, and then the 1 of $\mathrm{End}_{\mathbb{C}}(W)$ is $\sum_j w_j \otimes \check{w}_j$.

Now let $\Phi \in \mathcal{H}(G,\rho)$, $w \in W$, $\check{w} \in \check{W}$. We define a function $\phi = \Upsilon(\Phi \otimes w \otimes \check{w})$ from G to $\mathbb{C}$ by

$$\begin{aligned}\phi(g) &= \dim(\rho)\langle w, \Phi(g)\check{w}\rangle,\\ &= \dim(\rho)\mathrm{tr}(w \otimes \Phi(g)\check{w}), \quad g \in G.\end{aligned}$$

This function ϕ is certainly smooth and of compact support. The identity element $\mathbf{e}\otimes\sum_j w_j \otimes \check{w}_j$ of $\mathcal{H}(G,\rho)\otimes\mathrm{End}_{\mathbb{C}}(W)$ gets mapped to the identity e_ρ of $\mathcal{H}_\rho(G) = e_\rho * \mathcal{H}(G) * e_\rho$. To show that ϕ lies in $\mathcal{H}_\rho(G)$, we have to check that $e_\rho * \phi * e_\rho = \phi$. This follows immediately from the identity (which follows from the Schur orthogonality relation)

$$\int_K e_\rho(k)\check{\rho}(k^{-1})\, dk = 1_{\check{W}}.$$

We next show that our map Υ preserves multiplication, i.e. if $\Phi_i \in \mathcal{H}(G,\rho)$, $w_i \in W$, $\check{w}_i \in \check{W}$, and $\phi_i = \Upsilon(\Phi_i \otimes w_i \otimes \check{w}_i)$, for $i = 1, 2$, then

$$\phi_1 * \phi_2 = \Upsilon((\Phi_1 * \Phi_2) \otimes w_1 \otimes \check{w}_2\langle w_2, \check{w}_1\rangle).$$

By linearity, it is enough to treat the case in which Φ_1 is supported on just one double coset KgK. Abbreviating $d = \dim(\rho)$, we have

$$\begin{aligned} d^{-2}\phi_1 * \phi_2(x) &= d^{-2}\int_G \phi_1(y)\phi_2(y^{-1}x)\,dy \\ &= \int_G \langle w_1, \Phi_1(y)\check{w}_1\rangle\langle w_2, \Phi_2(y^{-1}x)\check{w}_2\rangle\,dy. \end{aligned}$$

The integrand is null unless $y \in KgK$. We write KgK as a disjoint union of cosets ℓgK, so this last integral comes down to

$$\begin{aligned} \sum_\ell \int_K &\langle \Phi_1(\ell g)\check{}.w_1, \check{\rho}(k)\check{w}_1\rangle\langle w_2, \check{\rho}(k^{-1})\Phi_2(g^{-1}\ell^{-1}x)\check{w}_2\rangle\,dk \\ &= \mu(K)d^{-1}\sum_\ell \langle \Phi_1(\ell g)\check{}.w_1, \Phi_2(g^{-1}\ell^{-1}x)\check{w}_2\rangle.\langle w_2, \check{w}_1\rangle \\ &= \mu(K)d^{-1}\sum_\ell \langle w_1, \Phi_1(\ell g)\Phi_2(g^{-1}\ell^{-1}x)\check{w}_2\rangle.\langle w_2, \check{w}_1\rangle \\ &= d^{-1}\int_G \langle w_1, \Phi_1(y)\Phi_2(y^{-1}x)\check{w}_2\rangle\langle w_2, \check{w}_1\rangle\,dy \\ &= d^{-2}\boldsymbol{\Upsilon}((\Phi_1 \otimes w_1 \otimes \check{w}_1)(\Phi_2 \otimes w_2 \otimes \check{w}_2))(x), \end{aligned}$$

as desired.

So, our map $\boldsymbol{\Upsilon}$ is an algebra homomorphism. We next show it is injective. Its kernel is a two-sided ideal of $\mathcal{H}(G,\rho) \otimes \mathrm{End}_{\mathbb{C}}(W)$, and is therefore of the form $J \otimes \mathrm{End}_{\mathbb{C}}(W)$, for some two-sided ideal J of $\mathcal{H}(G,\rho)$. Suppose that J is nonzero, and choose a nonzero $\Phi \in J$. Choose $g \in G$ so that $\Phi(g) \neq 0$, $\check{w} \in \check{W}$ with $\Phi(g)\check{w} \neq 0$, and $w \in W$ so that $\langle w, \Phi(g)\check{w}\rangle \neq 0$. This says that $\boldsymbol{\Upsilon}(\Phi \otimes w \otimes \check{w})$ is not zero. Therefore $J = 0$, and $\boldsymbol{\Upsilon}$ is injective.

To prove it is surjective, we take $\phi \in e_\rho * \mathcal{H}(G) * e_\rho$. We choose a basis $\{w_1, w_2, \dots, w_n\}$ of W, and the dual basis $\{\check{w}_1, \check{w}_2, \dots, \check{w}_n\}$ of $\check{W}$ (i.e. $\langle w_i, \check{w}_j\rangle = \delta_{ij}$ (Kronecker delta)). For each pair i, j and $g \in G$, we define an operator $\Phi_{ij}(g) \in \mathrm{End}_{\mathbb{C}}(\check{W})$ by

$$\Phi_{ij}(g)\check{v} = \int_K\int_K \langle \rho(\ell)w_i, \check{v}\rangle\check{\rho}(k^{-1})\check{w}_j\phi(kg\ell)\,dk.d\ell.$$

One verifies immediately that the function $g \mapsto \Phi_{ij}(g)$ lies in $\mathcal{H}(G,\rho)$. Consider the element

$$\Psi = d\,\mu^{-2}\sum_{i,j}\Phi_{ij} \otimes w_j \otimes \check{w}_i$$

of $\mathcal{H}(G,\rho) \otimes \mathrm{End}_{\mathbb{C}}(W)$, where we abbreviate $d = \dim(\rho)$, $\mu = \mu(K)$. We have

$$\begin{aligned}
\Upsilon(\Psi)(g) &= d^2\mu^{-2} \sum_{i,j} \langle w_j, \Phi_{ij}(g)\check{w}_i \rangle \\
&= d^2\mu^{-2} \sum_{i,j} \int_K \int_K \langle w_j, \check{\rho}(k^{-1})\check{w}_j \rangle \langle \rho(\ell)w_i, \check{w}_i \rangle \phi(kg\ell) \, dk \, d\ell \\
&= d^2\mu^{-2} \int_K \int_K \mathrm{tr}(\rho(k))\mathrm{tr}(\rho(\ell))\phi(kg\ell) \, dk \, d\ell \\
&= \int_K \int_K e_\rho(k^{-1}) e_\rho(\ell^{-1}) \phi(kg\ell) \, dk \, d\ell \\
&= e_\rho * \phi * e_\rho(g) = \phi(g).
\end{aligned}$$

This proves that Υ is surjective, and we are done. ■

Put another way, the algebras $\mathcal{H}(G,\rho)$, $e_\rho * \mathcal{H}(G) * e_\rho$ are Morita equivalent, and so have equivalent module categories. In particular, we can identify their sets of isomorphism classes of simple modules. This correspondence is given explicitly as follows:— if $\mathcal{M}$ is a simple left $\mathcal{H}(G,\rho)$-module, then $\mathcal{M} \otimes_{\mathbb{C}} W$ is the corresponding simple module over $e_\rho * \mathcal{H}(G) * e_\rho \cong \mathcal{H}(G,\rho) \otimes_{\mathbb{C}} \mathrm{End}_{\mathbb{C}}(W)$. To describe the inverse of this correspondence, we take $\check{W}$ and view it as a right $\mathrm{End}_{\mathbb{C}}(W)$-module in the natural way. Then, if M is an $e_\rho * \mathcal{H}(G) * e_\rho$-module, we form $\check{W} \otimes_{\mathrm{End}_{\mathbb{C}}(W)} M$, which is an $\mathcal{H}(G,\rho)$-module via the second factor and the canonical embedding $\mathcal{H}(G,\rho) \to e_\rho * \mathcal{H}(G) * e_\rho$. Since $\check{W} \otimes_{\mathrm{End}_{\mathbb{C}}(W)} W \cong \mathbb{C}$, this gives the desired inverse. Alternatively, one can view M as a K-space via the action $(k,m) \mapsto \phi_k m$, $k \in K$, $m \in M$, where ϕ_k is the function $e_\rho * {}^k e_\rho$, and ${}^k e_\rho$ is the function $x \mapsto e_\rho(k^{-1}x)$. If we now view $\check{W}$ as a left K-space in the usual way, the space $\check{W} \otimes_{\mathrm{End}_{\mathbb{C}}(W)} M$ above is just the space of K-fixed vectors in $\check{W} \otimes_{\mathbb{C}} M$.

Either way, this and (4.2.3) together yield:

(4.2.5) Corollary: *Let K be a compact open subgroup of G, and ρ an irreducible smooth representation of K. There is a canonical bijection between the set of isomorphism classes of simple $\mathcal{H}(G,\rho)$-modules and the set of equivalence classes of irreducible smooth representations $\mathcal{V}$ of G for which $\mathcal{V}^\rho \neq \{0\}$ (i.e. which* contain ρ).

(4.2.6) Remark: The results and arguments of **(4.1)** and **(4.2)** also hold in a slightly more general situation (where we will later use them without comment). We take a closed subgroup Z of the centre $\mathcal{Z}(G)$ such that $\mathcal{Z}(G)/Z$ is compact. We assume that K is open, contains Z and K/Z is compact. We replace the Haar measure dg by a Haar measure $d\dot{g}$ on G/Z. In **(4.1)**, we have to assume that ρ is a continuous finite-dimensional representation of K such that $\rho \mid Z$ is a multiple of a fixed continuous quasicharacter $\chi : Z \to \mathbb{C}^\times$. In **(4.2)**, we replace $\mathcal{H}(G)$ by the convolution algebra $\mathcal{H}_\chi(G)$ of smooth functions $\Phi : G \to \mathbb{C}$ which are compactly supported modulo Z and satisfy $\Phi(zg) = \chi(z)^{-1}\Phi(G)$, $z \in Z$, $g \in G$.

It will be useful later to have some account of how the isomorphism $\boldsymbol{\Upsilon}$ of (4.2.4) behaves under induction. Of course, the machinery of (4.2.4) only makes sense in the context of irreducible representations, so we have to assume that the induced representation is irreducible.

Let K be some compact open subgroup of G, and (ρ, V) an irreducible smooth representation of K. Let H be another compact open subgroup of G with $H \supset K$. Write (σ, W) for the induced representation $\mathrm{Ind}(\rho : K, H)$. *We assume that σ is irreducible.* As in **(4.1)**, we identify V with the subspace of W consisting of functions which vanish identically outside K, and write $\varpi : W \to V$ for the canonical K-projection. Thus $\mathrm{End}_\mathbb{C}(V)$ gets identified with the subalgebra $\varpi \mathrm{End}_\mathbb{C}(W)\varpi$ of $\mathrm{End}_\mathbb{C}(W)$. Of course, this subalgebra has unit element ϖ, which is different from that of $\mathrm{End}_\mathbb{C}(W)$. We therefore get a canonical embedding of algebras

$$\mathcal{H}(G,\rho) \otimes_\mathbb{C} \mathrm{End}_\mathbb{C}(V) \to \mathcal{H}(G,\sigma) \otimes_\mathbb{C} \mathrm{End}_\mathbb{C}(W)$$

by tensoring the isomorphism of (4.1.3) with the inclusion $\mathrm{End}_\mathbb{C}(V) \to \mathrm{End}_\mathbb{C}(W)$.

On the other hand, if we view $\mathcal{H}(G)$ as a left G-module by left translation, convolution with e_ρ (resp. e_σ) is the canonical projection of $\mathcal{H}(G)$ onto its space of ρ- (resp. σ-) isotypic vectors. We have $\mathcal{H}(G)^\rho \subset \mathcal{H}(G)^\sigma$, so we conclude

$$e_\rho * e_\sigma = e_\sigma * e_\rho = e_\rho .$$

Thus $e_\rho * \mathcal{H}(G) * e_\rho$ is a subalgebra of $e_\sigma * \mathcal{H}(G) * e_\sigma$.

(4.2.7) Proposition: *In the situation above, the diagram*

$$\begin{array}{ccc} \mathcal{H}(G,\rho) \otimes_\mathbb{C} \mathrm{End}_\mathbb{C}(V) & \longrightarrow & \mathcal{H}(G,\sigma) \otimes_\mathbb{C} \mathrm{End}_\mathbb{C}(W) \\ \boldsymbol{\Upsilon}_\rho \big\downarrow & & \big\downarrow \boldsymbol{\Upsilon}_\sigma \\ e_\rho * \mathcal{H}(G) * e_\rho & \longrightarrow & e_\sigma * \mathcal{H}(G) * e_\sigma \end{array}$$

commutes, where $\boldsymbol{\Upsilon}_\rho$, $\boldsymbol{\Upsilon}_\sigma$ are the isomorphisms given by (4.2.4).

Proof: First we identify $\check{V}$ with a subspace of $\check{W}$: we take $\check{v} \in \check{V}$ and extend it to a linear form on W by making it trivial on the canonical complement $\mathrm{Ker}(\varpi)$ of V in W. This is the same as the embedding given by the identification $\check{\sigma} = \mathrm{Ind}(\check{\rho})$. The implied embedding $V \otimes \check{V} \to W \otimes \check{W}$ is then the same as the above embedding $\mathrm{End}_{\mathbb{C}}(V) \to \mathrm{End}_{\mathbb{C}}(W)$, when we make our standard identifications.

Now take $\phi \in \mathcal{H}(G, \rho)$, and let Φ denote its image in $\mathcal{H}(G, \sigma)$, so that $\phi = (H : K)\check{\varpi}\Phi\check{\varpi}$, by (4.1.4). Let $v \in V$, $\check{v} \in \check{V}$. Then $\Upsilon_\sigma(\Phi \otimes v \otimes \check{v})$ is the function

$$g \mapsto \dim(\sigma)\langle v, \Phi(g)\check{v}\rangle, \quad g \in G.$$

However, we have $v = \varpi v$, $\check{v} = \check{\varpi}\check{v}$, so

$$\begin{aligned}\Upsilon_\sigma(\Phi \otimes v \otimes \check{v})(g) &= \dim(\sigma)\langle v, \check{\varpi}\Phi(g)\check{\varpi}\check{v}\rangle \\ &= \dim(\rho)\langle v, \phi(g)\check{v}\rangle \\ &= \Upsilon_\rho(\phi \otimes v \otimes \check{v})(g),\end{aligned}$$

as required. ■

The effect of this on module categories is also easy to work out, and worthy of note. If $\mathcal{A}$ temporarily denotes an associative $\mathbb{C}$-algebra with 1, we write $\mathcal{A}$-$\mathfrak{Mod}$ for the category of left $\mathcal{A}$-modules M which are unital in the sense that $1_{\mathcal{A}}m = m$, $m \in M$. In the situation above, we have an equivalence

$$\mathrm{End}_{\mathbb{C}}(W)\text{-}\mathfrak{Mod} \xrightarrow{\approx} \mathrm{End}_{\mathbb{C}}(V)\text{-}\mathfrak{Mod}$$

given by $M \mapsto \varpi M$ with inverse $L \mapsto \mathrm{End}_{\mathbb{C}}(W) \otimes_{\mathrm{End}_{\mathbb{C}}(V)} L$. Combining this with (4.2.7), we get

(4.2.8) Proposition: *Let G be a locally profinite group with compact open subgroups $K \subset H$. Let ρ be an irreducible smooth representation of K and write $\sigma = \mathrm{Ind}(\rho : K, H)$. Assume that σ is irreducible. Then*

$$\begin{gathered}M \mapsto e_\sigma * \mathcal{H}(G) * e_\sigma \otimes_{e_\rho * \mathcal{H}(G) * e_\rho} M, \quad M \in |e_\rho * \mathcal{H}(G) * e_\rho\text{-}\mathfrak{Mod}|, \\ L \mapsto e_\rho L, \quad L \in |e_\sigma * \mathcal{H}(G) * e_\sigma\text{-}\mathfrak{Mod}|,\end{gathered}$$

are mutually inverse equivalences

$$e_\rho * \mathcal{H}(G) * e_\rho\text{-}\mathfrak{Mod} \cong e_\sigma * \mathcal{H}(G) * e_\sigma\text{-}\mathfrak{Mod}.$$

(4.3) Unitary structures

We now assume that our given representation (ρ, W) is *irreducible*. Since K is compact, there is a positive definite K-invariant Hermitian form h on W. Thus $h : W \times W \to \mathbb{C}$, h is linear in the first argument, antilinear in the second, $h(\rho(k)w_1, \rho(k)w_2) = h(w_1, w_2)$, $h(w,w) > 0$, for $k \in K$, $w_1, w_2, w \in W$, $w \neq 0$.

All of this applies equally in the more general context of (4.2.5), provided we assume that the central quasicharacter χ of ρ is unitary, i.e. $|\chi| = 1$. Again, we work only with the compact case, but the results hold with the same proofs in the compact mod centre case.

The form h gives rise to a $\mathbb{C}$-antilinear isomorphism $\Theta = \Theta_h : W \to \check{W}$ by $\langle w, \Theta(v)\rangle = h(w,v)$, $w, v \in W$. Immediately, we have $\Theta \circ \rho(k) = \check{\rho}(k) \circ \Theta$, $k \in K$. Since W, $\check{W}$ are irreducible, there is at most one antilinear isomorphism $W \cong \check{W}$, up to scalar multiple. The form h is therefore uniquely determined up to a positive constant factor.

We can use the form h to define an involution on the $\mathbb{C}$-algebra $\mathrm{End}_{\mathbb{C}}(W)$. For $f \in \mathrm{End}_{\mathbb{C}}(W)$, define $\overline{f} \in \mathrm{End}_{\mathbb{C}}(W)$ by

$$h(w_1, f(w_2)) = h(\overline{f}(w_1), w_2), \quad w_i \in W,\ f \in \mathrm{End}_{\mathbb{C}}(W).$$

One verifies immediately that $f \mapsto \overline{f}$ is $\mathbb{C}$-antilinear and satisfies $\overline{\overline{f}} = f$, $\overline{f_1 \circ f_2} = \overline{f}_2 \circ \overline{f}_1$, for $f, f_i \in \mathrm{End}_{\mathbb{C}}(W)$. In other words, "bar" is indeed an involution on $\mathrm{End}_{\mathbb{C}}(W)$.

If we identify $\mathrm{End}_{\mathbb{C}}(W)$ with $W \otimes_{\mathbb{C}} \check{W}$, as in **(4.2)**, we get

$$\overline{(w_1 \otimes \check{w}_2)} = \Theta^{-1}(\check{w}_2) \otimes \Theta(w_1), \quad w_1 \in W,\ \check{w}_2 \in \check{W}.$$

We next take $\Phi \in \mathcal{H}(G, \check{\rho})$, and define a function $\overline{\Phi} : G \to \mathrm{End}_{\mathbb{C}}(W)$ by

$$\overline{\Phi}(g) = \overline{\Phi(g^{-1})}, \quad g \in G.$$

Then $\overline{\Phi} \in \mathcal{H}(G, \check{\rho})$ and we have $\overline{\overline{\Phi}} = \Phi$, $\overline{\Phi_1 * \Phi_2} = \overline{\Phi}_2 * \overline{\Phi}_1$, for $\Phi, \Phi_i \in \mathcal{H}(G, \check{\rho})$. Thus again, "bar" is an involution on $\mathcal{H}(G, \check{\rho})$.

We can use these involutions to define some more hermitian forms. First, we define a form h^2 on $\mathrm{End}_{\mathbb{C}}(W)$ by

(4.3.1) $$h^2(f_1, f_2) = \mathrm{tr}_W(f_1 \circ \overline{f}_2), \quad f_i \in \mathrm{End}_{\mathbb{C}}(W).$$

The form h^2 is positive definite. Likewise, we define a positive definite hermitian form $\check{\boldsymbol{h}}$ on $\mathcal{H}(G, \check{\rho})$ by

(4.3.2) $$\check{\boldsymbol{h}}(\Phi_1, \Phi_2) = \int_G h^2(\Phi_1(g), \Phi_2(g))\, dg, \quad \Phi_i \in \mathcal{H}(G, \check{\rho}).$$

Taking account of the definitions, we can rewrite this in the form

$$\begin{aligned}\check{\boldsymbol{h}}(\Phi_1, \Phi_2) &= \int_G \mathrm{tr}_W(\Phi_1(g) \circ \overline{\Phi_2(g)})\, dg \\ &= \mathrm{tr}_W(\Phi_1 * \overline{\Phi}_2(1_G)), \quad \Phi_i \in \mathcal{H}(G, \check{\rho}).\end{aligned}$$

We emphasise that all of these definitions are independent of the choice of the original K-invariant positive definite hermitian form h on W.

We could start instead with a positive definite K-invariant hermitian form $\check{h}$ on $\check{W}$, for example

$$\check{h}(\check{w}_1, \check{w}_2) = h(\Theta^{-1}(\check{w}_2), \Theta^{-1}(\check{w}_1)) = \langle \Theta^{-1}(\check{w}_2), \check{w}_1 \rangle.$$

This gives rise to canonical involutions and positive definite hermitian forms on $\mathrm{End}_{\mathbb{C}}(\check{W})$ and $\mathcal{H}(G, \rho)$, in exactly the same way.

(4.3.3) Remarks: *(i)* We can carry out the same constructions without the hypothesis that ρ is irreducible. However, we have to start by choosing a positive definite K-invariant hermitian form h on W and, when ρ is not irreducible, this h is not at all uniquely determined. Different choices of h may lead to different involutions on $\mathrm{End}_{\mathbb{C}}(W)$, and hence to different involutions and different forms on the Hecke algebras.

(ii) Return to the situation and notation of (4.1.3). Assume further that the induced representation ρ_* of H is *irreducible*. Thus the algebras $\mathrm{End}_{\mathbb{C}}(W_*)$, $\mathcal{H}(G, \rho_*)$ have canonical involutions. If we embed $\mathrm{End}_{\mathbb{C}}(W)$ in $\mathrm{End}_{\mathbb{C}}(W_*)$ as in the proof of (4.1.3), then it is easy to see that the canonical involution on $\mathrm{End}_{\mathbb{C}}(W_*)$ extends that on $\mathrm{End}_{\mathbb{C}}(W)$. The same applies on taking contragredients, and the isomorphism $\mathcal{H}(G, \rho) \cong \mathcal{H}(G, \rho_*)$ of (4.1.3) is an isomorphism of *algebras with involution*. To compare the forms on the Hecke algebras, take $\phi, \psi \in \mathcal{H}(G, \rho)$. We write

$$\phi * \overline{\psi} = \alpha \mathbf{e} + \xi,$$

where $\mathbf{e}$ is the unit element of $\mathcal{H}(G, \rho)$, $\alpha \in \mathbb{C}$, and $\xi \in \mathcal{H}(G, \rho)$ is some function whose support does not meet K. Then

$$\boldsymbol{h}(\phi, \psi) = \mathrm{tr}_W(\phi * \overline{\psi}(1)) = \mu(K)^{-1} \alpha \dim(\rho).$$

Now suppose $\phi \mapsto \Phi$, $\psi \mapsto \Psi$ under the isomorphism $\mathcal{H}(G, \rho) \cong \mathcal{H}(G, \rho_*)$ of (4.1.3). Then $\phi * \overline{\psi} \mapsto \Phi * \overline{\Psi}$. Further, $\mathbf{e} \mapsto \mathbf{E}$, the unit element of $\mathcal{H}(G, \rho_*)$. Let the function ξ above map to Ξ. The support of Ξ is contained in $H\mathrm{supp}(\xi)H$. However, since ρ induces irreducibly to H, the support of Ξ cannot meet K, so we have $\Phi * \overline{\Psi} = \alpha \mathbf{E} + \Xi$, and, if we write $\boldsymbol{h}_*$ for the canonical hermitian form on $\mathcal{H}(G, \rho_*)$, we get

$$\begin{aligned} \boldsymbol{h}_*(\Phi, \Psi) = \mathrm{tr}_{\check{W}}(\Phi * \overline{\Psi}(1)) &= \mu(H)^{-1} \alpha \dim(\rho_*) \\ &= \mu(K)^{-1} \alpha \dim(\rho) = \boldsymbol{h}(\phi, \psi). \end{aligned}$$

We deduce that *the algebra isomorphism of* (4.1.3) *preserves the canonical involutions, and is an isometry for the canonical hermitian forms on the algebras* $\mathcal{H}(G, \rho)$, $\mathcal{H}(G, \rho_*)$.

Now we pass to the situation of **(4.2)** and the algebra isomorphism

$$\Upsilon : \mathcal{H}(G,\rho) \otimes \mathrm{End}_{\mathbb{C}}(W) \cong e_\rho * \mathcal{H}(G) * e_\rho.$$

(Here, of course, we must have ρ irreducible.) The algebra $\mathcal{H}(G,\rho) \otimes \mathrm{End}_{\mathbb{C}}(W)$ carries a canonical involution, which is the tensor product of the involutions defined above. Moreover, inspired by **[FM]**, we can combine the canonical hermitian forms $\boldsymbol{h}$, h^2 into a canonical positive definite hermitian form $\boldsymbol{h}\otimes h^2$ on the tensor product. Explicitly, we have a linear form $\ell_1 : \mathcal{H}(G,\rho) \otimes \mathrm{End}_{\mathbb{C}}(W) \to \mathbb{C}$ by

$$\ell_1(\Phi \otimes w \otimes \check{w}) = \dim(\rho).\langle w, \Phi(1_G)\check{w}\rangle, \quad \Phi \in \mathcal{H}(G,\rho),\ w \in W,\ \check{w} \in \check{W}.$$

We then define

$$\boldsymbol{h} \otimes h^2(\Xi_1, \Xi_2) = \ell_1(\Xi_1.\overline{\Xi}_2), \quad \Xi_i \in \mathcal{H}(G,\rho) \otimes \mathrm{End}_{\mathbb{C}}(W).$$

Observe that the restriction of $\boldsymbol{h} \otimes h^2$ to $\mathcal{H}(G,\rho) \otimes 1_W \cong \mathcal{H}(G,\rho)$ is just $\dim(\rho)\boldsymbol{h}$, and its restriction to $\mathbf{e} \otimes \mathrm{End}_{\mathbb{C}}(W) \cong \mathrm{End}_{\mathbb{C}}(W)$ is $\mu(K)^{-1}\dim(\rho)h^2$.

Likewise, the algebra $e_\rho * \mathcal{H}(G) * e_\rho$ carries a canonical involution $\phi \mapsto \overline{\phi}$ given by

$$\overline{\phi}(g) = \overline{\phi(g^{-1})}, \quad g \in G,$$

where the last of these bars denotes complex conjugation. There is also a canonical positive definite hermitian form $(\ ,\ )$ on $e_\rho * \mathcal{H}(G) * e_\rho$ given by

$$(\phi, \psi) = \int_G \phi(g)\overline{\psi(g)}\,dg = \phi * \overline{\psi}(1_G).$$

(4.3.4) Proposition: *The algebra isomorphism*

$$\Upsilon : \mathcal{H}(G,\rho) \otimes_{\mathbb{C}} \mathrm{End}_{\mathbb{C}}(W) \xrightarrow{\approx} e_\rho * \mathcal{H}(G) * e_\rho$$

of (4.2.4) *is an isomorphism of algebras with involution, and is an isometry for the canonical positive definite hermitian forms* $\boldsymbol{h} \otimes h^2$, $(\ ,\ )$.

Proof: Take $\Phi \in \mathcal{H}(G,\rho)$, $w \in W$, $\check{w} \in \check{W}$, and consider

$$\Upsilon\left(\overline{\Phi \otimes w \otimes \check{w}}\right) = \Upsilon(\overline{\Phi} \otimes \Theta^{-1}(\check{w}) \otimes \Theta(w)).$$

As above, let $\check{h}$ denote the form $(\check{w}_1, \check{w}_2) \mapsto h(\Theta^{-1}(\check{w}_2), \Theta^{-1}(\check{w}_1))$ on $\check{W}$. We evaluate this last function at $g \in G$ to get

$$\begin{aligned}\Upsilon\left(\overline{\Phi \otimes w \otimes \check{w}}\right)(g) &= \dim(\rho)\langle \Theta^{-1}(\check{w}), \overline{\Phi}(g)\Theta(w)\rangle \\ &= \dim(\rho)\langle \Theta^{-1}(\check{w}), \overline{\Phi(g^{-1})}\Theta(w)\rangle.\end{aligned}$$

On the other hand,

$$\begin{aligned}\overline{\Upsilon(\Phi\otimes w\otimes \check{w})(g)} &= \overline{\Upsilon(\Phi\otimes w\otimes \check{w})(g^{-1})}\\ &= \dim(\rho)\overline{\langle w, \Phi(g^{-1})\check{w}\rangle}\\ &= \dim(\rho)\overline{\check{h}(\Phi(g^{-1})\check{w}, \Theta(w))}\\ &= \dim(\rho)\check{h}(\Theta(w), \Phi(g^{-1})\check{w})\\ &= \dim(\rho)\check{h}(\overline{\Phi(g^{-1})}\Theta(w), \check{w})\\ &= \dim(\rho)\langle\Theta^{-1}(\check{w}), \overline{\Phi(g^{-1})}\Theta(w)\rangle.\end{aligned}$$

We conclude that

$$\Upsilon\left(\overline{\Phi\otimes w\otimes \check{w}}\right) = \overline{\Upsilon(\Phi\otimes w\otimes \check{w})},$$

for $\Phi\in\mathcal{H}(G,\rho)$, $w\in W$, $\check{w}\in\check{W}$, as required.

Now we compare the hermitian forms. The form $(\ ,\ )$ on $e_\rho*\mathcal{H}(G)*e_\rho$ is also given by a linear form,

$$(\phi_1,\phi_2) = \ell_2(\phi_1*\overline{\phi}_2),\quad \phi_i\in e_\rho*\mathcal{H}(G)*e_\rho,$$

where $\ell_2(\phi)=\phi(1_G)$, $\phi\in e_\rho*\mathcal{H}(G)*e_\rho$. By the first part, it is enough to check that $\ell_2\circ\Upsilon=\ell_1$, which is immediate from the definition of Υ. ∎

5. SIMPLE TYPES

After the diversions of §4, we return to the mainstream of the paper, with the same notation as before. In particular, $A = \mathrm{End}_F(V)$, $G = \mathrm{Aut}_F(V) \cong GL(N,F)$.

The present chapter is the keystone of the whole structure. We take a simple stratum $[\mathfrak{A}, n, 0, \beta]$ in A and set, as usual, $E = F[\beta]$, $B =$ the A-centraliser of E (so that $B = \mathrm{End}_E(V)$), $\mathfrak{B} = \mathfrak{A} \cap B$. We take one of the simple characters $\theta \in \mathcal{C}(\mathfrak{A}, 0, \beta)$ as in §3, and consider certain representations of the group $J(\beta, \mathfrak{A})$ which contain θ when restricted to $H^1(\beta, \mathfrak{A})$. Among these, when the order $\mathfrak{A}$ is *principal*, will be the "simple types" which we use to analyse the representations of G.

Initially, we work with no restriction on $\mathfrak{A}$. The first step is easy:— there is a unique irreducible representation $\eta(\theta)$ of $J^1(\beta, \mathfrak{A})$ containing θ. It is when we try to extend this to a representation of $J(\beta, \mathfrak{A})$ that things become interesting. In most, but not all, cases, general theory predicts the existence of such an extension with little effort. However, there are usually many such, almost all of which are unsuitable for our purposes. This leads us to invoke a richer structure. Since the orders $\mathfrak{A}$, $\mathfrak{B}$ determine each other, we tend to regard the groups H^i, J^i and the character θ as functions of $\mathfrak{B}$ (the element β being fixed once for all). Because of the correspondences $\boldsymbol{\tau}$ of (3.6.2), the character θ varies with $\mathfrak{B}$ in a canonical and coherent way. The same applies to the representation $\eta(\theta)$. This behaviour is best expressed in terms of a triple $\mathfrak{B}_{\mathrm{M}} \supset \mathfrak{B} \supset \mathfrak{B}_{\mathrm{m}}$ of hereditary $\mathfrak{o}_E$-orders in B. We discuss this in **(5.1)**.

If we take the order $\mathfrak{B}_{\mathrm{M}}$ maximal and $\mathfrak{B}_{\mathrm{m}}$ minimal, the relation between the various η's provides the crucial first step in extending $\eta(\theta)$ to a representation $\kappa(\theta)$ of $J(\beta, \mathfrak{A})$. This approach obviates the exceptional cases which arise in the general theory, and the coherence condition determines the appropriate class of extensions κ much more clearly. The representation $\kappa(\theta)$ is still not uniquely determined, but it is determined up to an essentially trivial twist. This is the subject of **(5.2)**.

These representations κ are only indirectly important to our theory. The significant ones are of the form $\kappa(\theta) \otimes \sigma$, where σ is an irreducible cuspidal representation of the finite reductive group (over k_E) $J(\beta, \mathfrak{A})/J^1(\beta, \mathfrak{A})$. Indeed, there are further restrictions on the representation, but the early stages of the argument are general. In **(5.3)**, we give the first steps in the computation of the intertwining of the representations $\kappa \otimes \sigma$. Of course, since we want eventually to analyse the representations of G containing $\kappa \otimes \sigma$, we do not just need to know the intertwining as a set, but rather the structure of the $(\kappa \otimes \sigma)$-spherical

Hecke algebra $\mathcal{H}(G, \kappa \otimes \sigma)$. In preparation for this, we recall in **(5.4)** a collection of basically well-known facts about Weyl groups, BN-pairs and affine Hecke algebras.

With this to hand we can, in **(5.5)**, push further the calculation of the intertwining of $\kappa \otimes \sigma$. Then we specialise to the case where $\kappa \otimes \sigma$ is a "simple type". In this situation, we prove our Main Theorem in **(5.6)**, which gives explicitly the structure of the spherical Hecke algebra of $\kappa \otimes \sigma$.

We conclude the section by showing that simple types inherit the "intertwining implies conjugacy" property of simple strata and simple characters. We shall see later that this implies a strong uniqueness property for representations of G.

(5.1) Heisenberg representations

Let $[\mathfrak{A}, n, 0, \beta]$ be a simple stratum in the algebra $A = \mathrm{End}_F(V)$, and write $E = F[\beta]$. Throughout this section, we write B for the A-centraliser of E, so that $B = \mathrm{End}_E(V)$, and we also put $\mathfrak{B} = \mathfrak{A} \cap B$, $\mathfrak{P} = \mathrm{rad}(\mathfrak{A})$, $\mathfrak{Q} = \mathfrak{P} \cap B = \mathrm{rad}(\mathfrak{B})$.

(5.1.1) Proposition: *Let $[\mathfrak{A}, n, 0, \beta]$ be a simple stratum in A, and $\theta \in \mathcal{C}(\mathfrak{A}, 0, \beta)$. There exists a unique irreducible representation $\eta(\theta)$ of the group $J^1(\beta, \mathfrak{A})$ such that $\eta(\theta) \mid H^1(\beta, \mathfrak{A})$ contains θ. Moreover, $\eta(\theta) \mid H^1(\beta, \mathfrak{A})$ is a multiple of θ, and*

$$\dim(\eta(\theta)) = (J^1(\beta, \mathfrak{A}) : H^1(\beta, \mathfrak{A}))^{\frac{1}{2}}.$$

The G-intertwining of $\eta(\theta)$ is $J^1(\beta, \mathfrak{A}) B^\times J^1(\beta, \mathfrak{A})$.

Proof: Given (3.4.1), the first two assertions are standard: see for example **[BF1]** §**8**. The intertwining of θ is $J^1(\beta, \mathfrak{A}) B^\times J^1(\beta, \mathfrak{A})$ by (3.3.2). The representation of $J^1(\beta, \mathfrak{A})$ induced by θ is a multiple of $\eta(\theta)$, so the last assertion follows from (4.1.5). ■

We now need an effective machine for comparing the dimensions of various representations η given by (5.1.1). For this purpose, it is useful to have the "generalised group index" notation. If K is a group with subgroups K_1, K_2, such that $\mathsf{K}_1 \cap \mathsf{K}_2$ is of finite index in both K_2 and K_1, then we can put

$$(\mathsf{K}_1 : \mathsf{K}_2) = \frac{(\mathsf{K}_1 : \mathsf{K}_1 \cap \mathsf{K}_2)}{(\mathsf{K}_2 : \mathsf{K}_1 \cap \mathsf{K}_2)}.$$

If K is a topological group with (say) left Haar measure μ, and the K_i are compact open subgroups, we of course have $(\mathsf{K}_1 : \mathsf{K}_2) = \mu(\mathsf{K}_1)/\mu(\mathsf{K}_2)$.

(5.1.2) Proposition: *For $i = 1, 2$, let $[\mathfrak{A}_i, n_i, 0, \beta]$ be simple strata in A (with the same β), and let $\theta_i \in \mathcal{C}(\mathfrak{A}_i, 0, \beta)$. Let B be the A-centraliser* *of* β, $\mathfrak{P}_i = \mathrm{rad}(\mathfrak{A}_i)$, $\mathfrak{B}_i = \mathfrak{A}_i \cap B$, $\mathfrak{Q}_i = \mathfrak{P}_i \cap B$. *Let* η_i *be the unique irreducible representation of* $J^1(\beta, \mathfrak{A}_i)$ *which contains* θ_i. *Then*

$$\dim(\eta_1)(\boldsymbol{U}^1(\mathfrak{B}_1) : \boldsymbol{U}^1(\mathfrak{B}_2)) = \dim(\eta_2)(J^1(\beta, \mathfrak{A}_1) : J^1(\beta, \mathfrak{A}_2)).$$

Proof: We start with some lemmas.

(5.1.3) Lemma: *For* $1 \leq i \leq n+1$, *let* V_i *be a finite-dimensional* F*-vector space, and let* μ_i *be a Haar measure on* V_i. *Suppose that we have an exact sequence of* F*-linear maps*

$$0 \to V_1 \xrightarrow{f_1} V_2 \xrightarrow{f_2} \dots \xrightarrow{f_n} V_{n+1} \to 0. \tag{5.1.4}$$

For each i, *let* L_i, L_i' *be* $\mathfrak{o}_F$*-lattices in* V_i *and suppose that* (5.1.4) *restricts to give exact sequences*

$$0 \to L_1 \xrightarrow{f_1} L_2 \xrightarrow{f_2} \dots \xrightarrow{f_n} L_{n+1} \to 0,$$
$$0 \to L_1' \xrightarrow{f_1} L_2' \xrightarrow{f_2} \dots \xrightarrow{f_n} L_{n+1}' \to 0.$$

Then

$$\prod_{i=1}^{n+1} \mu_i(L_i)^{(-1)^i} = \prod_{i=1}^{n+1} \mu_i(L_i')^{(-1)^i}.$$

Proof: By induction from Fubini's theorem. ■

(5.1.5) Lemma: *Let* L_1, L_2 *be* $\mathfrak{o}_F$*-lattices in* A, *and define* L_1^*, L_2^* *by* (1.1.4). *Let* μ *be some Haar measure on* A. *Then*

$$\mu(L_1)\mu(L_1^*) = \mu(L_2)\mu(L_2^*).$$

Proof: It is enough to treat the case $L_1 \supset L_2$. The Pontrjagin dual of the finite abelian group L_1/L_2 is then isomorphic to L_2^*/L_1^*. Thus $(L_1 : L_2) = (L_2^* : L_1^*)$, and the assertion follows. ■

Now we prove (5.1.2). From (3.1.16), we have exact sequences

$$0 \to \mathfrak{Q}_i \to \mathfrak{J}^1(\beta, \mathfrak{A}_i) \xrightarrow{a_\beta} (\mathfrak{H}^1(\beta, \mathfrak{A}_i))^* \xrightarrow{s} \mathfrak{B}_i \to 0.$$

If μ_A is a Haar measure on A and μ_B a Haar measure on B, then (5.1.3) gives us

$$\frac{\mu_A(\mathfrak{J}^1(\beta, \mathfrak{A}_1))\mu_B(\mathfrak{B}_1)}{\mu_B(\mathfrak{Q}_1)\mu_A((\mathfrak{H}^1(\beta, \mathfrak{A}_1)^*)} = \frac{\mu_A(\mathfrak{J}^2(\beta, \mathfrak{A}_2))\mu_B(\mathfrak{B}_2)}{\mu_B(\mathfrak{Q}_2)\mu_A((\mathfrak{H}^2(\beta, \mathfrak{A}_2)^*)} \tag{5.1.6}$$

Using the obvious abbreviations, (5.1.5) gives $\mu_A(\mathfrak{H}_i^*) = c_A/\mu(\mathfrak{H}_i)$, for a constant c_A depending only on μ_A. On the other hand, the factor $\mu_B(\mathfrak{B}_i)/\mu_B(\mathfrak{Q}_i)$ can be rewritten $c_B\mu_B(\mathfrak{Q}_i)^{-2}$, for a constant c_B depending only on μ_B. Then we recall that

$$\mu_A(\mathfrak{J}^1(\beta,\mathfrak{A}_i))\mu_A(\mathfrak{H}^1(\beta,\mathfrak{A}_i)) = \mu_A(\mathfrak{J}^1(\beta,\mathfrak{A}_i))^2.\dim(\eta_i)^{-2}.$$

We substitute all this in (5.1.6) to get

$$\frac{c_B\mu_A(\mathfrak{J}^1(\beta,\mathfrak{A}_1))^2}{c_A\mu_B(\mathfrak{Q}_1)^2\dim(\eta_1)^2} = \frac{c_B\mu_A(\mathfrak{J}^1(\beta,\mathfrak{A}_2))^2}{c_A\mu_B(\mathfrak{Q}_2)^2\dim(\eta_2)^2}$$

The result follows. ■

We now record a more precise statement concerning the intertwining of the representations η constructed in (5.1.1). First, we need some notation. If ρ is a representation of a compact open subgroup K of G, and $g \in G$, we write

(5.1.7) $$I_g(\rho \mid K) = I_g(\rho) = \mathrm{Hom}_{K^g\cap K}(\rho, \rho^g),$$

using the notations of §**4**. Of course, by (4.1.2), we can identify $I_g(\rho)$ with the subspace of $\mathcal{H}(G,\rho)$ consisting of functions supported only on the coset KgK.

(5.1.8) Proposition: *Let $[\mathfrak{A}, n, 0, \beta]$ be a simple stratum in A, and use the notation above. Let $\theta \in \mathcal{C}(\mathfrak{A}, 0, \beta)$, and let $\eta = \eta(\theta)$ be the unique irreducible representation of $J^1(\beta,\mathfrak{A})$ containing θ, as in (5.1.1). Abbreviate $J^1 = J^1(\beta,\mathfrak{A})$, $H^1 = H^1(\beta,\mathfrak{A})$. Then, for $g \in G$, we have*

$$\dim_{\mathbb{C}}(I_g(\eta \mid J^1)) = \begin{cases} 1 & \textit{if } g \in J^1B^\times J^1, \\ 0 & \textit{otherwise.} \end{cases}$$

Proof: Write $d = \dim(\eta) = (J^1 : H^1)^{\frac{1}{2}}$. Then $\mathrm{Ind}(\theta : H^1, J^1)$ is a sum of d copies of η, whence $\dim(I_g(\mathrm{Ind}(\theta))) = d^2\dim(I_g(\eta))$. Now we use (4.1.5). We have

$$\dim(I_g(\theta)) = \begin{cases} 1 & \textit{if } g \in J^1B^\times J^1, \\ 0 & \textit{otherwise.} \end{cases}$$

by (3.3.2). Thus, for $g \in J^1B^\times J^1$, the dimension of $I_g(\mathrm{Ind}(\theta))$ is the number of (H^1, H^1)-double cosets contained in J^1gJ^1. If $g \notin J^1B^\times J^1$, this dimension is zero. The proposition now follows from:—

(5.1.9) Lemma: *Let $y \in B^\times$. Then J^1yJ^1 is the union of $(J^1 : H^1)$ distinct (H^1, H^1) double cosets.*

Proof: Choose a Haar measure $\mu^\times$ on G. Then, for $y \in B^\times$, we have $\mu^\times(J^1yJ^1) = \mu^\times(J^1).(J^1 : J^1 \cap (J^1)^y)$. On the other hand, since H^1 is a normal subgroup of J^1, we have $\mu^\times(H^1y'H^1) = \mu^\times(H^1yH^1) = \mu^\times(H^1).(H^1 : H^1 \cap (H^1)^y)$, for any $y' \in J^1yJ^1$. Therefore it is enough to prove:

(5.1.10) Lemma: *For $y \in B^\times$, we have $(J^1 : J^1 \cap (J^1)^y) = (H^1 : H^1 \cap (H^1)^y)$.*

Proof: Temporarily write $J^1 = 1+\mathfrak{j}$, $H^1 = 1+\mathfrak{h}$. Choose a Haar measure μ on A so that $\mu^\times(J^1) = \mu(\mathfrak{j})$, $\mu^\times(H^1) = \mu(\mathfrak{h})$. According to (3.1.16), we have an exact sequence

$$0 \longrightarrow \mathfrak{Q} \longrightarrow \mathfrak{j} \xrightarrow{a_\beta} \mathfrak{h}^* \xrightarrow{s} \mathfrak{B} \longrightarrow 0,$$

where, as usual, $\mathfrak{B} = \mathfrak{A} \cap B$, $\mathfrak{Q} = \mathfrak{P} \cap B = \mathrm{rad}(\mathfrak{B})$. This sequence remains exact when conjugated by y, and we assert that the sequence

$$0 \longrightarrow \mathfrak{Q} + \mathfrak{Q}^y \longrightarrow \mathfrak{j} + \mathfrak{j}^y \xrightarrow{a_\beta} \mathfrak{h}^* + (\mathfrak{h}^*)^y \xrightarrow{s} \mathfrak{B} + \mathfrak{B}^y \longrightarrow 0$$

is exact. The map induced by the tame corestriction s is certainly surjective. According to (3.1.19), we have $\mathfrak{h}^* = a_\beta(\mathfrak{j}) + \mathfrak{A}$, whence $\mathfrak{h}^* + (\mathfrak{h}^*)^y = a_\beta(\mathfrak{j} + \mathfrak{j}^y) + \mathfrak{A} + \mathfrak{A}^y$. The intersection of this with $\mathrm{Ker}(s) = a_\beta(A)$ is $(\mathfrak{A} \cap a_\beta(A)) + (\mathfrak{A} \cap a_\beta(A))^y + a_\beta(\mathfrak{j} + \mathfrak{j}^y)$ by (1.4.16). We have $(\mathfrak{A} \cap a_\beta(A)) = a_\beta(\mathfrak{Q}^r\mathfrak{N}(\beta, \mathfrak{A})) \subset a_\beta(\mathfrak{J}^{[\frac{r+1}{2}]}(\beta, \mathfrak{A}))$ by (3.1.10), where $r = -k_0(\beta, \mathfrak{A})$. It follows that the sequence is exact at $\mathfrak{h}^* + (\mathfrak{h}^*)^y$. At the remaining place, we have

$$\mathfrak{Q} + \mathfrak{Q}^y \subset (\mathfrak{j} + \mathfrak{j}^y) \cap \mathrm{Ker}(a_\beta) \subset (\mathfrak{P} + \mathfrak{P}^y) \cap \mathrm{Ker}(a_\beta) = \mathfrak{Q} + \mathfrak{Q}^y,$$

by (1.3.16). Thus our sequence is exact. We therefore have a commutative diagram with exact rows and columns

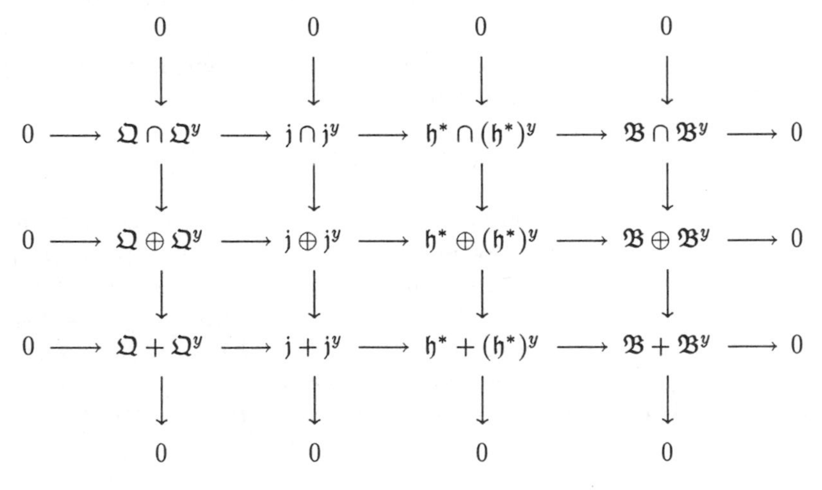

to which we can apply (5.1.3).

We fix Haar measures on A and B. To abbreviate the notation in the present computation only, we write $\{S\}_A$ for the measure of a subset S of A, and likewise for B. We omit the subscript when it is clear which space we are working in.

The first row of the diagram gives the relation

$$\{\mathfrak{j}\cap\mathfrak{j}^y\}/\{\mathfrak{h}^*\cap(\mathfrak{h}^*)^y\} = c_1\{\mathfrak{Q}\cap\mathfrak{Q}^y\}/\{\mathfrak{B}\cap\mathfrak{B}^y\}$$

for some constant c_1 which is given by the horizontal sequences via (5.1.3). We can rewrite the left hand side of this as $\{\mathfrak{j}\cap\mathfrak{j}^y\}\{\mathfrak{h}^* + \mathfrak{h}^{*y}\}/\{\mathfrak{h}^*\}^2$, using the third column. The right hand side is $\{\mathfrak{Q}\}^2/\{\mathfrak{Q}+\mathfrak{Q}^y\}\{\mathfrak{B}\cap\mathfrak{B}^y\}$. However, $\mathfrak{B}\cap\mathfrak{B}^y$ is the dual of $\mathfrak{Q}+\mathfrak{Q}^y$, so this reduces to $\{\mathfrak{Q}\}/\{\mathfrak{B}\}$. Now we use (5.1.5) to get $\{\mathfrak{h}^*+\mathfrak{h}^{*y}\} = \{\mathfrak{h}\}\{\mathfrak{h}^*\}/\{\mathfrak{h}\cap\mathfrak{h}^y\}$. However, we also have $\{\mathfrak{j}\}/\{\mathfrak{h}^*\} = c_1\{\mathfrak{Q}\}/\{\mathfrak{B}\}$. Plugging all this in, our relation above becomes

$$c_1\{\mathfrak{j}\cap\mathfrak{j}^y\}\{\mathfrak{h}\}\{\mathfrak{Q}\}/\{\mathfrak{B}\}\{\mathfrak{j}\}\{\mathfrak{h}\cap\mathfrak{h}^y\} = c_1\{\mathfrak{Q}\}/\{\mathfrak{B}\}.$$

Since c_1 is positive, the result follows. ■

This also completes the proof of (5.1.8). ■

It is sometimes convenient to view this result a little differently.

(5.1.11) Corollary: *Let η, θ be as in* (5.1.8), *and let $y \in B^\times$. Then the representations $\eta \mid (J^1\cap(J^1)^y)$, $\eta^y \mid (J^1\cap(J^1)^y)$ have a unique irreducible component in common (counted according to multiplicities).*

We now investigate various coherence properties exhibited by certain families of these representations η, reflecting the coherence properties of the characters θ given by the correspondences $\boldsymbol{\tau}$ defined in (3.6.2). Let $[\mathfrak{A}, n, 0, \beta]$ be a simple stratum in A as above, with $E = F[\beta]$, $B=\text{End}_E(V)$, $\mathfrak{B} = \mathfrak{A}\cap B$ as always. We choose (and fix) hereditary $\mathfrak{o}_E$-orders $\mathfrak{B}_{\text{M}}$, $\mathfrak{B}_{\text{m}}$ in B such that

(5.1.12) $$\mathfrak{B}_{\text{M}} \supset \mathfrak{B} \supset \mathfrak{B}_{\text{m}}.$$

Later, we will sometimes need to take $\mathfrak{B}_{\text{M}}$ maximal and $\mathfrak{B}_{\text{m}}$ minimal but, for the moment, we work in greater generality. We let $\mathfrak{A}_{\text{M}}$ (resp. $\mathfrak{A}_{\text{m}}$) denote the unique hereditary $\mathfrak{o}_F$-order in A which is stable under conjugation by $E^\times$ and such that $\mathfrak{A}_{\text{M}}\cap B = \mathfrak{B}_{\text{M}}$ (resp. $\mathfrak{A}_{\text{m}}\cap B = \mathfrak{B}_{\text{m}}$). We use the obvious associated notations, such as $\mathfrak{P}_{\text{M}} = \text{rad}(\mathfrak{A}_{\text{M}})$, $\mathfrak{Q}_{\text{M}} = \text{rad}(\mathfrak{B}_{\text{M}}) = \mathfrak{P}_{\text{M}}\cap B$, etc. It is often convenient to regard $\mathfrak{B}_{\text{M}}$, $\mathfrak{B}_{\text{m}}$ as fixed, while $\mathfrak{B}$ is allowed to vary over all hereditary $\mathfrak{o}_E$-orders between $\mathfrak{B}_{\text{M}}$ and $\mathfrak{B}_{\text{m}}$.

We now give some further notation, which will remain fixed throughout the present chapter.

(5.1.13) Notation: *(i) Let* $\theta_{\mathrm{M}} \in \mathcal{C}(\mathfrak{A}_{\mathrm{M}}, 0, \beta)$, $\theta_{\mathrm{m}} \in \mathcal{C}(\mathfrak{A}_{\mathrm{m}}, 0, \beta)$, $\theta \in \mathcal{C}(\mathfrak{A}, 0, \beta)$, *and assume that these characters are all related via the correspondences* $\boldsymbol{\tau}$ *of* (3.6.2)*:*

$$\theta_{\mathrm{M}} = \boldsymbol{\tau}_{\mathfrak{A}_{\mathrm{m}},\mathfrak{A}_{\mathrm{M}},\beta,0}(\theta_{\mathrm{m}}), \qquad \theta = \boldsymbol{\tau}_{\mathfrak{A}_{\mathrm{m}},\mathfrak{A},\beta,0}(\theta_{\mathrm{m}}).$$

(ii) Write $J^t = J^t(\beta, \mathfrak{A})$, $J^t_{\mathrm{m}} = J^t(\beta, \mathfrak{A}_{\mathrm{m}})$, $J^t_{\mathrm{M}} = J^t(\beta, \mathfrak{A}_{\mathrm{M}})$, $t \geq 1$, $J = J^0(\beta, \mathfrak{A})$, *and so on, with similar conventions for the groups* H.

(iii) Write η *(resp.* η_{m}, *resp.* η_{M}*) for the unique irreducible representation of* J^1 *(resp.* J^1_{m}, *resp.* J^1_{M}*) containing* θ *(resp.* θ_{m}, *resp.* θ_{M}*) given by* (5.1.1).

The containment relations $\mathfrak{B}_{\mathrm{M}} \supset \mathfrak{B} \supset \mathfrak{B}_{\mathrm{m}}$ give relations $\boldsymbol{U}(\mathfrak{B}_{\mathrm{M}}) \supset \boldsymbol{U}(\mathfrak{B}) \supset \boldsymbol{U}(\mathfrak{B}_{\mathrm{m}})$, $\boldsymbol{U}^1(\mathfrak{B}_{\mathrm{m}}) \supset \boldsymbol{U}^1(\mathfrak{B}) \supset \boldsymbol{U}^1(\mathfrak{B}_{\mathrm{M}})$, but of course there are no straightforward incidence relations between the various J-groups. However, since, e.g., $\boldsymbol{U}(\mathfrak{B}_{\mathrm{M}})$ normalises J^1_{M} and contains $\boldsymbol{U}^1(\mathfrak{B}_{\mathrm{m}})$, we can form groups like $\boldsymbol{U}^1(\mathfrak{B}_{\mathrm{m}})J^1_{\mathrm{M}} \subset J_{\mathrm{M}}$.

We now start comparing representations above various of the η's. This leads to some extremely useful auxiliary representations. The following three results are essentially identical, so we only prove the first of them.

(5.1.14) Proposition: *There exists a unique irreducible representation* $\tilde{\eta}_{\mathrm{M}}$ *of* $\boldsymbol{U}^1(\mathfrak{B}_{\mathrm{m}})J^1_{\mathrm{M}}$ *such that:*

(i) $\tilde{\eta}_{\mathrm{M}} \mid J^1_{\mathrm{M}} = \eta_{\mathrm{M}}$, *and*

(ii) the representations $\tilde{\eta}_{\mathrm{M}}$ *and* η_{m} *induce equivalent irreducible representations of* $\boldsymbol{U}^1(\mathfrak{A}_{\mathrm{m}})$.

(5.1.15) Proposition: *There exists a unique irreducible representation* $\tilde{\eta}$ *of* $\boldsymbol{U}^1(\mathfrak{B}_{\mathrm{m}})J^1$ *such that*

(i) $\tilde{\eta} \mid J^1 = \eta$, *and*

(ii) the representations $\tilde{\eta}$, η_{m} *induce equivalent irreducible representations of* $\boldsymbol{U}^1(\mathfrak{A}_{\mathrm{m}})$.

(5.1.16) Proposition: *There exists a unique irreducible representation* $\hat{\eta}_{\mathrm{M}}$ *of* $\boldsymbol{U}^1(\mathfrak{B})J^1_{\mathrm{M}}$ *such that*

(i) $\hat{\eta}_{\mathrm{M}} \mid J^1_{\mathrm{M}} = \eta_{\mathrm{M}}$, *and*

(ii) the representations $\hat{\eta}_{\mathrm{M}}$ *and* η *induce equivalent irreducible representations of* $\boldsymbol{U}^1(\mathfrak{A})$.

Proof: We start by noting that, in this situation, (5.1.2) gives us the relation:

(5.1.17) $\dim(\eta_{\mathrm{M}}).(\boldsymbol{U}^1(\mathfrak{A}_{\mathrm{m}}) : \boldsymbol{U}^1(\mathfrak{B}_{\mathrm{m}})J^1_{\mathrm{M}}) = \dim(\eta_{\mathrm{m}}).(\boldsymbol{U}^1(\mathfrak{A}_{\mathrm{m}}) : J^1_{\mathrm{m}}).$

Consider the representation $\lambda = \mathrm{Ind}(\eta_{\mathrm{m}} : J^1_{\mathrm{m}}, \boldsymbol{U}^1(\mathfrak{A}_{\mathrm{m}}))$. By (3.3.2), $\boldsymbol{U}^1(\mathfrak{A}_{\mathrm{m}}) \cap I_G(\eta_{\mathrm{m}}) = J^1_{\mathrm{m}}$, so λ is irreducible. We now consider the restriction $\lambda \mid J^1_{\mathrm{M}}$. The representation $\lambda' = \mathrm{Ind}(\theta_{\mathrm{m}} : H^1_{\mathrm{m}}, \boldsymbol{U}^1(\mathfrak{A}_{\mathrm{m}}))$ is a multiple of λ by (5.1.1). Mackey theory shows that $\lambda' \mid H^1_{\mathrm{M}}$ contains the representation of H^1_{M} induced by $\theta_{\mathrm{m}} \mid H^1_{\mathrm{m}} \cap H^1_{\mathrm{M}}$. By (3.6.1), it therefore contains θ_{M}. Therefore $\lambda \mid H^1_{\mathrm{M}}$ contains θ_{M}, and $\lambda \mid J^1_{\mathrm{M}}$ contains η_{M}. Let ν be an irreducible component of $\lambda \mid \boldsymbol{U}^1(\mathfrak{B}_{\mathrm{m}})J^1_{\mathrm{M}}$ such that $\nu \mid J^1_{\mathrm{M}}$ contains η_{M}. Then $\nu \mid J^1_{\mathrm{M}}$ is a multiple of η_{M}, and, computing intertwining again, the representation $\mathrm{Ind}(\nu : \boldsymbol{U}^1(\mathfrak{B}_{\mathrm{m}})J^1_{\mathrm{M}}, \boldsymbol{U}^1(\mathfrak{A}_{\mathrm{m}}))$ is irreducible, hence equal to λ. Comparing dimensions, (5.1.17) implies $\nu \mid J^1_{\mathrm{M}} = \eta_{\mathrm{M}}$. Thus we may take $\tilde{\eta}_{\mathrm{M}} = \nu$.

To prove uniqueness for $\tilde{\eta}_{\mathrm{M}}$, we need only show that η_{M} occurs in $\mathrm{Ind}(\eta_{\mathrm{m}} : J^1_{\mathrm{m}}, \boldsymbol{U}^1(\mathfrak{A}_{\mathrm{m}})) \mid J^1_{\mathrm{M}}$ with multiplicity one. To do this, we use the Mackey formula to compute the restriction

$$\mathrm{Ind}(\tilde{\eta}_{\mathrm{M}} : \boldsymbol{U}^1(\mathfrak{B}_{\mathrm{m}})J^1_{\mathrm{M}}, \boldsymbol{U}^1(\mathfrak{A}_{\mathrm{m}})) \mid J^1_{\mathrm{M}}.$$

If $x \in \boldsymbol{U}^1(\mathfrak{A}_{\mathrm{m}})$ intertwines $\tilde{\eta}_{\mathrm{M}}$ with η_{M}, then it intertwines η_{M} with itself, and so lies in $\boldsymbol{U}^1(\mathfrak{B}_{\mathrm{m}})J^1_{\mathrm{M}}$. Thus the multiplicity of η_{M} in $\mathrm{Ind}(\eta_{\mathrm{m}} : J^1_{\mathrm{m}}, \boldsymbol{U}^1(\mathfrak{A}_{\mathrm{m}}))$ is equal to its multiplicity in $\tilde{\eta}_{\mathrm{M}}$, which we know to be one. ■

Now we have a basic consistency property.

(5.1.18) Proposition: *In the notation above, we have*

$$\tilde{\eta}_{\mathrm{M}} \mid \boldsymbol{U}^1(\mathfrak{B})J^1_{\mathrm{M}} = \hat{\eta}_{\mathrm{M}}.$$

Proof: Consider the representation $\mathrm{Ind}(\tilde{\eta} : \boldsymbol{U}^1(\mathfrak{B}_{\mathrm{m}})J^1, \boldsymbol{U}^1(\mathfrak{B}_{\mathrm{m}})\boldsymbol{U}^1(\mathfrak{A}))$. This is certainly irreducible, and its restriction to $\boldsymbol{U}^1(\mathfrak{A})$ contains the representation induced by η. We have $\dim(\eta) = \dim(\tilde{\eta})$ and also

$$(\boldsymbol{U}^1(\mathfrak{B}_{\mathrm{m}})\boldsymbol{U}^1(\mathfrak{A}) : \boldsymbol{U}^1(\mathfrak{B}_{\mathrm{m}})J^1) = (\boldsymbol{U}^1(\mathfrak{A}) : J^1)$$

since $\boldsymbol{U}^1(\mathfrak{B}_{\mathrm{m}}) \cap \boldsymbol{U}^1(\mathfrak{A}) = \boldsymbol{U}^1(\mathfrak{B}) \subset J^1$. Therefore, comparing dimensions, we see that

$$\mathrm{Ind}(\tilde{\eta} : \boldsymbol{U}^1(\mathfrak{B}_{\mathrm{m}})J^1, \boldsymbol{U}^1(\mathfrak{B}_{\mathrm{m}})\boldsymbol{U}^1(\mathfrak{A})) \mid \boldsymbol{U}^1(\mathfrak{A}) = \mathrm{Ind}(\eta : J^1, \boldsymbol{U}^1(\mathfrak{A})).$$

Now restrict this representation to $\boldsymbol{U}^1(\mathfrak{B}_{\mathrm{m}})J^1_{\mathrm{M}}$. When restricted further to the subgroup $\boldsymbol{U}^1(\mathfrak{B})J^1_{\mathrm{M}}$, it contains $\hat{\eta}_{\mathrm{M}}$, so let λ be an irreducible component of

$$\mathrm{Ind}(\tilde{\eta} : \boldsymbol{U}^1(\mathfrak{B}_{\mathrm{m}})J^1, \boldsymbol{U}^1(\mathfrak{B}_{\mathrm{m}})\boldsymbol{U}^1(\mathfrak{A})) \mid \boldsymbol{U}^1(\mathfrak{B}_{\mathrm{m}})J^1_{\mathrm{M}}$$

which contains $\hat{\eta}_{\mathrm{M}}$. The restriction of λ to J^1_{M} contains, and is therefore a multiple of, η_{M}. The $\boldsymbol{U}^1(\mathfrak{A}_{\mathrm{m}})$-intertwining of λ is therefore contained in that of η_M, which equals $J^1_M B^\times J^1_M \cap \boldsymbol{U}^1(\mathfrak{A}_m) = \boldsymbol{U}^1(\mathfrak{B}_m)J^1_M$. Therefore λ induces irreducibly to $\boldsymbol{U}^1(\mathfrak{A}_m)$, and indeed

$$\mathrm{Ind}(\lambda : \boldsymbol{U}^1(\mathfrak{B}_m)J^1_M, \boldsymbol{U}^1(\mathfrak{A}_m)) = \mathrm{Ind}(\eta_m : J^1_m, \boldsymbol{U}^1(\mathfrak{A}_m)).$$

From the uniqueness property of $\tilde{\eta}_M$ given by (5.1.14), it follows that $\lambda = \tilde{\eta}_M$, as required. ∎

To conclude this section, we compute the intertwining of these new representations.

(5.1.19) Proposition: *Let* $\tilde{\eta}_M$ *be as in* (5.1.14), $\tilde{\eta}$ *as in* (5.1.15). *Then*

$$I_G(\tilde{\eta}_M, \tilde{\eta}_M) = J^1_M B^\times J^1_M,$$
$$I_G(\tilde{\eta}, \tilde{\eta}) = J^1 B^\times J^1.$$

Proof: The first statement is a special case of the second. In the second, we certainly have $I_G(\tilde{\eta}, \tilde{\eta}) \subset I_G(\eta, \eta) = J^1 B^\times J^1$. For the other containment, let $x \in B^\times$. Then x certainly intertwines

$$\mathrm{Ind}(\tilde{\eta} : \boldsymbol{U}^1(\mathfrak{B}_m)J^1, \boldsymbol{U}^1(\mathfrak{A}_m)) = \mathrm{Ind}(\eta_m : J^1_m, \boldsymbol{U}^1(\mathfrak{A}_m))$$

by (4.1.1), (4.1.3) and (5.1.1) (applied to η_m). We deduce that there exist $u, v \in \boldsymbol{U}^1(\mathfrak{A}_m)$ such that uxv intertwines $\tilde{\eta}$. In particular, uxv intertwines η. Thus there exist $j_1, j_2 \in J^1$ such that $j_1 uxv j_2 \in B^\times$, by (5.1.1) again. Now we recall from (1.6.1) that

$$\textbf{(5.1.20)} \qquad \boldsymbol{U}^1(\mathfrak{A}_m)x\boldsymbol{U}^1(\mathfrak{A}_m) \cap B^\times = \boldsymbol{U}^1(\mathfrak{B}_m)x\boldsymbol{U}^1(\mathfrak{B}_m),$$

for $x \in B^\times$. Therefore, we can find $u', v' \in \boldsymbol{U}^1(\mathfrak{B}_m)$ such that $u'xv' = j_1 uxv j_2$, and this element certainly intertwines $\tilde{\eta}$. Therefore so does x, as required. ∎

(5.2) Extending to level zero

We now come to the more interesting problem of extending the representation $\eta(\theta)$ constructed in (5.1.1) to a representation of $J(\beta, \mathfrak{A})$. This sort of process has often been, rather vaguely, subsumed under the general heading "Weil representation". See **[Wa]** p.127 for a discussion of this. There are certain difficulties associated with this approach. First, the "Mackey obstruction", in certain circumstances which can arise here, *even when all ramification is tame,* is taken from a non-vanishing cohomology group. (The authors are indebted to Harry Reimann for pointing out the fact that $\mathbf{H}^2(GL(3, \mathbb{F}_2), \mathbb{F}_2)$ is not trivial.) Moreover, once one has proved that an extension exists, one still has to be careful about which extension to choose. In general, many of the possible extensions are not at all suitable for our purposes, and it is not easy to specify the correct ones when trying to treat the order $\mathfrak{A}$, or the character θ, in isolation. We therefore treat a consistent family of extensions of the $\eta(\theta)$, as θ ranges over a collection of characters related by the bijections $\boldsymbol{\tau}$ of (3.6.2). As we shall see, this extra structure leads to an entirely different, and we think entirely new, method for dealing with this extension problem.

So, let $[\mathfrak{A}, n, 0, \beta]$ be a simple stratum in A, let $\theta \in \mathcal{C}(\mathfrak{A}, 0, \beta)$, and let η be the unique irreducible representation of $J^1(\beta, \mathfrak{A})$ such that $\eta \mid H^1(\beta, \mathfrak{A})$ contains θ, as in (5.1.1).

(5.2.1) Definition: *A β-extension of η is a representation κ of $J(\beta, \mathfrak{A})$ such that*

(i) $\kappa \mid J^1(\beta, \mathfrak{A}) = \eta$, and

(ii) κ is intertwined by the whole of $B^\times$.

Here, as usual, B denotes the A-centraliser of β.

Remark: As we observed in **(3.5)**, the character $\theta \in \mathcal{C}(\mathfrak{A}, 0, \beta)$ does not determine the simple stratum $[\mathfrak{A}, n, 0, \beta]$. So, suppose we have another simple stratum $[\mathfrak{A}, n, 0, \beta']$ such that $\mathcal{C}(\mathfrak{A}, 0, \beta') = \mathcal{C}(\mathfrak{A}, 0, \beta)$. Implicit in this is $H^1(\beta', \mathfrak{A}) = H^1(\beta, \mathfrak{A})$. We get $J(\beta', \mathfrak{A}) = J(\beta, \mathfrak{A})$, since this is the $\boldsymbol{U}(\mathfrak{A})$-intertwining of the character θ. It follows that $J^1(\beta', \mathfrak{A}) = J^1(\beta, \mathfrak{A})$. The representation η of $J^1(\beta, \mathfrak{A})$ is defined independent of β but, at first sight, the two elements β, β' might give rise to distinct families of "interesting" extensions to J. This is not the case. For, writing down the G-intertwining of the character θ by (3.3.2), we get the identity

$$J^1 B^\times J^1 = J^1 B'^\times J^1,$$

where B' is the A-centraliser of β'. Thus, if κ is a β-extension of η, it is intertwined by all of $B'^\times$, and is therefore also a β'-extension.

It follows that the notion of β-extension actually only depends on the underlying simple character θ.

The main result of this section is the following.

(5.2.2) Theorem: *Let $[\mathfrak{A}, n, 0, \beta]$ be a simple stratum in A, and put $E = F[\beta]$, $B =$ the A-centraliser of β. Let $\theta \in \mathcal{C}(\mathfrak{A}, 0, \beta)$, and let η be the unique irreducible representation of $J^1(\beta, \mathfrak{A})$ containing θ.*

(i) There exists a β-extension κ of η.

(ii) An irreducible representation κ' of $J(\beta, \mathfrak{A})$ is a β-extension of η if and only if $\kappa' = \kappa \otimes \chi \circ \det_B$, for some character χ of $\boldsymbol{U}(\mathfrak{o}_E)/\boldsymbol{U}^1(\mathfrak{o}_E)$.

(iii) As χ ranges over the characters of $\boldsymbol{U}(\mathfrak{o}_E)/\boldsymbol{U}^1(\mathfrak{o}_E)$, the representations $\kappa \otimes \chi \circ \det_B$ are distinct, and, further, distinct representations in this family do not intertwine.

The proof of this result will occupy the rest of this section (and part of it must wait until **(7.2)**). We now re-adopt the notation (5.1.12), (5.1.13). The first stage of the argument is summarised in the following result.

(5.2.3) Theorem: *Use the notation* (5.1.12), (5.1.13), *and assume further that* $\mathfrak{B}_{\mathrm{M}}$ *(resp.* $\mathfrak{B}_{\mathrm{m}}$*) is a maximal (resp. minimal) hereditary* $\mathfrak{o}_E$*-order in* B.

(i) There exists an irreducible representation κ_{M} *of* J_{M} *such that* $\kappa_{\mathrm{M}} \mid J^1_{\mathrm{M}} = \eta_{\mathrm{M}}$*, and the induced representations*

$$\mathrm{Ind}(\kappa_{\mathrm{M}} \mid \boldsymbol{U}^1(\mathfrak{B}_{\mathrm{m}})J^1_{\mathrm{M}} : \boldsymbol{U}^1(\mathfrak{B}_{\mathrm{m}})J^1_{\mathrm{M}}, \boldsymbol{U}^1(\mathfrak{A}_{\mathrm{m}})), \quad \mathrm{Ind}(\eta_{\mathrm{m}} : J^1_{\mathrm{m}}, \boldsymbol{U}^1(\mathfrak{A}_{\mathrm{m}}))$$

are equivalent and irreducible. These conditions determine κ_{M} *uniquely up to tensoring with a character* $\chi \circ \det_B$*, where* χ *is a character of* $\boldsymbol{U}(\mathfrak{o}_E)/\boldsymbol{U}^1(\mathfrak{o}_E)$.

(ii) Fix a representation κ_{M} *satisfying (i). Then there exists a unique representation* κ *of* J *such that* $\kappa \mid J^1 = \eta$ and $\mathrm{Ind}(\kappa : J, \boldsymbol{U}(\mathfrak{A})) \cong \mathrm{Ind}(\kappa_{\mathrm{M}} \mid \boldsymbol{U}(\mathfrak{B})J^1_{\mathrm{M}} : \boldsymbol{U}(\mathfrak{B})J^1_{\mathrm{M}}, \boldsymbol{U}(\mathfrak{A}))$. *These representations are irreducible.*

(iii) If κ *is as in (ii), then*

$$\mathrm{Ind}(\eta_{\mathrm{m}} : J^1_{\mathrm{m}}, \boldsymbol{U}^1(\mathfrak{A}_{\mathrm{m}})) \cong \mathrm{Ind}(\kappa \mid \boldsymbol{U}^1(\mathfrak{B}_{\mathrm{m}})J^1 : \boldsymbol{U}^1(\mathfrak{B}_{\mathrm{m}})J^1, \boldsymbol{U}^1(\mathfrak{A}_{\mathrm{m}})),$$

and these representations are irreducible.

Remark: (5.2.3) applies equally to the case $\mathfrak{A} = \mathfrak{A}_{\mathrm{m}}$. This gives us a unique representation κ_{m} of J^1_{m} satisfying *(ii)*. Of course, *(iii)* is empty in this case.

Remark: In (5.2.3)*(i)*, the representations $\kappa_{\mathrm{M}} \otimes \chi \circ \det_B$, as χ ranges over the characters of $\boldsymbol{U}(\mathfrak{o}_E)/\boldsymbol{U}^1(\mathfrak{o}_E)$, are all distinct. Indeed, they are distinguished by their determinants, since the order of χ divides $(q_E - 1)$ (hence is relatively prime to the residual characteristic p), while the dimension of κ_{M} is a power of p.

Proof of (5.2.3): Let $\tilde{\eta}_{\mathrm{M}}$ be the representation of $\boldsymbol{U}^1(\mathfrak{B}_{\mathrm{m}})J^1_{\mathrm{M}}$ constructed in (5.1.14). The first step of the proof is:—

(5.2.4) Proposition: *There exists a representation* κ_{M} *of* J_{M} *such that* $\kappa_{\mathrm{M}} \mid \boldsymbol{U}^1(\mathfrak{B}_{\mathrm{m}})J^1_{\mathrm{M}} = \tilde{\eta}_{\mathrm{M}}$. *If* κ'_{M} *is any other representation of* J_{M} *which restricts to* $\tilde{\eta}_{\mathrm{M}}$*, then* $\kappa'_{\mathrm{M}} = \kappa_{\mathrm{M}} \otimes \chi \circ \det_B$*, for some character* χ *of* $\boldsymbol{U}(\mathfrak{o}_E)/\boldsymbol{U}^1(\mathfrak{o}_E)$.

Proof: Let p be the residual characteristic of our base field F. The group J_{M} normalises the representation η_{M}, so we can extend η_{M} to a *projective* representation ξ_{M} of J_{M}. Thus, if η_{M} acts on the space W, then ξ_{M} is a map $J_M \to \mathrm{Aut}_{\mathbb{C}}(W)$ such that $\xi_M(x) = \eta_M(x)$ for $x \in J^1_M$, while in general, $\xi_M(x)^{-1}\eta_M(y)\xi_M(x) = \eta_M(x^{-1}yx)$, $x \in J_M$, $y \in J^1_M$. Moreover, we have $\xi_M(x)\xi_M(y) = \boldsymbol{\alpha}(x,y)\xi_M(xy)$, $x, y \in J_M$, for some 2-cocycle $\boldsymbol{\alpha}$ of J_M in $\mathbb{C}^\times$.

We can adjust the automorphisms $\xi_M(x)$ by scalars, at the cost of replacing $\boldsymbol{\alpha}$ by a cohomologous cocycle. Since J^1_M is a pro-p group, the determinants $\det_W(\eta_M(x))$, $x \in J^1_M$, are p-power roots of unity. Choose a set of coset representatives $\{x_i\}$ for J_M/J^1_M, with the convention that if $x_i \in J^1_M$, then $x_i = 1$. Choose the $\xi_M(x_i)$ to have determinant 1, and, for $x \in J^1_M$, set $\xi_M(x_ix) = \xi_M(x_i)\eta_M(x)$. Then, for $y_1, y_2 \in J_M$, we have

$$(\boldsymbol{\alpha}(y_1, y_2))^{\dim(\eta_M)} = \det_W(\xi_M(y_1)\xi_M(y_2)\xi_M(y_1y_2)^{-1}),$$

which is a p-power root of unity. The dimension of η_M is a power of p, so the values of $\boldsymbol{\alpha}$ are p-power roots of unity. It follows that the order of the cohomology class of $\boldsymbol{\alpha}$ in $\mathbf{H}^2(J_M, \mathbb{C}^\times)$ is a power of p.

However, we know that there exists an extension of η_M to a linear representation of $\boldsymbol{U}^1(\mathfrak{B}_m)J^1_M$, namely the representation $\tilde{\eta}_M$ constructed in (5.1.14). It follows that the restriction of $\boldsymbol{\alpha}$ to $\boldsymbol{U}^1(\mathfrak{B}_m)J^1_M$ is cohomologous to zero. Now we observe that $\boldsymbol{U}^1(\mathfrak{B}_m)J^1_M$ is a Sylow pro-p subgroup of J_M. Therefore, by **[Br]** III.10.3, the original cohomology class of $\boldsymbol{\alpha}$ in $\mathbf{H}^2(J_M, \mathbb{C}^\times)$ is trivial, and so η_M extends to a linear representation λ of J_M, as required.

Now let us compare $\lambda \mid \boldsymbol{U}^1(\mathfrak{B}_m)J^1_M$ with $\tilde{\eta}_M$. This restriction is of the form $\tilde{\eta}_M \otimes \phi$, for some abelian character ϕ of $\boldsymbol{U}^1(\mathfrak{B}_m)/\boldsymbol{U}^1(\mathfrak{B}_M)$. The representation λ (of J_M) is certainly intertwined by $\boldsymbol{U}(\mathfrak{B}_M)$. By (5.1.19), the representation $\tilde{\eta}_M$ is intertwined by the whole of $B^\times$. It is therefore intertwined by $\boldsymbol{U}(\mathfrak{B}_M)$, so the character ϕ is also intertwined by $\boldsymbol{U}(\mathfrak{B}_M)$. We can identify J_M/J^1_M with $GL(t, \mathrm{k}_E)$, for some integer t, where k_E is the residue class field of E. This identifies the image of $\boldsymbol{U}^1(\mathfrak{B}_m)$ with the unipotent radical $\mathcal{N}$ of a Borel subgroup of $\mathcal{G} = GL(t, \mathrm{k}_E)$. Thus we may think of ϕ as a character of $\mathcal{N}$ which is intertwined by every element of $\mathcal{G}$. We will show that such a character ϕ extends to a character ϕ' of $\mathcal{G}$. Inflating ϕ' to a character (also denoted ϕ') of J_M, we put $\kappa_M = \lambda \otimes \phi'^{-1}$, and this extends $\tilde{\eta}_M$, as required.

We now show that the character ϕ extends to $\mathcal{G}$. Conjugating as appropriate, we can assume that we are given a character ϕ of the group $\mathcal{N}$ of all upper triangular unipotent matrices in $\mathcal{G} = GL(t, \mathrm{k}_E)$, with the property that ϕ is intertwined by every element of $\mathcal{G}$. The commutator group of $\mathcal{N}$ consists of all matrices $\mathfrak{n} = (\mathfrak{n}_{ij}) \in \mathcal{N}$ such that $\mathfrak{n}_{i,i+1} = 0$, $1 \le i \le t-1$. Thus ϕ is of the form $\mathfrak{n} \mapsto \vartheta(\sum a_i \mathfrak{n}_{i,i+1})$, where ϑ is a nontrivial character of the additive group of k_E and $a_1, a_2, \dots, a_{t-1}$ are uniquely determined elements of k_E. Suppose first that k_E has at least three elements. The character ϕ is stable under conjugation by any diagonal matrix in $\mathcal{G}$, and this implies that all a_i are zero. Therefore ϕ is trivial, and certainly extends to a character of $\mathcal{G}$. We therefore assume that k_E has just two elements. If $t = 2$, the group $\mathcal{G}$ is the symmetric group on three symbols, and $\mathcal{N}$ is a quotient of $\mathcal{G}$. Thus ϕ again extends to a character of $\mathcal{G}$. So, we may assume also that $t \geq 3$. Let c be the permutation matrix in $\mathcal{G}$ corresponding to the cycle $(1, 2, \ldots, t)$. The fact that c intertwines ϕ implies $a_1 = a_2 = \ldots = a_{t-1}$, in the notation above. On the other hand, the fact that the permutation matrix corresponding to the transposition $(1, 2)$ intertwines ϕ implies that $a_2 = 0$. Hence again ϕ is trivial, and extends to $\mathcal{G}$, as required.

Now suppose we have another representation κ' of J_{M} with $\kappa' \mid \boldsymbol{U}^1(\mathfrak{B}_{\mathrm{m}})J^1_{\mathrm{M}} = \tilde{\eta}_{\mathrm{M}}$. Thus κ' is of the form $\kappa' = \kappa_{\mathrm{M}} \otimes \phi$, where we may now view ϕ as a character of $\mathcal{G} = GL(t, k_E)$ which is null on the image $\mathcal{N}$ of $\boldsymbol{U}^1(\mathfrak{B}_{\mathrm{m}})$ in $\mathcal{G}$. It follows that ϕ is null on the subgroup of $\mathcal{G}$ generated by all $\mathcal{G}$-conjugates of $\mathcal{N}$. We again think of $\mathcal{N}$ as the group of upper triangular unipotent matrices in $GL(t, k_E)$. The character ϕ is null on $\mathcal{N}$ and on every lower triangular unipotent matrix. Elementary row and column operations show readily that $SL(t, k_E)$ is generated by the upper and lower triangular unipotent subgroups. Thus ϕ is trivial on $SL(t, k_E)$, and factors through the determinant, as required. ∎

This technique of first extending to a p-Sylow subgroup was suggested by **[Wa]**.

(5.2.5) Proposition: *Let κ_{M} be as in* (5.2.4). *There is a representation κ of J which is uniquely determined by the following properties:*

(i) $\kappa \mid J^1 = \eta$;

(ii) κ and $\kappa_{\mathrm{M}} \mid \boldsymbol{U}(\mathfrak{B})J^1_{\mathrm{M}}$ induce equivalent irreducible representations of $\boldsymbol{U}(\mathfrak{A})$.

Moreover, this representation κ satisfies

(iii)

$$\mathrm{Ind}(\kappa : J, \boldsymbol{U}(\mathfrak{B})\boldsymbol{U}^1(\mathfrak{A})) \cong \mathrm{Ind}(\kappa_{\mathrm{M}} \mid \boldsymbol{U}(\mathfrak{B})J^1_{\mathrm{M}} : \boldsymbol{U}(\mathfrak{B})J^1_{\mathrm{M}}, \boldsymbol{U}(\mathfrak{B})\boldsymbol{U}^1(\mathfrak{A})).$$

Proof: The representation

$$\lambda = \mathrm{Ind}(\kappa_{\mathrm{M}} \mid \boldsymbol{U}(\mathfrak{B})J^1_{\mathrm{M}} : \boldsymbol{U}(\mathfrak{B})J^1_{\mathrm{M}}, \boldsymbol{U}(\mathfrak{B})\boldsymbol{U}^1(\mathfrak{A}))$$

is irreducible. When restricted to $\boldsymbol{U}^1(\mathfrak{A})$, it has a component

$$\begin{aligned}\mathrm{Ind}(\tilde{\eta}_{\mathrm{M}} : \boldsymbol{U}^1(\mathfrak{B})J^1_{\mathrm{M}}, \boldsymbol{U}^1(\mathfrak{A})) &= \mathrm{Ind}(\hat{\eta}_{\mathrm{M}} : \boldsymbol{U}^1(\mathfrak{B})J^1_{\mathrm{M}}, \boldsymbol{U}^1(\mathfrak{A})) \\ &= \mathrm{Ind}(\eta : J^1, \boldsymbol{U}^1(\mathfrak{A}))\end{aligned}$$

by (5.1.18) and (5.1.16). Thus $\lambda \mid J^1$ contains η. Let κ be an irreducible component of $\lambda \mid J$ such that $\kappa \mid J^1$ contains η. Then $\kappa \mid J^1$ is a multiple of η. The $\boldsymbol{U}(\mathfrak{B})\boldsymbol{U}^1(\mathfrak{A})$-intertwining of η is $\boldsymbol{U}(\mathfrak{B})\boldsymbol{U}^1(\mathfrak{A}) \cap JB^\times J = J$, so this implies that κ induces irreducibly to $\boldsymbol{U}(\mathfrak{B})\boldsymbol{U}^1(\mathfrak{A})$. Therefore $\mathrm{Ind}(\kappa : J, \boldsymbol{U}(\mathfrak{B})\boldsymbol{U}^1(\mathfrak{A})) = \lambda$. Comparing dimensions using (5.1.2), we get $\dim(\kappa) = \dim(\eta)$, whence the representation κ satisfies all the required conditions.

We have to show that κ is uniquely determined by *(i)* and *(ii)*. To do this, it is enough to show that η occurs with multiplicity one in the representation of $\boldsymbol{U}(\mathfrak{A})$ induced by κ, since this representation only depends on κ_{M}. We restrict this representation to J^1. By (5.1.19) and Mackey theory, we see that η occurs in $\mathrm{Ind}(\kappa : J, \boldsymbol{U}(\mathfrak{A}))$ with the same multiplicity as it occurs in κ, i.e. with multiplicity one. ■

We shall need one more property of this representation.

(5.2.6) Proposition: *Let κ be the representation of J constructed in* (5.2.5). *Then $\kappa \mid \boldsymbol{U}^1(\mathfrak{B}_{\mathrm{m}})J^1 = \tilde{\eta}$ (as in* (5.1.15)*).*

Proof: We have already noted that the representation

$$\mathrm{Ind}(\kappa_{\mathrm{M}} : \boldsymbol{U}(\mathfrak{B})J^1_{\mathrm{M}}, \boldsymbol{U}(\mathfrak{A})) = \mathrm{Ind}(\kappa : J, \boldsymbol{U}(\mathfrak{A}))$$

contains η with multiplicity one. The same applies to the representation

$$\mathrm{Ind}(\eta_{\mathrm{m}} : J^1_{\mathrm{m}}, \boldsymbol{U}^1(\mathfrak{A}_{\mathrm{m}})) = \mathrm{Ind}(\kappa_{\mathrm{M}} : \boldsymbol{U}^1(\mathfrak{B}_{\mathrm{m}})J^1_{\mathrm{M}}, \boldsymbol{U}^1(\mathfrak{A}_{\mathrm{m}}))$$

and it follows that the irreducible representation $\kappa \mid \boldsymbol{U}^1(\mathfrak{B}_{\mathrm{m}})J^1$ occurs in $\mathrm{Ind}(\eta_{\mathrm{m}} : J^1_{\mathrm{m}}, \boldsymbol{U}^1(\mathfrak{A}_{\mathrm{m}}))$. However, this last representation contains η with multiplicity one, as the restriction to J^1 of the component $\tilde{\eta}$. Therefore $\kappa \mid \boldsymbol{U}^1(\mathfrak{B}_{\mathrm{m}})J^1 = \tilde{\eta}$, as required. ■

Now we have proved (5.2.3) The three parts are (5.2.4), (5.2.5), (5.2.6) respectively. ■

Remark: We could have produced an extension κ of η to J by following the same procedure used to get κ_{M}. However, the resulting representation would only have been defined up to twisting with a character of $\boldsymbol{U}(\mathfrak{B})/\boldsymbol{U}^1(\mathfrak{B}) = (\mathfrak{B}/\mathfrak{Q})^\times$, lifted from the centre of the algebra $\mathfrak{B}/\mathfrak{Q}$ via the determinant. Of course, this centre is of the form $(\mathsf{k}_E)^e$, where $e = e(\mathfrak{B}|\mathfrak{o}_E)$.

We now work out the intertwining of these representations.

(5.2.7) Proposition: *Let κ be the representation of J constructed in* (5.2.3). *Then $I_G(\kappa) = JB^\times J = J^1B^\times J^1$.*

Proof: We surely have $I_G(\kappa) \subset I_G(\eta) = JB^\times J$ by (5.1.1), so we need only show that κ is intertwined by every $y \in B^\times$. Further, by (5.2.5)*(ii)*, it is enough to treat the case $\mathfrak{B} = \mathfrak{B}_{\mathrm{M}}$ is a maximal order, $\kappa = \kappa_{\mathrm{M}}$.

So, we fix an element $y \in B^\times$ and show that it intertwines κ_{M}. We choose a basis to identify $\mathfrak{B}_{\mathrm{M}}$ with $\mathbb{M}(R, \mathfrak{o}_E)$, where $R = \dim_E(V)$. We are at liberty to vary y in its $(\boldsymbol{U}(\mathfrak{B}_{\mathrm{M}}), \boldsymbol{U}(\mathfrak{B}_{\mathrm{M}}))$-double coset, so we assume

$$y = \operatorname{diag}(\pi_E^{a_1}, \ldots, \pi_E^{a_R}), \tag{5.2.8}$$

where π_E is some fixed prime element of E and $a_1 \geq a_2 \geq \ldots \geq a_R$. More precisely, suppose we have

$$\begin{aligned} a_1 = a_2 = \ldots = a_{n_1} > a_{n_1+1} = \ldots = a_{n_1+n_2} \\ > a_{n_1+n_2+1} = \ldots\ldots = a_{n_1+\ldots n_e}, \\ n_1 + n_2 + \ldots + n_e = R. \end{aligned}$$

Let $\mathfrak{B}$ be the hereditary $\mathfrak{o}_E$-order of all block matrices $(x_{ij})_{1 \leq i,j \leq e}$, in which the (i,j)-block x_{ij} has dimensions $n_i \times n_j$ and entries in $\mathfrak{o}_E$ if $i \leq j$ and in $\mathfrak{p}_E$ otherwise. Let $\mathfrak{M}(\mathfrak{B})$ denote the subring of block matrices $(x_{ij}) \in \mathfrak{B}$ such that $x_{ij} = 0$ whenever $i \neq j$. Then

(5.2.9) *(i) y centralises $\mathfrak{M}(\mathfrak{B})$ (and $\mathfrak{B}$ is maximal for this property);*
(ii) $\mathfrak{B}_{\mathrm{M}} \cap \mathfrak{B}_{\mathrm{M}}^y \subset \mathfrak{p}_E \mathfrak{B}_{\mathrm{M}} + (\mathfrak{B} \cap \mathfrak{B}^y)$.

We can now treat $\mathfrak{B}$ on the same footing as the order $\mathfrak{B}$ in (5.1.12).

There is a certain symmetry here. Let $\mathfrak{B}'$ be the transpose of $\mathfrak{B}$. Thus, in the obvious notation, we have $\mathfrak{M}(\mathfrak{B}') = \mathfrak{M}(\mathfrak{B})$, and $\mathfrak{B}'$ has properties relative to y^{-1} the same as those of $\mathfrak{B}$ relative to y. Explicitly,

(5.2.9)′ *(i) y^{-1} centralises $\mathfrak{M}(\mathfrak{B}')$ (and $\mathfrak{B}'$ is maximal for this property);*
(ii) $\mathfrak{B}_{\mathrm{M}} \cap {}^y\mathfrak{B}_{\mathrm{M}} \subset \mathfrak{p}_E + (\mathfrak{B}' \cap {}^y\mathfrak{B}')$.

Before proceeding, let us record one special case.

(5.2.10) Lemma: *Suppose, in the above situation, that $\mathfrak{B} = \mathfrak{B}_{\mathrm{M}}$ (or, equivalently, that $y \in E^\times$). Then y intertwines κ_{M}.*

Proof: Let $\mathfrak{B}_{\mathrm{m}}$ denote the obvious (upper triangular mod $\mathfrak{p}_E$) minimal hereditary $\mathfrak{o}_E$-order contained in $\mathfrak{B}$. Let $\tilde{\eta}_{\mathrm{M}}$ be the extension of $\eta = \eta_{\mathrm{M}}$ to $\boldsymbol{U}^1(\mathfrak{B}_{\mathrm{m}})J_{\mathrm{M}}^1$ given by (5.1.14). The element y normalises the groups J_{M}, $\boldsymbol{U}^1(\mathfrak{B}_{\mathrm{m}})J_{\mathrm{M}}^1$, and stabilises the representation $\tilde{\eta}_{\mathrm{M}}$, by (5.1.19). Since, by construction, κ_{M} is an extension of $\tilde{\eta}_{\mathrm{M}}$, so is κ_{M}^y. Thus $\kappa_{\mathrm{M}}^y = \kappa_{\mathrm{M}} \otimes \phi$, where ϕ is a character of $J_{\mathrm{M}}/J_{\mathrm{M}}^1 = \boldsymbol{U}(\mathfrak{B}_{\mathrm{M}})/\boldsymbol{U}^1(\mathfrak{B}_{\mathrm{M}})$ which is null on the image of $\boldsymbol{U}^1(\mathfrak{B}_{\mathrm{m}})$. This implies that ϕ has order prime to p, and, arguing as in the proof of (5.2.4), it factors through the determinant map $\det_B : \boldsymbol{U}(\mathfrak{B}_{\mathrm{M}}) \to \boldsymbol{U}(\mathfrak{o}_E)$. Temporarily write D for the group of diagonal matrices in $\boldsymbol{U}(\mathfrak{B}_{\mathrm{M}})$. Then y centralises D, with the result that the representations $\kappa_{\mathrm{M}}^y = \kappa_{\mathrm{M}} \otimes \phi$ and κ_{M} must agree on D. Taking determinants on the representation space, this implies

$$\det(\kappa_{\mathrm{M}}(d)) = \det(\kappa_{\mathrm{M}}(d))\phi(d)^{\dim(\kappa_{\mathrm{M}})}, \quad d \in D.$$

The dimension of κ_{M} is a power of p, so ϕ is null on D. Since D maps onto $\boldsymbol{U}(\mathfrak{o}_E)$ under $\det_B$, this implies that ϕ is trivial, as required. ∎

Now we return to the situation above. Consider the (irreducible) representation $\kappa_{\mathrm{M}} \mid \boldsymbol{U}(\mathfrak{B})J_{\mathrm{M}}^1$. We have $\boldsymbol{U}(\mathfrak{B}) = \mathfrak{M}(\mathfrak{B})^\times \boldsymbol{U}^1(\mathfrak{B})$ and we know from (5.1.18) that $\kappa_{\mathrm{M}} \mid \boldsymbol{U}^1(\mathfrak{B})J_{\mathrm{M}}^1 = \hat{\eta}_{\mathrm{M}}$, in the notation of (5.1.16). Further, by (5.1.18) and (5.1.19), the element y intertwines $\hat{\eta}_{\mathrm{M}}$. Indeed, the space $I_y(\hat{\eta}_{\mathrm{M}})$ is contained in $I_y(\eta_{\mathrm{M}})$ and therefore has dimension 1 by (5.1.8).

(5.2.11) Lemma: *Let G be a locally profinite group, H a compact open subgroup of G, and ρ an irreducible smooth representation of H. Let g be some element of G such that the intertwining space $I_g(\rho)$ has dimension one. Let N be a closed subgroup of G which normalises H and suppose that ρ admits extension to a linear representation of NH. Suppose also that*

(i) g centralises N, and

(ii) $(NH) \cap (NH)^g = N(H \cap H^g)$.

Let $\tilde{\rho}$ be an extension of ρ to NH. There exists a unique abelian character ϕ of N, which is trivial on $N \cap H$, such that g intertwines $\tilde{\rho}$ with $\tilde{\rho} \otimes \phi$. Moreover,

(a) y intertwines ρ' with $\rho' \otimes \phi$, for any extension ρ' of ρ to NH, and

(b) if ϕ' is a character of $N/N \cap H$ different from ϕ and if ρ' is an extension of ρ to NH, then y does not intertwine ρ' with $\rho' \otimes \phi'$.

Proof: Let ρ, $\tilde{\rho}$ act on the space W. We view ρ^g as also acting on W. Thus $\tilde{\rho}^g$ also acts on W, and $\tilde{\rho}^g(n) = \tilde{\rho}(n)$, $n \in N$. Let $\Phi \in \mathrm{Hom}_{H \cap H^g}(\rho, \rho^g)$ be a nonzero intertwining operator. Thus $\rho^g(h) \circ \Phi = \Phi \circ \rho(h)$, $h \in H \cap H^g$. Now consider the operator $\Phi(n) = \tilde{\rho}(n) \circ \Phi \circ \tilde{\rho}(n^{-1})$, for some $n \in N$. This is nonzero and independent of the choice of $\tilde{\rho}$. We find readily that $\Phi(n) \in \mathrm{Hom}_{H \cap H^g}(\rho, \rho^g)$. Thus $\Phi(n) = \Phi.\phi(n)$, for some character ϕ of $N/N \cap H \cap H^g$. We note however that $N \cap H \cap H^g = N \cap H$, and the first assertion of the lemma follows.

This construction of ϕ is independent of the choice of extension $\tilde{\rho}$, which implies *(a)*. In *(b)*, if y intertwines ρ' with $\rho' \otimes \phi'$ then, in the notation above, we have $\Phi.\phi(n) = \Phi.\phi'(n)$, for all $n \in N$. We can choose $n \in N$ such that $\phi(n) \neq \phi'(n)$, whence $\Phi = 0$, which is nonsense. ∎

The next step is to check that the hypotheses of (5.2.11) apply with $H = \boldsymbol{U}^1(\mathfrak{B})J_{\mathrm{M}}^1$, $\rho = \hat{\eta}_{\mathrm{M}}$, $g = y$, $N = \mathfrak{M}(\mathfrak{B})^\times$. The only non-obvious one is *(ii)*, which asserts that

$$(\mathfrak{M}(\mathfrak{B})^\times \boldsymbol{U}^1(\mathfrak{B})J^1_{\mathrm{M}}) \cap (\mathfrak{M}(\mathfrak{B})^\times \boldsymbol{U}^1(\mathfrak{B})J^1_{\mathrm{M}})^y \\ = \mathfrak{M}(\mathfrak{B})^\times(\boldsymbol{U}^1(\mathfrak{B})J^1_{\mathrm{M}} \cap (\boldsymbol{U}^1(\mathfrak{B})J^1_{\mathrm{M}})^y).$$

To do this, we temporarily write $J^1_{\mathrm{M}} = 1 + \mathfrak{j}$, and prove the additive assertion

$$(\mathfrak{M}(\mathfrak{B}) + \mathfrak{Q} + \mathfrak{j}) \cap (\mathfrak{M}(\mathfrak{B}) + \mathfrak{Q} + \mathfrak{j})^y = \mathfrak{M}(\mathfrak{B}) + (\mathfrak{Q} + \mathfrak{j}) \cap (\mathfrak{Q} + \mathfrak{j})^y,$$

where $\mathfrak{Q} = \mathrm{rad}(\mathfrak{B})$. This is an equality of $\mathfrak{o}_F$-orders, which implies equality between their unit groups, which is the desired relation. We surely have the containment

$$(\mathfrak{M}(\mathfrak{B}) + \mathfrak{Q} + \mathfrak{j}) \cap (\mathfrak{M}(\mathfrak{B}) + \mathfrak{Q} + \mathfrak{j})^y \supset \mathfrak{M}(\mathfrak{B}) + (\mathfrak{Q} + \mathfrak{j}) \cap (\mathfrak{Q} + \mathfrak{j})^y,$$

If we apply the adjoint map a_β to each side, the image of the left hand side is contained in $a_\beta(\mathfrak{j}) \cap a_\beta(\mathfrak{j}^y) = a_\beta(\mathfrak{j} \cap \mathfrak{j}^y)$. (This equality follows from the big commutative diagram in the proof of (5.1.10).) This is certainly contained in the image of the right hand side. Thus the two sides have the same image under a_β. We now compute the intersections of the two sides with $\mathrm{Ker}(a_\beta) = B$. The intersection of the left hand side with B is $\mathfrak{M}(\mathfrak{B}) + \mathfrak{Q} \cap \mathfrak{Q}^y$, which is contained in the right hand side, so we have the desired equality.

The lemma therefore applies, and we deduce that y must intertwine $\kappa_{\mathrm{M}} \mid \boldsymbol{U}(\mathfrak{B})J^1_{\mathrm{M}}$ with $\kappa_{\mathrm{M}} \mid \boldsymbol{U}(\mathfrak{B})J^1_{\mathrm{M}} \otimes \phi$, for some abelian character ϕ of $\mathfrak{M}(\mathfrak{B})^\times / \mathfrak{M}(\mathfrak{B})^\times \cap \boldsymbol{U}^1(\mathfrak{B})J^1_{\mathrm{M}}$. Let $\mathfrak{B}_{\mathrm{m}}$ be a minimal hereditary $\mathfrak{o}_E$-order in B contained in $\mathfrak{B}$, so that $\kappa_{\mathrm{M}} \mid \boldsymbol{U}^1(\mathfrak{B}_{\mathrm{m}})J^1_{\mathrm{M}} = \tilde{\eta}_{\mathrm{M}}$, in the notation of (5.1.14). Thus, in particular, y intertwines $\tilde{\eta}_{\mathrm{M}}$ with $\tilde{\eta}_{\mathrm{M}} \otimes \phi \mid \boldsymbol{U}^1(\mathfrak{B}_{\mathrm{M}})J^1_{\mathrm{M}}$. However, any intertwining operator here must lie in $\mathrm{Hom}_{J^1_{\mathrm{M}} \cap (J^1_{\mathrm{M}})^y}(\eta_{\mathrm{M}}, \eta^y_{\mathrm{M}})$, which is equal to the corresponding intertwining space for $\tilde{\eta}_{\mathrm{M}}$, by (5.1.19) and (5.1.8). It follows readily that ϕ is trivial on $\boldsymbol{U}^1(\mathfrak{B}_{\mathrm{m}})J^1_{\mathrm{M}}$. Since $\boldsymbol{U}^1(\mathfrak{B}_{\mathrm{m}})J^1_{\mathrm{M}}$ is a Sylow pro-p subgroup of $\boldsymbol{U}(\mathfrak{B})J^1_{\mathrm{M}}$, it follows that ϕ has order relatively prime to p. Further, it factors through the determinant in the following sense. We have, by definition;

$$\mathfrak{M}(\mathfrak{B}) = \prod_{i=1}^{e} \mathbb{M}(n_i, \mathfrak{o}_E),$$

so that $\phi = \phi_1 \otimes \ldots \otimes \phi_e$, where ϕ_i is a character of $GL(n_i, \mathfrak{o}_E)$. Moreover, each ϕ_i is trivial in $\boldsymbol{U}^1(\mathfrak{B}_{\mathrm{m}}) \cap GL(n_i \mathfrak{o}_E)$ and, in particular, on the principal congruence subgroup $1 + \mathfrak{p}_E \mathbb{M}(n_i, \mathfrak{o}_E)$. It follows (again as in the proof of (5.2.4)) that ϕ_i factors through the determinant map $GL(n_i, \mathfrak{o}_E) \to \boldsymbol{U}(\mathfrak{o}_E)$.

Now let κ be the representation of $J = J(\beta, \mathfrak{A})$ given by (5.2.5). We have $J = \mathfrak{M}(\mathfrak{B})^\times J^1$, and $\mathfrak{M}(\mathfrak{B})^\times \cap J^1 = \mathfrak{M}(\mathfrak{B})^\times \cap \boldsymbol{U}^1(\mathfrak{B}) = \mathfrak{M}(\mathfrak{B})^\times \cap J^1_{\mathrm{M}}$, so we may regard ϕ as a character of J which is trivial on J^1, or, equally, as a character of $\boldsymbol{U}(\mathfrak{B})\boldsymbol{U}^1(\mathfrak{A})$ which is null on $\boldsymbol{U}^1(\mathfrak{A})$. We can therefore form the representation $\kappa \otimes \phi$, and (5.2.5) gives us the relation

$$\begin{aligned}&\mathrm{Ind}(\kappa \otimes \phi : J, \boldsymbol{U}(\mathfrak{B})\boldsymbol{U}^1(\mathfrak{A}))\\ &\qquad\cong \mathrm{Ind}((\kappa_{\mathrm{M}} \mid \boldsymbol{U}(\mathfrak{B})J^1_{\mathrm{M}}) \otimes \phi : \boldsymbol{U}(\mathfrak{B})J^1_{\mathrm{M}}, \boldsymbol{U}(\mathfrak{B})\boldsymbol{U}^1(\mathfrak{A})).\end{aligned}$$

We now see that some element $z \in \boldsymbol{U}(\mathfrak{B})\boldsymbol{U}^1(\mathfrak{A})y\boldsymbol{U}(\mathfrak{B})\boldsymbol{U}^1(\mathfrak{A})$ intertwines κ with $\kappa \otimes \phi$. This element z must intertwine the representation $\eta = \kappa \mid J^1$, and (5.1.1), (1.6.1) now imply $z \in JyJ$. Therefore

(5.2.12) *The element y intertwines κ with $\kappa \otimes \phi$.*

It is easy to check that the hypotheses of (5.2.11) apply in the case $H = J^1$, $N = \mathfrak{M}(\mathfrak{B})^\times$, $g = y$, $\rho = \eta$, so we deduce

(5.2.13) *Let μ be a representation of J such that $\mu \mid J^1 = \eta$. Then y intertwines μ with $\mu \otimes \phi$.*

We will show in (7.2.11) below that there exists an extension μ of η to J which is intertwined by y. (5.2.11) now implies that ϕ is trivial. So, we finally see that y must intertwine the representation $\kappa_{\mathrm{M}} \mid \boldsymbol{U}(\mathfrak{B})J^1_{\mathrm{M}}$. For the rest of the proof, we abbreviate

$$K = \boldsymbol{U}(\mathfrak{B})J^1_{\mathrm{M}}.$$

We note that the definition of the representation κ_{M} is independent of the choice of the minimal hereditary $\mathfrak{o}_E$-order $\mathfrak{B}_{\mathrm{m}}$ used in its construction, since any two such orders $\mathfrak{B}_{\mathrm{m}}$ are conjugate by an element of $\boldsymbol{U}(\mathfrak{B}_{\mathrm{M}})$. This means that we may repeat the argument above, using the pair $(\mathfrak{B}', y^{-1})$ in place of $(\mathfrak{B}, y)$. We deduce that y^{-1} intertwines $\kappa_{\mathrm{M}} \mid \boldsymbol{U}(\mathfrak{B}')J^1_{\mathrm{M}}$. Intertwining is closed under inversion, so y also intertwines $\kappa_{\mathrm{M}} \mid \boldsymbol{U}(\mathfrak{B}')J^1_{\mathrm{M}}$. Abbreviate

$$K' = \boldsymbol{U}(\mathfrak{B}')J^1_{\mathrm{M}}.$$

Now let W denote the representation space of κ_{M}, and W^y that of κ^y_{M}. Choose a nonzero linear map $T : W \to W^y$ such that

$$T \circ \eta_{\mathrm{M}}(x) = \eta^y_{\mathrm{M}}(x) \circ T, \quad x \in J^1_{\mathrm{M}} \cap J^{1\,y}_{\mathrm{M}}.$$

This condition determines T uniquely, up to a scalar factor (by (5.1.8)). Therefore we have

$$T \circ \kappa_{\mathrm{M}}(x) = \kappa^y_{\mathrm{M}}(x) \circ T,$$

for $x \in (K \cap K^y) \cup (K' \cap K'^y)$.

Consider the group $J_{\mathrm{M}} \cap J_{\mathrm{M}}^y$. Arguing as before, the decomposition $J_{\mathrm{M}} = \boldsymbol{U}(\mathfrak{B}_{\mathrm{M}})J_{\mathrm{M}}^1$ gives us

$$J_{\mathrm{M}} \cap J_{\mathrm{M}}^y = (\boldsymbol{U}(\mathfrak{B}_{\mathrm{M}}) \cap \boldsymbol{U}(\mathfrak{B}_{\mathrm{M}})^y)(J_{\mathrm{M}}^1 \cap J_{\mathrm{M}}^{1\,y}).$$

By (5.2.9), this is the same as

$$J_{\mathrm{M}} \cap J_{\mathrm{M}}^y = (\boldsymbol{U}^1(\mathfrak{B}_{\mathrm{M}}) \cap \boldsymbol{U}(\mathfrak{B}_{\mathrm{M}})^y)(K \cap K^y).$$

We can reverse the order of factors here. Take $x \in J_{\mathrm{M}} \cap J_{\mathrm{M}}^y$, and write $x = ab$, $a \in K \cap K^y$, $b \in \boldsymbol{U}^1(\mathfrak{B}_{\mathrm{M}}) \cap \boldsymbol{U}(\mathfrak{B}_{\mathrm{M}})^y$. Elementary matrix manipulations now give

$$\boldsymbol{U}^1(\mathfrak{B}_{\mathrm{M}}) \cap \boldsymbol{U}(\mathfrak{B}_{\mathrm{M}})^y \subset (\boldsymbol{U}^1(\mathfrak{B}_{\mathrm{M}}) \cap \boldsymbol{U}^1(\mathfrak{B}_{\mathrm{M}})^y).(\boldsymbol{U}(\mathfrak{B}') \cap \boldsymbol{U}(\mathfrak{B}')^y).$$

We can therefore write $b = cd$, with $c \in \boldsymbol{U}^1(\mathfrak{B}_{\mathrm{M}}) \cap \boldsymbol{U}^1(\mathfrak{B}_{\mathrm{M}})^y$, $d \in K' \cap K'^y$. However, the restriction $\kappa_{\mathrm{M}} \mid \boldsymbol{U}^1(\mathfrak{B}_{\mathrm{M}})$ is a multiple of $\theta_{\mathrm{M}} \mid \boldsymbol{U}^1(\mathfrak{B}_{\mathrm{M}})$, for some simple character $\theta_{\mathrm{M}} \in \mathcal{C}(\mathfrak{A}_{\mathrm{M}}, 0, \beta)$. Therefore $\kappa_{\mathrm{M}} \mid \boldsymbol{U}^1(\mathfrak{B}_{\mathrm{M}}) \cap \boldsymbol{U}^1(\mathfrak{B}_{\mathrm{M}})^y$ is a multiple of $\chi \circ \det_B$, for some character χ of $\boldsymbol{U}^1(\mathfrak{o}_E)$. The same applies to $\kappa_{\mathrm{M}}^y \mid \boldsymbol{U}^1(\mathfrak{B}_{\mathrm{M}}) \cap \boldsymbol{U}^1(\mathfrak{B}_{\mathrm{M}})^y$, for the same character χ, since y intertwines θ_{M}. Thus, taking $x \in J_{\mathrm{M}} \cap J_{\mathrm{M}}^y$, written in this form $x = acd$, we get

$$T \circ \kappa_{\mathrm{M}}(x) = \kappa_{\mathrm{M}}^y(x) \circ T.$$

This says that y intertwines κ_{M}, as required, and we have completed the proof of (5.2.7). ∎

We can now prove (5.2.2). Part *(i)* is given by (5.2.3)*(ii)* and (5.2.7). In part *(ii)*, one implication is obvious, so let κ' be another β-extension of η. Then $\kappa' = \kappa \otimes \xi$, for some character ξ of $J/J^1 \cong \boldsymbol{U}(\mathfrak{B})/\boldsymbol{U}^1(\mathfrak{B})$. The character ξ is intertwined by the whole of $B^\times$, hence so is $\xi \mid \boldsymbol{U}(\mathfrak{B})$. It follows that $\xi = \chi \circ \det_B$, as required. Just as in the second remark following the statement of (5.2.3), the various representations $\kappa \otimes \chi \circ \det_B$ have distinct determinants, and hence are distinct. This proves the first part of (5.2.2)*(iii)*. To prove the remaining part, it is enough to show that κ and $\kappa \otimes \chi \circ \det_B$ do not intertwine, for any nontrivial character χ of $\boldsymbol{U}(\mathfrak{o}_E)/\boldsymbol{U}^1(\mathfrak{o}_E)$. This is given by (5.2.7) and (5.2.11).

This finally completes the proof of (5.2.2) (modulo (7.2.11) below). ∎

We observe that the construction given by (5.2.3) further gives a canonical bijection between β-extensions κ of η and β-extensions κ_{M} of η_{M}. This can be regarded as an extension of the correspondences $\boldsymbol{\tau}$ between the simple characters θ.

It is worth examining this a little more closely. While the hypothesis that $\mathfrak{B}_{\mathrm{M}}$ is maximal and $\mathfrak{B}_{\mathrm{m}}$ minimal is crucial in the constructions of (5.2.4), many of the other arguments depend only on containment relations and are valid more generally. Taking (5.2.7) into account, one can simply imitate the proof of (5.2.5) to get the following result.

(5.2.14) *Let $\mathfrak{B}_1 \supset \mathfrak{B}_2$ be hereditary $\mathfrak{o}_E$-orders in B with associated $\mathfrak{o}_F$-orders $\mathfrak{A}_1, \mathfrak{A}_2$. Let $\theta_i \in \mathcal{C}(\mathfrak{A}_i, 0, \beta)$ be related by (3.6.1), let η_i be the unique irreducible representation of $J^1(\beta, \mathfrak{A}_i)$ which contains θ_i. There is a canonical bijection between β-extensions κ_1 of η_1 and β-extensions κ_2 of η_2 expressed as follows:— given κ_1 (resp. κ_2), there exists a unique κ_2 (resp. κ_1) such that $\kappa_1 \mid \boldsymbol{U}(\mathfrak{B}_2)J^1(\beta, \mathfrak{A}_1)$ and κ_2 induce the same irreducible representation of $\boldsymbol{U}(\mathfrak{B}_2)\boldsymbol{U}^1(\mathfrak{A}_2)$.*

(5.3) A bound on intertwining

In this section, we just need a simple stratum $[\mathfrak{A}, n, 0, \beta]$ and a character $\theta \in \mathcal{C}(\mathfrak{A}, 0, \beta)$. The notations E, B, $\mathfrak{B}$, $J = J(\beta, \mathfrak{A})$ etc. are as before. Write η for the unique irreducible representation of J^1 which contains θ, and fix a β-extension κ of η.

We shall be interested in representations of the form

$$\lambda = \kappa \otimes \sigma, \tag{5.3.1}$$

where σ is (the inflation to J of) an irreducible cuspidal representation of the reductive algebraic group (over k_E) $J/J^1 = \boldsymbol{U}(\mathfrak{B})/\boldsymbol{U}^1(\mathfrak{B})$. Indeed, there will be further restrictions on the representation σ, but there is no harm in starting the discussion at this level of generality.

The aim of this section is to give a bound on the intertwining of this representation λ. However, this preliminary result has nothing to do with the special properties of σ, and it is useful to have it more generally.

(5.3.2) Proposition: *Let κ be a β-extension of η as above, and let ξ be the inflation to J of some irreducible representation of $J/J^1 = \boldsymbol{U}(\mathfrak{B})/\boldsymbol{U}^1(\mathfrak{B})$. Then:*

(i) $I_G(\kappa \otimes \xi) = J I_{B^\times}(\xi \mid \boldsymbol{U}(\mathfrak{B}))J$;

(ii) for $y \in B^\times$, the intertwining spaces $I_y(\kappa \otimes \xi)$, $I_y(\xi)$ have the same dimension;

(iii) the representation $\kappa \otimes \xi$ is irreducible.

Proof: Let $g \in G$ intertwine $\kappa \otimes \xi$. Then it also intertwines the representation $(\kappa \otimes \xi) \mid J^1$, which is a multiple of η. Thus, by (5.1.1), we may as well take $g \in B^\times$. Let X denote the representation space of κ, and Y that of ξ. Let $\phi \in I_g(\kappa \otimes \xi)$. We may write $\phi = \sum_j S_j \otimes T_j$, where $S_j \in \mathrm{End}_{\mathbb{C}}(X)$, $T_j \in \mathrm{End}_{\mathbb{C}}(Y)$, and where the set $\{T_j\}$ is linearly independent. Take $h \in J^1 \cap (J^1)^g$, so that we have

$$\kappa \otimes \xi(h) \circ \phi = \phi \circ (\kappa \otimes \xi)^g(h).$$

Since $J^1 \subset \mathrm{Ker}(\xi)$, this relation reads

$$\sum_j (\kappa(h) \circ S_j - S_j \circ \kappa^g(h)) \otimes T_j = 0.$$

Since the T_j are linearly independent, this implies $S_j \in I_g(\eta)$ for all j. The spaces $I_g(\eta)$, $I_g(\kappa)$ are equal and one-dimensional, by (5.1.8) and (5.2.7). Thus we may take $j = 1$, so that $\phi = S \otimes T$, with $S \in I_g(\kappa)$, $T \in \mathrm{End}_{\mathbb{C}}(Y)$. Now taking $h \in J \cap J^g$, the intertwining relation reads

$$(S \circ \kappa^g(h)) \otimes (T \circ \xi^g(h)) = (\kappa(h) \circ S) \otimes (T \circ \xi^g(h)) = (\kappa(h) \circ S) \otimes (\xi(h) \circ T).$$

This says that $T \in I_g(\xi)$. This argument is reversible:— if $y \in B^\times$ and $T \in I_y(\xi)$, then $S \otimes T \in I_g(\kappa \otimes \xi)$. This proves *(ii)*, which implies *(i)*. To get *(iii)*, we take $y = 1$. By *(ii)*, $\dim(I_1(\kappa \otimes \xi)) = 1$, so $\kappa \otimes \xi$ is indeed irreducible. ■

This is as far as we can get with our present, essentially lattice-theoretic, methods. We return to the intertwining of the representation λ after we introduce some new machinery in the next section.

(5.4) Affine Hecke algebras and Weyl groups

We now recall, in the context of A and G, some standard material concerning generalised Tits systems and affine Hecke algebras. See **[Cr]**, **[Iw]**, **[IM]** for this.

We start by choosing a minimal hereditary $\mathfrak{o}_F$-order $\mathfrak{C}$ in A, and an $\mathfrak{o}_F$-basis $\{u_1, u_2, \ldots, u_N\}$ of the lattice chain defining $\mathfrak{C}$, as in **(1.1)**. We use this basis to identify A with $\mathbb{M}(N, F)$ (and G with $GL(N, F)$), so that $\mathfrak{C}$ becomes identified with the subring of $\mathbb{M}(N, \mathfrak{o}_F)$ consisting of matrices which are upper triangular modulo $\mathfrak{p}_F$. Thus $\boldsymbol{U}(\mathfrak{C})$ is an Iwahori subgroup of G. We also fix a prime element π_F, or just π, of F. We define

(5.4.1) *(i)* $\boldsymbol{D}$ *= the group of all diagonal matrices in G whose eigenvalues are powers of π;*

(ii) $\boldsymbol{W_0}$*= the group of permutation matrices in G (i.e. $\boldsymbol{W_0}$ consists of all permutations of the basis $\{u_j\}$);*

(iii) $\widetilde{\boldsymbol{W}}$*= the group generated by $\boldsymbol{W_0}$ and $\boldsymbol{D}$.*

Thus, in particular, $\boldsymbol{W}_0$ normalises $\boldsymbol{D}$, and $\widetilde{\boldsymbol{W}}$ is the semidirect product $\boldsymbol{W}_0 \ltimes \boldsymbol{D}$ of $\boldsymbol{W}_0$ and $\boldsymbol{D}$.

We define some special elements of $\widetilde{\boldsymbol{W}}$.

(5.4.2) *(i) For $1 \le i \le N-1$, let $\boldsymbol{s}_i$ denote the transposition $\boldsymbol{s}_i(u_i) = u_{i+1}$, $\boldsymbol{s}_i(u_{i+1}) = u_i$, $\boldsymbol{s}_i(u_j) = u_j$ if $j \ne i, i+1$;*

(ii)

$$\Pi = \Pi(\mathfrak{C}) = \begin{pmatrix} 0 & 1 & 0 & & \dots & \dots & 0 \\ 0 & 0 & 1 & 0 & \dots & \dots & 0 \\ \vdots & \vdots & \ddots & \ddots & \ddots & & \vdots \\ \vdots & \vdots & & \ddots & \ddots & \ddots & \vdots \\ & & & & & & 0 \\ 0 & 0 & & \dots & \dots & 0 & 1 \\ \pi & 0 & & \dots & \dots & & 0 \end{pmatrix}.$$

Thus Π normalises $\boldsymbol{U}(\mathfrak{C})$ and $\Pi\mathfrak{C} = \mathrm{rad}(\mathfrak{C})$. Moreover, $\mathfrak{K}(\mathfrak{C})$ is generated by Π and $\boldsymbol{U}(\mathfrak{C})$.

The elements $\boldsymbol{s}_i$, $1 \le i \le N-1$, and Π generate the group $\widetilde{\boldsymbol{W}}$. Set $\boldsymbol{s}_0 = \Pi \boldsymbol{s}_1 \Pi^{-1}$, and let $\boldsymbol{W}$ be the group generated by $\{\boldsymbol{s}_0, \boldsymbol{s}_1, \dots, \boldsymbol{s}_{N-1}\}$. Then we have another semidirect product decomposition

$$\widetilde{\boldsymbol{W}} = \langle \Pi \rangle \ltimes \boldsymbol{W}.$$

The group $\boldsymbol{W}$ is a Coxeter group whose fundamental set of involutions is $\{\boldsymbol{s}_i : 0 \le i \le N-1\}$, and it carries the usual length function ℓ. We extend this to a function on $\widetilde{\boldsymbol{W}}$ by setting $\ell(\Pi^a \boldsymbol{w}) = \ell(\boldsymbol{w})$, $a \in \mathbb{Z}$, $\boldsymbol{w} \in \boldsymbol{W}$. We note a few more well-known facts:—

(5.4.3) *(i) The map $\boldsymbol{w} \mapsto \boldsymbol{U}(\mathfrak{C})\boldsymbol{w}\boldsymbol{U}(\mathfrak{C})$ gives a bijection*

$$\widetilde{\boldsymbol{W}} \xrightarrow{\approx} \boldsymbol{U}(\mathfrak{C})\backslash G/\boldsymbol{U}(\mathfrak{C}).$$

(ii) For a Haar measure μ on G, we have

$$\mu(\boldsymbol{U}(\mathfrak{C})\boldsymbol{w}\boldsymbol{U}(\mathfrak{C})) = \mu(\boldsymbol{U}(\mathfrak{C})).q^{\ell(\boldsymbol{w})}, \quad \boldsymbol{w} \in \widetilde{\boldsymbol{W}},$$

where $q = q_F$ is the cardinality of the residue class field of F.

(iii) We have

$$\bigcup_{\boldsymbol{w} \in \boldsymbol{W}_0} \boldsymbol{U}(\mathfrak{C})\boldsymbol{w}\boldsymbol{U}(\mathfrak{C}) = GL(N, \mathfrak{o}_F).$$

We shall be much concerned with the Hecke algebra $\mathcal{H}(G, \mathbf{1}_{U(\mathfrak{C})})$ of G relative to the trivial character $\mathbf{1}_{U(\mathfrak{C})}$ of the Iwahori subgroup $\boldsymbol{U}(\mathfrak{C})$.

To avoid awkward normalising factors, *we assume that the convolution in $\mathcal{H}(G, \mathbf{1}_{U(\mathfrak{C})})$ is defined relative to the Haar measure μ on G for which $\mu(\boldsymbol{U}(\mathfrak{C})) = 1$.* We pick out some special elements of this algebra. For $\boldsymbol{w} \in \widetilde{\boldsymbol{W}}$, define $\mathbf{f}_{\boldsymbol{w}} \in \mathcal{H}(G, \mathbf{1}_{U(\mathfrak{C})})$ by

$$\mathbf{f}_{\boldsymbol{w}} = \begin{cases} 1 & \text{if } x \in \boldsymbol{U}(\mathfrak{C})\boldsymbol{w}\boldsymbol{U}(\mathfrak{C}), \\ 0 & \text{otherwise.} \end{cases} \tag{5.4.4}$$

As a space of complex valued functions, the algebra $\mathcal{H}(G, \mathbf{1}_{U(\mathfrak{C})})$ carries a natural Hermitian inner product

$$(\phi, \psi) = \int_G \phi(g)\overline{\psi(g)}\, dg, \qquad \phi, \psi \in \mathcal{H}(G, \mathbf{1}_{U(\mathfrak{C})}), \tag{5.4.5}$$

where "bar" denotes complex conjugation, just as in **(4.3)**.

Now let m be a positive integer and $z \in \mathbb{C}^\times$. We recall the definition of the *affine Hecke algebra* $\mathcal{H}(m, z)$. This is the associative $\mathbb{C}$-algebra with 1 given by the following presentation (although others are known).

(5.4.6) Definition: *$\mathcal{H}(m, z)$ is generated by elements $[s_i]$, $1 \le i \le m-1$, and a pair of elements $[\zeta]$, $[\zeta']$, subject to the following relations:*

(i) $[\zeta][\zeta'] = [\zeta'][\zeta] = 1$ (so we always put $[\zeta'] = [\zeta]^{-1}$);

(ii) $([s_i] + 1)([s_i] - z) = 0$, $1 \le i \le m-1$;

(iii) $[\zeta]^2[s_1] = [s_{m-1}][\zeta]^2$;

(iv) $[\zeta][s_i] = [s_{i-1}][\zeta]$, $2 \le i \le m-1$;

(v) $[s_i][s_{i+1}][s_i] = [s_{i+1}][s_i][s_{i+1}]$, $1 \le i \le m-2$;

(vi) $[s_i][s_j] = [s_j][s_i]$, $1 \le i, j \le m-1$, $|i-j| \ge 2$.

(5.4.7) *The map $[s_i] \mapsto \mathbf{f}_{\boldsymbol{s}_i}$, $1 \le i \le N-1$, $[\zeta] \mapsto \mathbf{f}_{\Pi(\mathfrak{C})}$ extends uniquely to an isomorphism of algebras*

$$\mathcal{H}(N, q) \xrightarrow{\approx} \mathcal{H}(G, \mathbf{1}_{U(\mathfrak{C})}),$$

where $q = q_F$ is the cardinality of the residue field of F.

We denote the inverse of this isomorphism by

$$\mathbf{f}_{\boldsymbol{w}} \mapsto [\boldsymbol{w}], \quad \boldsymbol{w} \in \widetilde{\boldsymbol{W}}. \tag{5.4.8}$$

The elements $[\boldsymbol{w}]$, $\boldsymbol{w} \in \widetilde{\boldsymbol{W}}$, then form a $\mathbb{C}$-basis of $\mathcal{H}(N, q)$.

We can describe the elements $[\boldsymbol{w}]$ of $\mathcal{H}(N, q)$, $\boldsymbol{w} \in \widetilde{\boldsymbol{W}}$, without explicit reference to the algebra $\mathcal{H}(G, \mathbf{1}_{U(\mathfrak{C})})$. If $\boldsymbol{s}_i$, $1 \le i \le N-1$, is a transposition as in (5.4.2), we first set $[\boldsymbol{s}_i] = [s_i]$. In general, if $\boldsymbol{w} \in \widetilde{\boldsymbol{W}}$, we take a minimal expression for $\boldsymbol{w}$,

$$\boldsymbol{w} = \Pi^a \boldsymbol{w}_1 \boldsymbol{w}_2 \ldots \boldsymbol{w}_\ell,$$

where $a \in \mathbb{Z}$, the $\boldsymbol{w}_j$ lie in $\{\boldsymbol{s}_0, \dots, \boldsymbol{s}_{N-1}\}$, and $\ell = \ell(\boldsymbol{w})$. We have

$$[\boldsymbol{w}] = [\zeta]^a [\boldsymbol{w}_1] \dots [\boldsymbol{w}_\ell]. \tag{5.4.9}$$

Of course, we use the convention $[s_0] = [\zeta][s_1][\zeta]^{-1}$. It now follows from (5.4.6)*(ii)* that the elements $[\boldsymbol{w}]$, $\boldsymbol{w} \in \widetilde{\boldsymbol{W}}$, are *invertible* in the algebra $\mathcal{H}(N, z)$.

We can use this isomorphism to transfer the canonical inner product on the algebra $\mathcal{H}(G, \mathbf{1}_{U(\mathfrak{C})})$ to one on $\mathcal{H}(N, q)$, which we denote by $\langle\, , \rangle$. Explicitly, the elements $[\boldsymbol{w}]$, $\boldsymbol{w} \in \widetilde{\boldsymbol{W}}$, form an orthogonal basis of $\mathcal{H}(N, q)$ for $\langle\, , \rangle$, and

$$\langle [\boldsymbol{w}], [\boldsymbol{w}] \rangle = q^{\ell(\boldsymbol{w})},$$

where ℓ is the length function on $\widetilde{\boldsymbol{W}}$ as above.

The algebra $\mathcal{H}(m, z)$ has a number of interesting subalgebras and quotient algebras. Let $\mathcal{H}_0(m, z)$ denote the subalgebra of $\mathcal{H}(m, z)$ generated by the elements $[s_i]$, $1 \le i \le m-1$.

(5.4.10) *Let $\mathfrak{D}$ be the maximal order $\mathbb{M}(N, \mathfrak{o}_F)$ containing $\mathfrak{C}$. The isomorphism* (5.4.7) *identifies the subalgebra $\mathcal{H}_0(N, q)$ of $\mathcal{H}(N, q)$ with the subalgebra $\mathcal{H}(\boldsymbol{U}(\mathfrak{D}), \mathbf{1}_{U(\mathfrak{C})})$ of $\mathcal{H}(G, \mathbf{1}_{U(\mathfrak{C})})$.*

Now fix an unramified quasicharacter ω of $F^\times$. Write $\omega_{\mathfrak{C}}$ for the quasicharacter of $F^\times \boldsymbol{U}(\mathfrak{C})$ which is null on $\boldsymbol{U}(\mathfrak{C})$ and agrees with ω on $F^\times$. Thus $\mathcal{H}(G, \omega_{\mathfrak{C}}^{-1})$ is the convolution algebra of functions $\phi : G \to \mathbb{C}$ which are compactly supported mod $F^\times$ and satisfy

(5.4.11) *(i)* $\phi(xgy) = \phi(g)$, $x, y \in \boldsymbol{U}(\mathfrak{C})$, $g \in G$;
(ii) $\phi(zg) = \omega(z)\phi(g)$, $g \in G$, $z \in F^\times$.

There is a canonical surjective algebra homomorphism $\mathcal{H}(G, \mathbf{1}_{U(\mathfrak{C})}) \to \mathcal{H}(G, \omega_{\mathfrak{C}}^{-1})$, which maps the function $\phi \in \mathcal{H}(G, \mathbf{1}_{U(\mathfrak{C})})$ to the function ϕ^ω given by

$$\phi^\omega(x) = \int_{F^\times} \omega(y)^{-1} \phi(yx)\, d^\times y, \qquad x \in G, \tag{5.4.12}$$

where $d^\times y$ is the Haar measure on $F^\times$ giving $\boldsymbol{U}(\mathfrak{o}_F)$ measure 1.

In $\mathcal{H}(m, z)$, the element $[\zeta]^m$ is central, and, for $\xi \in \mathbb{C}^\times$, we write $\mathcal{H}^\xi(m, z)$ for the quotient of $\mathcal{H}(m, z)$ by the ideal generated by $[\zeta]^m - \xi$. Then:

(5.4.13) *The isomorphism* (5.4.7) *induces an isomorphism*

$$\mathcal{H}^{\omega(\pi)}(N, q) \xrightarrow{\approx} \mathcal{H}(G, \omega_{\mathfrak{C}}^{-1}).$$

If the quasicharacter ω is unitary, i.e. $|\omega| = 1$, then these quotient algebras inherit unitary structures, which are preserved by this isomorphism.

(5.5) Intertwining and Weyl groups

We now return to the configuration $\mathfrak{B}_{\mathrm{M}} \supset \mathfrak{B} \supset \mathfrak{B}_{\mathrm{m}}$ of $\mathfrak{o}_E$-orders in (5.1.12), where E is the field $F[\beta]$ attached to our simple stratum $[\mathfrak{A}, n, 0, \beta]$. *We assume throughout that $\mathfrak{B}_{\mathrm{M}}$ is maximal and $\mathfrak{B}_{\mathrm{m}}$ is minimal.* Write $R = \dim_E(V)$ and $e = e(\mathfrak{B}|\mathfrak{o}_E)$. Let $\mathcal{L} = \{L_i : i \in \mathbb{Z}\}$ be the $\mathfrak{o}_E$-lattice chain defining $\mathfrak{B}$, numbered so that $\mathfrak{B}_{\mathrm{M}} = \mathrm{End}_{\mathfrak{o}_E}(L_0)$. We choose an $\mathfrak{o}_E$-basis $\mathcal{V} = \{v_1, v_2, \dots, v_R\}$ of the lattice chain defining $\mathfrak{B}_{\mathrm{m}}$. In particular,

(5.5.1) *For $0 \leq i \leq e-1$, L_i is the $\mathfrak{o}_E$-linear span of the set*

$$\{v_1, v_2, \dots v_{d(i)}, \pi_E v_{d(i)+1}, \dots, \pi_E v_R\},$$

where $d(i) = \dim_{k_E}(L_i/L_e)$ and π_E is some prime element of E.

This choice of $\mathcal{V}$ identifies $U(\mathfrak{B}_{\mathrm{m}})$ with the standard Iwahori subgroup of $B^\times$ (as above), and $U(\mathfrak{B}_{\mathrm{M}})$ with $GL(R, \mathfrak{o}_E)$. For $1 \leq i \leq e$, let V^i be the E-linear span of the set $\{v_{d(i)+1}, \dots, v_{d(i-1)}\}$, with the convention $d(e) = 0$. Then V is the direct sum of the V^i and further:

(5.5.2) *(i) $L_j = \coprod_i L_j^i$, where $L_j^i = L_j \cap V^i$, $1 \leq i \leq e$, $j \in \mathbb{Z}$;*
(ii) $L_i^i = L_{i+1}^i = \dots = L_{i+e-1}^i$, $1 \leq i \leq e$;
(iii) the set $\{L_{ej}^i : j \in \mathbb{Z}\}$ is an $\mathfrak{o}_E$-lattice chain in V^i of period 1.

It will be convenient to relabel our chosen basis $\mathcal{V}$, by setting

(5.5.3) $$\mathcal{V}^i = \mathcal{V} \cap V^i = \{v_j^i : 1 \leq j \leq R(i)\}, \quad 1 \leq i \leq e,$$

where $R(i) = \dim_E(V^i)$. The numbering here is arranged so that our original ordered basis $\mathcal{V}$ appears as the union of the $\mathcal{V}^i$ ordered lexicographically. Put $B^i = \mathrm{End}_E(V^i)$, $M(B) = \prod_i B^i$, viewed as a subalgebra of B. We think of $M(B)$ as the algebra of "diagonal block matrices" b with the i-th block b_i having dimension $R(i)$:

$$b = \begin{pmatrix} b_1 & 0 & \dots & \dots & 0 \\ 0 & b_2 & 0 & \dots & 0 \\ \vdots & & \ddots & & \\ \vdots & & & \ddots & 0 \\ 0 & & \dots & 0 & b_e \end{pmatrix}.$$

We put $\mathfrak{B}^i = \mathfrak{B} \cap B^i = \mathrm{End}_{\mathfrak{o}_E}(L_i^i)$, which is a maximal $\mathfrak{o}_E$-order in B^i. Indeed, in our present pictorial setup, $\mathfrak{B}^i$ is just $\mathbb{M}(R(i), \mathfrak{o}_E)$, and

$J/J^1 = \boldsymbol{U}(\mathfrak{B})/\boldsymbol{U}^1(\mathfrak{B}) = \prod_i(\mathfrak{B}^i/\mathfrak{p}_E\mathfrak{B}^i)$. We also write $\mathfrak{M}(\mathfrak{B}) = \prod \mathfrak{B}^i$, which is a maximal $\mathfrak{o}_E$-order in the semisimple E-algebra $M(B)$.

We fix a prime element π_E, or just π, of the field E. We incorporate the ideas of **(5.4)** into the present situation, with base field E in place of F, and $\mathfrak{B}_{\mathrm{m}}$ in place of the order $\mathfrak{C}$. So we define

(5.5.4) *(i)* $\boldsymbol{D}$ = *the group of matrices in* $B^\times$ *which are diagonal relative to the basis* $\mathcal{V}$ *of* V, *and whose eigenvalues are all powers of* π_E;

(ii) $\boldsymbol{W}_0$ = *the group of permutations of the basis* $\mathcal{V}$;

(iii) $\widetilde{\boldsymbol{W}} = \boldsymbol{D} \rtimes \boldsymbol{W}_0$.

Thus, in particular, we have

$$B^\times = \boldsymbol{U}(\mathfrak{B}_{\mathrm{m}})\widetilde{\boldsymbol{W}}\boldsymbol{U}(\mathfrak{B}_{\mathrm{m}}).$$

Now let σ_i be an irreducible cuspidal representation of $(\mathfrak{B}^i/\mathfrak{p}_E\mathfrak{B}^i)^\times \cong GL(R(i), \mathrm{k}_E)$ for $1 \le i \le e$, and put $\sigma = \sigma_1 \otimes \ldots \otimes \sigma_e$, viewed as a representation of J via inflation. The following result, in the special case of a principal order, is left as an exercise in **[HM1]** Ch. 2 Lemma (1.1).

(5.5.5) Proposition: *In the above situation, let* $w \in \widetilde{\boldsymbol{W}}$ *intertwine* $\sigma \mid \boldsymbol{U}(\mathfrak{B})$. *Then* w *normalises the group* $\mathfrak{M}(\mathfrak{B})^\times$.

Proof: We need a sequence of lemmas.

(5.5.6) Lemma: *For* $j \in \mathbb{Z}$, *define* $i(j) \in \mathbb{Z}$ *by* $1 \le i(j) \le e$ *and* $j+1 \equiv i(j) \pmod e$. *Let* M *be an* $\mathfrak{o}_E$-*submodule of* V *generated by a set of scalar multiples of elements of* $\mathcal{V}$. *Let* j *be maximal for the property* $M \subset L_j$. *Then either*

(i) $\mathfrak{p}_E M$ *is strictly contained in* $L_{j+1} \cap M$, *or*

(ii) $M \subset V^{i(j)}$ *and* $\dim_{\mathrm{k}_E}(M/M \cap \mathfrak{p}_E L_j^{i(j)}) = \mathrm{rank}_{\mathfrak{o}_E}(M)$.

Proof: Abbreviate $i = i(j)$, and set $V' = \coprod_{k \ne i} V^k$. Then we observe that $L_{j+1} = \mathfrak{p}_E L_j^i \oplus (L_j \cap V')$ and further $M = (M \cap V^i) \oplus (M \cap V')$. Thus

$$L_{j+1} \cap M = (\mathfrak{p}_E L_j^i \cap M) \oplus (V' \cap M).$$

If *(i)* fails, we have $\mathfrak{p}_E M = L_{j+1} \cap M$ and the factor $M \cap V'$ is zero. Therefore *(ii)* holds. ■

(5.5.7) Lemma: *Let* $w \in \widetilde{\boldsymbol{W}}$. *Then either* w *permutes the set* $\{L_j^i : i, j \in \mathbb{Z}\}$, *or else there exist integers* i, j, k *such that*

$$\mathfrak{p}_E L_j^i \subsetneq L_j^i \cap w L_{k+1} \subset L_j^i \cap w L_k = L_j^i.$$

Proof: We fix a pair (i,j). Choose k maximal for the property $w^{-1}L_j^i \subset L_k$. We apply (5.5.6). If the first alternative holds for some pair (i,j), we have the desired containments. Otherwise, for each (i,j), there exist integers k and ℓ such that $w^{-1}L_j^i \subset L_\ell^k$, and

$$\dim_{k_E}(w^{-1}L_j^i / w^{-1}L_j^i \cap \mathfrak{p}_E L_\ell^k) = \operatorname{rank}_{\mathfrak{o}_E}(L_j^i) = \dim_E(V^i).$$

Since w permutes the V^i, it follows that $w^{-1}V^i = V^k$ and also that $w^{-1}L_j^i = L_\ell^k$, as asserted. ■

Now we can prove the proposition. Let $w \in \widetilde{W}$, and suppose that it does not normalise $\mathfrak{M}(\mathfrak{B})^\times$. If w permutes the lattices L_j^i, then it surely normalises $\mathfrak{M}(\mathfrak{B})^\times$. We therefore assume that w does not permute the L_j^i. So, by (5.5.7), there exist integers i, j, k such that

$$\mathfrak{p}_E L_j^i \subsetneq L_j^i \cap w L_{k+1} \subset L_j^i \cap w L_k = L_j^i.$$

We define a lattice chain $\overline{\mathcal{L}}$ in V^i of period 2 by setting $\overline{L}_0 = L_j^i$, $\overline{L}_1 = L_j^i \cap w L_{k+1}$. The lattice $\overline{L}_1$ is spanned by multiples of elements of our basis $\mathcal{V}$. Write $\overline{\mathfrak{B}}$ for the hereditary order in B^i defined by the chain $\overline{\mathcal{L}}$ and $\overline{\mathfrak{Q}} = \operatorname{rad}(\overline{\mathfrak{B}})$. We have $w\overline{\mathfrak{Q}}w^{-1} \subset \mathfrak{Q}$ since $L_k = (L_k \cap wV^i) \oplus (L_k \cap wV')$ for all k. Therefore, if $T \in I_w(\sigma \mid U(\mathfrak{B}))$, we must have $T \circ \sigma(h) = \sigma^w(h) \circ T$ for all $h \in 1+\overline{\mathfrak{Q}}$. However, $whw^{-1} \in U^1(\mathfrak{B}) \subset \operatorname{Ker}(\sigma)$, whence $T \circ \sigma(h) = T$. Therefore $\sigma(U^1(\overline{\mathfrak{B}})) = \sigma_i(U^1(\overline{\mathfrak{B}}))$ acts trivially on the image of T. The group $U(\overline{\mathfrak{B}})/U^1(\mathfrak{B}^i)$ is a proper parabolic subgroup of $U(\mathfrak{B}^i)/U^1(\mathfrak{B}^i)$, and its unipotent radical is $U^1(\overline{\mathfrak{B}})/U^1(\mathfrak{B}^i)$. Since σ_i is cuspidal as a representation of $U(\mathfrak{B}^i)/U^1(\mathfrak{B}^i)$, we conclude that $T = 0$, whence w cannot intertwine $\sigma \mid U(\mathfrak{B})$. This proves the result. ■

We now specialise to the case where the order $\mathfrak{B}$ is *principal.* Thus, in particular, the spaces V^i all have the same E-dimension $f = R/e$, $e = e(\mathfrak{B}|\mathfrak{o}_E)$. Let Σ_e denote the group of permutations of the set $\{1, 2, \ldots, e\}$. We define a homomorphism

(5.5.8)
$$w_{\mathfrak{B}} : \Sigma_e \to W_0 \quad by$$
$$w_{\mathfrak{B}}(s) : v_j^i \mapsto v_j^{s(i)},$$

for $s \in \Sigma_e$, $1 \le i \le e$, $1 \le j \le f$, where W_0 is as in (5.5.4). Define

(5.5.9) *(i)* $W_0(\mathfrak{B}) = w_{\mathfrak{B}}(\Sigma_e)$;
(ii) $D(\mathfrak{B})$ = *the* D-*normaliser (= the* D-*centraliser) of* $\mathfrak{M}(\mathfrak{B})^\times$;
(iii) $\widetilde{W}(\mathfrak{B}) = W_0(\mathfrak{B}) \ltimes D(\mathfrak{B})$.

We also have the special element

$$\Pi(\mathfrak{B}) = \begin{pmatrix} 0 & 1 & 0 & & \dots & \dots & 0 \\ 0 & 0 & 1 & 0 & \dots & \dots & 0 \\ \vdots & \vdots & \ddots & \ddots & \ddots & & \vdots \\ \vdots & \vdots & & \ddots & \ddots & \ddots & \vdots \\ & & & & & & 0 \\ 0 & & & & & 0 & 1 \\ \pi & 0 & & \dots & \dots & & 0 \end{pmatrix} \in \widetilde{\boldsymbol{W}}(\mathfrak{B}),$$

where the entries in this matrix are $f \times f$ blocks. Thus the bottom left hand entry means $\pi I_{f\times f}$, where π is our chosen prime of E. We then get a semidirect product decomposition

$$\widetilde{\boldsymbol{W}}(\mathfrak{B}) = \langle \Pi(\mathfrak{B}) \rangle \ltimes \boldsymbol{W}(\mathfrak{B}),$$

where $\boldsymbol{W}(\mathfrak{B})$ is the subgroup of $\widetilde{\boldsymbol{W}}(\mathfrak{B})$ generated by $\boldsymbol{W}_0(\mathfrak{B})$ and all conjugates of it by powers of $\Pi(\mathfrak{B})$. It is easy to see that $\widetilde{\boldsymbol{W}}(\mathfrak{B})$ is precisely the $\widetilde{\boldsymbol{W}}$-normaliser of $\mathfrak{M}(\mathfrak{B})^\times$ (and of $M(B)^\times$).

We now take a representation $\sigma = \sigma_1 \otimes \ldots \otimes \sigma_e$ of J as above, but we continue with the assumption that $\mathfrak{A}$ (equivalently $\mathfrak{B}$) is a principal order. Thus the algebras $\mathfrak{B}^i/\mathfrak{p}_E\mathfrak{B}^i$ are all isomorphic to each other (indeed they are all isomorphic to $\mathbb{M}(N/e, \mathrm{k}_E)$.) We further impose the condition

$$\sigma_1 \cong \sigma_2 \cong \ldots \cong \sigma_e.$$

Under these hypotheses, we say that the representation $\lambda = \kappa \otimes \sigma$ is a *simple type in* G. (This definition is meant to include the degenerate case where $\beta \in \mathfrak{o}_F$, $E = F$, and θ is the trivial character of $\boldsymbol{U}^1(\mathfrak{A}) = H^1(\beta, \mathfrak{A})$.) To be quite precise:—

(5.5.10) Definition: *A simple type in G is one of the following (a) or (b):*

(a) *an irreducible representation $\lambda = \kappa \otimes \sigma$ of $J = J(\beta, \mathfrak{A})$ where:*

(i) *$\mathfrak{A}$ is a principal $\mathfrak{o}_F$-order in A and $[\mathfrak{A}, n, 0, \beta]$ is a simple stratum;*

(ii) *for some $\theta \in \mathcal{C}(\mathfrak{A}, 0, \beta)$, κ is a β-extension of the unique irreducible representation η of $J^1(\beta, \mathfrak{A})$ which contains θ;*

(iii) *if we write $E = F[\beta]$, $\mathfrak{B} = \mathfrak{A} \cap \mathrm{End}_E(V)$, so that*

$$J(\beta, \mathfrak{A})/J^1(\beta, \mathfrak{A}) \cong \boldsymbol{U}(\mathfrak{B})/\boldsymbol{U}^1(\mathfrak{B}) \cong GL(f, \mathrm{k}_E)^e,$$

for certain integers e, f, then σ is the inflation of a representation $\sigma_0 \otimes \ldots \otimes \sigma_0$, where σ_0 is an irreducible cuspidal representation of $GL(f, \mathrm{k}_E)$.

(b) an irreducible representation σ of $\boldsymbol{U}(\mathfrak{A})$, where

(i) $\mathfrak{A}$ is a principal $\mathfrak{o}_F$-order in A;

(ii) if we write $\boldsymbol{U}(\mathfrak{A})/\boldsymbol{U}^1(\mathfrak{A}) \cong GL(f, \mathrm{k}_F)^e$, for certain integers e, f, then σ is the inflation of a representation $\sigma_0 \otimes \ldots \otimes \sigma_0$, where σ_0 is an irreducible cuspidal representation of $GL(f, \mathrm{k}_F)$.

This distinction into cases is quite superficial:— it derives from the fact that our machinery of simple strata does not allow us to recognise the trivial character of $\boldsymbol{U}^1(\mathfrak{A})$ as a simple character (and it would be inconvenient elsewhere to arrange this). In practice, we can invariably treat *(b)* as a special case of *(a)*, simply by setting $E = F$, $\mathfrak{B} = \mathfrak{A}$, $J^t(\beta, \mathfrak{A}) = \boldsymbol{U}^t(\mathfrak{A})$, and θ, η, κ all trivial.

(5.5.11) Proposition: *Let λ be a simple type in G, as in* (5.5.10). *Let $g \in G$ intertwine λ. Then*

$$g \in J\widetilde{\boldsymbol{W}}(\mathfrak{B})J,$$

where $\widetilde{\boldsymbol{W}}(\mathfrak{B})$ is defined by (5.5.9).

Proof: Let $g \in G$ intertwine λ. By (5.3.2), we have $g \in JyJ$, for some $y \in B^\times$ which intertwines $\sigma \mid \boldsymbol{U}(\mathfrak{B})$. Since J contains the Iwahori subgroup $\boldsymbol{U}(\mathfrak{B}_{\mathrm{m}})$, we may as well take $y \in \widetilde{\boldsymbol{W}}$. We have already remarked that $\widetilde{\boldsymbol{W}}(\mathfrak{B})$ is the $\widetilde{\boldsymbol{W}}$-normaliser of $\mathfrak{M}(\mathfrak{B})^\times$, so the result follows from (5.5.5). (This argument applies equally to the case (5.5.10)*(b)*, by taking $E = F$, $\mathfrak{B} = \mathfrak{A}$, and so on.) ∎

One could prove directly and without difficulty at this stage that the converse of (5.5.11) holds:— any $g \in J\widetilde{\boldsymbol{W}}(\mathfrak{B})J$ does intertwine λ. However, we subsume this within the proof of (5.6.6) below.

Again let (J, λ) be a simple type, $\lambda = \kappa \otimes \sigma$. Attached to this, we have an auxiliary representation λ' of $\boldsymbol{U}(\mathfrak{B})J^1_{\mathrm{M}}$ defined as follows. We can identify $\boldsymbol{U}(\mathfrak{B})J^1_{\mathrm{M}}/J^1_{\mathrm{M}}$ with $\boldsymbol{U}(\mathfrak{B})/\boldsymbol{U}^1(\mathfrak{B}_{\mathrm{M}})$, which has $\boldsymbol{U}(\mathfrak{B})/\boldsymbol{U}^1(\mathfrak{B})$ as a quotient. We can therefore view σ as a representation of $\boldsymbol{U}(\mathfrak{B})J^1_{\mathrm{M}}$, where it is irreducible. We put

$$\lambda' = (\kappa_{\mathrm{M}} \mid \boldsymbol{U}(\mathfrak{B})J^1_{\mathrm{M}}) \otimes \sigma, \tag{5.5.12}$$

where κ_{M} is the representation of J_{M} associated with κ, as in (5.2.3). Observe that in the case (5.5.10)*(b)*, the representations λ, λ' coincide.

(5.5.13) Proposition: *Let (J, λ) be a simple type in G as above, and define the representation λ' by* (5.5.12). *Then we have a canonical algebra isomorphism*

$$\mathcal{H}(G, \lambda) \cong \mathcal{H}(G, \lambda').$$

This isomorphism preserves support of functions:— if $\phi \in \mathcal{H}(G,\lambda)$ is supported on JyJ, for some $y \in B^\times$, the corresponding function $\phi' \in \mathcal{H}(G,\lambda')$ is supported on $\boldsymbol{U}(\mathfrak{B})J^1_{\mathrm{M}}y\boldsymbol{U}(\mathfrak{B})J^1_{\mathrm{M}}$.

Proof: Write

$$\nu = \mathrm{Ind}(\lambda : J, \boldsymbol{U}(\mathfrak{B})\boldsymbol{U}^1(\mathfrak{A})),$$
$$\nu' = \mathrm{Ind}(\lambda' : \boldsymbol{U}(\mathfrak{B})J^1_{\mathrm{M}}, \boldsymbol{U}(\mathfrak{B})\boldsymbol{U}^1(\mathfrak{A})).$$

Then

$$\nu = \mathrm{Ind}(\kappa : J, \boldsymbol{U}(\mathfrak{B})\boldsymbol{U}^1(\mathfrak{A})) \otimes \sigma,$$
$$\nu' = \mathrm{Ind}(\kappa_{\mathrm{M}} \mid \boldsymbol{U}(\mathfrak{B})J^1_{\mathrm{M}} : \boldsymbol{U}(\mathfrak{B})J^1_{\mathrm{M}}, \boldsymbol{U}(\mathfrak{B})\boldsymbol{U}^1(\mathfrak{A})) \otimes \sigma,$$

and these representations are equivalent by (5.2.5). The first assertion now follows from (4.1.3).

For the second, we observe that the isomorphism between $\mathcal{H}(G,\nu)$ and $\mathcal{H}(G,\nu')$ certainly preserves support, being given by an equivalence of representations. Let $\phi \in \mathcal{H}(G,\lambda)$ have support JyJ, for some $y \in B^\times$. The corresponding function $\phi' \in \mathcal{H}(G,\lambda')$ is then supported on a union of double cosets $\boldsymbol{U}(\mathfrak{B})J^1_{\mathrm{M}}x\boldsymbol{U}(\mathfrak{B})J^1_{\mathrm{M}}$ such that

$$\boldsymbol{U}(\mathfrak{B})J^1_{\mathrm{M}}x\boldsymbol{U}(\mathfrak{B})J^1_{\mathrm{M}} \subset \boldsymbol{U}(\mathfrak{B})\boldsymbol{U}^1(\mathfrak{A})y\boldsymbol{U}(\mathfrak{B})\boldsymbol{U}^1(\mathfrak{A}).$$

However, since x intertwines λ', it surely intertwines $\lambda' \mid J^1_{\mathrm{M}}$, which is a multiple of η_{M}. Thus $x \in J^1_{\mathrm{M}}B^\times J^1_{\mathrm{M}}$, and we may as well assume $x \in B^\times$. Now, $\boldsymbol{U}(\mathfrak{B})\boldsymbol{U}^1(\mathfrak{A})x\boldsymbol{U}(\mathfrak{B})\boldsymbol{U}^1(\mathfrak{A}) \cap B^\times = \boldsymbol{U}(\mathfrak{B})x\boldsymbol{U}(\mathfrak{B})$, by (1.6.1), and it follows that $\boldsymbol{U}(\mathfrak{B})x\boldsymbol{U}(\mathfrak{B}) = \boldsymbol{U}(\mathfrak{B})y\boldsymbol{U}(\mathfrak{B})$. Therefore ϕ' has support $\boldsymbol{U}(\mathfrak{B})J^1_{\mathrm{M}}y\boldsymbol{U}(\mathfrak{B})J^1_{\mathrm{M}}$, as required. ■

Let us now return to the group $\widetilde{\boldsymbol{W}}(\mathfrak{B})$, with its associated subgroups $\boldsymbol{D}(\mathfrak{B})$, $\boldsymbol{W}(\mathfrak{B})$, $\boldsymbol{W}_0(\mathfrak{B})$ and the element $\Pi(\mathfrak{B})$ as defined in (5.5.9). We relate this to the affine Hecke algebra $\mathcal{H}(e, q_E^f)$, with $e = e(\mathfrak{B}|\mathfrak{o}_E)$, $ef = \dim_E(V)$, $q_E = \#\mathrm{k}_E$, as before.

(5.5.14) Proposition: *Let K/E be an unramified field extension of degree f with $K^\times \subset \mathfrak{K}(\mathfrak{B})$. Put $C = \mathrm{End}_K(V)$, $\mathfrak{C} = \mathfrak{B} \cap C = \mathfrak{A} \cap C$, and view π_E as a prime element of K. Then*

(i) $\boldsymbol{U}(\mathfrak{C})$ is an Iwahori subgroup of $C^\times \cong GL(e,K)$;

(ii) $\{v_1^i : 1 \le i \le e\}$ (notation of (5.5.3)) is a basis of $\mathcal{L}$ as $\mathfrak{o}_K$-lattice chain;

(iii) $\widetilde{\boldsymbol{W}}(\mathfrak{B})$ is the affine Weyl group $\widetilde{\boldsymbol{W}}_K$ of $C^\times$ relative to the Iwahori subgroup $\boldsymbol{U}(\mathfrak{C})$, the basis $\{v_1^i\}$ and the prime element π_E of K. Further, the subgroups $\boldsymbol{W}_0(\mathfrak{B})$, $\boldsymbol{D}(\mathfrak{B})$, $\boldsymbol{W}(\mathfrak{B})$ of $\widetilde{\boldsymbol{W}}(\mathfrak{B})$ are equal to the corresponding subgroups $\boldsymbol{W}_{0,K}$, $\boldsymbol{D}_K$, $\boldsymbol{W}_K$ of $\widetilde{\boldsymbol{W}}_K$.

The proof is immediate. We can combine this with the algebra isomorphism $\mathcal{H}(C^\times, \mathbf{1}_{U(\mathfrak{C})}) \cong \mathcal{H}(e, q_E^f)$ of (5.4.7). (Observe that $q_E^f = \#\mathrm{k}_K$.) This allows us to view the standard basis $\{[w] : w \in \widetilde{W}_K\}$ of $\mathcal{H}(e, q_E^f)$ as being parametrised by the elements of $\widetilde{W}(\mathfrak{B})$. In terms of the standard generators, this amounts to:

(5.5.15) *(i)* $[\zeta] = [\Pi(\mathfrak{B})]$;
(ii) $[s_i] = [w_{\mathfrak{B}}(\tau_i)]$, *where* $\tau_i \in \Sigma_e$ *is the transposition* $(i, i+1)$, $1 \le i \le e-1$.

Let us put this more formally.

(5.5.16) Proposition: *Composition of the canonical isomorphism*

$$\mathcal{H}(e, q_E^f) \cong \mathcal{H}(C^\times, \mathbf{1}_{U(\mathfrak{C})})$$

and the identification (5.5.14) *between* $\widetilde{W}_K$ *and* $\widetilde{W}(\mathfrak{B})$ *induces an injection from the standard basis* $\{[w] : w \in \widetilde{W}_K\}$ *of* $\mathcal{H}(e, q_E^f)$ *to* $J\backslash G/J$, $J = J(\beta, \mathfrak{A})$, *whose image consists of the cosets* JwJ, $w \in \widetilde{W}(\mathfrak{B})$.

Proof: It remains only to prove that this map is injective, i.e., that $JwJ = Jw'J$, $w, w' \in \widetilde{W}(\mathfrak{B})$, implies $w = w'$. We have $J(\beta, \mathfrak{A}) = U(\mathfrak{B})J^1(\beta, \mathfrak{A})$, so $JwJ = Jw'J$ implies $J^1wJ^1 = J^1xw'J^1$, for some $x \in U(\mathfrak{B})$ (abbreviating $J^1(\beta, \mathfrak{A}) = J^1$). (1.6.1) now implies that $U^1(\mathfrak{B})wU^1(\mathfrak{B}) = U^1(\mathfrak{B})xw'U^1(\mathfrak{B})$ and therefore

$$U(\mathfrak{B})wU(\mathfrak{B}) = U(\mathfrak{B})w'U(\mathfrak{B}).$$

Write $\widetilde{W}_{\mathrm{m}}$ for the affine Weyl group of $B^\times$ relative to the basis $\{v_j^i\}$, the Iwahori subgroup $U(\mathfrak{B}_{\mathrm{m}})$ and the prime element π_E, so that $\widetilde{W}(\mathfrak{B}) \subset \widetilde{W}_{\mathrm{m}}$. We have a canonical bijection between $U(\mathfrak{B})\backslash B^\times/U(\mathfrak{B})$ and $\widetilde{W}_{\mathrm{m}} \cap U(\mathfrak{B})\backslash \widetilde{W}_{\mathrm{m}}/\widetilde{W}_{\mathrm{m}} \cap U(\mathfrak{B})$, and $\widetilde{W}(\mathfrak{B})$ injects (by **[IM]**(2.34)) into this double coset space. Thus $w = w'$, as required. ∎

The injection of (5.5.16) is largely independent of all the choices we have made. It does depend on the choice of a "base-point" L_0 in the lattice chain $\mathcal{L}$ which defines $\mathfrak{A}$. This, however, is not at all serious, since changing L_0 amounts to conjugating by some power of $\Pi(\mathfrak{B})$. Beyond that, we observe that any two $\mathfrak{o}_E$-bases of $\mathcal{L}$ are conjugate by an element of $U(\mathfrak{B})$. They therefore give rise to $U(\mathfrak{B})$-conjugate Weyl groups, and such a conjugation has no effect on (J, J)-double cosets. Changing the choice of prime π_E likewise has no effect on cosets. Moreover, any two choices of the field K in (5.5.14) are $U(\mathfrak{B})$ conjugate. Altogether,

(5.5.17) *For a fixed choice of* $L_0 \in \mathcal{L}$, *the inverse of the correspondence in* (5.5.16) *gives a canonical injection from the standard basis of* $\mathcal{H}(e, q_E^f)$ *to* $J\backslash G/J$, *which extends* (5.5.15) *and has image* $J\widetilde{W}(\mathfrak{B})J$.

It will follow from (5.6.6) below that this image is precisely the G-intertwining of any simple type λ attached to the simple stratum $[\mathfrak{A}, n, 0, \beta]$. One might further enquire as to whether this correspondence depends on the choice of the element β. If, however, we have a simple stratum $[\mathfrak{A}, n, 0, \beta']$ with $\mathcal{C}(\mathfrak{A}, 0, \beta') = \mathcal{C}(\mathfrak{A}, 0, \beta)$, it is not hard to see that an $\mathfrak{o}_E$-basis of $\mathcal{L}$ is automatically an $\mathfrak{o}_{E'}$-basis, where $E' = F[\beta']$, (this follows from (3.5.1)) and that β' gives rise to the same correspondence. We will not, however, need to be concerned with this.

(5.6) The Hecke algebra of a simple type

We now take a simple type λ in G, as in (5.5.10). Thus λ is an irreducible representation of $J = J(\beta, \mathfrak{A})$ of the form $\lambda = \kappa \otimes \sigma$, where κ is a β-extension of some η containing some $\theta \in \mathcal{C}(\mathfrak{A}, 0, \beta)$ as in (5.2.1), and σ is the inflation of a "homogeneous" cuspidal representation of $J/J^1 = \boldsymbol{U}(\mathfrak{B})/\boldsymbol{U}^1(\mathfrak{B}) \cong (GL(f, k_E))^e$, i.e. $\sigma = \sigma_0 \otimes \ldots \otimes \sigma_0$ for some irreducible cuspidal representation σ_0 of $GL(f, k_E)$. The other notations of the preceding sections also remain in force. In particular, we have the basis $\{[w] : w \in \widetilde{\boldsymbol{W}}(\mathfrak{B})\}$ of the affine Hecke algebra $\mathcal{H}(e, q_E^f)$ described by (5.5.17). Notice here that the subset $\{[w] : w \in \boldsymbol{W}_0(\mathfrak{B})\}$ is a basis of the subalgebra $\mathcal{H}_0(e, q_E^f)$.

Recall that $R = ef = \dim_E(V)$, $e = e(\mathfrak{B}|\mathfrak{o}_E)$. Let $\mathfrak{B}_{\mathrm{M}} \supset \mathfrak{B} \supset \mathfrak{B}_{\mathrm{m}}$ as in **(5.5)** (so that $\mathfrak{B}_{\mathrm{M}}$ is maximal and $\mathfrak{B}_{\mathrm{m}}$ minimal). Write $\overline{G} = GL(R, k_E)$ and identify $\overline{G}$ with $\boldsymbol{U}(\mathfrak{B}_{\mathrm{M}})/\boldsymbol{U}^1(\mathfrak{B}_{\mathrm{M}})$. Write $x \mapsto \bar{x}$ for the quotient map $\boldsymbol{U}(\mathfrak{B}_{\mathrm{M}}) \to \overline{G}$. The image $\overline{\boldsymbol{W}}_0$ of the group $\boldsymbol{W}_0$ of permutations of our original E-basis $\mathcal{V}$ of V is then the ordinary Weyl group of $\overline{G}$ with respect to the Borel subgroup $\boldsymbol{U}(\mathfrak{B}_{\mathrm{m}})/\boldsymbol{U}^1(\mathfrak{B}_{\mathrm{M}})$ and the diagonal torus in $\overline{G}$. Note that $\overline{\boldsymbol{W}}_0 \cong \boldsymbol{W}_0$. Write $\overline{P}$ for the image $\boldsymbol{U}(\mathfrak{B})/\boldsymbol{U}^1(\mathfrak{B}_{\mathrm{M}})$ of $\boldsymbol{U}(\mathfrak{B})$ in $\overline{G}$. Thus $\overline{P}$ is the standard (upper triangular) parabolic subgroup of $\overline{G}$ with all blocks of size $f \times f$. We can further identify $\boldsymbol{U}(\mathfrak{B})/\boldsymbol{U}^1(\mathfrak{B})$ with the Levi subgroup $(GL(f, k_E))^e$ embedded in $\overline{G}$ as blocks on the diagonal. Write $\bar{\sigma}$ for σ viewed as a representation of $\overline{P}$.

Now let $\overline{\boldsymbol{W}}_0(\mathfrak{B})$ denote the image of $\boldsymbol{W}_0(\mathfrak{B})$ in $\overline{G}$. Thus $\overline{\boldsymbol{W}}_0(\mathfrak{B})$ is the group of permutations in $\overline{\boldsymbol{W}}_0$ which preserve, and act trivially on, the blocks in $(GL(f, k_E))^e$. Then one knows (see **[HM1]** Ch. 1 Th. 5.1 or **[HL]**) that the intertwining of the representation $\bar{\sigma}$ in $\overline{G}$ is the set $\overline{P}.\overline{\boldsymbol{W}}_0(\mathfrak{B}).\overline{P}$. More precisely, we have

(5.6.1) *There is a unique algebra isomorphism*

$$\bar{g} : \mathcal{H}_0(e, q_E^f) \xrightarrow{\approx} \mathcal{H}(\overline{G}, \bar{\sigma})$$
$$[w] \mapsto \bar{g}_{[w]}, \quad w \in \boldsymbol{W}_0(\mathfrak{B}),$$

such that the function $\bar{g}_{[w]}$ has support $\overline{P}.\overline{w}.\overline{P}$ for all $w \in \boldsymbol{W}_0(\mathfrak{B})$.

With $\lambda = \kappa \otimes \sigma$, we let κ_{M} be the associated representation of J_{M} as in (5.2.3), and λ' the representation

$$\lambda' = (\kappa_{\mathrm{M}} \mid \boldsymbol{U}(\mathfrak{B})J^1_{\mathrm{M}}) \otimes \sigma$$

of $\boldsymbol{U}(\mathfrak{B})J^1_{\mathrm{M}}$ as in (5.5.12). Reduction modulo the normal subgroup J^1_{M} of J_{M} induces a canonical algebra isomorphism

$$\mathcal{H}(J_{\mathrm{M}}, \sigma \mid \boldsymbol{U}(\mathfrak{B})J^1_{\mathrm{M}}) \cong \mathcal{H}(\overline{G}, \bar{\sigma}). \tag{5.6.2}$$

(5.6.3) Lemma: *Let κ_{M} act on the space X, σ on Y, and view σ as a representation of $\boldsymbol{U}(\mathfrak{B})J^1_{\mathrm{M}}$. For $\phi \in \mathcal{H}(J_{\mathrm{M}}, \sigma)$, define a function ϕ' from J_{M} to $\mathrm{End}_{\mathbb{C}}(\check{X} \otimes \check{Y})$ by $\phi'(g) = \check{\kappa}_{\mathrm{M}}(g) \otimes \phi(g)$, $g \in G$. Then $\phi \mapsto \phi'$ is an algebra isomorphism $\mathcal{H}(J_{\mathrm{M}}, \sigma) \cong \mathcal{H}(J_{\mathrm{M}}, \lambda')$.*

Proof: This is a purely formal consequence of the fact that κ_{M} restricts irreducibly to $\boldsymbol{U}(\mathfrak{B})J^1_{\mathrm{M}}$. The details are quite straightforward. ∎

Putting (5.6.1) and (5.6.3) together, we get:—

(5.6.4) Proposition: *There is a canonical algebra isomorphism*

$$g' : \mathcal{H}_0(e, q_E^f) \xrightarrow{\approx} \mathcal{H}(J_{\mathrm{M}}, \lambda')$$

which preserves support in the sense that, for $w \in \boldsymbol{W}_0(\mathfrak{B})$, the support of the function $g'[w]$ is $\boldsymbol{U}(\mathfrak{B})J^1_{\mathrm{M}}w\boldsymbol{U}(\mathfrak{B})J^1_{\mathrm{M}}$.

Remark: The map g' of (5.6.4) is uniquely determined by the property that it preserves supports. For, it is certainly uniquely determined by its values on the generators $[s_i] = [w_{\mathfrak{B}}(\tau_i)]$, $1 \le i \le e-1$, where $\tau_i \in \Sigma_e$ is the transposition $(i, i+1)$. However, since g' is an isomorphism, the space $I_{w_{\mathfrak{B}}(\tau_i)}(\lambda')$ has dimension one, whence $g'[s_i]$ is uniquely determined up to a constant scalar factor. This factor must be 1, because of the relation (5.4.6)*(ii)*.

Thus, so far, we have algebra maps

$$\mathcal{H}_0(e, q_E^f) \cong \mathcal{H}(J_{\mathrm{M}}, \lambda') \hookrightarrow \mathcal{H}(G, \lambda') \cong \mathcal{H}(G, \lambda),$$

the last of these being given by (5.5.13). Write $\boldsymbol{\Psi}_0$ for the composite of these maps:

$$\begin{aligned} \boldsymbol{\Psi}_0 : \mathcal{H}_0(e, q_E^f) &\to \mathcal{H}(G, \lambda) \\ [w] &\mapsto \boldsymbol{\Psi}_0[w], \quad w \in \boldsymbol{W}_0(\mathfrak{B}). \end{aligned} \tag{5.6.5}$$

By its construction, this map preserves supports in the sense that, if $w \in \boldsymbol{W}_0(\mathfrak{B})$, then the function $\boldsymbol{\Psi}_0[w]$ has support JwJ.

(5.6.6) Main Theorem: *Let (J, λ) be a simple type in $G = \mathrm{Aut}_F(V)$, attached to a simple stratum $[\mathfrak{A}, n, 0, \beta]$ in $A = \mathrm{End}_F(V)$. Write $E = F[\beta]$, $B =$ the A-centraliser of E, $\mathfrak{B} = \mathfrak{A} \cap B$, as usual. Let $e = e(\mathfrak{B}|\mathfrak{o}_E)$, $R = \dim_E(V)$, $ef = R$. Then*

(i) there exists a nonzero element $\psi \in \mathcal{H}(G, \lambda)$ supported on $J\Pi(\mathfrak{B}) = J\Pi(\mathfrak{B})J$ (notation of (5.5.9)), and this is unique up to scalar multiple;

(ii) for any element ψ as in (i), there is a unique algebra map $\boldsymbol{\Psi} : \mathcal{H}(e, q_E^f) \to \mathcal{H}(G, \lambda)$, which extends $\boldsymbol{\Psi}_0$ and such that $\boldsymbol{\Psi}[\zeta] = \psi$.

Every map $\boldsymbol{\Psi}$ in (ii) is an isomorphism of algebras and preserves support of functions, in the sense that, for $w \in \widetilde{\boldsymbol{W}}(\mathfrak{B})$, the function $\boldsymbol{\Psi}[w]$ has support JwJ. Moreover, the restriction of $\boldsymbol{\Psi}$ to the subalgebra of $\mathcal{H}(e, q_E^f)$ spanned by $\{[w] : w \in \boldsymbol{W}(\mathfrak{B})\}$ is uniquely determined, independent of the choice of ψ.

Proof: As a matter of convenience, we assume that the algebra structure of $\mathcal{H}(G, \lambda)$ is defined relative to a Haar measure μ on G for which $\mu(J) = 1$.

(5.6.7) Lemma: *The space $I_{\Pi(\mathfrak{B})}(\lambda)$ has dimension 1, and any nonzero element of it is an automorphism of the representation space of λ.*

Proof: The element $\Pi = \Pi(\mathfrak{B})$ normalises the group J and λ is irreducible, so $\dim_{\mathbb{C}}(I_\Pi(\lambda)) \leq 1$, and any nonzero intertwining operator here is an automorphism of the representation space. Now we observe that $\Pi(\mathfrak{B})$ intertwines κ (by (5.2.7)) and also $\sigma \mid \boldsymbol{U}(\mathfrak{B})$ by the definition of σ. Indeed, $\Pi(\mathfrak{B})$ operates by permuting the (equivalent) tensor factors σ_i of σ. Thus, by (5.3.2), $\Pi(\mathfrak{B})$ does intertwine λ, and the lemma follows. ∎

(5.6.8) Lemma: *Let $\psi \in \mathcal{H}(G, \lambda)$ be nonzero with support $J\Pi(\mathfrak{B})J$. Then ψ is invertible in $\mathcal{H}(G, \lambda)$, and its inverse has support $J\Pi(\mathfrak{B})^{-1}J$. Moreover, if $a \in \mathbb{Z}$ and $\phi \in \mathcal{H}(G, \lambda)$ has support $J\Pi(\mathfrak{B})^a J$, then ϕ is a scalar multiple of ψ^a.*

Remark: If a is positive, ψ^a of course means the a-fold convolution of ψ with itself. If a is negative, $\psi^a = (\psi^{-1})^{-a}$.

Proof: On the double coset $J\Pi(\mathfrak{B})^{-1}J = \Pi(\mathfrak{B})^{-1}J$, (5.6.7) allows us to define a function ψ' by $\psi'(\Pi(\mathfrak{B})^{-1}j) = \psi(\Pi(\mathfrak{B}))^{-1}\check{\lambda}(j)$, $j \in J$. Extending ψ' trivially to a function on G, we see that $\psi' \in \mathcal{H}(G, \lambda)$ and the support of ψ' is $J\Pi(\mathfrak{B})^{-1}J$. We compute the convolutions $\psi * \psi'$, $\psi' * \psi$ directly. Using the fact that J has measure 1, we find that both of these functions have support J, and their value at 1 is the identity operator. Thus $\psi * \psi' = \psi' * \psi$ is the unit element of $\mathcal{H}(G, \lambda)$.

The same computation shows that if $\psi'' \in \mathcal{H}(G,\lambda)$ has support $J\Pi(\mathfrak{B})^{-1}J$, then $\psi * \psi''$ is supported on J, whence is a multiple of 1. Thus ψ'' is a scalar multiple of $\psi' = \psi^{-1}$. This proves the second statement when $|a| = 1$. The general case is a straightforward induction. ■

Again take ψ as in (5.6.8). The fact that $\Pi(\mathfrak{B})$ normalises J makes it very easy to compute convolutions with ψ. Indeed, we have:

(5.6.9) Lemma: *Let $\phi \in \mathcal{H}(G,\lambda)$ be nonzero and supported on JwJ, for some $w \in \widetilde{\boldsymbol{W}}(\mathfrak{B})$. Let $a \in \mathbb{Z}$, ψ be as in (5.6.8). Then $\phi * \psi^a$ (resp. $\psi^a * \phi$) is nonzero and has support $Jw\Pi(\mathfrak{B})^a J$ (resp. $J\Pi(\mathfrak{B})^a wJ$).*

Proof: Trivial. ■

If τ_i is the transposition $(i, i+1)$ in the symmetric group Σ_e, $1 \le i \le e-1$, write $\boldsymbol{s}_i = w_{\mathfrak{B}}(\tau_i)$, as in (5.5.8). Likewise, if $\tau_0 = (1,e)$, put $\boldsymbol{s}_0 = \Pi(\mathfrak{B})\boldsymbol{s}_1\Pi(\mathfrak{B})^{-1}$. Thus, according to (5.5.15), the standard generators $[s_i]$ of $\mathcal{H}(e, q_E^f)$ are given by $[s_i] = [\boldsymbol{s}_i]$.

(5.6.10) Lemma: *For $0 \le i \le e-1$, the space $I_{\boldsymbol{s}_i}(\lambda)$ has dimension one. Moreover, any element ϕ of $\mathcal{H}(G,\lambda)$ with support $J\boldsymbol{s}_iJ$ is invertible.*

Proof: In the first assertion, the cases $i \ne 0$ follow from (5.6.5). The remaining one follows on conjugating by a function ψ as in (5.6.8). The second assertion follows from the fact that $[s_i]$ is invertible in $\mathcal{H}(e, q_E^f)$, by (5.4.6)*(ii)*. ■

Now we prove the theorem. The map $\boldsymbol{\Psi}$ extends to an algebra homomorphism provided the elements $\boldsymbol{\Psi}[\zeta]$, $\boldsymbol{\Psi}[s_i]$, $1 \le i \le e-1$, satisfy the relations (5.4.6). Setting $\boldsymbol{\Psi}[\zeta'] = (\boldsymbol{\Psi}[\zeta])^{-1}$, as we may by (5.6.8), takes care of *(i)*. Relations *(ii)*, *(v)* and *(vi)* hold by the construction (5.6.5) of $\boldsymbol{\Psi}_0$. We just have to check *(iii)* and *(iv)*. These have identical proofs, so we just do *(iv)*. Take $2 \le i \le e-1$, and put $\phi = \boldsymbol{\Psi}[s_i]$, $\psi = \boldsymbol{\Psi}[\zeta]$. The convolution $\psi * \phi * \psi^{-1}$ has support $J\Pi(\mathfrak{B})J\boldsymbol{s}_iJ\Pi(\mathfrak{B})^{-1}J = J\pi(\mathfrak{B})\boldsymbol{s}_i\Pi(\mathfrak{B})^{-1}J$ by (5.6.9), and this equals $J\boldsymbol{s}_{i-1}J$. Therefore, by (5.6.10), we have $\psi * \phi * \psi^{-1} = \alpha\boldsymbol{\Psi}[s_{i-1}]$, for a complex constant α. We have already observed that the relation (5.4.6)*(ii)* is preserved by the restriction $\boldsymbol{\Psi}_0$ of $\boldsymbol{\Psi}$, so the functions $\boldsymbol{\Psi}[s_j]$ are "roots" of the polynomial $(X - q_E^f)(X+1)$. The same applies to the conjugate $\psi * \phi * \psi^{-1}$ of $\boldsymbol{\Psi}[s_i]$, so we have $\alpha = 1$, as required.

Thus indeed $\boldsymbol{\Psi}_0$ extends to an algebra homomorphism $\boldsymbol{\Psi}$ from $\mathcal{H}(e, q_E^f)$ to $\mathcal{H}(G,\lambda)$.

(5.6.11) Lemma: *Let $w \in \widetilde{\boldsymbol{W}}(\mathfrak{B})$. Then $\boldsymbol{\Psi}[w]$ has support JwJ, and is an invertible element of $\mathcal{H}(G,\lambda)$.*

Proof: We work by induction on the length $\ell(w)$ of w. By (5.6.8), (5.6.9), we may as well take $w \in \boldsymbol{W}(\mathfrak{B})$. If $\ell(w) = 1$, the assertion is given by

(5.6.10). In general, we write $w = w_1 w_2 \dots w_\ell$, $\ell = \ell(w)$, with the $w_j \in \{s_0, s_1, \dots, s_{e-1}\}$. Put $w' = w_1 \dots w_{\ell-1}$, so that $[w] = [w'][w_\ell]$ in $\mathcal{H}(e, q_E^f)$, and $\Psi[w] = \Psi[w'] * \Psi[w_\ell]$.

The function $\Psi[w]$ is certainly invertible as an element of $\mathcal{H}(G, \lambda)$, since $[w]$ is invertible in $\mathcal{H}(e, q_E^f)$ and Ψ is at least a ring homomorphism. In particular, it is nonzero, so we only have to show that $\Psi[w] = \Psi[w'] * \Psi[w_\ell]$ has support contained in JwJ. To abbreviate the notation, write $w_\ell = s = s_i$, for some i. By (5.6.9), we can conjugate by $\Pi(\mathfrak{B})$ without significantly changing anything to ensure that $i \neq 0$.

We now pass to the auxiliary representation $\lambda' = \kappa_{\mathrm{M}} \mid U(\mathfrak{B}) J^1_{\mathrm{M}} \otimes \sigma$ of $U(\mathfrak{B}) J^1_{\mathrm{M}}$, as in (5.5.12). Write

$$\Xi : \mathcal{H}(G, \lambda) \xrightarrow{\approx} \mathcal{H}(G, \lambda')$$

for the isomorphism given by (5.5.13), and put $\Psi' = \Psi \circ \Xi$. The algebra isomorphism Ξ preserves support, so it is enough to show, in the above situation, that $\Psi'[w] = \Psi'[w'] * \Psi'[s]$ has support contained in $U(\mathfrak{B}) J^1_{\mathrm{M}} w U(\mathfrak{B}) J^1_{\mathrm{M}}$. Observe here that $U(\mathfrak{B})$ and the element s are contained in $U(\mathfrak{B}_{\mathrm{M}})$, so they normalise J^1_{M}. The convolution $\Psi'[w'] * \Psi'[s]$ has support contained in $U(\mathfrak{B}) J^1_{\mathrm{M}} w' U(\mathfrak{B}) J^1_{\mathrm{M}} s U(\mathfrak{B}) J^1_{\mathrm{M}}$ $= U(\mathfrak{B}) J^1_{\mathrm{M}} w' U(\mathfrak{B}) s U(\mathfrak{B}) J^1_{\mathrm{M}}$, and consists of a union of double cosets of the form $U(\mathfrak{B}) J^1_{\mathrm{M}} v U(\mathfrak{B}) J^1_{\mathrm{M}}$ with $v \in W(\mathfrak{B})$. It follows from (1.6.1) that

$$U(\mathfrak{B}) J^1_{\mathrm{M}} w' U(\mathfrak{B}) s U(\mathfrak{B}) J^1_{\mathrm{M}} \cap B^\times = U(\mathfrak{B}) w' U(\mathfrak{B}) s U(\mathfrak{B}).$$

So, we only have to prove the following lemma.

(5.6.12) Lemma: *Let w_1, $w_2 \in \widetilde{W}(\mathfrak{B})$, and suppose that $\ell(w_1 w_2) = \ell(w_1) + \ell(w_2)$. Then $U(\mathfrak{B}) w_1 U(\mathfrak{B}) w_2 U(\mathfrak{B}) = U(\mathfrak{B}) w_1 w_2 U(\mathfrak{B})$.*

Proof: Inductively, there is no harm in assuming that w_2 is a basic involution s_i, and we can again conjugate by $\Pi(\mathfrak{B})$ to ensure that $i \neq 0$. To simplify notation, we write $w_2 = s$, $w_1 = w$, $U = U(\mathfrak{B})$.

So, in this situation, we just need the Tits system-like property

$$wUs \subset UwsU. \tag{5.6.13}$$

We can embed the whole setup in the Tits system on $B^\times$ relative to our original E-basis $\mathcal{V}$ of V and the Iwahori subgroup $U_{\mathrm{m}} = U(\mathfrak{B}_{\mathrm{m}})$. Write $\widetilde{W}(\mathfrak{B})$ for the associated Weyl group, and ℓ_{m} for its length function. For $v \in \widetilde{W}_{\mathrm{m}}$, $\ell = \ell_{\mathrm{m}}(v)$, we have $q_E^\ell = (U_{\mathrm{m}} : U_{\mathrm{m}} \cap (U_{\mathrm{m}})^v)$. Then, appealing to our block pictures, we see that

$$\ell_{\mathrm{m}}(w) = f\ell(w), \quad w \in \widetilde{W}(\mathfrak{B}). \tag{5.6.14}$$

We have $\boldsymbol{U} = \mathfrak{M}(\mathfrak{B})^\times \boldsymbol{U}_{\mathrm{m}}$, where $\mathfrak{M}(\mathfrak{B})^\times$ is the group introduced in **(5.5)**. We know that $\widetilde{\boldsymbol{W}}(\mathfrak{B})$ normalises $\mathfrak{M}(\mathfrak{B})^\times$, and by (5.6.14), we have $\ell_{\mathrm{m}}(\boldsymbol{ws}) = \ell_{\mathrm{m}}(\boldsymbol{w}) + \ell_{\mathrm{m}}(\boldsymbol{s})$. Thus

$$\begin{aligned} \boldsymbol{wUs} &= \boldsymbol{w}\mathfrak{M}(\mathfrak{B})^\times \boldsymbol{U}_{\mathrm{m}}\boldsymbol{s} = \mathfrak{M}(\mathfrak{B})^\times \boldsymbol{w}\boldsymbol{U}_{\mathrm{m}}\boldsymbol{s} \\ &\subset \mathfrak{M}(\mathfrak{B})^\times \boldsymbol{U}_{\mathrm{m}}\boldsymbol{ws}\boldsymbol{U}_{\mathrm{m}} \subset \boldsymbol{UwsU}, \end{aligned}$$

as required. ∎

This also completes the proof of (5.6.11). ∎

So, our map $\boldsymbol{\Psi}$ preserves supports. In particular, each $\boldsymbol{\Psi}[\boldsymbol{w}]$, $\boldsymbol{w} \in \widetilde{\boldsymbol{W}}(\mathfrak{B})$, is nonzero, and the set $\{\boldsymbol{\Psi}[\boldsymbol{w}] : \boldsymbol{w} \in \widetilde{\boldsymbol{W}}(\mathfrak{B})\}$ is linearly independent. This shows that $\boldsymbol{\Psi}$ is injective. To show it is surjective, we have to prove:

(5.6.15) Lemma: *For $\boldsymbol{w} \in \widetilde{\boldsymbol{W}}(\mathfrak{B})$, the space $I_{\boldsymbol{w}}(\lambda)$ has dimension at most one.*

Proof: By (5.3.2), we have $\dim(I_{\boldsymbol{w}}(\lambda)) = \dim(I_{\boldsymbol{w}}(\sigma \mid \boldsymbol{U}(\mathfrak{B})))$. However, $I_{\boldsymbol{w}}(\sigma \mid \boldsymbol{U}(\mathfrak{B})) \subset I_{\boldsymbol{w}}(\sigma \mid \mathfrak{M}(\mathfrak{B})^\times)$. Since σ restricts irreducibly to $\mathfrak{M}(\mathfrak{B})^\times$ and $\widetilde{\boldsymbol{W}}(\mathfrak{B})$ normalises $\mathfrak{M}(\mathfrak{B})^\times$, this last space has dimension at most one. ∎

We have finally completed the proof of the Main Theorem (5.6.6). ∎

Continuing in the same situation, we know from **(4.3)** that the algebra $\mathcal{H}(G, \lambda)$ carries a naturally defined Hermitian inner product $\boldsymbol{h}$. In **(5.4)**, we introduced an inner product $\langle\, , \rangle$ on $\mathcal{H}(e, q_E^f)$. We now consider the the way these inner products are related via the isomorphisms $\boldsymbol{\Psi}$ of (5.6.6).

(5.6.16) Definition: *Let $\boldsymbol{\Psi} : \mathcal{H}(e, q_E^f) \to \mathcal{H}(G, \lambda)$ be an isomorphism as in (5.6.6). Call $\boldsymbol{\Psi}$ unitary if $\boldsymbol{\Psi}[\zeta] * \overline{\boldsymbol{\Psi}[\zeta]} = 1$, where "bar" denotes the canonical involution of $\mathcal{H}(G, \lambda)$, as in **(4.3)**.*

(5.6.17) Corollary: *Suppose in (5.6.6) that $\boldsymbol{\Psi}[\zeta]$ is chosen so that the isomorphism $\Psi : \mathcal{H}(e, q_E^f) \cong \mathcal{H}(G, \lambda)$ is unitary. Then Ψ is an isometry with respect to the canonical inner products $\langle\, , \rangle$, $\dim(\lambda)^{-1}\boldsymbol{h}$ on $\mathcal{H}(e, q_E^f)$, $\mathcal{H}(G, \lambda)$:—*

$$\dim(\lambda)^{-1}\boldsymbol{h}(\boldsymbol{\Psi}(\boldsymbol{x}), \boldsymbol{\Psi}(\boldsymbol{y})) = \langle \boldsymbol{x}, \boldsymbol{y} \rangle, \quad \boldsymbol{x}, \boldsymbol{y} \in \mathcal{H}(e, q_E^f).$$

Proof: We define an involution "bar" on $\mathcal{H}(e, q_E^f)$ as follows:

(i) $\overline{\alpha x} = \overline{\alpha}.\overline{x}$, $x \in \mathcal{H}(e, q_E^f)$, $\alpha \in \mathbb{C}$;

(ii) $\overline{[\zeta]} = [\zeta]^{-1}$, $\overline{[s_i]} = [s_i]$, $1 \le i \le e-1$.

This does indeed extend to an involution (i.e. involutary anti-automorphism) of $\mathcal{H}(e, q_E^f)$. In terms of our standard basis $\{[\boldsymbol{w}] : \boldsymbol{w} \in \widetilde{\boldsymbol{W}}(\mathfrak{B})\}$ of $\mathcal{H}(e, q_E^f)$, it is given by

$$\overline{[\boldsymbol{w}]} = [\boldsymbol{w}^{-1}].$$

(5.6.18) Lemma: *Let $\langle\, , \rangle$ be the canonical inner product on $\mathcal{H}(e, q_E^f)$, and "bar" the involution defined above. Then*

$$\langle [\boldsymbol{w}_1], [\boldsymbol{w}_2] \rangle = \langle [\boldsymbol{w}_1]\overline{[\boldsymbol{w}_2]}, \mathbf{1} \rangle,$$

where $\mathbf{1}$ *denotes the unit element of* $\mathcal{H}(e, q_E^f)$.

Proof: As in **(5.5)**, let K/E be an unramified field extension of degree f, with $K^\times \subset \mathfrak{K}(\mathfrak{B})$. Let C be the A-centraliser of K, and $\mathfrak{C} = C \cap \mathfrak{B}$, so that $\boldsymbol{U}(\mathfrak{C})$ is an Iwahori subgroup of $C^\times$. The canonical algebra isomorphism $\mathcal{H}(e, q_E^f) \cong \mathcal{H}(C^\times, \mathbf{1}_{U(\mathfrak{C})})$ described in **(5.4)** preserves inner products. Comparing the effect on generators, it is easy to see that it also preserves the canonical involutions. When we use this isomorphism to transport the assertion of the lemma to $\mathcal{H}(C^\times, \mathbf{1}_{U(\mathfrak{C})})$, it becomes immediate. ∎

We next show that our unitary algebra isomorphism $\boldsymbol{\Psi}$ carries this involution on $\mathcal{H}(e, q_E^f)$ to the canonical involution on $\mathcal{H}(G, \lambda)$, as described in **(4.3)**:

$$\textbf{(5.6.19)} \qquad \boldsymbol{\Psi}(\overline{[\boldsymbol{w}]}) = \overline{\boldsymbol{\Psi}[\boldsymbol{w}]}, \quad \boldsymbol{w} \in \widetilde{\boldsymbol{W}}(\mathfrak{B}).$$

We only need check this on the generators $[\zeta] = [\Pi(\mathfrak{B})]$, $[s_i]$, $1 \le i \le e-1$. The first case is just the definition of a unitary homomorphism $\boldsymbol{\Psi}$. Take the second case. The functions $\boldsymbol{\Psi}\left(\overline{[\boldsymbol{s}_i]}\right)$, $\overline{\boldsymbol{\Psi}[\boldsymbol{s}_i]}$ both have support $J\boldsymbol{s}_iJ$, and therefore differ by a constant, $\boldsymbol{\Psi}\left(\overline{[\boldsymbol{s}_i]}\right) = \alpha_i\overline{\boldsymbol{\Psi}[\boldsymbol{s}_i]}$ say. Consider the map $\omega : \mathcal{H}(e, q_E^f) \to \mathcal{H}(e, q_E^f)$ given by

$$\omega(x) = \boldsymbol{\Psi}^{-1}\left(\overline{\boldsymbol{\Psi}(\overline{x})}\right), \quad x \in \mathcal{H}(e, q_E^f).$$

This is an algebra automorphism of $\mathcal{H}(e, q_E^f)$ which fixes $[\zeta]$ and maps $[s_i]$ to $\overline{\alpha}_i[s_i]$. Then (5.4.6)*(ii)* implies that $\alpha_i = 1$. This proves (5.6.19).

Now we prove the corollary. By linearity, we have to show

$$\dim(\lambda)^{-1}\boldsymbol{h}(\boldsymbol{\Psi}[\boldsymbol{w}_1], \boldsymbol{\Psi}[\boldsymbol{w}_2]) = \langle [\boldsymbol{w}_1], [\boldsymbol{w}_2] \rangle, \quad \boldsymbol{w}_i \in \widetilde{\boldsymbol{W}}(\mathfrak{B}).$$

If $w_1 \neq w_2$, then $\langle [w_1], [w_2] \rangle = 0$, and the functions $\Psi[w_1]$, $\Psi[w_2]$ have disjoint supports, so surely $h(\Psi[w_1], \Psi[w_2]) = 0$ also. We may therefore assume that $w_1 = w_2 = w$, say. We have $\langle [w], \overline{[w]} \rangle = q_E^{f\ell}$, $\ell = \ell(w)$. On the other hand, $h(\Psi[w], \Psi[w]) = \mathrm{tr}_{\check{W}} \Psi[w] * \overline{\Psi[w]}(1)) = \mathrm{tr}_{\check{W}}(\Psi([w]\overline{[w]})(1))$. The element $[w]\overline{[w]} \in \mathcal{H}(e, q_E^f)$ is of the form $\alpha \mathbf{1} + x$, where $\alpha \in \mathbb{C}$, and $x \in \mathcal{H}(e, q_E^f)$ is a linear combination of various $[v]$, $v \in \widetilde{W}(\mathfrak{B})$ with $v \neq 1$. Indeed, $\alpha = \langle [w]\overline{[w]}, \mathbf{1} \rangle = \langle [w], [w] \rangle = q_E^{f\ell}$. Thus $\Psi([w]\overline{[w]}) = \alpha \Psi(\mathbf{1}) + \Psi(x)$, and the function $\Psi(x)$ vanishes at 1_G. Therefore $\Psi([w]\overline{[w]})(1) = \alpha \mathrm{tr}_{\check{W}}(e(1)) = \alpha \dim(\lambda)$, as required. ■

Remark: We have imposed throughout the condition $\mu(J) = 1$ on our Haar measure μ on G. Varying this has no effect at all on the construction of support preserving algebra isomorphisms $\Psi : \mathcal{H}(e, q_E^f) \xrightarrow{\approx} \mathcal{H}(G, \lambda)$. Indeed, the two algebra structures on $\mathcal{H}(G, \lambda)$ given by different Haar measures are related by a unique support-preserving isomorphism.

Changing the measure does have some slight effect on unitary structures, given the way we defined the inner product on $\mathcal{H}(G, \lambda)$. The involution "bar" on $\mathcal{H}(G, \lambda)$ is defined independently of Haar measure. We can still define an isomorphism Ψ to be unitary if $\Psi[\zeta] * \overline{\Psi[\zeta]}$ is the unit element of $\mathcal{H}(G, \lambda)$. With this definition, unitary isomorphisms still take the involution on $\mathcal{H}(e, q_E^f)$ to that on $\mathcal{H}(G, \lambda)$. However, with our definition of the inner product

$$h(\phi, \psi) = \int \mathrm{tr}(\phi(x).\overline{\psi(x)})\, dx = \mathrm{tr}(\phi * \overline{\psi}(1_G)),$$

where tr denotes the trace of operators on $\check{W}$, a unitary isomorphism Ψ gives the relation

$$\langle [w_1], [w_2] \rangle = \mu(J) \dim(\lambda)^{-1} h(\Psi[w_1], \Psi[w_2]). \tag{5.6.20}$$

(5.7) Intertwining and conjugacy for simple types

Now we effectively return to the situation of **(5.5)**, and prove an "intertwining implies conjugacy" theorem for simple types in G, of which the corresponding results for simple strata (2.6.1) and simple characters (3.5.11) may be regarded as preliminary steps.

(5.7.1) Theorem: *Let (J_1, λ_1), (J_2, λ_2) be simple types in G, attached to principal $\mathfrak{o}_F$-orders $\mathfrak{A}_1$, $\mathfrak{A}_2$, respectively. Suppose that $\mathfrak{A}_1 \cong \mathfrak{A}_2$ as $\mathfrak{o}_F$-orders, and that the representations λ_1, λ_2 intertwine in G. Then there exists $x \in G$ such that $J_2 = x^{-1} J_1 x$ and λ_2 is equivalent to λ_1^x.*

Proof: Suppose first that (J_1, λ_1) is defined by some simple stratum $[\mathfrak{A}_1, n_1, 0\beta_1]$ with $n_1 \geq 1$. Then, we assert, (J_2, λ_2) is defined by a simple stratum $[\mathfrak{A}_2, n_2, 0, \beta_2]$ with $n_2 \geq 1$. For, if λ_2 is of the sort (5.5.10)*(b)*, then λ_2 is trivial on $\boldsymbol{U}^1(\mathfrak{A}_2)$, and the nonsplit fundamental stratum $[\mathfrak{A}_1, n_1, n_1, -1, \beta_1]$ must intertwine with the null stratum $[\mathfrak{A}_2, 1, 0, 0]$, contrary to (2.6.2).

We therefore assume initially that (J_i, λ_i) is defined by a simple stratum $[\mathfrak{A}_i, n_i, 0, \beta_i]$, with $n_i \geq 1$, for $i = 1, 2$. There is no harm in replacing (J_1, λ_1) by a G-conjugate and assuming further that $\mathfrak{A}_1 = \mathfrak{A}_2 = \mathfrak{A}$, say. Since the nonsplit fundamental strata $[\mathfrak{A}, n_i, n_i - 1, \beta_i]$ must intertwine, we also have $n_1 = n_2$. Let $\theta_i \in \mathcal{C}(\mathfrak{A}, 0, \beta_i)$ be the simple character occurring in $\lambda_i \mid H^1(\beta_i, \mathfrak{A})$, $i = 1, 2$. Abbreviate $H_i = H^1(\beta_i, \mathfrak{A})$. The characters θ_1, θ_2 intertwine, so we can apply (3.5.11): there exists $x \in \boldsymbol{U}(\mathfrak{A})$ such that $H_1^x = H_2$ and $\theta_2 = \theta_1^x$. Replacing λ_1 by a conjugate, we may take $x = 1$ here. The $\boldsymbol{U}(\mathfrak{A})$-intertwining of θ_i is J_i, so we also have $J_1 = J_2$. Moreover, we get $\eta_1 = \eta_2$, where η_i is the unique irreducible representation of J_i^1 which contains $\theta_1 = \theta_2$ on $H_1 = H_2$. We can now write our simple types in the form $\lambda_i = \kappa_i \otimes \sigma_i$, where κ_i is some β_i-extension of η_i. However, the remark following (5.2.1) shows that κ_1 is also a β_2-extension of $\eta_1 = \eta_2$, so we may as well take $\kappa_1 = \kappa_2$ here.

Now consider the representations σ_i of J/J^1, where $J = J(\beta_1, \mathfrak{A}) = J(\beta_2, \mathfrak{A})$. Each of these is irreducible and, if we view J/J^1 as an algebraic group over k_{E_i}, $E_i = F[\beta_i]$, then σ_i is cuspidal. However, this condition can be described in purely group-theoretic terms: a Borel subgroup of J/J^1 is just the normaliser of a p-Sylow subgroup, a parabolic subgroup is a subgroup containing a Borel subgroup, and the unipotent radical of a parabolic subgroup is its unique maximal normal p-subgroup. Thus the cuspidality condition on σ_i is independent of β_i. The homogeneity condition can be similarly described:— it is equivalent to the condition that σ_i is stable under conjugation by $\mathfrak{K}(\mathfrak{B}_i)J_i$, $\mathfrak{B}_i = \mathfrak{A} \cap \mathrm{End}_{E_i}(V)$. We have $\mathfrak{K}(\mathfrak{B}_1)J_1 = \mathfrak{K}(\mathfrak{B}_2)J_2$ by (3.3.17). Therefore, from now on, we may as well take $\beta_1 = \beta_2 = \beta$, say.

Write $E = F[\beta]$, $B = \mathrm{End}_E(V)$, $\mathfrak{B} = \mathfrak{A} \cap B$, as usual. The representations λ_i of $J = J(\beta, \mathfrak{A})$ are intertwined by some $x \in G$. This element x must intertwine $\lambda_1 \mid J^1 = \lambda_2 \mid J^1$, and this is a multiple of $\eta_1 = \eta_2 = \kappa_i \mid J^1$. We deduce ((5.1.1)) that $x \in J^1 B^\times J^1$. Therefore we may assume $x \in B^\times$. Exactly as in (5.3.2), it follows that x intertwines the representations $\sigma_i \mid \boldsymbol{U}(\mathfrak{B})$.

At this stage, therefore, we have irreducible representations σ_1, σ_2 of $\boldsymbol{U}(\mathfrak{B})$ which are inflated from "homogeneous" cuspidal representations of $\boldsymbol{U}(\mathfrak{B})/\boldsymbol{U}^1(\mathfrak{B})$ ($\cong GL(f, \mathrm{k}_E)^e$, say) and which intertwine in $B^\times$. We show that they are equivalent, and this will prove the theorem. To do this, we proceed as in **(5.5)**. We choose a basis of the lattice chain defining $\mathfrak{B}$ exactly as in (5.5.2), which gives rise to an affine Weyl group $\widetilde{\boldsymbol{W}}$. With $\mathfrak{M}(\mathfrak{B})$ as in **(5.5)**, the proof of (5.5.5) applies without change to show that any $\boldsymbol{w} \in \widetilde{\boldsymbol{W}}$ which intertwines the σ_i must normalise $\mathfrak{M}(\mathfrak{B})^\times$. We can push this further. The proof of (5.5.11) applies without change to show that any such $\boldsymbol{w}$ lies in $\widetilde{\boldsymbol{W}}(\mathfrak{B})$. Such an element $\boldsymbol{w}$ acts by permuting the blocks $GL(f, \mathfrak{o}_E)$ of $\mathfrak{M}(\mathfrak{B})^\times$. If we write $\sigma_i = \tau_i \otimes \ldots \otimes \tau_i$, for some cuspidal representation τ_i of $GL(f, k_E)$ (inflated in the usual way), then any $\boldsymbol{w} \in \widetilde{\boldsymbol{W}}(\mathfrak{B})$ which intertwines σ_1 with σ_2 effectively identifies τ_1 with τ_2. We deduce that $\sigma_1 \cong \sigma_2$, and the result follows.

This leaves us with the case in which the simple types (J_i, λ_i) are each of the form (5.5.10)*(b)*. The proof in this case is exactly the same as the last step in the first case, which shows that the σ_i are equivalent. ∎

6. MAXIMAL TYPES

In this and the following chapter, we analyse the set of (equivalence classes of) irreducible smooth representations π of G which contain a given simple type (J, λ), in the sense that the restriction $\pi \mid J$ has λ as a factor. In the present chapter, we give a complete account of the irreducible *supercuspidal* representations which contain a simple type. It is reassuring to know in advance that the scope of these results is actually wider. In (8.4.1) we will prove :—

(6.0.1) *Let π be an irreducible supercuspidal representation of G. Then π contains some simple type in G.*

Therefore this chapter contains a complete account of all the irreducible supercuspidal representations of G. However, we will not directly appeal to (6.0.1) here.

So, we now fix a simple type (J, λ) in our group $G = \mathrm{Aut}_F(V) \cong GL(N, F)$, and consider those irreducible smooth representations $(\pi, \mathcal{V})$ of G which contain λ. We observed in **(4.2)** that such representations are classified by the simple modules over the spherical Hecke algebra $\mathcal{H}(G, \lambda)$, or, equivalently via a canonical Morita equivalence, the simple modules over the subalgebra $e_\lambda * \mathcal{H}(G) * e_\lambda$ of the algebra $\mathcal{H}(G)$ of locally constant compactly supported functions on G.

Initially, the type (J, λ) carries along with it all the associated notations as in §**5**. In particular, we have a simple stratum $[\mathfrak{A}, n, 0, \beta]$ in $A = \mathrm{End}_F(V)$, such that the order $\mathfrak{A}$ is principal. We put $E = F[\beta]$, $B =$ the A-centraliser of $E = \mathrm{End}_E(V)$, $\mathfrak{B} = \mathfrak{A} \cap B$, $e = e(\mathfrak{B}|\mathfrak{o}_E)$, $J = J(\beta, \mathfrak{A})$ and so on.

(6.1) Extension by a central character

When analysing certain types of representation π of G, it is convenient to have control of the central quasicharacter ω_π of π. We therefore start by giving a refinement of our Main Theorem (5.6.6) incorporating this extra structure.

For our fixed simple type (J, λ), we have $J \cap F^\times = \mathfrak{o}_F^\times$, and $\lambda \mid \mathfrak{o}_F^\times$ is a multiple of a character ω_λ of $\mathfrak{o}_F^\times$. Let ω be some quasicharacter of $F^\times$ with $\omega \mid \mathfrak{o}_F^\times = \omega_\lambda$. We can then define a representation $\lambda\omega$ of $F^\times J$ by

$$\lambda\omega(zj) = \omega(z)\lambda(j), \quad j \in J,\ z \in F^\times. \tag{6.1.1}$$

Here, of course, we are identifying $F^\times$ with the centre of G. The intertwining of this representation $\lambda\omega$ is, of course, the same as that of λ, namely $J\widetilde{\boldsymbol{W}}(\mathfrak{B})J$ in the notation of §**5**. In particular, the space $I_{\boldsymbol{w}}(\lambda\omega)$ has dimension one, for $\boldsymbol{w} \in \widetilde{\boldsymbol{W}}(\mathfrak{B})$.

We also have the group $E^\times J$, which normalises $F^\times J$ and the representation $\lambda\omega$. Moreover, since the quotient $E^\times J/F^\times J$ is cyclic of order $e' = e(E|F)$, it follows that $\lambda\omega$ extends to a representation Λ of $E^\times J$.

(6.1.2) Proposition: *Let Λ be an irreducible representation of $E^\times J$ such that $\Lambda \mid J$ contains λ. Then*

(i) there is a unique quasicharacter ω of $F^\times$ such that $\omega \mid \mathfrak{o}_F^\times = \omega_\lambda$ and $\Lambda \mid F^\times J = \lambda\omega$;

(ii) given $\lambda\omega$ as in (6.1.1), *there exist precisely $e(E|F)$ distinct extensions Λ of $\lambda\omega$ to $E^\times J$;*

(iii) given $\boldsymbol{w} \in \widetilde{\boldsymbol{W}}(\mathfrak{B})$ and an extension Λ of $\lambda\omega$ to $E^\times J$, there is a unique extension Λ' of $\lambda\omega$ such that $\boldsymbol{w}$ intertwines Λ with Λ'.

In particular, if $\mathfrak{B}$ is a maximal order in B, distinct extensions of $\lambda\omega$ to $E^\times J$ do not intertwine in G.

Proof: Part *(i)* is immediate, given our earlier remark that any $\lambda\omega$ does extend to $E^\times J$. In part *(ii)*, consider the induced representation $\mathrm{Ind}(\lambda\omega : F^\times J, E^\times J) = \Lambda \otimes \mathrm{Ind}(\mathbf{1} : F^\times J, E^\times J)$, for any extension Λ and where $\mathbf{1}$ denotes the trivial representation of $F^\times J$. This is the sum of the $e(E|F)$ irreducible representations $\Lambda \otimes \chi$, where χ ranges over the characters of $E^\times/F^\times\mathfrak{o}_E^\times$. The intertwining of this induced representation is $E^\times J\widetilde{\boldsymbol{W}}(\mathfrak{B})JE^\times = J\widetilde{\boldsymbol{W}}(\mathfrak{B})J$, and we can compute the spaces $I_{\boldsymbol{w}}(\mathrm{Ind}(\lambda\omega))$ of intertwining operators, $\boldsymbol{w} \in \widetilde{\boldsymbol{W}}(\mathfrak{B})$.

(6.1.3) Lemma: *For $\boldsymbol{w} \in \widetilde{\boldsymbol{W}}(\mathfrak{B})$, the space $I_{\boldsymbol{w}}(\mathrm{Ind}(\lambda\omega))$ has dimension $e(E|F)$.*

Proof: We have already noted that $I_{\boldsymbol{w}}(\lambda\omega)$ has dimension one, so the required dimension is the number of $F^\times J$-double cosets $F^\times J\boldsymbol{v}J$ contained in $E^\times J\boldsymbol{w}J$, $\boldsymbol{v} \in \widetilde{\boldsymbol{W}}(\mathfrak{B})$. This number is clearly $e(E|F)$. ■

The lemma immediately implies part *(iii)* of the proposition, and part *(ii)* is the special case $\boldsymbol{w} = 1$. When $\mathfrak{B}$ is a maximal order in B, $\widetilde{\boldsymbol{W}}(\mathfrak{B})$ is the cyclic group generated by π_E, and the final assertion follows. ■

Remark: The final assertion of (6.1.2) holds without restriction on the order $\mathfrak{B}$. The proof of this fact, however, requires the machinery of §**7**, and we have no immediate use for it.

We now work out the structure of the Hecke algebra $\mathcal{H}(G, \lambda\omega)$. As usual, we take $\mathcal{H}(G, \lambda)$ to be defined relative to the Haar measure on G which gives $\mu(J) = 1$. There is a canonical surjective algebra homomorphism

$$P_\omega : \mathcal{H}(G, \lambda) \to \mathcal{H}(G, \lambda\omega)$$

given by

$$P_\omega(\phi)(x) = \int_{F^\times} \omega(z)\phi(zx)\,dz,$$

where $\phi \in \mathcal{H}(G,\lambda)$ and dz is the Haar measure on $F^\times$ giving $\mathfrak{o}_F^\times$ measure 1.

Now fix a prime element π_F of F. There is a unique function $\phi_1 \in \mathcal{H}(G,\lambda)$ such that

(6.1.4) *(i)* ϕ_1 *has support* $J\pi_F J = \pi_F J$;
(ii) $\phi_1(\pi_F j) = \omega(\pi_F)^{-1}\check{\lambda}(j)$, $j \in J$.

Let $\mathbf{e}$ denote the identity element of $\mathcal{H}(G,\lambda)$, so that $\mathbf{e}$ has support J and $\mathbf{e}(j) = \check{\lambda}(j)$, $j \in J$.

(6.1.5) Lemma: *The kernel of the map* P_ω *is the ideal* $\mathcal{I}$ *of* $\mathcal{H}(G,\lambda)$ *generated by* $\mathbf{e} - \phi_1$.

Proof: Surely $\mathcal{I} \subset \mathrm{Ker}(P_\omega)$, so take $\phi \in \mathrm{Ker}(P_\omega)$. If $x \in \mathrm{supp}(\phi)$, then the restriction of ϕ to $F^\times JxJ$ (i.e. the function in $\mathcal{H}(G,\lambda)$ agreeing with ϕ on $F^\times JxJ$ and null elsewhere) still lies in $\mathrm{Ker}(P_\omega)$, and ϕ is a finite sum of functions of this form. In other words, we may as well assume that $\mathrm{supp}(\phi) \subset F^\times JxJ$, for some $x \in G$ with $\phi(x) \neq 0$. We have

$$P_\omega(\phi)(g) = \sum_{n=-\infty}^{\infty} \omega(\pi_F)^n \phi(\pi_F^n g), \qquad g \in G, \tag{6.1.6}$$

so there are at least two values of n for which $\phi(\pi_F^n x) \neq 0$. Choose m maximal and n minimal for the property

$$\mathrm{supp}(\phi) \subset \bigcup_{m \le i \le n} \pi_F^i JxJ.$$

If $n - m = 1$, we have (by (5.6.8), (5.6.9)) $\phi = \psi * (\mathbf{e} - \phi_1)$, for some $\psi \in \mathcal{H}(G,\lambda)$ with support $\pi_F^m J$, so $\phi \in \mathcal{I}$. In general, we can add to ϕ an element of $\mathcal{I}$ of this same form $\psi * (\mathbf{e} - \phi_1)$ to reduce $n - m$. The result follows. ■

Now we choose an isomorphism $\boldsymbol{\Psi} : \mathcal{H}(e, q_E^f) \to \mathcal{H}(G,\lambda)$ as in (5.6.6), and put $\psi = \boldsymbol{\Psi}[\zeta]$, using the same notation. With $e = e(\mathfrak{B}|\mathfrak{o}_E)$, $e' = e(E|F)$ as before, we have $\psi^{ee'} = \alpha\phi_1$, for some $\alpha \in \mathbb{C}^\times$, since the ee'-fold convolution $\psi^{ee'}$ has support $\pi_F J$. Thus $\boldsymbol{\Psi}$ induces an isomorphism

$$\frac{\mathcal{H}(e, q_E^f)}{([\zeta]^{ee'} - \alpha 1)} \cong \mathcal{H}(G, \lambda\omega). \tag{6.1.7}$$

Observe that changing $\boldsymbol{\Psi}$ (i.e. the choice of ψ) changes α to $\alpha\alpha_1^{ee'}$, for some α_1, so there is a choice of $\boldsymbol{\Psi}$ (in fact ee' distinct choices) for which

$$\frac{\mathcal{H}(e, q_E^f)}{([\zeta]^{ee'} - 1)} \cong \mathcal{H}(G, \lambda\omega). \tag{6.1.8}$$

Observe also that, when $\mathfrak{B}$ is a maximal order, (6.1.2) gives an algebra decomposition of $\mathcal{H}(G, \lambda\omega)$:

$$\mathcal{H}(G, \lambda\omega) = \prod_{\Lambda} \mathcal{H}(G, \Lambda), \tag{6.1.9}$$

where Λ ranges over the irreducible representations of $E^\times J$ with $\Lambda \mid F^\times J = \lambda\omega$.

The element $[\zeta]^e$ is central in $\mathcal{H}(e, q_E^f)$, so the two-sided ideal of $\mathcal{H}(e, q_E^f)$ generated by $[\zeta]^{ee'} - 1$ is just

$$([\zeta]^{ee'} - 1)\mathcal{H}(e, q_E^f) = \mathcal{H}(e, q_E^f)([\zeta]^{ee'} - 1).$$

If we temporarily write $\mathcal{H}_1 = \mathcal{H}(e, q_E^f)/([\zeta]^{ee'} - 1)$, it follows that the set $\{[\zeta]^i[\boldsymbol{w}] : 0 \leq i \leq ee' - 1, \boldsymbol{w} \in \boldsymbol{W}(\mathfrak{B})\}$ is a basis of $\mathcal{H}_1$.

(6.1.10) Proposition: *With the notation above, suppose that $\mathcal{H}_1$ has a nonzero finite-dimensional left ideal $\mathfrak{a}$. Then $e = 1$.*

Proof: Assume $e > 1$ so that $\boldsymbol{W}(\mathfrak{B})$ is an infinite group. We define a subset $\boldsymbol{W}_\mathfrak{a}$ of $\boldsymbol{W}(\mathfrak{B})$ by the condition: $\boldsymbol{w} \in \boldsymbol{W}_\mathfrak{a}$ if there exists an element

$$[\boldsymbol{w}][\zeta]^i + \alpha_1[\boldsymbol{w}_1][\zeta]^{i_1} + \ldots + \alpha_r[\boldsymbol{w}_r][\zeta]^{i_r} \in \mathfrak{a},$$

where $\alpha_i \in \mathbb{C}$, $\boldsymbol{w}_i \in \boldsymbol{W}(\mathfrak{B})$, $\boldsymbol{w}_i \neq \boldsymbol{w}$. Since $\mathfrak{a}$ is not zero, the set $\boldsymbol{W}_\mathfrak{a}$ is nonempty. Since $\mathfrak{a}$ is finite-dimensional, the set $\boldsymbol{W}_\mathfrak{a}$ is finite. Take $\boldsymbol{w} \in \boldsymbol{W}_\mathfrak{a}$ of maximal length, and an element in $\mathfrak{a}$ involving $\boldsymbol{w}$ as above. We take a fundamental involution $\boldsymbol{s} \in \boldsymbol{W}(\mathfrak{B})$ such that $\ell(\boldsymbol{sw}) > \ell(\boldsymbol{w})$. It follows easily that $\boldsymbol{sw} \in \boldsymbol{W}_\mathfrak{a}$, and we have a contradiction. ■

(6.2) Supercuspidal representations

We continue with our fixed simple type (J, λ) in G, and consider the irreducible *supercuspidal* representations of G which contain λ. The picture is very simple, given by the following two results.

(6.2.1) Theorem: *Let (J, λ) be a simple type in G, and suppose there exists an irreducible supercuspidal representation $(\pi, \mathcal{V})$ of G such that $\pi \mid J$ contains λ. Then the $\mathfrak{o}_E$-order $\mathfrak{B}$ attached to (J, λ) is maximal.*

(6.2.2) Theorem: *Suppose that the $\mathfrak{o}_E$-order $\mathfrak{B}$ attached to (J, λ) is maximal. Then any irreducible representation π of G containing λ is supercuspidal. Moreover, for any such representation π, there is a uniquely determined representation Λ of $E^\times J$ such that $\Lambda \mid J = \lambda$ and*

$$\pi = \text{c-Ind}(\Lambda : E^\times J, G).$$

Remark: In (6.2.2), c-Ind denotes the functor "smooth induction with compact supports". In fact, we do not need compact support here:— see **[Bu2]**.

We start by proving (6.2.2). So, suppose that $\mathfrak{B}$ is a maximal order. The group $\widetilde{\boldsymbol{W}}(\mathfrak{B})$ is then just the cyclic group generated by $\Pi(\mathfrak{B}) = \pi_E 1_G$, and it is contained in $E^\times$. Thus, for any irreducible representation Λ of $E^\times J$ containing λ, we have $I_G(\Lambda) = E^\times J$. It follows (**[Ca]** (1.5)) that c-Ind$(\Lambda : E^\times J, G)$ is irreducible and supercuspidal. Any irreducible representation π of G containing λ contains such an extension Λ, and therefore $\pi = \text{c-Ind}(\Lambda)$.

Next let Λ, Λ' be irreducible representations of $E^\times J$ containing λ. We have to show that c-Ind(Λ) = c-Ind(Λ') implies $\Lambda = \Lambda'$. The representations Λ, Λ' certainly agree on $F^\times$, so $\Lambda \mid F^\times J = \Lambda' \mid F^\times J = \lambda\omega$, where $\omega = \omega_\pi$. Moreover, the representations Λ, Λ' intertwine in G, so $\Lambda = \Lambda'$ by (6.1.2). This proves (6.2.2). ■

Now let us prove (6.2.1). Let $(\pi, \mathcal{V})$ be an irreducible supercuspidal representation of G containing λ. If $\omega = \omega_\pi$ is the central quasicharacter of π, then $\pi \mid F^\times J$ contains $\lambda\omega$. Write $\mathcal{H}(G, \omega)$ for the convolution algebra of smooth functions ϕ on G which are compactly supported mod $F^\times$ and satisfy $\phi(zg) = \omega^{-1}(z)\phi(g)$, $g \in G$, $z \in F^\times$. As in **(4.2)**, the $\lambda\omega$-isotypic component $\mathcal{V}^{\lambda\omega}$ of $\mathcal{V}$ is a simple module over the ring $e_{\lambda\omega} * \mathcal{H}(G, \omega) * e_{\lambda\omega}$, where $e_{\lambda\omega}$ is the function defined by (4.2.1) (with $\rho = \lambda\omega$). Now we refer to **[Cr]** p.122. If $v \in \mathcal{V}$, the map $\mathcal{H}(G, \omega) \to \mathcal{V}$ given by $\phi \mapsto \pi(\phi)v$ is a split surjection. If we choose $v \in \mathcal{V}^{\lambda\omega}$, this restricts to a split surjection $e_{\lambda\omega} * \mathcal{H}(G, \omega) * e_{\lambda\omega} \to \mathcal{V}^{\lambda\omega}$. In particular, the ring $e_{\lambda\omega} * \mathcal{H}(G, \omega) * e_{\lambda\omega}$ has a non-trivial, finite-dimensional projective simple left ideal, namely the image of any splitting here. Thus the Morita equivalent ring $\mathcal{H}(G, \lambda\omega)$ has an ideal of the same type. The same therefore applies to the ring $\mathcal{H}(e, q_E^f)/([\zeta]^{ee'} - 1)$ by (6.1.8). By (6.1.10), we get $e = 1$, so $\mathfrak{B}$ is maximal, as required. ■

Remark: We shall give a different proof of (6.2.1) below, using the Jacquet functor:— see the remark following (7.3.2). Moreover, when we take (6.0.1) into account, Theorems (6.2.1), (6.2.2) give a complete account of the irreducible supercuspidal representations of G.

In the light of these results, we say that the simple type (J, λ) is *maximal* if the associated invariant $e(\mathfrak{B}|\mathfrak{o}_E)$ is 1, i.e. if λ occurs only in supercuspidal irreducible representations of G. We summarise:

(6.2.3) Corollary: *Let (J, λ) be a simple type in G, attached to the simple stratum $[\mathfrak{A}, n, 0, \beta]$. Write $E = F[\beta]$, and let $\mathfrak{B}$ denote the $\mathfrak{A}$-centraliser of β. The following are equivalent:*

(i) $e(\mathfrak{B}|\mathfrak{o}_E) = 1$;

(ii) there exists an irreducible supercuspidal representation $(\pi, \mathcal{V})$ of G such that $\pi \mid J$ contains λ;

(iii) any irreducible representation $(\pi, \mathcal{V})$ of G such that $\pi \mid J$ contains λ is supercuspidal.

Suppose these conditions hold, and let π be an irreducible representation of G which contains λ. Then an irreducible representation π' of G contains λ if and only if $\pi' = \pi \otimes \chi \circ \det_A$, for some unramified quasicharacter χ of $F^\times$.

Proof: The equivalence of the three conditions is already established. For the final assertion, we note that one implication is obvious. Conversely, if π' contains λ, then the central quasicharacters ω_π, $\omega_{\pi'}$ agree on $\mathfrak{o}_F^\times$. We may therefore tensor π' with an unramified quasicharacter of G to achieve $\omega_\pi = \omega_{\pi'}$. Therefore $\pi = \mathrm{Ind}(\Lambda)$, $\pi' = \mathrm{Ind}(\Lambda')$, for representations Λ, Λ' of $E^\times J$ extending $\lambda\omega$, $\omega = \omega_\pi$. Immediately from (6.1.2), there is an unramified quasicharacter χ of $F^\times$ such that $\Lambda' = \Lambda \otimes \chi \circ \det_A \mid E^\times J$. (We could equally well have deduced this part from the Hecke algebra isomorphism.) ■

Now we turn to uniqueness properties.

(6.2.4) Theorem: *Let π be an irreducible supercuspidal representation of G containing two simple types (J, λ), (J_1, λ_1). Then (J, λ), (J_1, λ_1) are conjugate in G.*

Proof: We start by showing that the representation π immediately yields a considerable amount of information concerning any simple type in π.

(6.2.5) Lemma: *With π as in* (6.2.4), *let $[\mathfrak{A}, n, 0, \beta]$ be the simple stratum defining (J, λ), and write $E = F[\beta]$. Let χ be an unramified quasicharacter of $F^\times$. Then $\pi \cong \pi \otimes (\chi \circ \det_A)$ if and only if χ has finite order dividing $N/e(E|F)$.*

Proof: Let $\omega = \omega_\pi$ be the central quasicharacter of π. The central quasicharacter of $\pi \otimes (\chi \circ \det_A)$ is $\omega.\chi^N$, so we may as well assume that χ has order dividing N. Let Λ be the unique extension of $\lambda\omega$ to $E^\times J$ such that $\pi = \mathrm{Ind}(\Lambda)$, as in (6.2.2). Then $\pi \cong \pi \otimes (\chi \circ \det_A)$ if and only if $\Lambda \cong \Lambda \otimes (\chi \circ \det_A)$. This last condition is equivalent to $(\chi \circ \det_A) \mid E^\times = 1$ by (6.1.2)*(ii)*. However, $(\chi \circ \det_A) \mid E^\times = (\chi \circ \boldsymbol{N}_{E/F})^R$, where $\boldsymbol{N}_{E/F}$

denotes the field norm, and $R = \dim_E(V)$. This character is trivial if and only if the order of χ divides $R.f(E|F) = N/e(E|F)$. ■

In the notation of (6.2.5), the order $\mathfrak{A}$ is principal and also maximal for the property of being normalised by $E^\times$. Thus, in particular, $e(\mathfrak{A}|\mathfrak{o}_F) = e(E|F) = N/g(\boldsymbol{\pi})$, where $g(\boldsymbol{\pi})$ is the number of unramified quasicharacters χ of $F^\times$ such that $\boldsymbol{\pi} \otimes \chi \circ \det \cong \boldsymbol{\pi}$. Likewise, if $\mathfrak{A}_1$ is the principal order underlying the simple type (J_1, λ_1), we also get $e(\mathfrak{A}_1|\mathfrak{o}_F) = N/g(\boldsymbol{\pi})$. It follows that $\mathfrak{A}_1 \cong \mathfrak{A}$ as $\mathfrak{o}_F$-orders. (6.2.4) now follows from (5.7.1). ■

7. TYPICAL REPRESENTATIONS

We now fix a simple type (J, λ) in $G = \mathrm{Aut}_F(V) \cong GL(N, F)$, attached to a simple stratum $[\mathfrak{A}, n, 0, \beta]$ in $A = \mathrm{End}_F(V)$ and a simple character $\theta \in \mathcal{C}(\mathfrak{A}, 0, \beta)$. We use our standard notation for this situation: $E = F[\beta]$, $B = \mathrm{End}_E(V)$, $\mathfrak{B} = \mathfrak{A} \cap B$, $e = e(\mathfrak{B}|\mathfrak{o}_E)$. If our simple type happens to be of the kind (5.5.10)*(b)*, we adopt our usual strategy of setting $E = F$, $\mathfrak{B} = \mathfrak{A}$, and taking θ, η, κ all trivial. The first object of this section is to characterise, in terms of standard classification theory, those irreducible representations of G which contain λ. Of course, we have already classified these in one sense:— they are the same as the simple modules over the ring $\mathcal{H}(G, \lambda)$, which is isomorphic via (5.6.6) to $\mathcal{H}(e, q_E^f)$, and the simple modules over this algebra are classified in **[KL]**.

We first have to establish an "Iwahori decomposition" for the group $J(\beta, \mathfrak{A})$, and investigate the behaviour of the representation λ relative to this decomposition. This preliminary material occupies the first two subsections but, once done, it puts us in the same position as **[Bo]** and **[Cs]** relative to the "trivial type" consisting of the trivial character of an Iwahori subgroup of G. In the process, we produce a maximal simple type (J', λ') (in the group $G' = GL(N/e, F)$) associated to (J, λ), and the main result shows that an irreducible representation π of G contains λ if and only if the supercuspidal support of π consists of representations of G' which contain λ' (and are hence known from (6.2.3)). When we take (6.0.1) into account, we therefore have a complete description, via (5.6.6), of the irreducible representations of G whose supercuspidal support consists of unramified twists of a single supercuspidal representation.

We can then refine this in **(7.5)** to get an equivalence of categories. With this in hand, we then show how various representation-theoretic properties are transmitted by these equivalences of categories. In particular, we show how to transfer all questions concerning decomposition of induced representations, and determination of the discrete series, to the case of **[Bo]** or **[Cs]** which deals with representations having Iwahori fixed vectors.

The methods we use ultimately go back to **[Cs]**, but we have often been inspired by the treatment of certain special cases in **[Wa]**. Our situation is more general and has a different starting point, but we are often reduced to checking details in parallel to **[Wa]**.

While we never actually appeal to (6.0.1) in this section, it is useful to bear it in mind for the sake of the extra force which it gives the results.

(7.1) Some Iwahori decompositions

We start by investigating some geometrical properties of the groups $J = J(\beta, \mathfrak{A})$, $H = H(\beta, \mathfrak{A})$ relative to certain parabolic subgroups of G. Ultimately, we shall only be interested in those parabolic subgroups suggested by the decomposition of V given by (5.5.2). Initially, however, we have to work in greater generality.

To start with, let $\mathfrak{A} = \mathrm{End}^0_{\mathfrak{o}_F}(\mathcal{L})$ be some hereditary $\mathfrak{o}_F$-order in A defined by an $\mathfrak{o}_F$-lattice chain $\mathcal{L} = \{L_k : k \in \mathbb{Z}\}$ in V. We take an integer $t \geq 1$ and an ordered decomposition $V = V^{(1)} \oplus V^{(2)} \oplus \ldots \oplus V^{(t)}$ of V as a direct sum of F-subspaces.

(7.1.1) Definition: *The F-decomposition $V = V^{(1)} \oplus V^{(2)} \oplus \ldots \oplus V^{(t)}$ of V is* subordinate to the $\mathfrak{o}_F$-chain $\mathcal{L}$ *(or the $\mathfrak{o}_F$-order $\mathfrak{A}$) if:*

(i) $L_k = \coprod_{i=1}^t L_k^{(i)}$, *where* $L_k^{(i)} = L_k \cap V^{(i)}$, $1 \leq i \leq t$, $k \in \mathbb{Z}$;

(ii) $L^{(i)}_{i+mt} = L^{(i)}_{i+mt+1} = \ldots = L^{(i)}_{i+(m+1)t-1} \neq L^{(i)}_{i+(m+1)t}$, $1 \leq i \leq t$, $m \in \mathbb{Z}$.

If this holds, then each $\{L^{(i)}_{tj} : j \in \mathbb{Z}\}$ is an $\mathfrak{o}_F$-lattice chain in $V^{(i)}$ of period $e(\mathfrak{A}|\mathfrak{o}_F)/t$. In particular, t must divide $e(\mathfrak{A}|\mathfrak{o}_F)$, and the case $t = 1$ is invariably trivial.

(7.1.2) Remarks: *(i)* It is not hard to show that, given $\mathfrak{A}$ (or $\mathcal{L}$) and a positive divisor t of $e = e(\mathfrak{A}|\mathfrak{o}_F)$, there exists a decomposition $V = V^{(1)} \oplus \ldots \oplus V^{(t)}$ of V subordinate to $\mathfrak{A}$. To do this, one takes some decomposition $V = V^{(1)} \oplus \ldots \oplus V^{(t)}$ of V, with each $V^{(i)}$ having dimension $\geq e/t$, and a lattice chain $\mathcal{L}'^{(i)}$ in $V^{(i)}$ of period e/t. One uses (7.1.1) to construct from the $\mathcal{L}'^{(i)}$ a lattice chain $\mathcal{L}'$ in V of period e, such that the chosen decomposition of V is subordinate to $\mathcal{L}'$. For suitable choices of $V^{(i)}$, $\mathcal{L}'^{(i)}$, one gets $\mathcal{L}' \cong \mathcal{L}$. Then one applies a suitable automorphism of V to achieve $\mathcal{L}' = \mathcal{L}$.

(ii) The decomposition in (5.5.2) is an E-decomposition of V subordinate to the $\mathfrak{o}_E$-order $\mathfrak{B}$, with $t = e(\mathfrak{B}|\mathfrak{o}_E)$. It is therefore also an F-decomposition subordinate to $\mathfrak{A}$.

(iii) Observe that the notion of a decomposition of V subordinate to a chain is quite different from that of a decomposition splitting the chain, in the sense of **(2.3)**, although the two notions do have several properties in common.

Before proceeding we note some identities concerning the enumeration of various of the lattices turning up in (7.1.1).

(7.1.3) Lemma: *In the situation of* (7.1.1), *let* $1 \leq i \leq t$. *Then*

$$L_j^{(i)} = L^{(i)}_{(k+1)t}, \quad \text{where } k = \left[\tfrac{j-i}{t}\right], \ j \in \mathbb{Z}.$$

Proof: Trivial. ∎

We now fix $\mathfrak{A} = \mathrm{End}^0_{\mathfrak{o}_F}(\mathcal{L})$ and an F-decomposition $V = \coprod_{i=1}^t V^{(i)}$ of V satisfying (7.1.1). We put

$$A^{(ij)} = \mathrm{Hom}_F(V^{(j)}, V^{(i)}), \quad 1 \le i, j \le t,$$

and view this as a subspace of A. Thus we get a "block decomposition" of the algebra A,

$$A = \coprod_{1 \le i,j \le t} A^{(ij)},$$

in which the spaces $A^{(i)} = A^{(ii)}$ are algebras. Write $\mathbf{1}^{(i)}$ for the projection $V \to V^{(i)}$ with kernel $\coprod_{j \ne i} V^{(j)}$, so that in fact $\mathbf{1}^{(i)}$ is the identity element of $A^{(i)}$, and

$$A^{(ij)} = \mathbf{1}^{(i)}.A.\mathbf{1}^{(j)}, \quad 1 \le i, j \le t.$$

The defining condition (7.1.1)*(i)* implies that $\mathbf{1}^{(i)} L_k = L_k^{(i)} \subset L_k$, $k \in \mathbb{Z}$, so that

(7.1.4)
$$\mathbf{1}^{(i)} \in \mathfrak{A}, \quad 1 \le i \le t.$$

This simple fact has many useful consequences.

(7.1.5) Lemma: *In the above situation, write* $\mathfrak{P} = \mathrm{rad}(\mathfrak{A})$, *and let* $m \in \mathbb{Z}$. *Then* $\mathfrak{P}^m \cap A^{(ij)} = \mathbf{1}^{(i)}.\mathfrak{P}^m.\mathbf{1}^{(j)}$ *and*

$$\mathfrak{P}^m = \coprod_{1 \le i,j \le t} \mathfrak{P}^m \cap A^{(ij)}.$$

Proof: (7.1.4) implies $\mathbf{1}^{(i)}.\mathfrak{P}^m.\mathbf{1}^{(j)} \subset \mathfrak{P}^m$, whence $\mathbf{1}^{(i)}.\mathfrak{P}^m.\mathbf{1}^{(j)} \subset \mathfrak{P}^m \cap A^{(ij)}$. The opposite containment is obvious, and the lemma follows. ∎

The sets $\mathfrak{P}^m \cap A^{(ij)}$ are $\mathfrak{o}_F$-lattices, and $\mathfrak{A}^{(i)} = \mathfrak{A} \cap A^{(i)}$ is an $\mathfrak{o}_F$-order in the algebra $A^{(i)}$. Also, $\mathfrak{P}^m \cap A^{(ij)}$ is an $(\mathfrak{A}^{(i)}, \mathfrak{A}^{(j)})$-bimodule. We can describe the orders $\mathfrak{A}^{(i)}$ completely.

(7.1.6) Proposition: *In the notation above, we have*

$$\mathfrak{A}^{(i)} = \mathrm{End}^0_{\mathfrak{o}_F}(\{L_{jt}^{(i)} : j \in \mathbb{Z}\}), \qquad 1 \le i \le t.$$

In particular, $\mathfrak{A}^{(i)}$ *is a hereditary* $\mathfrak{o}_F$*-order in* $A^{(i)} = \mathrm{End}_F(V^{(i)})$ *and* $e(\mathfrak{A}^{(i)}|\mathfrak{o}_F) = e(\mathfrak{A}|\mathfrak{o}_F)/t$.

Proof: It is clear that $\mathfrak{A}^{(i)} \subset \mathrm{End}^0_{\mathfrak{o}_F}(\{L_{jt}^{(i)}\})$. For the reverse containment, take $x \in \mathrm{End}^0_{\mathfrak{o}_F}(\{L_{jt}^{(i)}\})$. Thus $x \in A^{(i)}$ and $xL_{jt}^{(i)} \subset L_{jt}^{(i)}$, $j \in \mathbb{Z}$.

We have to show $x \in \mathfrak{A}$, which amounts to showing $xL_k \subset L_k$, $k \in \mathbb{Z}$. Now, $x = \mathbf{1}^{(i)}x\mathbf{1}^{(i)}$, so $xL_k = xL_k^{(i)}$. We use (7.1.3) to get $L_k^{(i)} = L_{(j+1)t}^{(i)}$, where $j = [(k-i)/t]$. Thus $xL_k^{(i)} = xL_{(j+1)t}^{(i)} \subset L_{(j+1)t}^{(i)} \subset L_{(j+1)t}$. However, $(j+1)t \geq k$, so $xL_k \subset L_k$, as required. ■

(7.1.7) Proposition: *In the situation above, put* $\mathfrak{P}^{(i)} = \mathrm{rad}(\mathfrak{A}^{(i)})$, $1 \leq i \leq t$, *and take* $m \in \mathbb{Z}$. *Then* $\mathfrak{P}^m \cap A^{(i)} = \mathbf{1}^{(i)}.\mathfrak{P}^m.\mathbf{1}^{(i)} = (\mathfrak{P}^{(i)})^{m'}$, *where*

$$m' = \left[\tfrac{m+t-1}{t}\right].$$

Proof: We start by showing that $\mathbf{1}^{(i)}.\mathfrak{P}^m.\mathbf{1}^{(i)} \subset \mathfrak{P}^{(i)m'}$. This amounts to showing that $\mathbf{1}^{(i)}.\mathfrak{P}^m.\mathbf{1}^{(i)}L_{jt}^{(i)} \subset L_{(j+m')t}^{(i)}$, $j \in \mathbb{Z}$. (7.1.3) gives $L_{jt}^{(i)} = L_{i+jt-1}^{(i)}$, so $\mathbf{1}^{(i)}.\mathfrak{P}^m.\mathbf{1}^{(i)}L_{jt}^{(i)} \subset L_{i+jt+m-1}^{(i)}$. On the other hand, $L_{(j+m')t}^{(i)} = L_{i+(j+m'-1)t}^{(i)}$, so we just have to check that $i+(j+m'-1)t \leq i+jt+m-1$, which is trivial.

To get the opposite containment, we need to show that $\mathfrak{P}^{(i)m'}L_j \subset L_{j+m}$, $j \in \mathbb{Z}$. This is the same as showing that $\mathfrak{P}^{(i)m'}L_j^{(i)} \subset L_{j+m}^{(i)}$. To start with, we have $\mathfrak{P}^{(i)m'}L_j^{(i)} = L_{i+kt-1}^{(i)}$, where $k = m'+1+[(j-i)/t]$. On the other hand, $L_{j+m}^{(i)} = L_{i+t\ell}^{(i)}$, with $\ell = [(j+m-i)/t]$. It is therefore enough to show that $t\ell \leq i+kt-1$, which is easy. ■

It is now appropriate to record another elementary identity.

(7.1.8) Lemma: *Let* $t \in \mathbb{Z}$, $t \geq 1$. *For* $m \in \mathbb{Z}$, *define* $m' \in \mathbb{Z}$ *by*

$$m' = \left[\tfrac{m+t-1}{t}\right].$$

Let $n \in \mathbb{Z}$ *be divisible by* t, *so that in particular* $n = n't$. *Then*

$$\left(\left[\tfrac{n}{2}\right]+1\right)' = \left[\tfrac{n'}{2}\right]+1 \quad \text{and} \quad \left[\tfrac{n+1}{2}\right]' = \left[\tfrac{n'+1}{2}\right].$$

Proof: Trivial. ■

The configuration of (7.1.1) behaves well, in certain circumstances, with respect to change of base field. We exhibit this formally.

(7.1.9) Proposition: *Let* $\mathfrak{A}$ *be a hereditary* $\mathfrak{o}_F$*-order in* A, *and* $V = V^{(1)} \oplus V^{(2)} \oplus \ldots \oplus V^{(t)}$ *an* F*-decomposition of* V *subordinate to* $\mathfrak{A}$. *Let* K/F *be a subfield of* A *with* $K^\times \subset \mathfrak{K}(\mathfrak{A})$. *Suppose also that the* $V^{(i)}$ *are all* K*-subspaces of* V. *Then* $V = \coprod_{i=1}^t V^{(i)}$ *is a* K*-decomposition of* V *subordinate to the hereditary* $\mathfrak{o}_K$*-order* $\mathfrak{A} \cap \mathrm{End}_K(V)$.

There is nothing to prove here.

Now suppose we have a simple stratum $[\mathfrak{A}, n, 0, \beta]$ in A. We use our customary notation $\mathfrak{P} = \mathrm{rad}(\mathfrak{A})$, $E = F[\beta]$, $B = \mathrm{End}_E(V)$, $\mathfrak{B} = \mathfrak{A} \cap B$, $\mathfrak{Q} = \mathrm{rad}(\mathfrak{B})$, etc. We also write $\mathcal{L} = \{L_k : k \in \mathbb{Z}\}$ for the $\mathfrak{o}_F$-lattice chain in V defining $\mathfrak{A}$. We let $V^{(1)} \oplus V^{(2)} \oplus \ldots \oplus V^{(t)}$ be an E-decomposition of V subordinate to $\mathfrak{B}$. We use the notation $A^{(ij)}$, $A^{(i)}$, $\mathbf{1}^{(i)}$ etc., as above. Note that here we have $\mathbf{1}^{(i)} \in \mathfrak{B}$, $1 \leq i \leq t$.

We now pick a (W, E)-decomposition $\mathfrak{A} = \mathfrak{A}(E) \otimes_{\mathfrak{o}_E} \mathfrak{B}$ of $\mathfrak{A}$, as in **(1.2)**, which respects this extra structure. To do this, we choose an $\mathfrak{o}_E$-basis $\mathcal{W}^{(i)}$ of each $\mathfrak{o}_E$-chain $\{L_{jt}^{(i)}\}$ in such a way that the union $\mathcal{W}$ of the $\mathcal{W}^{(i)}$ is an $\mathfrak{o}_E$-basis of the lattice L_0. When suitably ordered, the set $\mathcal{W}$ is an $\mathfrak{o}_E$-basis of the chain $\mathcal{L}$. Let $W^{(i)}$ denote the F-span of $\mathcal{W}^{(i)}$, and $W = \coprod W^{(i)}$. Thus $V = E \otimes_F W$, and this restricts to identify $V^{(i)} = E \otimes_F W^{(i)}$, for each i. We view V and $V^{(i)}$ as left modules over the algebra $A(E) = \mathrm{End}_F(E)$, via the natural action of $A(E)$ on the tensor factor E. (This corresponds to the algebra embedding $A(E) \to A$ implied by the choice of W.) Moreover, the projections $\mathbf{1}^{(i)}$ commute with this left action of $A(E)$. Thus, if K/F is a subfield of $A(E)$, then the $V^{(i)}$ are K-subspaces of V. Further, our choice of bases gives a (W, E)-decomposition $\mathfrak{A} = \mathfrak{A}(E) \otimes_{\mathfrak{o}_E} \mathfrak{B}$, where $\mathfrak{A}(E) = \mathrm{End}^0_{\mathfrak{o}_F}(\{\mathfrak{p}_E^j : j \in \mathbb{Z}\})$, and therefore

$$\mathfrak{A} \cap A^{(ij)} = \mathbf{1}^{(i)}.\mathfrak{A}.\mathbf{1}^{(j)} = \mathfrak{A}(E) \otimes_{\mathfrak{o}_E} \mathfrak{B}^{(ij)},$$

where $\mathfrak{B}^{(ij)} = \mathfrak{B} \cap A^{(ij)} = \mathbf{1}^{(i)}\mathfrak{B}\mathbf{1}^{(j)}$, $1 \leq i, j \leq t$. We use (1.2.11) to view $\mathfrak{K}(\mathfrak{A}(E))$ as a subgroup of $\mathfrak{K}(\mathfrak{A})$, so in all we have:

(7.1.10) Proposition: *Let $V = \coprod_{i=1}^t V^{(i)}$ be an E-decomposition of V subordinate to $\mathfrak{B}$. Let $\mathfrak{A} = \mathfrak{A}(E) \otimes_{\mathfrak{o}_E} \mathfrak{B}$ be a (W, E)-decomposition of $\mathfrak{A}$ given by an $\mathfrak{o}_E$-basis of the chain $\mathcal{L}$ which is the union of $\mathfrak{o}_E$-bases of the chains $\{L_{kt}^{(i)}\}$. Let K/F be a subfield of $A(E)$ such that $K^\times \subset \mathfrak{K}(\mathfrak{A}(E))$, acting on V via the embedding $A(E) \to A$ implied by the choice of basis. Then $V = \coprod_{i=1}^t V^{(i)}$ is a K-decomposition of V subordinate to the $\mathfrak{o}_K$-order $\mathfrak{A} \cap \mathrm{End}_K(V)$.*

The purpose of this apparatus is to enable us to perform, in the present context, the sort of inductive argument necessary to treat the rings $\mathfrak{H}(\beta, \mathfrak{A})$, $\mathfrak{J}(\beta, \mathfrak{A})$. We can now express and prove the results we seek. First we summarise our notation.

(7.1.11) Notation: $[\mathfrak{A}, n, 0, \beta]$ is a simple stratum in $A = \mathrm{End}_F(V)$, $E = F[\beta]$, $B = \mathrm{End}_E(V)$, $\mathfrak{B} = \mathfrak{A} \cap B$. Also, $V = V^{(1)} \oplus V^{(2)} \oplus \ldots \oplus V^{(t)}$ is an E-decomposition of V subordinate to the $\mathfrak{o}_E$-order $\mathfrak{B}$. We put $A^{(ij)} = \mathrm{Hom}_F(V^{(j)}, V^{(i)})$, $A^{(i)} = A^{(ii)}$, $\mathfrak{A}^{(i)} = \mathfrak{A} \cap A^{(i)}$, $\mathbf{1}^{(i)} = 1_{A^{(i)}}$, $1 \leq i, j \leq t$, and use similar notations attached to B.

(7.1.12) Proposition: *Using the notation* (7.1.11), *let* $k \in \mathbb{Z}$, $k \geq 0$, *and set* $k' = \left[\frac{k+t-1}{t}\right]$. *Then:*

(i)

$$\mathfrak{H}^k(\beta,\mathfrak{A}) \cap A^{(ij)} = \mathbf{1}^{(i)}.\mathfrak{H}^k(\beta,\mathfrak{A}).\mathbf{1}^{(j)}, \quad 1 \leq i,j \leq t,$$

and similarly for $\mathfrak{J}(\beta,\mathfrak{A})$.

(ii)

$$\mathfrak{H}^k(\beta,\mathfrak{A}) = \coprod_{1\leq i,j\leq t} \mathfrak{H}^k(\beta,\mathfrak{A}) \cap A^{(ij)},$$

and likewise for $\mathfrak{J}(\beta,\mathfrak{A})$.

(iii) $\mathfrak{H}^k(\beta,\mathfrak{A}) \cap A^{(i)} = \mathfrak{H}^{k'}(\beta,\mathfrak{A}^{(i)})$ *and* $\mathfrak{J}^k(\beta,\mathfrak{A}) \cap A^{(i)} = \mathfrak{J}^{k'}(\beta,\mathfrak{A}^{(i)})$, $1 \leq i \leq t$.

Proof: The projections $\mathbf{1}^{(i)}$ lie in $\mathfrak{B}$, and $\mathfrak{H}^k(\beta,\mathfrak{A})$ is a $(\mathfrak{B},\mathfrak{B})$-bimodule. Thus $\mathbf{1}^{(i)}.\mathfrak{H}^k(\beta,\mathfrak{A}).\mathbf{1}^{(j)}$ is contained in $\mathfrak{H}^k(\beta,\mathfrak{A}) \cap A^{(ij)}$, and the assertions *(i)*, *(ii)* (for $\mathfrak{H}$) follow immediately. The corresponding assertions for $\mathfrak{J}$ are proved similarly.

To prove *(iii)*, we first observe that n is divisible by t, since it is divisible by $e = e(\mathfrak{B}|\mathfrak{o}_E)$, so that $n' = n/t$, and the strata $[\mathfrak{A}^{(i)}, n', 0, \beta]$ are all simple by (1.4.13). We proceed by induction along β, treating only the $\mathfrak{H}$-case, the $\mathfrak{J}$-case being exactly the same.

Suppose to start with that β is minimal over F. We may as well take $k \leq [n/2]$, since otherwise $\mathfrak{H}^k(\beta,\mathfrak{A}) = \mathfrak{P}^k$, $\mathfrak{H}^{k'}(\beta,\mathfrak{A}^{(i)}) = \mathfrak{P}^{(i)k'}$ by (7.1.8), and the assertion follows from (7.1.7). Setting $\mathfrak{Q} = \mathrm{rad}(\mathfrak{B})$, we therefore have $\mathfrak{H}^k(\beta,\mathfrak{A}) = \mathfrak{Q}^k + \mathfrak{P}^{[n/2]+1}$, and

$$\begin{aligned}\mathbf{1}^{(i)}.\mathfrak{H}^k(\beta,\mathfrak{A}).\mathbf{1}^{(i)} &= \mathbf{1}^{(i)}\mathfrak{Q}^k.\mathbf{1}^{(i)} + \mathbf{1}^{(i)}\mathfrak{P}^{[n/2]+1}.\mathbf{1}^{(i)} \\ &= \mathfrak{Q}^{(i)k'} + \mathfrak{P}^{(i)[n'/2]+1} = \mathfrak{H}^{k'}(\beta,\mathfrak{A}^{(i)}),\end{aligned}$$

where $\mathfrak{Q}^{(i)} = \mathrm{rad}(\mathfrak{B}^{(i)})$, by (7.1.7), (7.1.8).

In the general case, we take a (W,E)-decomposition $\mathfrak{A} = \mathfrak{A}(E)\otimes_{\mathfrak{o}_E}\mathfrak{B}$ given by an $\mathfrak{o}_E$-basis of $\mathcal{L}$ which is the union of $\mathfrak{o}_E$-bases of the $\{L_{jt}^{(i)}\}$, and use (7.1.10). Let $r = -k_0(\beta,\mathfrak{A})$, $r(E) = -k_0(\beta,\mathfrak{A}(E)) = r/e$, $r' = -k_0(\beta,\mathfrak{A}^{(i)}) = r/t$. Choose a simple stratum $[\mathfrak{A}(E), n/e, r(E), \gamma]$ equivalent to $[\mathfrak{A}(E), n/e, r(E), \beta]$. The strata $[\mathfrak{A}, n, r, \gamma]$, $[\mathfrak{A}^{(i)}, n', r', \gamma]$ are then simple and equivalent respectively to $[\mathfrak{A}, n, r, \beta]$, $[\mathfrak{A}^{(i)}, n', r', \beta]$. Again we may as well take $k \leq [r/2]$, so that $\mathfrak{H}^k(\beta,\mathfrak{A}) = \mathfrak{Q}^k + \mathfrak{H}^{[r/2]+1}(\gamma,\mathfrak{A})$. This gives $\mathbf{1}^{(i)}.\mathfrak{H}^k(\beta,\mathfrak{A}).\mathbf{1}^{(i)} = \mathfrak{Q}^{(i)k'} + \mathbf{1}^{(i)}\mathfrak{H}^{[r/2]+1}(\gamma,\mathfrak{A}).\mathbf{1}^{(i)}$. However, by (7.1.10), $V = \coprod_i V^{(i)}$ is an $F[\gamma]$-decomposition of V subordinate to $\mathfrak{A} \cap \mathrm{End}_{F[\gamma]}(V)$. We can therefore apply our inductive hypothesis to get

$$\mathbf{1}^{(i)}\mathfrak{H}^{[r/2]+1}(\gamma,\mathfrak{A})\mathbf{1}^{(i)} = \mathfrak{H}^{([r/2]+1)'}(\gamma,\mathfrak{A}^{(i)}) = \mathfrak{H}^{[r'/2]+1}(\gamma,\mathfrak{A}^{(i)})$$

by (7.1.8). The result follows. ■

We continue using the notation (7.1.11), and define

(7.1.13) *(i)* $P = G \cap \prod_{1 \le i \le j \le t} A^{(ij)}$;

(ii) $M = M(A)^\times$, *where* $M(A) = \prod_{i=1}^{t} A^{(i)}$;

(iii) $\mathbb{N} = \prod_{1 \le i < j \le t} A^{(ij)}$, $U = 1 + \mathbb{N}$;

(iv) $\mathbb{N}^- = \prod_{1 \le j < i \le t} A^{(ij)}$, $U^- = 1 + \mathbb{N}^-$.

Thus $M(A)$ is a subalgebra of A, and P is a parabolic subgroup of G with unipotent radical U and Levi decomposition $P = MU$. We also have the opposite parabolic subgroup $P^- = MU^-$. We shall only use the group P^- very occasionally, but it should be observed that, in everything which follows, the groups P, P^- play symmetric roles.

The additive decompositions of the lattices $\mathfrak{H}^k$, $\mathfrak{J}^k$ given by (7.1.12) imply multiplicative decompositions ("Iwahori decompositions") for the groups $H(\beta, \mathfrak{A})$, $J(\beta, \mathfrak{A})$.

(7.1.14) Theorem: *Using the notation* (7.1.11), (7.1.13), *and letting* $\mathcal{G}$ *denote any of the groups* $H(\beta, \mathfrak{A})$, $H^1(\beta, \mathfrak{A})$, $J(\beta, \mathfrak{A})$, $J^1(\beta, \mathfrak{A})$, *we have*

$$\mathcal{G} = (\mathcal{G} \cap U^-).(\mathcal{G} \cap M).(\mathcal{G} \cap U);$$
$$\mathcal{G} \cap P = (\mathcal{G} \cap M).(\mathcal{G} \cap U).$$

Moreover, the decomposition of an element of $\mathcal{G}$ *implied by these expressions is unique.*

Proof: We start with a lemma.

(7.1.15) Lemma: *We have* $\mathfrak{A} \cap \mathbb{N} = \mathfrak{P} \cap \mathbb{N}$ *and* $\mathfrak{A} \cap \mathbb{N}^- = \mathfrak{P} \cap \mathbb{N}^-$. *In particular,* $J(\beta, \mathfrak{A}) \cap U = J^1(\beta, \mathfrak{A}) \cap U$, $J(\beta, \mathfrak{A}) \cap U^- = J^1(\beta, \mathfrak{A}) \cap U^-$, *and likewise for* $H(\beta, \mathfrak{A})$.

Proof: By (7.1.5), it is enough to show that $\mathfrak{A} \cap A^{(ij)} \subset \mathfrak{P}$ whenever $i \ne j$. So, we take $i \ne j$, let $x \in \mathfrak{A} \cap A^{(ij)}$, and we have to show that $xL_k \subset L_{k+1}$ for all $k \in \mathbb{Z}$. Now, $xL_k^{(j)} \subset L_k^{(i)}$ while $xL_k^{(\ell)} = \{0\}$ if $\ell \ne j$. If $k \not\equiv j - 1 \pmod t$, we have $L_k^{(j)} = L_{k+1}^{(j)}$ by (7.1.3), so here $xL_k = xL_k^{(j)} = L_{k+1}^{(j)} \subset L_{k+1}$. If, however, $k \equiv j - 1 \pmod t$, we must have $k \not\equiv i - 1 \pmod t$, so $xL_k \subset L_k^{(i)} = L_{k+1}^{(i)} \subset L_{k+1}$, as required. ■

Now we prove (7.1.14) in the case $\mathcal{G} = J = J(\beta, \mathfrak{A})$. Take $x \in J$, and write $x = (x_{ij})$, where $x_{ij} \in A^{(ij)}$. We have $x_{ij} \in \mathfrak{J}^1(\beta, \mathfrak{A})$ if $i \ne j$ by (7.1.15), while $x_{ii} \in \mathfrak{J}(\beta, \mathfrak{A}^{(i)})$. The same applies to $x^{-1} \in J$, and so $(x^{-1})_{ii} x_{ii} \equiv 1 \pmod{\mathfrak{J}^1}$. Thus $x_{ii} \in \mathfrak{J}(\beta, \mathfrak{A}^{(i)}) \cap \boldsymbol{U}(\mathfrak{A}^{(i)}) = J(\beta, \mathfrak{A}^{(i)})$. The decomposition for J now follows easily using "elementary row and column operations". The decomposition for J^1 now follows from that for J and (7.1.15). The cases H, H^1 are exactly the same. ■

There are various useful and easily derived refinements of this. Immediately, we have

(7.1.16)

$$\text{(i)} \qquad J \cap M(A) = J \cap M = \prod_{i=1}^{t} J(\beta, \mathfrak{A}^{(i)});$$

$$\text{(ii)} \qquad J^1 \cap M = \prod_{i=1}^{t} J^1(\beta, \mathfrak{A}^{(i)}),$$

while the module structures implied by, e.g., (3.1.13) give

(7.1.17)

$$\text{(i)} \qquad J \cap M.H^1 = (H^1 \cap U^-).(J \cap M).(H^1 \cap U),$$

$$\text{(ii)} \qquad J \cap P.H^1 = (H^1 \cap U^-).(J \cap M).(J^1 \cap U).$$

Here, we abbreviate $J = J(\beta, \mathfrak{A})$, $J^1 = J^1(\beta, \mathfrak{A})$, and so on. Further, standard theory gives results like:

(7.1.18) *The product map*

$$(J^1 \cap U^-) \times (J \cap M) \times (J^1 \cap U) \to J$$

is a homeomorphism.

To conclude this section, we start the analysis of various representations defined in terms of these decompositions.

(7.1.19) Proposition: *Use the notation* (7.1.11), (7.1.13). *Let* $\theta \in \mathcal{C}(\mathfrak{A}, 0, \beta)$. *Then the characters* $\theta \mid (H^1(\beta, \mathfrak{A}) \cap U^-)$, $\theta \mid (H^1(\beta, \mathfrak{A}) \cap U)$ *are both null. After identifying* $H^1(\beta, \mathfrak{A}) \cap M$ *with* $\prod_i H^1(\beta, \mathfrak{A}^{(i)})$, *we have*

$$\theta \mid (H^1(\beta, \mathfrak{A}) \cap M) = \theta^{(1)} \otimes \theta^{(2)} \otimes \ldots \otimes \theta^{(t)},$$

where $\theta^{(i)} \in \mathcal{C}(\mathfrak{A}^{(i)}, 0, \beta)$, *and indeed*

$$\theta^{(i)} = \boldsymbol{\tau}_{\mathfrak{A}, \mathfrak{A}^{(i)}, \beta, 0}(\theta),$$

in the notation of (3.6.13).

Proof: We proceed by induction along β. Assume first that β is minimal over F, and take $x \in \mathfrak{H}^1(\beta, \mathfrak{A}) \cap A^{(ij)}$ for some pair $i \neq j$. Writing $\mathfrak{Q} = \mathrm{rad}(\mathfrak{B})$ we have $x \in \mathbf{1}^{(i)}\mathfrak{H}^1(\beta, \mathfrak{A})\mathbf{1}^{(j)} = \mathbf{1}^{(i)}\mathfrak{Q}.\mathbf{1}^{(j)} + \mathbf{1}^{(i)}\mathfrak{P}^{[n/2]+1}.\mathbf{1}^{(j)}$. We can therefore write $1 + x = (1 + y)(1 + z)$ with $y \in \mathbf{1}^{(i)}\mathfrak{Q}.\mathbf{1}^{(j)}$, $z \in \mathbf{1}^{(i)}\mathfrak{P}^{[n/2]+1}.\mathbf{1}^{(j)}$. There is a character χ of $E^\times$ such that $\theta(1+y)(1+z) = \chi(\det_B(1 + y)).\psi_\beta(1 + z)$. Since $(1 + y)$ is a unipotent element of $B^\times$, we have $\det_B(1+y) = 1$, so the first factor here is null. The second factor is $\psi_F(\mathrm{tr}_{A/F}(\beta z))$, where ψ_F is our preordained character of F. The element βz lies in $A^{(ij)}$, whence is nilpotent and has zero trace. Therefore θ is null on $H^1 \cap 1 + A^{(ij)}$ when $i \neq j$. The groups $1 + \mathfrak{H} \cap A^{(ij)}$, for $i > j$, generate $H^1 \cap U^-$, so $\theta \mid H^1 \cap U^-$ is null, and likewise for $\theta \mid H^1 \cap U$.

For the second assertion in this case, take $x \in H^1 \cap M$, so that $x = (x_1, \dots, x_t)$, with $x_i \in H^1(\beta, \mathfrak{A}^{(i)})$. Put $n' = n/t$, so that $x_i = y_i z_i$ with $y_i \in 1 + \mathfrak{Q} \cap A^{(i)}$, $z_i \in 1 + \mathfrak{P}^{(i)[n'/2]+1}$. Let χ be a character of $E^\times$ such that $\theta \mid (1+\mathfrak{Q}) = \chi \circ \det_B$. Then

$$\theta(x) = \prod_i \chi(\det_{B^{(i)}}(y_i)).\prod_i \psi_{A^{(i)},\beta}(z_i) = \prod_i \theta^{(i)}(x_i),$$

where $\theta^{(i)} = \tau_{\mathfrak{A},\mathfrak{A}^{(i)}}(\theta)$, as required.

In the general case, we choose a common (W, E)-decomposition of the $\mathfrak{A}^{(i)}$, as in (7.1.10). Put $r = -k_0(\beta, \mathfrak{A})$ and choose a simple stratum $[\mathfrak{A}, n, r, \gamma]$ equivalent to $[\mathfrak{A}, n, r, \beta]$ with $F[\gamma]^\times \subset \mathfrak{K}(\mathfrak{A}(E))$. For $i \neq j$, we have $H^1 \cap (1+A^{(ij)}) = (1+\mathfrak{Q}\cap A^{(ij)}).(1+\mathfrak{H}^{[r'/2]+1}(\gamma, \mathfrak{A})\cap A^{(ij)})$, where we put $r' = r/t = -k_0(\beta, \mathfrak{A}^{(i)})$. As in the minimal case, θ is null on the first factor. On the second, θ takes the form $\theta_0 \psi_c$, where $\theta_0 \in \mathcal{C}(\mathfrak{A}, [r/2], \gamma)$ and $c = \beta - \gamma$. The character θ_0 is null on $(1+\mathfrak{H}^{[r'/2]+1}(\gamma, \mathfrak{A}) \cap A^{(ij)})$ by inductive hypothesis, while the fact that $c \in A(E)$ implies $cA^{(ij)} \subset A^{(ij)}$. Thus ψ_c is also null on this factor, and the assertions follow as before.

To get the factorisation of $\theta \mid H^1 \cap M$ in this case, we write $\theta \mid U^1(\mathfrak{B}) = \chi \circ \det_B$, and $\theta \mid H^{[r/2]+1} = \theta_0 \psi_c$, where $\theta_0 \in \mathcal{C}(\mathfrak{A}, [r/2], \gamma)$. By (3.3.21), θ_0 is the restriction of some $\theta_1 \in \mathcal{C}(\mathfrak{A}, 0, \gamma)$. We can therefore use induction to show that $\theta_1 \mid H^1(\gamma, \mathfrak{A}) \cap M$ factors in the required manner. The same therefore applies to its restriction θ_0. The other components $\chi \circ \det_B$, ψ_c factor for the same reasons as in the first case. The result now follows. ■

(7.2) Iwahori factorisation of a simple type

The main function of the groups $J(\beta, \mathfrak{A})$, $J^1(\beta, \mathfrak{A})$ has been to carry various representations attached to a character $\theta \in \mathcal{C}(\mathfrak{A}, 0, \beta)$. We now investigate the behaviour of these representations on various subgroups suggested by the decompositions of **(7.1)**. We continue to use the notation (7.1.11), (7.1.13), but we *sometimes* additionally impose:

(7.2.1) Notation: *$\mathfrak{A}$ is a principal order and $t = e = e(\mathfrak{B}|\mathfrak{o}_E)$.*

This condition only operates in this section when it is explicitly invoked. When it does, we are exactly in the situation of **(5.5)**, so we also employ the notation introduced there.

We start, without (7.2.1), by considering the nondegenerate alternating form $\boldsymbol{k}_\theta$ introduced in **(3.4)**:

$$\boldsymbol{k}_\theta(x,y) = \theta[x,y], \quad x,y \in J^1(\beta,\mathfrak{A}).$$

(7.1.14) gives a direct product decomposition (using the obvious abbreviations):

$$\textbf{(7.2.2)} \qquad J^1/H^1 = \frac{J^1\cap U^-}{H^1\cap U^-} \times \frac{J^1\cap M}{H^1\cap M} \times \frac{J^1\cap U}{H^1\cap U},$$

and we can identify

$$\frac{J^1\cap M}{H^1\cap M} = \prod_{i=1}^{t} \frac{J^1(\beta,\mathfrak{A}^{(i)})}{H^1(\beta,\mathfrak{A}^{(i)})}.$$

(7.1.19) now gives us:

(7.2.3) Proposition: *(i) The subspaces $J^1\cap U^-/H^1\cap U^-$, $J^1\cap U/H^1\cap U$ of J^1/H^1 are both totally isotropic for the form $\boldsymbol{k}_\theta$, and orthogonal to the subspace $(J^1\cap M)/(H^1\cap M)$.*

(ii) The restriction of $\boldsymbol{k}_\theta$ to the group

$$(J^1\cap M)/(H^1\cap M) = \prod_i J^1(\beta,\mathfrak{A}^{(i)})/H^1(\beta,\mathfrak{A}^{(i)})$$

is the orthogonal sum of the pairings $\boldsymbol{k}_{\theta^{(i)}}$, $1\le i\le t$, (where $\theta^{(i)}$ is as in (7.1.19)*) and is, in particular, nondegenerate.*

(iii) We have an orthogonal sum decomposition

$$\frac{J^1}{H^1} = \frac{J^1\cap M}{H^1\cap M} \perp \left(\frac{J^1\cap U^-}{H^1\cap U^-} \times \frac{J^1\cap U}{H^1\cap U}\right)$$

In particular, the restriction of $\boldsymbol{k}_\theta$ to $(J^1\cap U^-)/(H^1\cap U^-)\times(J^1\cap U)/(H^1\cap U))$ is nondegenerate.

Proof: Part *(i)* is a direct consequence of (7.1.19) and the fact that $J\cap M$ normalises both $J\cap U$ and $J\cap U^-$. The various blocks $J^1(\beta,\mathfrak{A}^{(i)})$ in $J^1\cap M$ commute with each other, and the pairings $\boldsymbol{k}_{\theta^{(i)}}$ are nondegenerate. This gives *(ii)*. Part *(iii)* is now just a formal property of alternating bilinear forms. ■

We also have many other canonical isomorphisms like

$$\frac{J^1\cap U}{H^1\cap U} \cong \frac{(J^1\cap U).H^1}{H^1},$$

which we use without specific mention.

The orthogonal sum decompositions of (7.2.3) translate directly into properties of various representations:

(7.2.4) Proposition: *Let η (resp. $\eta^{(i)}$) be the unique irreducible representation of $J^1(\beta, \mathfrak{A})$ (resp. $J^1(\beta, \mathfrak{A}^{(i)})$) containing θ (resp. $\theta^{(i)}$). Write $\mathcal{G}$ for the finite abelian group $(J^1(\beta,\mathfrak{A}) \cap U)/(H^1(\beta,\mathfrak{A}) \cap U)$, and let $\phi \in \widehat{\mathcal{G}}$ (viewed as a character of $J^1 \cap U$ null on $H^1 \cap U$).*

(i) There is a unique irreducible representation η_ϕ of $(J^1(\beta,\mathfrak{A}) \cap P)H^1(\beta,\mathfrak{A})$ such that $\eta_\phi \mid J^1 \cap M = \eta^{(1)} \otimes \eta^{(2)} \otimes \ldots \otimes \eta^{(t)}$, $\eta_\phi \mid H^1(\beta,\mathfrak{A})$ is a multiple of θ, and $\eta_\phi \mid J^1(\beta,\mathfrak{A}) \cap U$ is a multiple of ϕ.

(ii) $\eta \mid (J^1(\beta,\mathfrak{A}) \cap P)H^1(\beta,\mathfrak{A}) = \prod_{\phi \in \widehat{\mathcal{G}}} \eta_\phi$.

(iii) $\mathrm{Ind}(\eta_\phi : (J^1 \cap P)H^1, J^1) = \eta$ for every $\phi \in \widehat{\mathcal{G}}$.

(iv) Given $\phi_1, \phi_2 \in \widehat{\mathcal{G}}$, there is a unique $x \in J^1 \cap U^- / H^1 \cap U^-$ such that $\eta_{\phi_2} = (\eta_{\phi_1})^x$.

Proof: This comes from the representation theory of extra-special p-groups of class two. Briefly, write $\mathcal{K} = \mathrm{Ker}(\theta \mid H^1)$. Then $\mathcal{K}$ is a normal subgroup of J^1, and (3.4.1) implies that $H^1/\mathcal{K}$ is the centre of $J^1/\mathcal{K}$. Moreover, θ defines a faithful character of $H^1/\mathcal{K}$. Temporarily write $\mathcal{V} = J^1/H^1$. The totally isotropic subspaces of $\mathcal{V}$ (relative to the form $\boldsymbol{k}_\theta$) are then the images in $\mathcal{V}$ of the abelian subgroups of $J^1/\mathcal{K}$ containing the centre $H^1/\mathcal{K}$. The representation η of **(5.1)** is constructed as follows. Take any maximal totally isotropic subspace $\mathcal{U}$ of $\mathcal{V}$ and any character ϑ of the inverse image $\overline{\mathcal{U}}$ of $\mathcal{U}$ in J^1 such that $\vartheta \mid H^1 = \theta$. Then η is the representation of J^1 induced by ϑ.

In the present situation, we take $\mathcal{U}$ of the form $\mathcal{U}' \oplus \mathcal{U}''$, where $\mathcal{U}'$ is a maximal totally isotropic subspace of the image of $J^1 \cap M$ in $\mathcal{V}$, and $\mathcal{U}''$ is the image of $J^1 \cap U$. The assertions of the Proposition are now given by elementary induction-restriction arguments. ■

In particular, using the same notation, we have

(7.2.5) *The representations $\eta_\phi \mid J^1(\beta,\mathfrak{A}) \cap P$, $\eta_\phi \mid J^1(\beta,\mathfrak{A}) \cap M$ are both irreducible, and*

$$\eta_\phi \mid J^1(\beta,\mathfrak{A}) \cap M = \eta^{(1)} \otimes \eta^{(2)} \otimes \ldots \otimes \eta^{(t)}.$$

The most interesting case for us will be the one where $\phi = \phi_0$ is the trivial character of $J^1 \cap U$. In this case, we write

(7.2.6)

$$\eta_P = \eta_{\phi_0} \quad \text{(as a representation of } (J^1(\beta,\mathfrak{A}) \cap P)H^1(\beta,\mathfrak{A})),$$

(7.2.7) $\qquad \eta_U = \eta_P \mid J^1(\beta,\mathfrak{A}) \cap M = \eta^{(1)} \otimes \eta^{(2)} \otimes \ldots \otimes \eta^{(t)}.$

We can equally regard η_U as a representation of $J^1 \cap P/J^1 \cap U$.

We can also describe these representations η_P, η_U somewhat differently. Observe that $(J^1 \cap P)H^1$ is a normal subgroup of J^1 since it contains H^1. Temporarily write θ_0 for that character of $(J^1 \cap U)H^1$ which is null on $J^1 \cap U$ and agrees with θ on H^1. This character is stabilised by the group $(J^1 \cap P)H^1$ (and even by $(J \cap P)H^1$), and the space of θ_0-vectors in η is identical with the space of $(J^1 \cap U)$-fixed vectors. Thus η_P can be viewed as the natural representation of $H^1(J \cap P)$ on the space of $(J^1 \cap U)$-fixed vectors in η. Restricting to $J^1 \cap M$, we get

(7.2.8) Corollary: *The natural representation of the group*

$$(J^1(\beta, \mathfrak{A}) \cap P)/(J^1(\beta, \mathfrak{A}) \cap U) = J^1(\beta, \mathfrak{A}) \cap M$$

on the space of $(J^1(\beta, \mathfrak{A}) \cap U)$-fixed vectors in η is irreducible, and equivalent to η_U.

It will later be necessary to have a fairly detailed knowledge of the intertwining of the various representations η_ϕ of $(J^1 \cap P)H^1$, $\phi \in \widehat{\mathcal{G}}$.

(7.2.9) Proposition: *Let $y \in B^\times$, $\phi \in \widehat{\mathcal{G}}$. There is a unique double coset $(J^1 \cap P)H^1x(J^1 \cap P)H^1 \subset J^1yJ^1$ which intertwines η_ϕ. For this x, we have $\dim_{\mathbb{C}}(I_x(\eta_\phi)) = 1$.*

Proof: Immediate from (5.1.8) and (7.2.4)*(iii)*. ■

(7.2.10) Corollary: *Let $y \in M(B)^\times = M \cap B^\times$. An element $x \in J^1yJ^1$ intertwines η_P if and only if $x \in (J^1 \cap P)H^1y(J^1 \cap P)H^1$. If (7.2.1) holds, the same applies to the element $y = \Pi(\mathfrak{B})$ (notation of (5.5.9)).*

Proof: Take first the case $y \in M(B)^\times$. Since y normalises the groups M, U, U^-, and intertwines the representation $\eta_U = \eta_P \mid J^1 \cap M$, one sees directly that it intertwines η_P. The uniqueness follows from (7.2.9).

Now assume that (7.2.1) holds, and take $y = \Pi(\mathfrak{B})$. Then y intertwines η, and normalises the group $J^1 \cap M$. Since $\eta \mid J^1 \cap M$ is a multiple of η_U, it follows that y intertwines η_U. Now we use the fact that η_P is null on $J^1 \cap U$ and $H^1 \cap U^-$ to show directly that y intertwines η_P. ■

In particular, if (7.2.1) holds, we note here that the representation η_U of $J^1(\beta, \mathfrak{A}) \cap M$ is *normalised* by the group $\boldsymbol{D}(\mathfrak{B})$ (see (5.5.9)) and the element $\Pi(\mathfrak{B})$.

We now embark on a small digression, with the the aim of completing the proof of (5.2.7). We continue with our representation η of $J^1 = J^1(\beta, \mathfrak{A})$. We take an E-decomposition of $V = V^{(1)} \oplus \ldots \oplus V^{(e)}$ of V which is subordinate to $\mathfrak{B}$, with $e = e(\mathfrak{B}|\mathfrak{o}_E)$ (that is, we are in the situation of (5.5.2)). Let P be the associated parabolic subgroup of G, and we use the other notations above. In this situation, the $\mathfrak{o}_E$-orders

$\mathfrak{B}^{(i)} = \mathfrak{B} \cap A^{(i)}$ are all maximal. We are given an element y of the centre of $M(B)^\times = M \cap B^\times$ (*cf.* (5.2.9)), and we have to show:

(7.2.11) Proposition: *There is exists an irreducible representation μ of J which is intertwined by y and such that $\mu \mid J^1 = \eta$.*

Proof: Consider first the representation $\eta_U = \eta^{(1)} \otimes \ldots \eta^{(e)}$. According to (5.2.10), there exists a representation $\mu^{(i)}$ of $J(\beta, \mathfrak{A}^{(i)})$ which extends $\eta^{(i)}$ and is intertwined by the component $y^{(i)}$ of y in $A^{(i)\times}$. We form the representation $\mu_U = \mu^{(1)} \otimes \ldots \otimes \mu^{(e)}$ of $J \cap M$, and this is certainly intertwined (indeed normalised) by y. We "inflate" this to a representation μ_P of $(J \cap P)H^1$ by setting

$$\mu_P(hmj) = \mu_U(m), \quad h \in H^1 \cap U^-, \ m \in J \cap M, \ j \in J^1 \cap U.$$

This gives a well-defined representation of $(J \cap P)H^1$, which is surely intertwined by y. Now we form $\mu = \mathrm{Ind}(\mu_P : (J \cap P)H^1, J)$. The restriction of this J^1 is exactly η, by (7.2.4)*(iii)* and the Mackey restriction formula. Thus μ extends η, and is intertwined by y, as required. ∎

We return to the main argument. With (7.2.11) established, we have the full β-extension machinery of §**5**. Let κ be some β-extension of η to a representation of $J = J(\beta, \mathfrak{A})$, as in (5.2.1). Since J normalises θ and $J \cap P$ normalises the character θ_0 used above, we can form the natural representation κ_P of $(J \cap P)H^1$ on the space of $(J \cap U) = (J^1 \cap U)$-fixed vectors in κ. Then

(7.2.12) $$\kappa_P \mid (J^1(\beta, \mathfrak{A}) \cap P)H^1(\beta, \mathfrak{A}) = \eta_P.$$

In particular, the representation κ_P is irreducible.

We can also form the representation $\kappa_U = \kappa_P \mid J(\beta, \mathfrak{A}) \cap M$ which we view, when convenient, as a representation of $(J \cap P)/(J \cap U)$. In either case, it is irreducible. Indeed,

(7.2.13) $$\kappa_U \mid J^1(\beta, \mathfrak{A}) \cap M = \eta_U = \eta^{(1)} \otimes \ldots \otimes \eta^{(t)}.$$

Thus, besides being irreducible, κ_U is of the form

(7.2.14) $$\kappa_U = \kappa^{(1)} \otimes \kappa^{(2)} \otimes \ldots \otimes \kappa^{(t)},$$

for a uniquely determined irreducible representation $\kappa^{(i)}$ of $J(\beta, \mathfrak{A}^{(i)})$ such that $\kappa^{(i)} \mid J^1(\beta, \mathfrak{A}^{(i)}) = \eta^{(i)}$, $1 \le i \le t$.

(7.2.15) Proposition: *Let κ_P be the natural representation of $(J \cap P)H^1$ on the space of $(J \cap U)$-fixed vectors in κ. Then*

$$\mathrm{Ind}(\kappa_P : (J^1 \cap P)H^1 : J) = \kappa.$$

Moreover, κ_P is intertwined by every element of the group $M(B)^\times = M \cap B^\times$ and, if (7.2.1) holds, by the element $\Pi(\mathfrak{B})$. The same applies to the representation κ_U, and moreover κ_U is normalised by $\Pi(\mathfrak{B})$ (when (7.2.1) holds).

Proof: Let $\tilde{\kappa}$ be the representation of J induced by κ_P. Then $\tilde{\kappa}$ certainly contains κ. However, the Mackey restriction formula and (7.2.4) show that $\tilde{\kappa} \mid J^1 = \eta$. Thus $\tilde{\kappa}$ is irreducible, whence $\tilde{\kappa} = \kappa$, as required.

Take $y \in M(B)^\times$ or (provided (7.2.1) holds) $y = \Pi(\mathfrak{B})$. Then y intertwines κ by (5.2.7), and indeed $I_y(\kappa) = I_y(\eta)$. By the first part, some double coset $(J \cap P)H^1x(J \cap P)H^1$ contained in JyJ intertwines κ_P. However, this coset must also intertwine η_P, and is therefore equal to $(J \cap P)H^1y(J \cap P)H^1$ by (7.2.10). Thus y intertwines κ_P, as required. The assertions concerning κ_U are now immediate. ∎

(7.2.16) Corollary: *For $1 \le i \le t$, the representation $\kappa^{(i)}$ of (7.2.14) is a β-extension of $\eta^{(i)}$. Moreover, if (7.2.1) holds, then for any pair i, j, any isomorphism $\mathfrak{B}^{(i)} \xrightarrow{\approx} \mathfrak{B}^{(j)}$ of $\mathfrak{o}_E$-orders induces an equivalence $\kappa^{(i)} \cong \kappa^{(j)}$.*

Proof: We have $\kappa^{(i)} \mid J^1(\beta, \mathfrak{A}^{(i)}) = \eta^{(i)}$, and $\kappa^{(i)}$ is intertwined by the whole of $B^{(i)\times} \subset M(B)^\times$ by the proposition. Thus $\kappa^{(i)}$ is a β-extension of $\eta^{(i)}$. Conjugation of κ_U by an appropriate power of $\Pi(\mathfrak{B})$ gives the assertion relative to some isomorphism $\mathfrak{B}^{(i)} \cong \mathfrak{B}^{(j)}$. Any two isomorphisms $\mathfrak{B}^{(i)} \cong \mathfrak{B}^{(j)}$ of $\mathfrak{o}_E$-orders differ by conjugation by an element of $\mathfrak{K}(\mathfrak{B}^{(i)})$, and such conjugations normalise $\kappa^{(i)}$. ∎

We now pass from the β-extension κ of η to a simple type $\lambda = \kappa \otimes \sigma$. The factor σ is, by definition, trivial on J^1, in particular, it is trivial on $J \cap U$ and $J \cap U^-$ by (7.1.15). This allows us to translate properties of κ directly to λ. We summarise:

(7.2.17) Theorem: *Let (J, λ) be a simple type in G, attached to a simple stratum $[\mathfrak{A}, n, 0, \beta]$. Use the notation (7.1.11), (7.1.13). Let λ_P denote the natural representation of $(J(\beta, \mathfrak{A}) \cap P)H^1(\beta, \mathfrak{A})$ on the space of $(J^1(\beta, \mathfrak{A}) \cap U)$-fixed vectors in λ, and put $\lambda_U = \lambda_P \mid J(\beta, \mathfrak{A}) \cap M$.*

(i) The representations λ_P, λ_U are irreducible and $\lambda \cong \mathrm{Ind}(\lambda_P : H^1(J \cap P), J)$.

(ii) For $1 \le i \le t$, there is a unique simple type $(J(\beta, \mathfrak{A}^{(i)}), \lambda^{(i)})$ in $A^{(i)\times}$ such that

$$\lambda_U = \lambda^{(1)} \otimes \lambda^{(2)} \otimes \ldots \otimes \lambda^{(t)},$$

after identifying $J(\beta, \mathfrak{A}) \cap M$ with $\prod_i J(\beta, \mathfrak{A}^{(i)})$.

Now suppose that (7.2.1) holds. Then:—

(iii) For any pair i, j, any isomorphism $\mathfrak{B}^{(i)} \cong \mathfrak{B}^{(j)}$ of $\mathfrak{o}_E$-orders induces an equivalence $\lambda^{(i)} \cong \lambda^{(j)}$.

(iv) The types $(J(\beta, \mathfrak{A}^{(i)}), \lambda^{(i)})$ are maximal.

All of this is immediate. In *(iv)*, the condition "$t = e$" which we imposed means that the orders $\mathfrak{B}^{(i)}$ are all maximal, whence the associated types are maximal.

Remark: (7.2.17)*(iii)* holds under the weaker hypothesis that $\mathfrak{A}$ is a principal order and that all the $V^{(i)}$ have the same dimension. Although we shall not need it, it is worth observing that we can weaken (7.2.1) (wherever it is used) to this hypothesis on $\mathfrak{A}$ and P, at the cost of replacing $\Pi(\mathfrak{B})$ by an appropriate power, namely $\Pi(\mathfrak{B})^{e'}$, where $e' = e(\mathfrak{B}|\mathfrak{o}_E)/t$. Of course, (7.2.17)*(iv)* does require the full force of (7.2.1).

Before proceeding, let us encapsulate some of these constructions in new terminology.

(7.2.18) Terminology: *(i)* Let $[\mathfrak{A}, n, 0, \beta]$ be a simple stratum in $A = \mathrm{End}_F(V)$ and put $E = F[\beta]$, $\mathfrak{B} =$ the $\mathfrak{A}$-centraliser of β. A parabolic subgroup P of G is *subordinate* to $[\mathfrak{A}, n, 0, \beta]$ if it is constructed, as in (7.1.13), from an E-decomposition $V = V^{(1)} \oplus \ldots \oplus V^{(t)}$ of V subordinate to $\mathfrak{A}$ (or $\mathfrak{B}$). We always view such a P as coming equipped with the distinguished Levi component $M = M(A)^\times = \prod_i \mathrm{Aut}_F(V^{(i)})$.

(ii) If (J, λ) is a simple type in G attached to a simple stratum $[\mathfrak{A}, n, 0, \beta]$, and P is a parabolic subgroup of G, we say that P is *subordinate to* (J, λ) if it is subordinate to $[\mathfrak{A}, n, 0, \beta]$.

(iii) Let (J, λ) be a simple type in $G = \mathrm{Aut}_F(V) \cong GL(N, F)$, attached to the simple stratum $[\mathfrak{A}, n, 0, \beta]$ with $E = F[\beta]$, $\mathfrak{B} = \mathfrak{A} \cap \mathrm{End}_E(V)$, $e = e(\mathfrak{B}|\mathfrak{o}_E)$ as always. Let P be a parabolic subgroup of G, subordinate to (J, λ) and minimal for this property (i.e. "$t = e$" in the earlier setup). Let U be the unipotent radical of P. By (7.2.17), there is a unique maximal simple type (J', λ') in $G' = GL(N/e, F)$ such that the natural representation λ_U of $(J \cap P)/(J \cap U)$ on the space of $J \cap U$-fixed vectors in λ is equivalent to $\lambda' \otimes \ldots \otimes \lambda'$ (e factors). We refer to (J', λ') as *the associated maximal type of* (J, λ).

To conclude this section, we record an intertwining property of these representations λ_P in a significant special case.

(7.2.19) Proposition: *Let (J, λ) be a simple type in G, defined by a simple stratum $[\mathfrak{A}, n, 0, \beta]$. Put $E = F[\beta]$, $B = \mathrm{End}_E(V)$, $\mathfrak{B} = \mathfrak{A} \cap B$. Choose an $\mathfrak{o}_E$-basis of the lattice chain defining $\mathfrak{B}$ as in (5.5.2), and let $P = MU$ be the parabolic subgroup of G determined by the associated decomposition $V = \coprod V^i$, as in (5.5.1). Define the affine Weyl group $\widetilde{\boldsymbol{W}}(\mathfrak{B})$ as in (5.5.9). Then, in the notation above, every $w \in \widetilde{\boldsymbol{W}}(\mathfrak{B})$ intertwines the representation λ_P, we have $I_w(\lambda_P) = I_w(\lambda_U)$, and this space has dimension 1.*

Proof: First we observe that the representations λ_P, λ_U act on the same vector space, so we can compare intertwining operators. Next, we observe that $\widetilde{\boldsymbol{W}}(\mathfrak{B})$ normalises $J \cap M$ and stabilises the representation λ_U. Thus the space $I_{\boldsymbol{w}}(\lambda_U)$ has dimension one, for $\boldsymbol{w} \in \widetilde{\boldsymbol{W}}(\mathfrak{B})$. We surely have $I_{\boldsymbol{w}}(\lambda_P) \subset I_{\boldsymbol{w}}(\lambda_U)$. On the other hand, we can write $K = (H^1 \cap U^-)(J \cap M)(J^1 \cap U)$ in the form $K = (J \cap M)X$, where $X = 1 + \mathfrak{H}^1 \cap \mathbb{N}^- + \mathfrak{J}^1 \cap \mathbb{N}$, and then $K^{\boldsymbol{w}} \cap K = (X^{\boldsymbol{w}} \cap X)(J \cap M)$. Since λ_P is null on X, the assertion follows immediately. ■

(7.3) Main theorems

If $(\boldsymbol{\pi}, \mathcal{V})$ is a smooth representation of G, K a compact subgroup of G and ρ an irreducible representation of K, we write $\mathcal{V}^\rho$ for the ρ-isotypic subspace of $\mathcal{V}$ and use the other notational conventions of §**4**. In particular, if K is also open, we have the idempotent $e_\rho \in \mathcal{H}(G)$ given by

$$e_\rho(x) = \begin{cases} \dfrac{\dim(\rho)}{\mu(K)} \operatorname{tr}(\rho(x^{-1})) \; \textit{if } x \in K, \\ 0 \; \textit{otherwise}, \end{cases} \qquad x \in G.$$

Here, μ is some preordained Haar measure on G and tr denotes the trace for operators on the representation space of ρ.

We now fix a simple type (J, λ) in G, along with its usual associated notations. We let P be a parabolic subgroup of G which is subordinate to G and minimal for this property. Indeed, we take the one given, via (7.1.13), by the E-decomposition of V defined by (5.5.2), and the notations U, U^-, M, λ_P, λ_U etc. are as above. We shall also need some of the notation of §**5**. In particular, we recall the group $\boldsymbol{D}(\mathfrak{B})$ of diagonal matrices (relative to the basis chosen to define P) whose eigenvalues are powers of some fixed prime element π_E of E, and which normalise (equivalently centralise) $\mathfrak{M}(\mathfrak{B})^\times = \mathfrak{B}^\times \cap M$.

(7.3.1) Principal Lemma: *Let $(\boldsymbol{\pi}, \mathcal{V})$ be a smooth representation of G and let $a \in \boldsymbol{D}(\mathfrak{B})$. Then the operator $\boldsymbol{\pi}(e_{\lambda_P}) \circ \boldsymbol{\pi}(a) \circ \boldsymbol{\pi}(e_{\lambda_P})$ is an isomorphism*

$$\mathcal{V}^{\lambda_P} \xrightarrow{\approx} \mathcal{V}^{\lambda_P}.$$

We postpone the proof of this result until the next section. Given this lemma, the results of this section become little more than formalities: indeed, they and their proofs are virtually lifted from **[Cs]** §**3.3** (with some hints from **[Bo]** and **[Wa]**).

Let $(\boldsymbol{\pi}, \mathcal{V})$ be a smooth representation of G. We write $(\boldsymbol{\pi}_U, \mathcal{V}_U)$ for the Jacquet module of $(\boldsymbol{\pi}, \mathcal{V})$ relative to the parabolic subgroup P. Thus $(\boldsymbol{\pi}_U, \mathcal{V}_U)$ is the natural smooth representation of P/U (or M) on the maximal U-fixed quotient of $\mathcal{V}$. One knows ([Cs] (3.3.1)) that if $(\pi, \mathcal{V})$ is admissible (in particular, if it is irreducible), then the representation $(\pi_U, \mathcal{V}_U)$ is admissible.

(7.3.2) Theorem: *Let $(\pi, \mathcal{V})$ be an admissible representation of G. The canonical map $\mathcal{V} \to \mathcal{V}_U$ induces an isomorphism*

$$\mathcal{V}^{\lambda_P} \xrightarrow{\approx} (\mathcal{V}_U)^{\lambda_U}.$$

Remark: Unless $e(\mathfrak{B}|\mathfrak{o}_E) = 1$, when the type (J, λ) is maximal, the parabolic subgroup P is proper. If, on the other hand, (J, λ) is not maximal and if π is irreducible containing λ, then π also contains λ_P, so that $\mathcal{V}^{\lambda_P}$ is nonzero. It follows from the theorem that $\mathcal{V}_U$ is not zero. Thus π cannot be supercuspidal, and we have another proof of (6.2.1).

This proof of (6.2.1) relies on the definition of an irreducible supercuspidal representation as one with only trivial Jacquet modules. On the other hand, the proof in §6 uses the characterisation by compactly supported (mod centre) matrix coefficients.

Remark: The hypothesis of admissibility in (7.3.2) is probably unnecessary, but we do not pursue this matter. However, our proof certainly gives usefully stronger results, as we remark below.

Remark: Both (7.3.2) and (7.3.1) are rather formal in nature, at least to the extent that they can be formulated axiomatically. It will be useful later for us to have this, so we outline the necessary ideas once we have completed the proof.

Proof of (7.3.2): Since $\lambda_P \mid J \cap M = \lambda_U$, $J = J(\beta, \mathfrak{A})$, the quotient map $\mathcal{V} \to \mathcal{V}_U$ surely maps $\mathcal{V}^{\lambda_P}$ to $(\mathcal{V}_U)^{\lambda_U}$, and this is a $J \cap P$ (or $J \cap M$) homomorphism.

We recall from [Cs](3.2.1) that a vector $v \in \mathcal{V}$ lies in the kernel $\mathcal{V}(U)$ of $\mathcal{V} \to \mathcal{V}_U$ if and only if there is a compact open subgroup U_1 of U such that

$$\int_{U_1} \pi(u)v\, du = 0, \tag{7.3.3}$$

where du is a Haar measure on U. Also, if the pair (v, U_1) satisfies (7.3.3) and U_2 is a compact open subgroup of U containing U_1, then

$$\int_{U_2} \pi(u)v\, du = 0$$

as well.

We first prove that $\mathcal{V}^{\lambda_P}$ maps injectively to $\mathcal{V}_U$. By the definition of λ_P, every $v \in \mathcal{V}^{\lambda_P}$ is fixed by the group $J \cap U = J^1 \cap U$, so

$$\textbf{(7.3.4)} \qquad \int_{J^1\cap U} \pi(u)v\,du \neq 0, \quad v \in \mathcal{V}^{\lambda_P},\ v \neq 0.$$

Let $v \in \mathcal{V}^{\lambda_P} \cap \mathcal{V}(U)$, $v \neq 0$. Choose an open compact subgroup U_1 of U so that (7.3.3) holds. There exists $a \in \boldsymbol{D}(\mathfrak{B})$ such that $a^{-1}U_1a \subset J^1 \cap U$. For such an element a, consider the vector $\pi(e_{\lambda_P})\pi(a)v$. By (7.3.1), this is a nonzero element of $\mathcal{V}^{\lambda_P}$. The operator $\pi(e_{\lambda_P})$ commutes with $\pi(u)$, $u \in J^1 \cap U$, so we get

$$\begin{aligned}\int_{J^1\cap U} \pi(u)\pi(e_{\lambda_P})\pi(a)v\,du &= \pi(e_{\lambda_P})\int_{J^1\cap U} \pi(u)\pi(a)v\,du \\ &= \pi(e_{\lambda_P})\pi(a)\int_{J^1\cap U} \pi(a^{-1}ua)v\,du \\ &= c\pi(e_{\lambda_P})\pi(a)\int_{a(J^1\cap U)a^{-1}} \pi(u)v\,du,\end{aligned}$$

for some constant $c > 0$. However, this last integral vanishes because $a(J^1 \cap U)a^{-1} \supset U_1$. This contradicts (7.3.4), and therefore the map $\mathcal{V}^{\lambda_P} \to (\mathcal{V}_U)^{\lambda_U}$ is injective.

To prove surjectivity, we first observe that $J \cap M$ normalises $H^1 \cap U^-$, so we have a surjection $(J \cap M).(H^1 \cap U^-) \to J \cap M$ which we use to inflate λ_U to a representation λ^- of $(J \cap M).(H^1 \cap U^-)$.

(7.3.5) Lemma: *The spaces $\mathcal{V}^{\lambda^-}$, $\mathcal{V}^{\lambda_P}$ have the same images in $\mathcal{V}_U$.*

Proof: Consider the operator $\boldsymbol{p}^+ : \mathcal{V}^{\lambda^-} \to \mathcal{V}$ given by

$$\boldsymbol{p}^+ : v \mapsto \int_{J^1\cap U} \pi(u)v\,du,$$

where du is the Haar measure on U giving $J^1 \cap U$ measure one. Then we have $\mathcal{V}^{\lambda_P} \subset \mathcal{V}^{\lambda^-}$, $\boldsymbol{p}^+(\mathcal{V}^{\lambda^-}) \subset \mathcal{V}^{\lambda_P}$, and $\boldsymbol{p}^+(v) = v$ if $v \in \mathcal{V}^{\lambda_P}$. Thus $\boldsymbol{p}^+$ projects $\mathcal{V}^{\lambda^-}$ onto $\mathcal{V}^{\lambda_P}$. However, for any $v \in \mathcal{V}$, the vectors v and $\boldsymbol{p}^+(v)$ have the same image in $\mathcal{V}_U$. ■

Now let $\bar{v} \in (\mathcal{V}_U)^{\lambda_U}$. Since $\mathcal{V}$ is semisimple as a $(J \cap M)$-space, $\bar{v}$ is the image of some element $v \in \mathcal{V}^{\lambda_U}$. This element v is fixed by some open compact subgroup U_1^- of U^-. Since $\mathcal{V}$ is admissible, so is $\mathcal{V}_U$, whence $(\mathcal{V}_U)^{\lambda_U}$ is finite-dimensional. We may therefore choose U_1^- independently of $\bar{v}$. That is, we may choose U_1 so that, for any $\bar{v} \in (\mathcal{V}_U)^{\lambda_U}$, there exists $v \in \mathcal{V}^{\lambda_U}$ which is fixed by U_1 and which maps to $\bar{v}$. If $U_1^- \supset J^1 \cap U^-$, we use (7.3.5) to show that $\bar{v}$ is the image of $\boldsymbol{p}^+(v) \in \mathcal{V}^{\lambda_P}$.

Otherwise, we choose $a \in \boldsymbol{D}(\mathfrak{B})$ so that $aU_1^- a^{-1} \supset H^1 \cap U^-$. It follows that $\boldsymbol{\pi}(a)\mathsf{v}$ is fixed by $H^1 \cap U^-$. Observe that $\boldsymbol{\pi}(a)\mathsf{v} \in \mathcal{V}^{\lambda_U}$ since $\boldsymbol{D}(\mathfrak{B})$ normalises λ_U. Thus, in particular, $\boldsymbol{\pi}(a)\mathsf{v} \in \mathcal{V}^{\lambda^-}$ and $\boldsymbol{\pi}_U(a)\bar{\mathsf{v}}$ is the image of $\boldsymbol{\pi}(a)\mathsf{v}$. However, $\boldsymbol{p}^+(\boldsymbol{\pi}(a)\mathsf{v}) \in \mathcal{V}^{\lambda_P}$, and this has the same image in $\mathcal{V}_U$ as $\boldsymbol{\pi}(a)\mathsf{v}$. This says that the element $\boldsymbol{\pi}_U(a)\bar{\mathsf{v}} \in (\mathcal{V}_U)^{\lambda_U}$ lies in the image of $\mathcal{V}^{\lambda_P}$. However, $\boldsymbol{\pi}_U(a)$ acts as an automorphism of $(\mathcal{V}_U)^{\lambda_U}$, so the result follows. ∎

(7.3.6) Remarks: *(i)* The proof above that $\mathcal{V}^{\lambda_P}$ maps injectively to $(\mathcal{V}_U)^{\lambda_U}$ is valid under the weaker hypothesis that $\boldsymbol{\pi}$ is smooth. While we did use the hypothesis of admissibility in the proof of surjectivity, its only function was to show that $(\mathcal{V}_U)^{\lambda_U}$ was finite-dimensional. *Thus in fact we have a bijection $\mathcal{V}^{\lambda_P} \cong (\mathcal{V}_U)^{\lambda_U}$ provided only that $(\boldsymbol{\pi}, \mathcal{V})$ is smooth and $(\mathcal{V}_U)^{\lambda_U}$ is finite-dimensional.*

(ii) Without even this weak hypothesis we get a useful result. For, the same argument shows that if we take a nonzero $\bar{\mathsf{v}} \in (\mathcal{V}_U)^{\lambda_U}$, there exists $a \in \boldsymbol{D}(\mathfrak{B})$ such that $\boldsymbol{\pi}_U(a)\bar{\mathsf{v}}$ is the image of some element of $\mathcal{V}^{\lambda_P}$. Therefore: *if $(\boldsymbol{\pi}, \mathcal{V})$ is smooth and if $(\mathcal{V}_U)^{\lambda_U} \neq \{0\}$, then $\mathcal{V}^{\lambda_P} \neq \{0\}$.*

(7.3.7) Remark: As we go through §**8** below, we shall need a number of variations on the theme of (7.3.2). It is therefore worth analysing carefully the structure of this proof. First, we need an opposite pair $P = MU$, $P^- = MU^-$ of proper parabolic subgroups of G, and a compact open subgroup K of G such that $\mathsf{K} = (\mathsf{K} \cap U^-)(\mathsf{K} \cap M)(\mathsf{K} \cap U)$. Suppose also that we have an irreducible smooth representation ρ of K such that the restrictions $\rho \mid \mathsf{K} \cap U^-$, $\rho \mid \mathsf{K} \cap U$ are both null (i.e. multiples of the trivial character), while $\rho_U = \rho \mid \mathsf{K} \cap M$ is irreducible. We also need a subgroup D of M with the following properties:

(a) For any smooth representation $(\boldsymbol{\pi}, \mathcal{V})$ of G and any $d \in \mathsf{D}$, the operator $\boldsymbol{\pi}(e_\rho) \circ \boldsymbol{\pi}(d) \circ \boldsymbol{\pi}(e_\rho)$ is an isomorphism $\mathcal{V}^\rho \cong \mathcal{V}^\rho$.

(b) The group D normalises $\mathsf{K} \cap M$ and stabilises the representation ρ_U.

(c) Given compact open subgroups U_1, U_2 of U (or U^-), there exists $d \in \mathsf{D}$ such that $d^{-1}U_1 d \subset U_2$.

Exactly the same argument then yields the analogue of (7.3.2):— *For any admissible representation $(\boldsymbol{\pi}, \mathcal{V})$ of G, the canonical map $\mathcal{V} \to \mathcal{V}_U$ induces an isomorphism $\mathcal{V}^\rho \cong \mathcal{V}_U^{\rho_U}$.*

The analogues of (7.3.6)*(i)* and *(ii)* above also hold.

We now write $P/U = G^{(1)} \times G^{(2)} \times \ldots \times G^{(e)}$, with $G^{(i)} = \mathrm{Aut}_F(V^{(i)})$, where $V = \coprod_i V^{(i)}$ is the E-decomposition of V defining P, and regard the tensor factor $\lambda^{(i)}$ of λ_U as a simple type in $G^{(i)}$, $1 \leq i \leq e$. Of course, the groups $G^{(i)}$ are all isomorphic (to $GL(N/e, F)$) and each $\lambda^{(i)}$ is equivalent to the maximal type λ' associated to (J, λ). Indeed, there is no harm in thinking of $\lambda_U = \lambda^{(1)} \otimes \ldots \otimes \lambda^{(e)}$ as a "maximal simple type in $M = \prod_i G^{(i)}$". We have:

(7.3.8) Lemma: *Let $\boldsymbol{\xi}$ be an irreducible smooth representation of $P/U = \prod_i G^{(i)}$ containing the representation $\lambda_U = \lambda^{(1)} \otimes \ldots \otimes \lambda^{(e)}$ of $\prod_i J(\beta, \mathfrak{A}^{(i)})$. Then $\boldsymbol{\xi}$ is supercuspidal, and there is a unique extension Λ_U of λ_U to the group $\prod_i \mathfrak{K}(\mathfrak{B}^{(i)})J(\beta, \mathfrak{A}^{(i)})$ such that $\boldsymbol{\xi} = \mathrm{Ind}(\Lambda_U)$.*

Proof: We have $\boldsymbol{\xi} = \boldsymbol{\xi}^{(1)} \otimes \ldots \otimes \boldsymbol{\xi}^{(e)}$, for a uniquely determined irreducible representation $\boldsymbol{\xi}^{(i)}$ of $G^{(i)}$, $1 \leq i \leq e$, and $\boldsymbol{\xi}^{(i)}$ contains $\lambda^{(i)}$. The assertion follows from (6.2.2). ∎

For a smooth representation $(\boldsymbol{\sigma}, \mathcal{W})$ of M, viewed as a representation of P via inflation, we write $\mathrm{Ind}(\boldsymbol{\sigma}) = \mathrm{Ind}(\boldsymbol{\sigma} : P, G)$ for the "ordinary" induced representation. Thus $\mathrm{Ind}(\boldsymbol{\sigma})$ is the space of right G-smooth functions $\phi : G \to \mathcal{W}$ such that $\phi(pg) = \boldsymbol{\sigma}(p)\phi(g)$, $p \in P$, $g \in G$, with G acting by right translation. This gives rise to a canonical isomorphism (*Frobenius Reciprocity*)

$$\mathrm{Hom}_P(\boldsymbol{\pi}, \boldsymbol{\sigma}) \cong \mathrm{Hom}_G(\boldsymbol{\pi}, \mathrm{Ind}(\boldsymbol{\sigma})),$$

for any smooth representation $\boldsymbol{\pi}$ of G.

We use the symbol Ind^u for the "normalised" induction functor

$$\mathrm{Ind}^u(\boldsymbol{\sigma}) = \mathrm{Ind}(\delta_P^{1/2} \otimes \boldsymbol{\sigma}),$$

where δ_P is the module of P. In the above situation, if $\boldsymbol{\pi}_i$ is an irreducible smooth representation of $G^{(i)}$, $1 \leq i \leq e$, we write

$$\boldsymbol{\pi}_1 \times \boldsymbol{\pi}_2 \times \ldots \times \boldsymbol{\pi}_e = \mathrm{Ind}^u(\boldsymbol{\pi}_1 \otimes \ldots \otimes \boldsymbol{\pi}_e).$$

In the present context, this detail makes very little difference. There are unramified quasicharacters χ_i of $F^\times$ such that

$$\boldsymbol{\pi}_1 \times \boldsymbol{\pi}_2 \times \ldots \times \boldsymbol{\pi}_e = \mathrm{Ind}(\boldsymbol{\pi}_1' \otimes \ldots \otimes \boldsymbol{\pi}_e'), \qquad \boldsymbol{\pi}_i' = \boldsymbol{\pi}_i \otimes \chi_i \circ \det,$$

and twisting with unramified quasicharacters does not affect the simple types.

(7.3.9) Theorem: *Let $(\boldsymbol{\pi}, \mathcal{V})$ be an irreducible smooth representation of G which contains the simple type λ. There is an irreducible supercuspidal representation $\boldsymbol{\pi}_i$ of $G^{(i)}$ which contains the type $\lambda^{(i)}$, $1 \leq i \leq e$, such that $(\boldsymbol{\pi}, \mathcal{V})$ is isomorphic to a G-subspace of $\boldsymbol{\pi}_1 \times \boldsymbol{\pi}_2 \times \ldots \times \boldsymbol{\pi}_e$.*

Proof: We start with a lemma.

(7.3.10) Lemma: *Let $(\boldsymbol{\xi}, \mathcal{W})$ be a smooth representation of $P/U = \prod_i G^{(i)}$ of finite composition length and which contains the representation $\lambda_U = \lambda^{(1)} \otimes \ldots \otimes \lambda^{(e)}$. Then $(\boldsymbol{\xi}, \mathcal{W})$ has an irreducible quotient containing λ_U.*

Proof: We proceed by induction on the composition length of $\mathcal{W}$, starting with the case where this length is 2. So, let

$$0 \to \mathcal{W}_1 \to \mathcal{W} \to \mathcal{W}_2 \to 0$$

be a composition series for $\mathcal{W}$ and assume that $\mathcal{W}_2$ does not contain λ_U. Therefore $\mathcal{W}_1$ does contain λ_U. Let $\mathcal{Z}$ denote the centre of P/U. If ξ possesses a central quasicharacter, i.e. if there is a quasicharacter ω of $\mathcal{Z}$ such that $\xi(z)w = \omega(z)w$, $z \in \mathcal{Z}$, $w \in \mathcal{W}$, then the exact sequence above splits. This is because $\mathcal{W}_1$ is supercuspidal by (7.3.8), and supercuspidal representations of P/U are injective in the category of smooth representations of P/U with fixed central quasicharacter ([**Cs**] (5.4.1)). The result follows in this case.

So, we assume that ξ does not have a central quasicharacter. Let ω_1 denote the central quasicharacter of the irreducible representation $\mathcal{W}_1$. Since, by hypothesis, $\mathcal{W}$ has no central quasicharacter, there exists $z \in \mathcal{Z}$ such that $\xi(z) \neq \omega_1(z)1_{\mathcal{W}}$. For such a z, consider the map $w \mapsto \xi(z)w - \omega_1(z)w$, $w \in \mathcal{W}$. This is a nonzero P/U-homomorphism $\mathcal{W} \to \mathcal{W}$ which is null on $\mathcal{W}_1$. We may therefore regard it as an embedding of $\mathcal{W}_2$ in $\mathcal{W}$, and its cokernel is the desired quotient of $\mathcal{W}$.

In the general case, we take a composition series for $\mathcal{W}$,

$$0 \subset \mathcal{W}_1 \subset \ldots \subset \mathcal{W}_{n-1} \subset \mathcal{W}_n = \mathcal{W}.$$

If $\mathcal{W}_n/\mathcal{W}_{n-1}$ contains λ_U, we are done. Otherwise, the subspace $\mathcal{W}_{n-1}$ contains λ_U and, by induction, we can further assume that $\mathcal{W}_{n-1}/\mathcal{W}_{n-2}$ contains λ_U. We apply the first step to $\mathcal{W}/\mathcal{W}_{n-2}$, and the result follows. ∎

Returning to (7.3.9) the irreducible representation $(\pi, \mathcal{V})$ is admissible (by [**J2**]), and it is certainly finitely generated. Therefore ([**Cs**] (3.3.1)) $(\pi_U, \mathcal{V}_U)$ is admissible and finitely generated (as a representation of P/U). It follows that $(\pi_U, \mathcal{V}_U)$ has finite composition length ([**Cs**] (6.3.10), but see also (7.3.19) below). We can now apply (7.3.10) to get a P/U-surjection from $\mathcal{V}$ to an irreducible quotient $(\xi, \mathcal{W})$ of $(\pi_U, \mathcal{V}_U)$ which contains λ_U. Therefore $\xi = \pi_1 \otimes \ldots \otimes \pi_e$, where π_i is an irreducible representation of $G^{(i)}$ containing $\lambda^{(i)}$, and which is supercuspidal by (6.2.3). The theorem now follows from Frobenius Reciprocity. ∎

For the moment, let π denote some irreducible smooth representation of G. It is an elementary consequence of the definitions and Frobenius Reciprocity that π is equivalent to a subspace of some representation $\pi_1 \times \ldots \times \pi_r$, with the π_i all irreducible and supercuspidal. Somewhat deeper is the following fact (see [**Cs**] (6.3.7), (6.3.11) (which hold for any reductive group G) or [**BZ**] (for $GL(N)$)):

(7.3.11) *Given irreducible supercuspidal representations $\sigma_1, \dots, \sigma_s$, then π is equivalent to a subquotient of $\sigma_1 \times \dots \times \sigma_s$ if and only if $r = s$ and $(\sigma_1, \dots, \sigma_s)$ is a permutation of $(\pi_1, \dots, \pi_r)$ (up to equivalence).*

The "multiset" or "set with multiplicity" of equivalence classes

$$\{\pi_1, \dots, \pi_r\}$$

is then called the *supercuspidal support of* π.

(7.3.12) Corollary: *Let (J, λ) be a simple type in G, and let $(\pi, \mathcal{V})$ be an irreducible smooth representation of G containing λ. The supercuspidal support of π then consists of unramified twists $\pi' \otimes \chi \circ \det$ of a single supercuspidal representation π' of a group $G' \cong GL(N/e, F)$, and this representation π' contains λ', where (J', λ') is the maximal simple type associated to (J, λ).*

We shall shortly see that the converse of this holds. When we have shown in §**8** that any supercuspidal representation contains a simple type, then we will also know that an irreducible representation of G contains a simple type if and only if its supercuspidal support consists of unramified twists of a single supercuspidal.

Returning to the situation in hand, we can use the notion of supercuspidal support to analyse the composition factors of the Jacquet module $(\pi_U, \mathcal{V}_U)$.

(7.3.13) Proposition: *Let $(\pi, \mathcal{V})$ be a smooth irreducible representation of G containing the simple type λ. Then every composition factor of $(\pi_U, \mathcal{V}_U)$ contains λ_U, and the composition length of $(\pi_U, \mathcal{V}_U)$ equals the multiplicity of λ in $(\pi, \mathcal{V})$.*

Proof: As in the proof of (7.3.9) above, the representation $(\pi_U, \mathcal{V}_U)$ of P/U has finite composition length. If $\mathcal{V}_U$ has a composition factor not containing λ_U, then $(\mathcal{V}_U)\check{} = (\check{\mathcal{V}})_{U^-}$ has a composition factor not containing $(\lambda_U)\check{} = (\check{\lambda})_{U^-}$. Applying (7.3.10) repeatedly, we see that $(\mathcal{V}_U)\check{}$ has an irreducible subspace not containing $(\lambda_U)\check{}$. It follows that $\mathcal{V}_U$ has an irreducible quotient $(\xi, \mathcal{W})$ not containing λ_U. This quotient takes the form $\xi = \xi^{(1)} \otimes \dots \otimes \xi^{(e)}$ and some $\xi^{(i)}$ does not contain $\lambda^{(i)}$. Therefore π embeds in a representation $\xi'^{(1)} \times \dots \times \xi'^{(e)}$, where $\xi'^{(i)} = \xi^{(i)} \otimes \chi_i \circ \det$ for various unramified quasicharacters χ_i of $F^\times$. Observe now that the representations $\xi'^{(i)}$ are all supercuspidal: otherwise, the supercuspidal support of π would have at least $e + 1$ elements (counted according to multiplicity), while (7.3.9) says that this support has exactly e elements. Therefore the supercuspidal support of π is precisely the multiset $\{\xi'^{(1)}, \dots, \xi'^{(e)}\}$, which must therefore be a permutation of $\{\pi_1, \dots, \pi_e\}$ (in the notation of (7.3.9)), by (7.3.11). However, each π_i contains a maximal simple type equivalent to the one associated to λ, while some $\boldsymbol{\xi}'^{(i)}$ does not. This contradiction shows that every composition factor of $\mathcal{V}_U$ contains λ_U.

Any irreducible representation $\boldsymbol{\xi}$ of P/U which contains λ_U is irreducibly induced from a uniquely determined extension of λ_U, by (7.3.8). It follows that λ_U occurs with multiplicity one in any such representation $\boldsymbol{\xi}$. Thus the composition length of $\mathcal{V}_U$ is the multiplicity of λ_U in $\boldsymbol{\pi}_U$. However, by (7.3.2), this multiplicity equals the multiplicity of λ_P in $\boldsymbol{\pi}$. Since λ_P induces irreducibly to λ, the representations λ, λ_P have the same multiplicities in $\boldsymbol{\pi}$. ■

(7.3.14) Theorem: *Let $\boldsymbol{\xi} = \boldsymbol{\xi}^{(1)} \otimes \ldots \otimes \boldsymbol{\xi}^{(e)}$ be a smooth irreducible representation of P/U which contains λ_U, and let $(\boldsymbol{\pi}, \mathcal{V})$ be a G-composition factor of $\boldsymbol{\xi}^{(1)} \times \ldots \times \boldsymbol{\xi}^{(e)}$. Then $\boldsymbol{\pi}$ contains the simple type λ.*

Proof: Suppose first that $(\boldsymbol{\pi}, \mathcal{V})$ is a subspace of $\boldsymbol{\xi}^{(1)} \times \ldots \times \boldsymbol{\xi}^{(e)}$. Put $\boldsymbol{\xi}' = \delta_P^{1/2} \otimes \boldsymbol{\xi}$ so that $\boldsymbol{\xi}^{(1)} \times \ldots \times \boldsymbol{\xi}^{(e)} = \mathrm{Ind}(\boldsymbol{\xi}')$. The representation $\boldsymbol{\xi}'$ also contains λ_U. Write $\mathcal{W}$ for the representation space of $\boldsymbol{\xi}'$. The G-embedding $\mathcal{V} \to \mathrm{Ind}(\boldsymbol{\xi}')$ leads, via Frobenius Reciprocity, to a surjective P-homomorphism $\mathcal{V} \to \mathcal{W}$ which must factor through $\mathcal{V} \to \mathcal{V}_U$. This gives us a surjection $\mathcal{V}_U \to \mathcal{W}$. Since $\mathcal{W}^{\lambda_U}$ is nonzero, it follows that $(\mathcal{V}_U)^{\lambda_U}$ is nonzero. (7.3.2) now shows that $\mathcal{V}^{\lambda_P}$ is nonzero, i.e. that $\boldsymbol{\pi}$ contains λ_P. It therefore also contains λ, as required.

Now let $(\boldsymbol{\pi}, \mathcal{V})$ be a composition factor of $\boldsymbol{\xi}^{(1)} \times \ldots \times \boldsymbol{\xi}^{(e)}$. By (7.3.11), there is a permutation $(\boldsymbol{\tau}^{(1)}, \ldots, \boldsymbol{\tau}^{(e)})$ of $(\boldsymbol{\xi}^{(1)}, \ldots, \boldsymbol{\xi}^{(e)})$ such that $\boldsymbol{\pi}$ is isomorphic to a subspace of $\boldsymbol{\tau}^{(1)} \times \ldots \times \boldsymbol{\tau}^{(e)}$ However, the representation $\boldsymbol{\tau}^{(1)} \otimes \ldots \otimes \boldsymbol{\tau}^{(e)}$ of P/U contains a representation isomorphic to λ_U. The result now follows from the first case. ■

We can use this same machinery of supercuspidal supports to analyse behaviour relative to more general parabolic subgroups of G. We continue with our fixed simple type (J, λ) in G. As before, P is a parabolic subgroup subordinate to the type (J, λ) and minimal for this property, and U denotes the unipotent radical of P.

(7.3.15) Proposition: *Let $(\boldsymbol{\pi}, \mathcal{V})$ be an irreducible smooth representation of G containing the simple type (J, λ). Let P_1 be a parabolic subgroup of G with unipotent radical U_1. The following are equivalent:*

(i) $\mathcal{V}_{U_1} \neq \{0\}$;

(ii) there exists an irreducible representation $\boldsymbol{\tau}$ of P_1/U_1 such that $\boldsymbol{\pi}$ is isomorphic to a G-subspace of $\mathrm{Ind}(\boldsymbol{\tau} : P_1, G)$;

(iii) some G-conjugate of P_1 contains P.

Proof: The equivalence of *(i)* and *(ii)* is standard. We know that $\mathcal{V}_U \neq \{0\}$, so transitivity of the Jacquet functor gives *(iii)$\Rightarrow$(i)*. Let us assume that *(ii)* holds. We can find a parabolic subgroup $\overline{P}_2$ of P_1/U_1, with unipotent radical $\overline{U}_2$, and an irreducible supercuspidal representation $\boldsymbol{\tau}_2$ of $\overline{P}_2/\overline{U}_2$ such that $\boldsymbol{\tau}$ embeds in $\mathrm{Ind}(\boldsymbol{\tau}_2 : \overline{P}_2, P_1)$. There is a parabolic subgroup P_2 of G, with unipotent radical U_2 and $P_2 \subset P_1$, whose image in P_1/U_1 is $\overline{P}_2$. By transitivity of induction, $\boldsymbol{\pi}$ embeds in $\mathrm{Ind}(\boldsymbol{\tau}_2 : P_2, G)$. If we write $P_2/U_2 = G_1 \times \ldots \times G_r$, with $G_i \cong GL(N_i, F)$, for various integers N_i, we have $\boldsymbol{\tau}_2 = \boldsymbol{\rho}_1 \otimes \ldots \otimes \boldsymbol{\rho}_r$, where $\boldsymbol{\rho}_i$ is an irreducible supercuspidal representation of G_i, and $\boldsymbol{\rho}_i$ is contained in the supercuspidal support of $\boldsymbol{\pi}$. This implies that $N_i = N/e$, $e = e(J, \lambda)$, so P_2 is G-conjugate to P. This proves *(iii)*. ■

In the situation of (7.3.15), we can use the same simple techniques to describe the representation $\boldsymbol{\tau}$.

(7.3.16) Proposition: *Let (J, λ) be a simple type in G, and P_1 a parabolic subgroup of G with unipotent radical U_1. Write $P_1/U_1 = G_1 \times \ldots \times G_r$, with $G_i \cong GL(N_i, F)$ say. Let $\boldsymbol{\tau} = \boldsymbol{\tau}_1 \otimes \ldots \otimes \boldsymbol{\tau}_r$ be an irreducible smooth representation of P_1/U_1. The following are equivalent:*

(i) $\boldsymbol{\tau}_1 \times \ldots \times \boldsymbol{\tau}_r$ contains λ;

(ii) every composition factor of $\boldsymbol{\tau}_1 \times \ldots \times \boldsymbol{\tau}_r$ contains λ;

(iii) for each i, $\boldsymbol{\tau}_i$ contains a simple type (J_i, λ_i) whose associated maximal type is equivalent to that of λ.

Proof: For each i, $1 \leq i \leq r$, we can find supercuspidal representations $\boldsymbol{\sigma}_{i1}, \ldots, \boldsymbol{\sigma}_{is_i}$ such that $\boldsymbol{\tau}_i$ embeds in $\boldsymbol{\sigma}_{i1} \times \ldots \times \boldsymbol{\sigma}_{is_i}$. Set

$$\boldsymbol{\Sigma} = \boldsymbol{\sigma}_{i1} \times \boldsymbol{\sigma}_{i2} \times \ldots \times \boldsymbol{\sigma}_{rs_r}.$$

Transitivity of induction implies that $\boldsymbol{\tau}_1 \times \ldots \times \boldsymbol{\tau}_r$ embeds in $\boldsymbol{\Sigma}$. Consider the statement

(iv) Each $\boldsymbol{\sigma}_{ij}$, $1 \leq i \leq r$, $1 \leq j \leq s_i$, contains a maximal simple type equivalent to that associated to λ.

Then *(iv)⇒(iii)* by (7.3.14), while *(iii)⇒(iv)* by (7.3.12). Next, *(iv)⇒(ii)* by (7.3.14), and certainly *(ii)⇒(i)*. The supercuspidal support of any composition factor of $\boldsymbol{\tau}_1 \times \ldots \times \boldsymbol{\tau}_r$ is $\{\boldsymbol{\sigma}_{ij} : 1 \leq i \leq r,\ 1 \leq j \leq s_r\}$, so *(i)⇒(iv)* by (7.3.12). ■

We conclude this section with a powerful uniqueness property.

(7.3.17) Theorem: *Let $\boldsymbol{\pi}$ be an irreducible smooth representation of G containing two simple types (J_1, λ_1), (J_2, λ_2). Then these types are conjugate:— there exists $x \in G$ such that $J_2 = x^{-1} J_1 x$ and λ_2 is equivalent to λ_1^x.*

Proof: Let $[\mathfrak{A}_i, n_i, 0, \beta_i]$ be a simple stratum in A defining the simple type (J_i, λ_i), and $\theta_i \in \mathcal{C}(\mathfrak{A}_i, 0, \beta_i)$ the simple character occurring in $\lambda_i \mid H^1(\beta_i, \mathfrak{A}_i)$, $i = 1, 2$. Write $E_i = F[\beta_i]$, $B_i = \mathrm{End}_{E_i}(V)$, $\mathfrak{B}_i = \mathfrak{A}_i \cap$

B_i. Any representation in the supercuspidal support of π contains the maximal simple type associated to (J_i, λ_i), for $i = 1, 2$, by (7.3.12). We apply (6.2.5) to these maximal simple types to deduce that $e(E_1|F) = e(E_2|F)$. Further, we have $e(\mathfrak{B}_1|\mathfrak{o}_{E_1}) = e(\mathfrak{B}_2|\mathfrak{o}_{E_2})$, since this integer is the cardinality of the supercuspidal support of π. Since $e(\mathfrak{A}_i|\mathfrak{o}_F) = e(\mathfrak{B}_i|\mathfrak{o}_{E_i})e(E_i|F)$, we deduce that $e(\mathfrak{A}_1|\mathfrak{o}_F) = e(\mathfrak{A}_2|\mathfrak{o}_F)$, and hence that the principal orders $\mathfrak{A}_i$ in A are isomorphic. Since the representations λ_i both occur in π, they intertwine in G, so we can apply (5.7.1) to get the result. ■

(7.3.18) Comment: In the foregoing, we have had to make frequent use of the fact [Cs] (6.3.10) that, for any reductive group G over F, a finitely generated admissible representation has finite composition length. We point out that, for $G = GL(N, F)$ (or the Levi component of a parabolic subgroup of such a G), this is an easy consequence of (2.6.2).

(7.4) Proof of the principal lemma

We now prove (7.3.1). Let us abbreviate $\lambda_P = \rho$, $K = (J \cap P)H^1$, and suppose ρ acts on the vector space W.

If $\phi \in \mathcal{H}(G)$ and $g \in G$, write ${}^g\phi$ for the function

$${}^g\phi : x \mapsto \phi(g^{-1}x), \quad x \in G.$$

Then, if $(\pi, \mathcal{V})$ is any smooth representation of G, it is trivial to check that the operator $\pi(e_\rho) \circ \pi(g) \circ \pi(e_\rho)$ on $\mathcal{V}$ (in fact on $\mathcal{V}^\rho$) is none other than $\pi(e_\rho * {}^g e_\rho)$. So, with $a \in \boldsymbol{D}(\mathfrak{B})$ as in (7.3.1), we have to examine the element $e_\rho * {}^a e_\rho \in e_\rho * \mathcal{H}(G) * e_\rho$. The desired result is implied by:

(7.4.1) *$e_\rho * {}^a e_\rho$ is an invertible element of the algebra $e_\rho * \mathcal{H}(G) * e_\rho$.*

We prove this by working out the image Ψ of the function $e_\rho * {}^a e_\rho$ under the inverse of the algebra isomorphism

$$\Upsilon : \mathcal{H}(G, \rho) \otimes_{\mathbb{C}} \mathrm{End}_{\mathbb{C}}(W) \xrightarrow{\approx} e_\rho * \mathcal{H}(G) * e_\rho$$

described in (4.2.4). For convenience, we assume that our Hecke algebras are defined relative to a Haar measure μ on G giving $\mu(K) = 1$. We see straightaway that the support of $e_\rho * {}^a e_\rho$ is contained in KaK, so the same applies to Ψ. By (7.2.9) and (7.2.10), there is a nonzero function $\psi \in \mathcal{H}(G, \rho)$ with support KaK, and this is unique up to scalar. The representation $\lambda_U = \rho \mid J \cap M$ is irreducible, and the element a normalises the group $J \cap M$. It follows that the representations $\rho \mid K \cap K^a$, $\rho^a \mid K \cap K^a$ are irreducible, and so the nonzero values of ψ are isomorphisms.

The main step is to show:

(7.4.2) *There is a nonzero* $\Phi \in I_a(\rho)$ *such that* $\Upsilon^{-1}(e_\rho * {}^a e_\rho) = \psi \otimes \Phi$.

Let us first deduce the result. We have to show that $\psi \otimes \Phi$ is an invertible element of the algebra $\mathcal{H}(G, \rho) \otimes \mathrm{End}_{\mathbb{C}}(W)$. This amounts to showing that ψ is invertible in $\mathcal{H}(G, \rho)$, since Φ is certainly invertible in $\mathrm{End}_{\mathbb{C}}(W)$. (7.2.17)*(i)* shows that $\mathrm{Ind}(\rho : K, J) = \lambda$. (4.1.3) now gives us an algebra isomorphism $\mathcal{H}(G, \rho) \cong \mathcal{H}(G, \lambda)$ which carries ψ to some nonzero function $\psi' \in \mathcal{H}(G, \lambda)$ with support JaJ. We know from (5.6.6) that such an ψ' is invertible in $\mathcal{H}(G, \lambda)$, so we have the result.

Now let us prove (7.4.2). We start by choosing dual bases $\{w_i\}$, $\{\check{w}_j\}$ of W, $\check{W}$, and write $\phi = e_\rho * {}^a e_\rho$. Then $\Upsilon^{-1}(\phi)$ is the sum of terms $w_i \otimes \check{w}_j \otimes \Psi_{ij}$, where Ψ_{ij} is the $\mathrm{End}_{\mathbb{C}}(\check{W})$-valued function given by

$$\Psi_{ij}(g)\check{v} = d \iint_{K \times K} \langle \rho(\ell) w_i, \check{v} \rangle \check{\rho}(k^{-1}) \check{w}_j . \phi(kg\ell)\, dk d\ell,$$

for $g \in G$, $\check{v} \in \check{W}$, and where $d = \dim(\rho)$. We substitute

$$\begin{aligned} \phi(g) &= \int_K e_\rho(x) e_\rho(a^{-1}x^{-1}g)\, dx \\ &= d^2 \sum_{b,c} \int_K \langle \rho(x^{-1}) w_b, \check{w}_b \rangle \langle \rho(g^{-1}xa) w_c, \check{w}_c \rangle\, dx, \end{aligned}$$

with the understanding that the symbol $\rho(y)$ means 0 if $y \notin K$. We use the Schur orthogonality relation to integrate out the variable ℓ, to get

$$\begin{aligned} \Psi_{ij}(g)\check{v} = d^2 \sum_{b,c} \iint_{K \times K} & \langle w_i, \check{w}_c \rangle \langle \rho(g^{-1}k^{-1}xa) w_c, \check{v} \rangle \\ & . \langle \rho(x^{-1}) w_b, \check{w}_b \rangle \check{\rho}(k^{-1}) \check{w}_j\, dk dx. \end{aligned}$$

Now put $g = a$, so that we have $k, x \in K$, $k^{-1}x \in aKa^{-1}$. Writing $H = K \cap aKa^{-1}$, we can put $x = kh$, with k allowed to range over K, h over H. Thus, for $u \in W$, $\check{v} \in \check{W}$, we have

$$\begin{aligned} & \langle u, \Psi_{ij}(a)\check{v} \rangle \\ & = d^2 \sum_{b,c} \int_K \int_H \langle w_i, \check{w}_c \rangle \langle \rho(a^{-1}ha) w_c, \check{v} \rangle \\ & \qquad \langle \rho(h^{-1}k^{-1}) w_b, \check{w}_b \rangle \langle u, \check{\rho}(k^{-1}) \check{w}_j \rangle\, dk dh. \end{aligned}$$

Since the element a intertwines ρ via an isomorphism, there is an automorphism Φ of W such that $\rho(a^{-1}ha) = \Phi^{-1} \circ \rho(h) \circ \Phi$, for $h \in H$.

Indeed, any nonzero $\Phi \in I_a(\rho)$ will do here:— we take $\Phi = (\psi(a)^{-1})\check{}$. We substitute this into our integral. Now we observe that ρ restricts irreducibly to H, since H contains $J \cap M(A)^\times$, and $\rho \mid J \cap M(A)^\times = \lambda_U$, which is irreducible. It follows that we can use the Schur orthogonality relation again to reduce our integral to

$$d(K:H)^{-1} \sum_{b,c} \int_K \langle w_i, \check{w}_c\rangle\langle \Phi w_c, \check{w}_b\rangle$$
$$\langle \rho(k^{-1})w_b, (\Phi^{-1})\check{}\,\check{v}\rangle\langle u, \check{\rho}(k^{-1})\check{w}_j\rangle\, dk.$$

We can reduce this again to

$$(K:H)^{-1} \sum_{b,c} \langle w_i, \check{w}_c\rangle\langle \Phi w_c, \check{w}_b\rangle\langle w_b, \check{w}_j\rangle\langle u, (\Phi^{-1})\check{}\,\check{v}\rangle$$
$$= (K:H)^{-1}\langle \Phi w_i, \check{w}_j\rangle\langle u, (\Phi^{-1})\check{}\,\check{v}\rangle.$$

Summing over i and j, we conclude that

$$\Upsilon^{-1}(e_\rho * {}^a e_\rho)(a) = (K:H)^{-1}(\Phi^{-1})\check{} \otimes \Phi = (K : K \cap {}^a K)^{-1}\psi(a) \otimes \Phi,$$

as required. ■

Remark: It is worth exhibiting the exact outcome of this computation. Observe that if we take $x, y \in K$, we get

$$\Upsilon^{-1}(e_\rho * {}^a e_\rho)(xay) = (K : K \cap {}^a K)^{-1}\psi(xay) \otimes \Phi,$$

as we should. Therefore

$$\text{(7.4.3)} \qquad \Upsilon^{-1}(e_{\lambda_P} * {}^a e_{\lambda_P}) = (K : K \cap {}^a K)^{-1}\psi \otimes (\psi(a)^{-1})\check{},$$

for $a \in \boldsymbol{D}(\mathfrak{B})$, and any nonzero $\psi \in \mathcal{H}(G, \lambda_P)$ with support KaK.

On the other hand, for $x, y \in K$, we get

(7.4.4)
$$\Upsilon^{-1}(e_{\lambda_P} * {}^{xay} e_{\lambda_P})(a) = (K : K \cap {}^a K)^{-1}\psi(a) \otimes (\lambda_P(x) \circ (\psi(a)^{-1})\check{} \circ \lambda_P(y)).$$

Thus, allowing x and y to vary over K, this accounts for the $\dim(\lambda_P)^2$ linearly independent functions in $e_\rho * \mathcal{H}(G) * e_\rho$ supported on KaK.

(7.4.5) Remark: (7.4.1) holds under substantially more general hypotheses than those used here. Let us return to the situation of (7.3.7). We have a pair of opposite parabolic subgroups $P = MU$, $P^- = MU^-$ of G, a compact open subgroup K of G such that $\mathsf{K} = (\mathsf{K} \cap U^-)(\mathsf{K} \cap M)(\mathsf{K} \cap U)$, and an irreducible representation ρ of K which is null on both $\mathsf{K} \cap U^-$ and $\mathsf{K} \cap U$. Further, the representation $\rho_U = \rho \mid \mathsf{K} \cap M$ is irreducible.

We are given a subgroup D of M which normalises $\mathsf{K} \cap M$ and stabilises the representation ρ_U. It follows readily that every $\mathsf{d} \in \mathsf{D}$ intertwines ρ. Indeed, the space $I_{\mathsf{d}}(\rho)$ has dimension one, and every nonzero element of it is an automorphism of the underlying vector space. *We further assume that, for every* $\mathsf{d} \in \mathsf{D}$, *the (essentially unique) nonzero function* $\phi_{\mathsf{d}} \in \mathcal{H}(G, \rho)$ *with support* KdK *is invertible in* $\mathcal{H}(G, \rho)$.

Under these hypotheses, the same proof as that given above shows:

(7.4.6) *For every* $\mathsf{d} \in \mathsf{D}$, *the function* $e_\rho * {}^{\mathsf{d}}e_\rho$ *is an invertible element of the algebra* $e_\rho * \mathcal{H}(G) * e_\rho$.

(7.5) The strong intertwining property

We start with a brief discussion of a general notion. For the moment, let G denote any locally profinite group, H a compact open subgroup of G and ρ an irreducible smooth representation of H. We say that the triple $(\mathsf{G}, \mathsf{H}, \rho)$ *has the strong intertwining property* if the following assertion holds:

(7.5.1) *Let* $(\pi, \mathcal{V})$ *be a smooth representation of* G, *and suppose*

(i) $\mathcal{V}$ *is generated as* G*-space by* $\mathcal{V}^\rho$;

(ii) $\mathcal{V}^\rho$ *is finite-dimensional.*

Then $\mathcal{V}$ *has finite composition length over* G, *and every composition factor of it contains* ρ.

At this level of generality, we need only one simple result. For a triple $(\mathsf{G}, \mathsf{H}, \rho)$ as above, let e_ρ denote the idempotent of $\mathcal{H}(\mathsf{G})$ defined by (4.2.1).

(7.5.2) Lemma: *(i) Let* $(\mathsf{G}, \mathsf{H}, \rho)$ *be a triple as above with the following property:*

any smooth representation $(\pi, \mathcal{V})$, such that $\mathcal{V}^\rho$ is a simple $e_\rho * \mathcal{H}(\mathsf{G}) * e_\rho$-module of finite $\mathbb{C}$-dimension, and which is generated (over G) by $\mathcal{V}^\rho$, is irreducible over G.

Then $(\mathsf{G}, \mathsf{H}, \rho)$ *has the strong intertwining property.*

(ii) For $1 \leq i \leq t$, *let* $(\mathsf{G}_i, \mathsf{H}_i, \rho_i)$ *be a triple with the strong intertwining property. Then* $(\mathsf{G}_1 \times \ldots \times \mathsf{G}_t, \mathsf{H}_1 \times \ldots \times \mathsf{H}_t, \rho_1 \otimes \ldots \otimes \rho_t)$ *has the strong intertwining property.*

Proof: In *(i)*, let $(\pi, \mathcal{V})$ be a smooth representation of G which is generated by $\mathcal{V}^\rho$, and such that $\mathcal{V}^\rho$ has finite dimension. We have to show that $\mathcal{V}$ has finite composition length over G, and that every composition factor contains ρ. We proceed by induction on the $e_\rho * \mathcal{H}(\mathsf{G}) * e_\rho$-composition length of $\mathcal{V}^\rho$. By hypothesis, the result holds when this length is one. We therefore assume that this length is at least two, and we choose a maximal $e_\rho * \mathcal{H}(\mathsf{G}) * e_\rho$-submodule M of $\mathcal{V}^\rho$. Let $\mathcal{W}$ be the G-subspace of $\mathcal{V}$

generated by M. Since $\mathcal{W}^\rho = e_\rho * \mathcal{W} = M$, we have $\mathcal{W} \neq \mathcal{V}$, and $\mathcal{W}$ satisfies (7.5.1)*(i)*, *(ii)*. By inductive hypothesis therefore, $\mathcal{W}$ has finite length and each of its composition factors contains ρ. Moreover, the quotient $\mathcal{V}/\mathcal{W}$ is generated, over G, by the image $\mathcal{V}^\rho + \mathcal{W}/\mathcal{W} = \mathcal{V}^\rho/M = (\mathcal{V}/\mathcal{W})^\rho$ of $\mathcal{V}^\rho$. Since $\mathcal{V}^\rho/M$ is a simple $e_\rho * \mathcal{H}(\mathrm{G}) * e_\rho$-module, it follows that $\mathcal{V}/\mathcal{W}$ is simple and the assertion follows.

In part *(ii)*, we need only treat the case $t = 2$. Write $\mathrm{G} = \mathrm{G}_1 \times \mathrm{G}_2$, $\mathrm{H} = \mathrm{H}_1 \times \mathrm{H}_2$, $\rho = \rho_1 \otimes \rho_2$. Let $(\pi, \mathcal{V})$ be a smooth representation of G satisfying (7.5.1)*(i)*, *(ii)*. By (7.5.2)*(i)*, we may as well assume that the $e_\rho * \mathcal{H}(\mathrm{G}) * e_\rho$-module $\mathcal{V}^\rho$ is *simple*. We have $\mathcal{H}(\mathrm{G}) \cong \mathcal{H}(\mathrm{G}_1) \otimes_{\mathbb{C}} \mathcal{H}(\mathrm{G}_2)$ as $\mathbb{C}$-algebras, and this identifies the subalgebras

$$e_\rho * \mathcal{H}(\mathrm{G}) * e_\rho = e_{\rho_1} * \mathcal{H}(\mathrm{G}_1) * e_{\rho_1} \otimes_{\mathbb{C}} e_{\rho_2} * \mathcal{H}(\mathrm{G}_2) * e_{\rho_2}.$$

Thus $\mathcal{V}^\rho = M_1 \otimes M_2$, where M_i is some simple $e_{\rho_i} * \mathcal{H}(\mathrm{G}_i) * e_{\rho_i}$-module.

By hypothesis, we have a G-surjection

$$\mathcal{H}(\mathrm{G}) \otimes_{e_\rho * \mathcal{H}(\mathrm{G}) * e_\rho} \mathcal{V}^\rho \to \mathcal{V}^\rho.$$

Abbreviate $\mathcal{H}(\mathrm{G}) \otimes_{e_\rho * \mathcal{H}(\mathrm{G}) * e_\rho} \mathcal{V}^\rho = \mathcal{H}(\mathrm{G}) \otimes_\rho \mathcal{V}^\rho$, and use similar notations in G_i. We may identify $\mathcal{H}(\mathrm{G}) \otimes_\rho \mathcal{V}^\rho = (\mathcal{H}(\mathrm{G}_1) \otimes_{\rho_1} M_1) \otimes_{\mathbb{C}} (\mathcal{H}(\mathrm{G}_2) \otimes_{\rho_2} M_2)$, and by hypothesis each factor $\mathcal{H}(\mathrm{G}_i) \otimes_{\rho_i} M_i$ is an irreducible representation of G_i, i.e. a simple $\mathcal{H}(\mathrm{G}_i)$-module. It follows that $\mathcal{H}(\mathrm{G}) \otimes_\rho \mathcal{V}^\rho$ is a simple $\mathcal{H}(\mathrm{G})$-module and so an irreducible representation of G which contains ρ. ∎

Continuing in this same general situation, we can now sharpen the result (4.2.3).

(7.5.3) Proposition: *Let* G *be a locally profinite group,* H *a compact open subgroup of* G, *and* ρ *an irreducible smooth representation of* H. *Suppose that the triple* $(\mathrm{G}, \mathrm{H}, \rho)$ *has the strong intertwining property. Let* $\mathfrak{C}$ *denote the category of all smooth representations* $(\pi, \mathcal{V})$ *of* G *such that* $\mathcal{V}$ *is generated over* G *by* $\mathcal{V}^\rho$ *and* $\mathcal{V}^\rho$ *is finite-dimensional. The process* $\mathcal{V} \mapsto \mathcal{V}^\rho$ *then gives an equivalence of categories between* $\mathfrak{C}$ *and the category* $\mathfrak{M}$ *of* $e_\rho * \mathcal{H}(\mathrm{G}) * e_\rho$*-modules of finite complex dimension.*

Proof: The operation $\mathcal{V} \mapsto \mathcal{V}^\rho$ surely gives a functor $\mathfrak{C} \to \mathfrak{M}$. In the opposite direction, let M be a left $e_\rho * \mathcal{H}(\mathrm{G}) * e_\rho$-module of finite dimension and consider the space $\mathcal{V} = \mathcal{H}(\mathrm{G}) \otimes_{e_\rho * \mathcal{H}(\mathrm{G}) * e_\rho} M$. This affords a smooth representation of G, with G acting on the factor $\mathcal{H}(\mathrm{G})$ by left translation. We have $\mathcal{V}^\rho = e_\rho * \mathcal{V} = e_\rho * \mathcal{H}(\mathrm{G}) \otimes M$. However, $M = e_\rho * M$ so this collapses to $\mathcal{V}^\rho = e_\rho * \mathcal{H}(\mathrm{G}) * e_\rho \otimes M = e_\rho \otimes M \cong M$. Thus $\mathcal{V}$ is generated over G by $\mathcal{V}^\rho$, which is of finite dimension. Thus $\mathcal{V}$ lies in the object class $|\mathfrak{C}|$ of $\mathfrak{C}$, and this process $M \mapsto \mathcal{H}(\mathrm{G}) \otimes_{e_\rho * \mathcal{H}(\mathrm{G}) * e_\rho} M$ gives a functor $\mathfrak{M} \to \mathfrak{C}$.

We have just shown $(\mathcal{H}(\mathrm{G}) \otimes M)^\rho \cong M$, so the composition $\mathfrak{M} \to \mathfrak{C} \to \mathfrak{M}$ is equivalent to the identity. In the opposite direction, for any $\mathcal{V} \in |\mathfrak{C}|$, we have a canonical G-surjection

$$\mathcal{H}(\mathrm{G}) \otimes_{e_\rho * \mathcal{H}(\mathrm{G}) * e_\rho} \mathcal{V}^\rho \to \mathcal{V}$$

by $\phi \otimes v \mapsto \phi * v$, $\phi \in \mathcal{H}(\mathrm{G})$, $v \in \mathcal{V}^\rho$. This induces the identity map on $\mathcal{V}^\rho$, so any kernel cannot meet $e_\rho \otimes \mathcal{V}^\rho$. However, any proper G-subspace of $\mathcal{V}$ does meet $\mathcal{V}^\rho$, by hypothesis. Therefore the canonical map $\mathcal{H}(\mathrm{G}) \otimes \mathcal{V}^\rho \to \mathcal{V}$ is an isomorphism. This shows that the composition $\mathfrak{C} \to \mathfrak{M} \to \mathfrak{C}$ is equivalent to the identity, and completes the proof. ■

Remark: In (4.2.3), we showed that one could obtain an irreducible representation of G containing ρ from a simple $e_\rho * \mathcal{H}(\mathrm{G}) * e_\rho$-module M by forming $\mathcal{V}$ as above, and factoring out a maximal G-subspace $\mathcal{V}_1$ of $\mathcal{V}$ with the property that $\mathcal{V}_1 \cap (e_\rho \otimes M) = \{0\}$. We just showed that this space $\mathcal{V}_1$ is trivial when we have the strong intertwining property .

Now we revert to our standard notation, with $G = \mathrm{Aut}_F(V)$.

(7.5.4) Theorem: *Let $G = \mathrm{Aut}_F(V)$ as before, and let (J, λ) be a simple type in G. Then the triple (G, J, λ) has the strong intertwining property.*

Proof: Let $(\pi, \mathcal{V})$ be a smooth representation of G such that $\mathcal{V}^\lambda$ is finite-dimensional and generates $\mathcal{V}$ over G. By (7.5.2), we need only treat the case in which $\mathcal{V}^\lambda$ is a simple $e_\lambda * \mathcal{H}(G) * e_\lambda$-module. Take some nonzero $v_0 \in \mathcal{V}^\lambda$, and let $\mathcal{W}$ be a G-subspace of $\mathcal{V}$ maximal for the property of not containing v_0. Zorn's Lemma guarantees the existence of such a $\mathcal{W}$. By the simplicity of $\mathcal{V}^\lambda$, the element v_0 generates $\mathcal{V}$ and so

(7.5.5) *(i) $\mathcal{V}/\mathcal{W}$ is irreducible;*
(ii) $\mathcal{W}^\lambda = \{0\}$;
(iii) the quotient map identifies $\mathcal{V}^\lambda = (\mathcal{V}/\mathcal{W})^\lambda$.

We have to show that $\mathcal{W} = \{0\}$.

We treat first the case in which the simple type (J, λ) is *maximal.* We start by showing that $\mathcal{V}$ admits a central quasicharacter. Let ω be the central quasicharacter of the irreducible representation $\mathcal{V}/\mathcal{W}$. For $z \in F^\times$ (the centre of G), the map $v \mapsto (\pi(z)v - \omega(z)v)$, $v \in \mathcal{V}$, is a G-endomorphism of $\mathcal{V}$ whose image is contained in $\mathcal{W}$. This image is generated by the image of the generating set $\mathcal{V}^\lambda$, and this image is contained in $\mathcal{W}^\lambda = \{0\}$. This map is therefore null, and ω is the central quasicharacter of $\mathcal{V}$. However, the irreducible representation $\mathcal{V}/\mathcal{W}$ is supercuspidal, since (J, λ) is maximal, so $\mathcal{W}$ is a quotient of $\mathcal{V}$ not containing λ. This implies that $\mathcal{W} = \{0\}$, as required.

Now let (J, λ) be a general simple type, and assume for a contradiction that $\mathcal{W} \neq \{0\}$. As before, let P denote a parabolic subgroup of G which is subordinate to the type (J, λ), and minimal for this property. Let U denote the unipotent radical of P. As in the first case, $\mathcal{V}$ admits a central quasicharacter, so any supercuspidal subquotient $\mathcal{U}$ of $\mathcal{W}$ is a direct summand of $\mathcal{V}$, hence a quotient of $\mathcal{V}$. Since $\mathcal{U}^\lambda \subset \mathcal{W}^\lambda = \{0\}$, this is impossible. We deduce that $\mathcal{W}$ has no supercuspidal subquotient. So, if $\mathcal{W}$ is not zero, there exists a proper parabolic subgroup P_1 of G, with unipotent radical U_1, say, so that $\mathcal{W}_{U_1} \neq \{0\}$. We assume that P_1 is minimal for this property. The Jacquet functor relative to U_1 is exact ([**Cs**] (3.2.3)), so we have an exact sequence

$$0 \to \mathcal{W}_{U_1} \to \mathcal{V}_{U_1} \to (\mathcal{V}/\mathcal{W})_{U_1} \to 0.$$

We first eliminate the possibility that $(\mathcal{V}/\mathcal{W})_{U_1} = \{0\}$. Suppose, for a contradiction, that $(\mathcal{V}/\mathcal{W})_{U_1}$ is zero. The representation $\mathcal{V}_{U_1} = \mathcal{W}_{U_1}$ of P_1/U_1 is finitely generated ([**Cs**] (3.3.1)), and so has an irreducible quotient $(\boldsymbol{\tau}, \mathcal{U})$. By the definition of P_1, this representation $\boldsymbol{\tau}$ is supercuspidal. We write $P_1/U_1 = G_1 \times \ldots \times G_r$, with $G_i \cong GL(N_i, F)$. The representation $\boldsymbol{\tau}$ then takes the form $\boldsymbol{\tau} = \boldsymbol{\tau}_1 \otimes \ldots \otimes \boldsymbol{\tau}_r$, where $\boldsymbol{\tau}_i$ is an irreducible supercuspidal representation of G_i. Let λ_0 be the maximal simple type associated to λ. The next step is to show that some $\boldsymbol{\tau}_i$ cannot contain a representation equivalent to λ_0. Suppose the contrary. By Frobenius Reciprocity, we have a nontrivial homomorphism $\mathcal{W} \to \text{Ind}(\boldsymbol{\tau} : P_1, G)$ and, by (7.3.15), every composition factor of this induced representation contains λ. It follows that $\mathcal{W}$ contains λ, contrary to (7.5.5). Therefore some $\boldsymbol{\tau}_i$ does not contain λ_0. Since $\boldsymbol{\tau}_i$ is supercuspidal, it therefore contains no simple type whose associated maximal type is λ_0. Now we use the fact that $\mathcal{V}_{U_1} = \mathcal{W}_{U_1}$. We have a nontrivial homomorphism $\mathcal{V} \to \text{Ind}(\boldsymbol{\tau})$. By (7.3.16), the representation $\text{Ind}(\boldsymbol{\tau})$ does not contain λ, so $\mathcal{V}$ has a quotient which does not contain λ, which is impossible.

We conclude that $(\mathcal{V}/\mathcal{W})_{U_1} \neq \{0\}$, and we can therefore assume that $P_1 \supset P$ by (7.3.15). Writing $P_1/U_1 = G_1 \times \ldots \times G_r$ as before, let (J_i, λ_i) be a simple type in G_i whose associated maximal type is that of (J, λ). Put $\lambda_{U_1} = \lambda_1 \otimes \ldots \otimes \lambda_r$. The representation $(\mathcal{V}/\mathcal{W})_{U_1}$ has finite length ([**Cs**] (3.3.1) and (6.3.10) again), and every composition factor of it contains λ_{U_1}. We assert, on the other hand, that $\mathcal{W}_{U_1}$ cannot contain λ_{U_1}. For, suppose it does contain λ_{U_1}. We can identify $\mathcal{W}_U$ with the Jacquet module of $\mathcal{W}_{U_1}$ relative to the parabolic subgroup P/U_1 of P_1/U_1. (7.3.6)*(i)* then shows that $\mathcal{W}_U$ contains λ_U, and hence ((7.3.6)*(ii)*) $\mathcal{W}$ contains λ, contrary to (7.5.5). Thus $\mathcal{W}_{U_1}$ does not contain λ_{U_1}. We can further assume that $(\mathcal{V}_{U_1})^{\lambda_{U_1}}$ generates $\mathcal{V}_{U_1}$. If it did not, $\mathcal{V}_{U_1}$ would have a simple quotient $\mathcal{X}$ not containing λ_{U_1}. There would be a nontrivial map $\mathcal{V} \to \mathrm{Ind}(\mathcal{X})$, and hence $\mathcal{V}$ would have a quotient not containing λ, by (7.3.2) again. We can now use (7.5.2) and induction on the rank of G to show that $\mathcal{W}_{U_1} = \{0\}$, which is the desired contradiction. ■

(7.5.6) Corollary: *Let (J, λ) be a simple type in G, and let M be a finite-dimensional $e_\lambda * \mathcal{H}(G) * e_\lambda$-module. Then the space*

$$\mathcal{V} = \mathcal{H}(G) \otimes_{e_\lambda * \mathcal{H}(G) * e_\lambda} M$$

affords a smooth representation of G (with G acting on the factor $\mathcal{H}(G)$ by left translation), which has the following properties:

(i) $\mathcal{V}^\lambda = e_\lambda * \mathcal{V} \cong M$ as $e_\lambda * \mathcal{H}(G) * e_\lambda$-module;

(ii) $\mathcal{V}$ is admissible, of finite length, and every composition factor of $\mathcal{V}$ contains λ.

Proof: Immediate from (7.5.3). ■

Write $\mathfrak{Ad}_\lambda(G)$ for the category whose objects are smooth representations $(\pi, \mathcal{V})$ of G which satisfy (7.5.6)*(ii)*. That is, $(\pi, \mathcal{V}) \in |\mathfrak{Ad}_\lambda(G)|$ if and only if $\mathcal{V}$ is admissible, of finite length, and every composition factor of $\mathcal{V}$ contains λ. (As is customary, we write $|\mathcal{C}|$ for the class of objects of a category $\mathcal{C}$.) Equivalently, we could say that $(\pi, \mathcal{V})$ is admissible and generated as G-space by $\mathcal{V}^\lambda$. Further, given $(\pi_i, \mathcal{V}_i) \in |\mathfrak{Ad}_\lambda(G)|$, $i = 1, 2$, the morphism set $\mathrm{Mor}_{\mathfrak{Ad}_\lambda(G)}(\mathcal{V}_1, \mathcal{V}_2)$ is to be $\mathrm{Hom}_G(\mathcal{V}_1, \mathcal{V}_2)$.

On the other hand, if $\mathcal{A}$ is an associative $\mathbb{C}$-algebra with 1, we write $\mathcal{A}$-$\mathfrak{Mod}_\mathfrak{f}$ for the category of all left $\mathcal{A}$-modules *of finite $\mathbb{C}$-dimension*, together with all $\mathcal{A}$-homomorphisms.

(7.5.7) Theorem: *Let (J, λ) be a simple type in $G = \mathrm{Aut}_F(V)$. Let $[\mathfrak{A}, n, 0, \beta]$ be a simple stratum in $A = \mathrm{End}_F(V)$ defining (J, λ), let $E = F[\beta]$, $B = \mathrm{End}_E(V)$, $\mathfrak{B} = B \cap \mathfrak{A}$, $e = e(\mathfrak{B}|\mathfrak{o}_E)$, $ef = \dim_E(V)$, $q_E = \#\mathrm{k}_E$. The following categories are equivalent:*

(i) $\mathfrak{Ad}_\lambda(G)$;

(ii) $e_\lambda * \mathcal{H}(G) * e_\lambda$-$\mathfrak{Mod}_\mathfrak{f}$;

(iii) $\mathcal{H}(G, \lambda)$-$\mathfrak{Mod}_\mathfrak{f}$;

(iv) $\mathcal{H}(e, q_E^f)$-$\mathfrak{Mod}_\mathfrak{f}$.

Proof: (5.6.6) gives us an algebra isomorphism

$$\Psi : \mathcal{H}(e, q_E^f) \xrightarrow{\approx} \mathcal{H}(G, \lambda),$$

which induces an equivalence of categories

$$\mathcal{H}(e, q_E^f)\text{-}\mathfrak{Mod}_\mathfrak{f} \approx \mathcal{H}(G, \lambda)\text{-}\mathfrak{Mod}_\mathfrak{f}$$

in the obvious way. (4.2.4) gives us an algebra isomorphism

$$\Upsilon : \mathcal{H}(G,\lambda) \otimes_{\mathbb{C}} \mathrm{End}_{\mathbb{C}}(W) \xrightarrow{\approx} e_\lambda * \mathcal{H}(G) * e_\lambda,$$

where W is the vector space on which λ acts. Thus the rings $\mathcal{H}(G,\lambda)$, $e_\lambda * \mathcal{H}(G) * e_\lambda$ are Morita equivalent, and their module categories are equivalent. Explicitly, we have a functor $\mathcal{H}(G,\lambda)\text{-}\mathfrak{Mod} \to e_\lambda * \mathcal{H}(G) * e_\lambda\text{-}\mathfrak{Mod}$ given by $M \mapsto M \otimes W$, $M \in |\mathcal{H}(G,\lambda)\text{-}\mathfrak{Mod}|$, which surely preserves finite-dimensionality. In the opposite direction, write W^* for the linear dual of W, viewed in the natural way as a right $\mathrm{End}_{\mathbb{C}}(W)$-module. Then we have a functor $e_\lambda * \mathcal{H}(G) * e_\lambda\text{-}\mathfrak{Mod} \to \mathcal{H}(G,\lambda)\text{-}\mathfrak{Mod}$ by $\mathcal{M} \mapsto W^* \otimes_{\mathrm{End}_{\mathbb{C}}(W)} \mathcal{M}$. This again preserves finite-dimensionality. The effect of these functors on morphisms is obvious, and one verifies that they provide mutually inverse equivalences between the categories $e_\lambda * \mathcal{H}(G) * e_\lambda\text{-}\mathfrak{Mod}_{\mathrm{f}}$ and $\mathcal{H}(G,\lambda)\text{-}\mathfrak{Mod}_{\mathrm{f}}$. (See also the remarks following the proof of (4.2.4).)

Finally, the equivalence between $\mathfrak{Ad}_\lambda(G)$ and $e_\rho * \mathcal{H}(G) * e_\rho\text{-}\mathfrak{Mod}_{\mathrm{f}}$ is given by (7.5.3). ■

We record some minor properties of these correspondences. Write $e' = e(E|F)$. Thus the element $[\Pi(\mathfrak{B})]^{ee'}$ is central in $\mathcal{H}(e, q_E^f)$ so, if M is a simple $\mathcal{H}(e, q_E^f)$-module, there exists $\alpha_M \in \mathbb{C}^\times$ such that

(7.5.8) $$[\Pi(\mathfrak{B})]^{ee'} m = \alpha_M m, \quad m \in M.$$

On the other hand, if $(\pi, \mathcal{V}) \in |\mathfrak{Ad}_\lambda(G)|$ is irreducible, its central quasicharacter ω_π is already specified on $\mathfrak{o}_F^\times$, since $\lambda \mid \mathfrak{o}_F^\times$ is a multiple of $\omega_\pi \mid \mathfrak{o}_F^\times$. If we fix a prime element π_F of F, and let $\phi_1 \in \mathcal{H}(G,\lambda)$ be the function with support $\pi_F J$ and $\phi_1(\pi_F) = 1$, then ω_π is the unique quasicharacter of $F^\times$ which agrees with λ on $\mathfrak{o}_F^\times$ and has $\omega_\pi(\pi_F)$ equal to the unique eigenvalue of ϕ_1 on the $\mathcal{H}(G,\lambda)$-module corresponding to π via (7.5.7).

(7.5.9) Proposition: *Let $\Psi : \mathcal{H}(e, q_E^f) \to \mathcal{H}(G,\lambda)$ be a support-preserving algebra isomorphism, and define $\alpha_\Psi \in \mathbb{C}^\times$ by $\Psi[\Pi(\mathfrak{B})]^{ee'} = \alpha_\Psi \phi_1$, using the notation above. Let M be a simple left $\mathcal{H}(e, q_E^f)$-module, and let $(\pi, \mathcal{V})$ be the irreducible representation of G corresponding to M via the equivalence* (7.5.7) *induced by Ψ. The central quasicharacter ω_π of π then satisfies $\omega_\pi(\pi_F) = \alpha_M \alpha_\Psi^{-1}$.*

Proof: Immediate. ■

We can also work out the effect of changing the isomorphism Ψ. Take $\epsilon \in \mathbb{C}^\times$, and let Ψ' be the unique support-preserving algebra isomorphism $\mathcal{H}(e, q_E^f) \to \mathcal{H}(G,\lambda)$ such that $\Psi'[\Pi(\mathfrak{B})] = \epsilon\Psi[\Pi(\mathfrak{B})]$.

(7.5.10) Proposition: *In the above situation, let M be a simple $\mathcal{H}(e, q_E^f)$-module, and let $(\pi, \mathcal{V})$ (resp. $(\pi', \mathcal{V}')$) be the irreducible representation of G corresponding to M under the equivalence $\mathcal{H}(e, q_E^f)$-$\mathfrak{Mod}_{\mathfrak{f}} \cong \mathfrak{Ad}_\lambda(G)$ induced by Ψ (resp. Ψ'). Let χ be any unramified quasicharacter of $F^\times$ such that $\chi(\det_A(\Pi(\mathfrak{B})) = \epsilon^{-1}$. Then $\pi' \cong \pi \otimes \chi \circ \det$.*

Proof: Straightforward. ■

Now let us recall another standard notation. Let $\|\cdot\|$ denote the standard normalised absolute value on F, so that $\|\pi_F\| = q_F^{-1}$ for any prime element π_F of F. For a smooth representation π of G and $t \in \mathbb{C}^\times$, we define a representation $\pi(t)$ of G by

$$\pi(t) : g \mapsto \pi(g)\|\det(g)\|^t, \quad g \in G.$$

There is an analogous construction for $\mathcal{H}(e, q_E^f)$-modules. If M is such and $t \in \mathbb{C}^\times$, we define another $\mathcal{H}(e, q_E^f)$-module $M(t)$ by

(7.5.11) *(i) $M \cong M(t)$ as $\mathbb{C}$-space, via a map $m \mapsto m_t$, $m \in M$;*
(ii) $[s_i]m_t = ([s_i]m)_t$, $1 \le i \le e-1$, and $[\zeta]m_t = (q_E^f)^{-t}([\zeta]m)_t$.

Here, $[s_i]$ and $[\zeta]$ are the standard generators of $\mathcal{H}(e, q_E^f)$ as in (5.4.6).

(7.5.12) Proposition: *Let $\Psi : \mathcal{H}(e, q_E^f) \to \mathcal{H}(G, \lambda)$ be a support-preserving algebra isomorphism, let $(\pi, \mathcal{V}) \in |\mathfrak{Ad}_\lambda(G)|$, and let M be the $\mathcal{H}(e, q_E^f)$-module corresponding to $(\pi, \mathcal{V})$ via Ψ. Then, for $t \in \mathbb{C}^\times$, the $\mathcal{H}(e, q_E^f)$-module corresponding to $(\pi(t), \mathcal{V})$ is $M(t)$.*

Proof: Again trivial. ■

Finally, we give a "splitting property", originally proved for the trivial simple type in **[Bo]**.

(7.5.13) Proposition: *Let (J, λ) be a simple type in G. Let $(\pi, \mathcal{V})$ be an admissible representation of G, and let $\mathcal{V}_1$ be the G-subspace of $\mathcal{V}$ generated by $\mathcal{V}^\lambda$. There is a unique G-subspace $\mathcal{V}_2$ of $\mathcal{V}$ such that $\mathcal{V} = \mathcal{V}_1 \oplus \mathcal{V}_2$.*

Proof: Since $\mathcal{V}$ is admissible, the space $\mathcal{V}^\lambda$ is finite-dimensional. It follows from (7.5.4) that $\mathcal{V}_1$ has finite length and all of its composition factors contain λ. On the other hand, $\mathcal{V}/\mathcal{V}_1$ is admissible and does not contain λ.

We pass to the contragredient $(\check{\pi}, \check{\mathcal{V}})$ of $(\pi, \mathcal{V})$. This is admissible, and contains the simple type $(J, \check{\lambda})$. We have an exact sequence

$$0 \to (\mathcal{V}/\mathcal{V}_1)^\vee \to \check{\mathcal{V}} \to \check{\mathcal{V}}_1 \to 0.$$

The space $\check{\mathcal{V}}_1$ has finite length and all its composition factors contain $\check{\lambda}$, while the space $(\mathcal{V}/\mathcal{V}_1)^\vee$ is admissible and does not contain $\check{\lambda}$. Let $\mathcal{U}$

denote the subspace of $\check{\mathcal{V}}$ generated by $\check{\mathcal{V}}^{\check{\lambda}}$. Again, this has finite length and all its composition factors contain $\check{\lambda}$. Thus $\mathcal{U} \cap (\mathcal{V}/\mathcal{V}_1)\check{} = \{0\}$, and $\mathcal{U}$ maps injectively to $\check{\mathcal{V}}_1$. Comparing multiplicities of $\check{\lambda}$, we see that $\mathcal{U}$ maps isomorphically to $\check{\mathcal{V}}_1$. In other words, this last sequence splits. Dualising (and identifying $(\check{\mathcal{V}})\check{}$ with $\mathcal{V}$ and so on), the sequence

$$0 \to \mathcal{V}_1 \to \mathcal{V} \to \mathcal{V}/\mathcal{V}_1 \to 0$$

splits. Since $\mathcal{V}/\mathcal{V}_1$ does not contain λ, we have $\mathrm{Hom}_G(\mathcal{V}/\mathcal{V}_1, \mathcal{V}_1) = \{0\}$, so the splitting is unique. ■

(7.6) Jacquet functors and Hecke algebra maps

We continue with our simple type (J, λ) in G and investigate further the equivalence $\mathfrak{Ad}_\lambda(G) \cong \mathcal{H}(G,\lambda)\text{-}\mathfrak{Mod}_{\mathrm{f}}$ given by (7.5.7). Let $P_1 = M_1U_1$ be a parabolic subgroup of G which is subordinate to λ. This gives rise to the representation λ_{U_1} of $J \cap M_1$, in the notation of (7.2.17). The triple $(M_1, J \cap M_1, \lambda_{U_1})$ then has the strong intertwining property by (7.2.17), (7.5.4) and (7.5.2). We can thus form the category $\mathfrak{Ad}_{\lambda_{U_1}}(M_1)$ by analogy with $\mathfrak{Ad}_\lambda(G)$. (7.3.16) implies that induction gives us a functor

$$\mathrm{Ind}_{G/P_1} : \mathfrak{Ad}_{\lambda_{U_1}}(M_1) \to \mathfrak{Ad}_\lambda(G).$$

Likewise, it is not difficult to show (from (7.3.13), (7.3.16)) that $\mathcal{V} \mapsto \mathcal{V}_{U_1}$ gives a functor

$$r_{U_1} : \mathfrak{Ad}_\lambda(G) \to \mathfrak{Ad}_{\lambda_{U_1}}(M_1).$$

We also have an equivalence $\mathfrak{Ad}_{\lambda_{U_1}}(M_1) \cong \mathcal{H}(M_1, \lambda_{U_1})\text{-}\mathfrak{Mod}_{\mathrm{f}}$ given by (7.5.3) and the standard Morita equivalence. The first aim of this section is to describe these functors in terms of an algebra homomorphism $\mathcal{H}(M_1, \lambda_{U_1}) \to \mathcal{H}(G, \lambda)$.

In fact, we only treat the case where $P_1 = P = MU$ is minimal for the property of subordination to (J, λ):— it is not too hard to see how the general case must go. Indeed, we may as well abandon all pretence at generality, and take P to be defined by the E-decomposition (5.5.2) of the underlying F-vector space V used to define $\widetilde{\boldsymbol{W}}(\mathfrak{B})$ in (5.5.9). Having once picked this affine Weyl group, we have algebra isomorphisms $\mathcal{H}(G,\lambda) \cong \mathcal{H}(e, q_E^f)$ and $\mathcal{H}(M, \lambda_U) \cong \mathcal{H}(1, q_E^f)^{\otimes e}$ (the e-fold tensor product of $\mathcal{H}(1, q_E^f)$ with itself). We show that the representation-theoretic functors r_U, $\mathrm{Ind}_{G/P}$ are then realised by a uniquely determined algebra map $\mathcal{H}(1, q_E^f)^{\otimes e} \to \mathcal{H}(e, q_E^f)$.

It turns out that this particular map does actually depend on λ. However, given careful identifications, there is a *canonical* map

$$\mathcal{H}(1, q_E^f)^{\otimes e} \to \mathcal{H}(e, q_E^f)$$

which realises the normalised induction functor $\mathrm{Ind}^u_{G/P}$. In the context of the trivial type, this is described in **[MW]** and **[Ro]**. It is partially generalised in **[Wa]**. Here we generalise to our arbitrary simple type (J, λ) in G.

As a minor technical convenience, we avoid the affine Hecke algebras here, going directly to the Hecke algebra $\mathcal{H}(C^\times, \mathbf{1}_{\mathcal{I}})$ of the trivial character of an Iwahori subgroup $\mathcal{I}$ in the G-centraliser $C^\times$ of a field K/E as in (5.5.14). This saves us a further epidemic of generators and relations, and reduces computation. The translation to affine Hecke algebras can then be read directly from **[MW]**.

Irrespective of the cosmetics, the results of this section transfer all questions concerning the structure of certain induced representations (namely those supported on the set of unramified twists of a single supercuspidal) to the case where the supporting supercuspidal is the trivial character of $GL(1)$, at least modulo (6.0.1). We shall later see (in **(8.5)**) that the qualifier "certain" can actually be removed. Of course, all of this is compatible with the standard classification **[Z]** of the non-supercuspidal representations of $GL(N)$ to which, we emphasise, we have made no appeal. Hence the results of this section give an alternative approach to the classification problem.

First let us have a general notation. If $\mathbb{A}$ and $\mathbb{B}$ are associative $\mathbb{C}$-algebras with 1, and $\phi : \mathbb{A} \to \mathbb{B}$ is a homomorphism (algebra homomorphisms here are always assumed to preserve the 1's), we get functors on the module categories

$$\phi^* : \mathbb{B}\text{-}\mathfrak{Mod} \to \mathbb{A}\text{-}\mathfrak{Mod},$$
$$\phi_* : \mathbb{A}\text{-}\mathfrak{Mod} \to \mathbb{B}\text{-}\mathfrak{Mod}.$$

Explicitly, if M is a $\mathbb{B}$-module, then $\phi^* M$ is the $\mathbb{A}$-module whose underlying abelian group is that of M, and on which $\mathbb{A}$ acts via ϕ. On the other hand, if L is an $\mathbb{A}$-module, we put

$$\phi_* L = \mathrm{Hom}_{\mathbb{A}}(\mathbb{B}, L).$$

Here we view $\mathbb{B}$ as a left $\mathbb{A}$-module via ϕ, and $\mathbb{B}$ acts on this Hom group by right translation: $bf : b' \mapsto f(b'b)$, $f \in \mathrm{Hom}_{\mathbb{A}}(\mathbb{B}, L)$, $b, b' \in \mathbb{B}$. The functors ϕ^*, ϕ_* form an adjoint pair, in that we have natural isomorphisms

$$\mathrm{Hom}_{\mathbb{A}}(\phi^* M, L) \cong \mathrm{Hom}_{\mathbb{B}}(M, \phi_* L), \quad L \in |\mathbb{A}\text{-}\mathfrak{Mod}|, \ M \in |\mathbb{B}\text{-}\mathfrak{Mod}|.$$

Of course, we can play the adjoint functors game the other way round by putting ${}_*\phi L = \mathbb{B} \otimes_{\mathbb{A}} L$, when we get

$$\mathrm{Hom}_{\mathbb{A}}(L, \phi^* M) \cong \mathrm{Hom}_{\mathbb{B}}({}_*\phi L, M).$$

Write

$$\mathrm{Mod}_\lambda : \mathfrak{Ad}_\lambda(G) \to \mathcal{H}(G,\lambda)\text{-}\mathfrak{Mod}_{\mathrm{f}}$$

for the functor obtained by composing $(\boldsymbol{\pi}, \mathcal{V}) \mapsto \mathcal{V}^\lambda$ with the canonical Morita equivalence between $e_\lambda * \mathcal{H}(G) * e_\lambda\text{-}\mathfrak{Mod}_{\mathrm{f}}$ and $\mathcal{H}(G,\lambda)\text{-}\mathfrak{Mod}_{\mathrm{f}}$, as in (7.5.7).

The representation λ_P of $(J \cap P)H^1$ defined in (7.2.17) induces to λ, hence a representation of G contains λ if and only if it contains λ_P. Thus, extending our notation in the obvious way, the categories $\mathfrak{Ad}_\lambda(G)$, $\mathfrak{Ad}_{\lambda_P}(G)$ are identical. (4.1.3) gives us an isomorphism $\mathcal{H}(G,\lambda_P) \cong \mathcal{H}(G,\lambda)$, and hence an equivalence $\mathcal{H}(G,\lambda_P)\text{-}\mathfrak{Mod}_{\mathrm{f}} \cong \mathcal{H}(G,\lambda)\text{-}\mathfrak{Mod}_{\mathrm{f}}$ of categories. (4.2.7), (4.2.8) then show that the diagram

$$\begin{array}{ccc} \mathfrak{Ad}_{\lambda_P}(G) & \xrightarrow{\mathrm{Mod}_{\lambda_P}} & \mathcal{H}(G,\lambda_P)\text{-}\mathfrak{Mod}_{\mathrm{f}} \\ \Vert & & \downarrow \approx \\ \mathfrak{Ad}_\lambda(G) & \xrightarrow{\mathrm{Mod}_\lambda} & \mathcal{H}(G,\lambda)\text{-}\mathfrak{Mod}_{\mathrm{f}} \end{array}$$

commutes. The effect of this is that we may interchange λ and λ_P as convenient in the ensuing arguments.

The representation λ_U of $J \cap M \subset M$ is a tensor product of simple types, so it has the strong intertwining property by (7.5.2), (7.5.4). Therefore, using the obvious notation, we have an equivalence

$$\mathrm{Mod}_{\lambda_U} : \mathfrak{Ad}_{\lambda_U}(M) \xrightarrow{\approx} \mathcal{H}(M,\lambda_U)\text{-}\mathfrak{Mod}_{\mathrm{f}}.$$

We also recall the group $\boldsymbol{D}(\mathfrak{B})$ of (5.5.9). This consists of all matrices $\boldsymbol{a}$ in $B = \mathbb{M}(R, E)$ of the form

$$\boldsymbol{a} = \mathrm{diag}(\pi_E^{a_1} I_f, \ldots, \pi_E^{a_e} I_f),$$

where I_f is the $f \times f$ identity matrix, π_E is a fixed prime element of E, and the a_i are integers. We write $\boldsymbol{D}^-(\mathfrak{B})$ for the subset consisting of $\boldsymbol{a} \in \boldsymbol{D}(\mathfrak{B})$ such that $a_1 \geq \ldots \geq a_e$. Invoking (3.1.9), (3.1.10), we have

$$\begin{array}{c} \boldsymbol{a}^{-1}(H^1 \cap U^-)\boldsymbol{a} \subset H^1 \cap U^-, \\ \boldsymbol{a}(J^1 \cap U)\boldsymbol{a}^{-1} \subset J^1 \cap U, \end{array} \qquad \boldsymbol{a} \in \boldsymbol{D}^-(\mathfrak{B}).$$

Let $\boldsymbol{r}_U : (\boldsymbol{\pi}, \mathcal{V}) \mapsto (\boldsymbol{\pi}_U, \mathcal{V}_U)$ be the Jacquet functor. According to (7.3.13), $\boldsymbol{r}_U$ gives a functor $\mathfrak{Ad}_\lambda(G) \to \mathfrak{Ad}_{\lambda_U}(M)$. By (7.3.14), the (unnormalised) induction process gives us a functor $\mathrm{Ind} : \mathfrak{Ad}_{\lambda_U}(M) \to \mathfrak{Ad}_\lambda(G)$. Frobenius Reciprocity then says exactly that the functors $\boldsymbol{r}_U$, Ind form an adjoint pair:

$$\mathrm{Hom}_M(\boldsymbol{r}_U(\mathcal{V}), \mathcal{W}) = \mathrm{Hom}_G(\mathcal{V}, \mathrm{Ind}(\mathcal{W})),$$

for $\mathcal{V} \in |\mathfrak{Ad}_\lambda(G)|$, $\mathcal{W} \in |\mathfrak{Ad}_{\lambda_U}(M)|$.

(7.6.1) Theorem: *There is a unique algebra homomorphism*

$$\boldsymbol{j}_{\delta_P}[\lambda] : \mathcal{H}(M, \lambda_U) \to \mathcal{H}(G, \lambda)$$

with the following properties.

(i) If $\phi \in \mathcal{H}(M, \lambda_U)$ has support $(J \cap M)\boldsymbol{a}$, $\boldsymbol{a} \in \mathbf{D}^-(\mathfrak{B})$, then $\boldsymbol{j}_{\delta_P}[\lambda](\phi)$ has support $J\boldsymbol{a}J$.

(ii) The following diagram is commutative:

$$\begin{array}{ccc} \mathfrak{Ad}_\lambda(G) & \xrightarrow{\mathrm{Mod}_\lambda} & \mathcal{H}(G,\lambda)\text{-}\mathfrak{Mod}_{\mathrm{f}} \\ {\scriptstyle \boldsymbol{r}_U}\downarrow & & \downarrow{\scriptstyle \boldsymbol{j}_{\delta_P}[\lambda]^*} \\ \mathfrak{Ad}_{\lambda_U}(M) & \xrightarrow{\mathrm{Mod}_{\lambda_U}} & \mathcal{H}(M,\lambda_U)\text{-}\mathfrak{Mod}_{\mathrm{f}} \end{array}$$

The map $\boldsymbol{j}_{\delta_P}[\lambda]$ is injective, and moreover we have a commutative diagram

$$\begin{array}{ccc} \mathfrak{Ad}_{\lambda_U}(M) & \xrightarrow{\approx} & \mathcal{H}(M,\lambda_U)\text{-}\mathfrak{Mod}_{\mathrm{f}} \\ {\scriptstyle \mathrm{Ind}}\downarrow & & \downarrow{\scriptstyle (\boldsymbol{j}_{\delta_P}[\lambda])_*} \\ \mathfrak{Ad}_\lambda(G) & \xrightarrow{\approx} & \mathcal{H}(G,\lambda)\text{-}\mathfrak{Mod}_{\mathrm{f}} \end{array}$$

Proof: Let us write $P/U = G'^e$, $G' = GL(N/e, F)$, and $\lambda_U = \lambda' \otimes \ldots \otimes \lambda'$, where (J', λ') is the maximal type associated to (J, λ). The G'-intertwining of λ' is $J'E^\times$, so the M-intertwining of λ_U is $\boldsymbol{D}(\mathfrak{B})J \cap M$. Also, since $\boldsymbol{D}(\mathfrak{B})$ normalises λ_U, any nonzero value of any $\phi \in \mathcal{H}(M, \lambda_U)$ is an automorphism of the representation space W of λ_U (or λ_P).

For the first step in the proof, we write $K = H^1(J \cap P)$, and define our Hecke algebras relative to measures μ_M on M, μ_G on G, such that $\mu_G(K) = \mu_M(J \cap M) = 1$.

(7.6.2) Proposition: *There is a unique algebra homomorphism*

$$\boldsymbol{j}[\lambda_P] = \boldsymbol{j} : \mathcal{H}(M, \lambda_U) \to \mathcal{H}(G, \lambda_P)$$

such that:

(i) if $\boldsymbol{a} \in \boldsymbol{D}^-(\mathfrak{B})$ and $\phi \in \mathcal{H}(M, \lambda_U)$ has support $\boldsymbol{a}(J \cap M)$, then $\boldsymbol{j}\phi$ has support $K\boldsymbol{a}K$;

(ii) if $\boldsymbol{a}$, ϕ are as in (i), we have $\boldsymbol{j}\phi(\boldsymbol{a}) = \phi(\boldsymbol{a})$ in $\mathrm{Aut}_{\mathbb{C}}(\check{W})$.

This homomorphism is injective.

Proof: We start with a lemma.

(7.6.3) Lemma: *Let $\boldsymbol{a}, \boldsymbol{b} \in \boldsymbol{D}(\mathfrak{B})$, and let $\phi, \psi \in \mathcal{H}(M, \lambda_U)$ have support $(J \cap M)\boldsymbol{a}$, $(J \cap M)\boldsymbol{b}$ respectively. Then $\phi * \psi$ has support $(J \cap M)\boldsymbol{ab}$, and $\phi * \psi(\boldsymbol{ab}) = \phi(\boldsymbol{a}) \circ \psi(\boldsymbol{b})$.*

Proof: This is a straightforward computation. ∎

As $\mathbb{C}$-algebras, we have $\mathcal{H}(M, \lambda_U) \cong \mathcal{H}(G', \lambda') \otimes \ldots \otimes \mathcal{H}(G', \lambda')$ (e factors), and, either directly from (7.6.3) or from (5.6.6), we have $\mathcal{H}(G', \lambda') \cong \mathbb{C}[x, x^{-1}]$ (Laurent polynomial ring), where we may think of the generator x as being some nonzero function with support $J'\pi_E$. Write x_i for the image of x in $\mathcal{H}(M, \lambda_U)$ embedded on the i-th tensor factor. Set

$$X_i = \prod_{k=1}^{i} x_k, \qquad 1 \leq i \leq e,$$

so that X_i is a nonzero function with support $(J \cap M)\boldsymbol{a}_i$, where $\boldsymbol{a}_i \in \boldsymbol{D}^-(\mathfrak{B})$ is the element whose first i diagonal blocks are $\pi_E I_f$, and all others I_f. Moreover, we still have $\mathcal{H}(M, \lambda_U) \cong \mathbb{C}[X_1, \ldots, X_e, X_1^{-1}, \ldots X_e^{-1}]$ (Laurent polynomial ring). Thus, in (7.6.2), if the homomorphism $\boldsymbol{j}$ exists, it is injective and is uniquely determined by the stated conditions.

Now we show it exists. We take $\phi, \psi \in \mathcal{H}(M, \lambda_U)$ with supports $(J \cap M)\boldsymbol{a}$, $(J \cap M)\boldsymbol{b}$ respectively, for elements $\boldsymbol{a}, \boldsymbol{b} \in \boldsymbol{D}^-(\mathfrak{B})$. We recall from (7.2.19) that every $\boldsymbol{a} \in \boldsymbol{D}^-(\mathfrak{B}) \subset \widetilde{\boldsymbol{W}}(\mathfrak{B})$ intertwines λ_P and that $I_{\boldsymbol{a}}(\lambda_P) = I_{\boldsymbol{a}}(\lambda_U)$. Therefore there exists a unique function $\boldsymbol{j}\phi = \Phi$ (resp. $\boldsymbol{j}\psi = \Psi$) in $\mathcal{H}(G, \lambda_P)$ with support $K\boldsymbol{a}K$ (resp. $K\boldsymbol{b}K$) such that $\Phi(\boldsymbol{a}) = \phi(\boldsymbol{a})$ (resp. $\Psi(\boldsymbol{b}) = \psi(\boldsymbol{b})$) in $\mathrm{Aut}_{\mathbb{C}}(\check{W})$. We proceed to compute the convolution $\Phi * \Psi$.

(7.6.4) Lemma: *In the notation above, the support of $\Phi * \Psi$ is $K\boldsymbol{ab}K$, and $\Phi * \Psi(\boldsymbol{ab}) = \Phi(\boldsymbol{a}) \circ \Psi(\boldsymbol{b})$.*

This lemma implies (7.6.2). To prove the lemma, we note first that the support of $\Phi * \Psi$ is contained in $K\boldsymbol{a}K\boldsymbol{b}K$. We write K in the form $(J^1 \cap U)(J \cap M)(H^1 \cap U^-)$. Since $\boldsymbol{a} \in \boldsymbol{D}^-(\mathfrak{B})$, we have $\boldsymbol{a}(J^1 \cap U)\boldsymbol{a}^{-1} \subset J^1 \cap U$ and $\boldsymbol{b}^{-1}(H^1 \cap U^-)\boldsymbol{b} \subset H^1 \cap U^-$. The elements $\boldsymbol{a}$,$\boldsymbol{b}$ normalise $J \cap M$, so in all we have $K\boldsymbol{a}K\boldsymbol{b}K \subset K\boldsymbol{ab}K$. Thus the support of $\Phi * \Psi$ is contained in $K\boldsymbol{ab}K$.

We now compute

$$\Phi * \Psi(\boldsymbol{ab}) = \int_G \Phi(x)\Psi(x^{-1}\boldsymbol{ab})\, dx.$$

The integral is effectively taken over $x \in K\boldsymbol{a}K$, so we write this double coset as a disjoint union of cosets $z\boldsymbol{a}K$, where z runs over $K/K \cap {}^{\boldsymbol{a}}K$. The integral reduces to

$$\Phi * \Psi(\boldsymbol{ab}) = \sum_z \Phi(z\boldsymbol{a})\Psi(\boldsymbol{a}^{-1}z^{-1}\boldsymbol{ab}).$$

Since $\boldsymbol{a} \in \boldsymbol{D}^-(\mathfrak{B})$, we can identify $K/K \cap {}^{\boldsymbol{a}}K$ with $(J^1 \cap U)/{}^{\boldsymbol{a}}(J^1 \cap U)$, noting that ${}^{\boldsymbol{a}}(J^1 \cap U) \subset (J^1 \cap U)$. In other words, we may as well take $z \in J^1 \cap U$. Thus $\check{\lambda}_P(z) = 1_{\check{W}}$, and $\Phi(z\boldsymbol{a}) = \Phi(\boldsymbol{a})$. On the other hand, the factor $\Psi(\boldsymbol{a}^{-1}z^{-1}\boldsymbol{ab})$ vanishes unless $\boldsymbol{a}^{-1}z^{-1}\boldsymbol{ab} \in K\boldsymbol{b}K$, or $\boldsymbol{a}^{-1}z^{-1}\boldsymbol{a} \in K\boldsymbol{b}K\boldsymbol{b}^{-1}$. Since $\boldsymbol{b} \in \boldsymbol{D}^-(\mathfrak{B})$, we have $K\boldsymbol{b}K\boldsymbol{b}^{-1} = K\boldsymbol{b}(H^1 \cap U^-)\boldsymbol{b}^{-1}$ which intersects U in $J^1 \cap U$. The factor in question therefore vanishes unless $\boldsymbol{a}^{-1}z^{-1}\boldsymbol{a} \in J^1 \cap U$, i.e. $z \in {}^{\boldsymbol{a}}(J^1 \cap U)$. Thus only the coset $\boldsymbol{a}K$ contributes to the integral, and we get $\Phi * \Psi(\boldsymbol{ab}) = \Phi(\boldsymbol{a}) \circ \Psi(\boldsymbol{b})$, as required. ■

This completes the proof of (7.6.2). ■

Now let χ be some unramified quasicharacter of M. Thus χ is a homomorphism $M \to \mathbb{C}^\times$ which is null on a maximal compact subgroup of M. In particular, χ is null on $J \cap M$. We can then define an automorphism τ_χ of the Hecke algebra $\mathcal{H}(M, \lambda_U)$ by

$$\tau_\chi \phi(x) = \chi(x)^{-1}\phi(x), \quad \phi \in \mathcal{H}(M, \lambda_U), \ x \in M.$$

We therefore get an injective algebra homomorphism

$$\textbf{(7.6.5)} \qquad \boldsymbol{j}_\chi[\lambda_P] = \boldsymbol{j}[\lambda_P] \circ \tau_\chi : \mathcal{H}(M, \lambda_U) \to \mathcal{H}(G, \lambda_P).$$

This only depends on $\chi \mid \boldsymbol{D}(\mathfrak{B})$.

(7.6.6) Remark: Suppose we have an algebra homomorphism $\boldsymbol{j}' : \mathcal{H}(M, \lambda_U) \to \mathcal{H}(G, \lambda_P)$ satisfying (7.6.2)*(i)*. Then it is easy to see that there is an unramified quasicharacter χ of M such that $\boldsymbol{j}' = \boldsymbol{j}_\chi[\lambda_P]$. Moreover, the restriction $\chi \mid \boldsymbol{D}(\mathfrak{B})$ is uniquely determined by this equality.

We can transport this structure to the algebra $\mathcal{H}(G, \lambda)$ by setting

$$\textbf{(7.6.7)} \qquad \boldsymbol{j}_\chi[\lambda] : \mathcal{H}(M, \lambda_U) \xrightarrow{\boldsymbol{j}_\chi[\lambda_P]} \mathcal{H}(G, \lambda_P) \xrightarrow{\approx} \mathcal{H}(G, \lambda).$$

We now abbreviate $e_U = e_{\lambda_U} \in \mathcal{H}(M)$, $e_P = e_{\lambda_P} \in \mathcal{H}(G)$, and write $\mathbf{e}_U$, $\mathbf{e}_P$ for the unit elements of $\mathcal{H}(M, \lambda_U)$, $\mathcal{H}(G, \lambda_P)$ respectively. We identify $e_U * \mathcal{H}(M) * e_U$, $e_P * \mathcal{H}(G) * e_P$ with $\mathcal{H}(M, \lambda_U) \otimes \mathrm{End}_{\mathbb{C}}(W)$, $\mathcal{H}(G, \lambda_P) \otimes \mathrm{End}_{\mathbb{C}}(W)$ respectively, via maps $\boldsymbol{\Upsilon}_M$, $\boldsymbol{\Upsilon}_G$ as in (4.2.4). This allows us to extend $\boldsymbol{j}_\chi[\lambda_P]$ to an algebra homomorphism

$$\textbf{(7.6.8)} \qquad \tilde{\boldsymbol{j}}_\chi[\lambda_P] = \boldsymbol{j}_\chi[\lambda_P] \otimes 1 : e_U * \mathcal{H}(M) * e_U \to e_P * \mathcal{H}(G) * e_P.$$

Strictly speaking, $\tilde{\boldsymbol{j}}_\chi[\lambda_P] = \boldsymbol{\Upsilon}_G \circ (\boldsymbol{j}_\chi[\lambda_P] \otimes 1) \circ \boldsymbol{\Upsilon}_M^{-1}$.

Now let $(\pi, \mathcal{V})$ be a smooth representation of G, and write p_U for the canonical map $\mathcal{V} \to \mathcal{V}_U$. Then we know ((7.3.6)(*i*)) that p_U restricts to an injective map of vector spaces

$$\mathcal{V}^{\lambda_P} \to (\mathcal{V}_U)^{\lambda_U}$$

which is surjective at least when $(\mathcal{V}_U)^{\lambda_U}$ is finite-dimensional.

Next, let $\delta = \delta_P$ denote the usual module character of P. Since it is trivial on U, we tend to regard δ_P as an (unramified) quasicharacter of M. Explicitly,

$$\textbf{(7.6.9)} \qquad \delta_P(m) = (L : mLm^{-1}), \quad m \in M,$$

for any compact open subgroup L of U. (We use the generalised group index here.) In particular, if $\boldsymbol{a} \in \boldsymbol{D}^-(\mathfrak{B})$ and $K = H^1(J \cap P)$ as before, we have

$$\textbf{(7.6.10)} \qquad \delta_P(\boldsymbol{a}) = (K : K \cap {}^{\boldsymbol{a}}K).$$

We now show that, for any $(\pi, \mathcal{V})$, we have

(7.6.11)

$$p_U(\pi(\tilde{j}_\delta(\phi))v) = \pi_U(\phi)p_U(v), \quad v \in \mathcal{V}^{\lambda_P}, \ \phi \in e_U * \mathcal{H}(M) * e_U,$$

where $\tilde{j}_\delta = \tilde{j}_{\delta_P}[\lambda_P]$ in the notation of (7.6.8). Of course, we need only check this identity for a family of algebra generators ϕ of $e_U * \mathcal{H}(M) * e_U$. Let us first find such a family. By (7.4.4), we have

$$\textbf{(7.6.12)} \quad \Upsilon_G^{-1}(e_P * {}^{xay}e_P) = (K : K \cap {}^{\boldsymbol{a}}K)^{-1}\phi \otimes \lambda_P(x) \circ \Phi \circ \lambda_P(y),$$

where $\boldsymbol{a} \in \boldsymbol{D}(\mathfrak{B})$, $x, y \in K$, $\phi \in \mathcal{H}(G, \lambda_P)$ is nonzero with support $K\boldsymbol{a}K$, and $\Phi = (\phi(\boldsymbol{a})^{-1})\check{} \in \mathrm{Aut}_{\mathbb{C}}(W)$. Exactly the same computation (*cf.* (7.4.5)) yields

$$\textbf{(7.6.13)} \qquad \Upsilon_M^{-1}(e_U * {}^{xay}e_U) = \psi \otimes \lambda_U(x) \circ \Psi \circ \lambda_U(y),$$

where $\boldsymbol{a} \in \boldsymbol{D}(\mathfrak{B})$, $x, y \in J \cap M$, $\psi \in \mathcal{H}(M, \lambda_U)$ is nonzero with support $(J \cap M)\boldsymbol{a}$, and $\Psi = (\psi(\boldsymbol{a})^{-1})\check{} \in \mathrm{Aut}_{\mathbb{C}}(W)$.

(7.6.14) Lemma: *The functions $e_M * {}^z e_M$, for $z \in (J \cap M) \cup \boldsymbol{D}^-(\mathfrak{B})$, together with their inverses, generate $e_U * \mathcal{H}(M) * e_U$ as $\mathbb{C}$-algebra.*

Proof: Let $\mathcal{A}$ be the subalgebra of $e_U * \mathcal{H}(M) * e_U$ generated by this family of functions. Consider first the function $e_U * {}^z e_U$ for $z \in J \cap M$. Then $\Upsilon_M^{-1}(e_U * {}^z e_U) = \mathrm{e}_U \otimes \lambda_U(z)$ by (7.6.13). The operators $\lambda_U(z)$, $z \in J \cap M$, span $\mathrm{End}_{\mathbb{C}}(W)$ since λ_U is irreducible. Thus $\Upsilon_M^{-1}(\mathcal{A}) \supset \mathrm{e}_U \otimes \mathrm{End}_{\mathbb{C}}(W)$. On the other hand, if $\boldsymbol{a} \in \boldsymbol{D}^-(\mathfrak{B})$, we write $\Upsilon_M^{-1}(e_U * {}^{\boldsymbol{a}}e_U) = \phi_{\boldsymbol{a}} \otimes \Phi$, where $\phi_{\boldsymbol{a}}$ has support $(J \cap M)\boldsymbol{a}$. We know that $\mathbf{e}_U \otimes \Phi^{-1} \in \Upsilon_M^{-1}(\mathcal{A})$, so $\phi_{\boldsymbol{a}} \otimes 1 \in \Upsilon_M^{-1}(\mathcal{A})$, and the functions $\phi_{\boldsymbol{a}}$, together with their inverses, generate the algebra $\mathcal{H}(M, \lambda_U)$. Thus $\mathcal{A} = e_U * \mathcal{H}(M) * e_U$, as required. ∎

We now check (7.6.11) in the case $\phi = e_U * {}^x e_U$, $x \in J \cap M$. Then $\tilde{\boldsymbol{j}}_\delta(\phi) = \Upsilon_G(\mathbf{e}_P \otimes \lambda_P(x)) = e_P * {}^x e_P$. Thus

$$\boldsymbol{p}_U(\boldsymbol{\pi}(\tilde{\boldsymbol{j}}_\delta(\phi))\mathrm{v}) = \boldsymbol{p}_U(\boldsymbol{\pi}(x)\mathrm{v}) = \boldsymbol{\pi}_U(x)\boldsymbol{p}_U(\mathrm{v}) = \boldsymbol{\pi}_U(\phi)\boldsymbol{p}_U(\mathrm{v}),$$

as required for (7.6.11).

Next take $\boldsymbol{a} \in \boldsymbol{D}^-(\mathfrak{B})$, $\phi = e_U * {}^{\boldsymbol{a}} e_U$. Then $\tilde{\boldsymbol{j}}_\delta(\phi) = e_P * {}^{\boldsymbol{a}} e_P$ by (7.6.9/10/12) and (7.6.2). Therefore

$$\boldsymbol{p}_U(\boldsymbol{\pi}(\tilde{\boldsymbol{j}}_\delta(\phi))\mathrm{v}) = \boldsymbol{p}_U(\boldsymbol{\pi}(e_P)\boldsymbol{\pi}(\boldsymbol{a})\mathrm{v}).$$

Since $\boldsymbol{a} \in \boldsymbol{D}^-(\mathfrak{B})$, the vector $\boldsymbol{\pi}(\boldsymbol{a})\mathrm{v}$ is fixed by $H^1 \cap U^-$, and

$$\boldsymbol{\pi}(e_P)\boldsymbol{\pi}(\boldsymbol{a})\mathrm{v} = \int_{J^1 \cap U} \boldsymbol{\pi}(u)\boldsymbol{\pi}(\boldsymbol{a})\mathrm{v}\, du,$$

where du is the Haar measure on U giving $J^1 \cap U$ measure 1. Thus $\boldsymbol{p}_U(\boldsymbol{\pi}(e_P)\boldsymbol{\pi}(\boldsymbol{a})\mathrm{v})$ is equal to $\boldsymbol{p}_U(\boldsymbol{\pi}(\boldsymbol{a})\mathrm{v}) = \boldsymbol{\pi}_U(\boldsymbol{a})\boldsymbol{p}_U(\mathrm{v}) = \boldsymbol{\pi}_U(\phi)\boldsymbol{p}_U(\mathrm{v})$, and (7.6.11) holds here.

Now we have the case where ϕ is the inverse of $\psi = e_U * {}^{\boldsymbol{a}} e_U$ in $\mathcal{H}(M, \lambda_U)$, with $\boldsymbol{a} \in \boldsymbol{D}^-(\mathfrak{B})$. Since $\tilde{\boldsymbol{j}}_\delta$ is a ring homomorphism, we have $\tilde{\boldsymbol{j}}_\delta(\psi^{-1}) = \tilde{\boldsymbol{j}}_\delta(\psi)^{-1}$, and $\tilde{\boldsymbol{j}}_\delta(\psi)$ acts as an automorphism of the space $\mathcal{V}^{\lambda_P}$. So, in (7.6.11), we may put $\mathrm{v} = \boldsymbol{\pi}(\tilde{\boldsymbol{j}}_\delta(\psi))\mathrm{w}$, for some $\mathrm{w} \in \mathcal{V}^{\lambda_P}$, to get

$$\boldsymbol{p}_U(\boldsymbol{\pi}(\tilde{\boldsymbol{j}}_\delta(\phi))\mathrm{v}) = \boldsymbol{p}_U(\mathrm{w}) = \boldsymbol{\pi}_U(\phi)\boldsymbol{\pi}_U(\psi)\boldsymbol{p}_U(\mathrm{w}) = \boldsymbol{\pi}_U(\phi)\boldsymbol{p}_U(\mathrm{v})$$

by the previous case. Thus (7.6.11) holds for all $\phi \in e_U * \mathcal{H}(M) * e_U$.

In the cases where $\boldsymbol{p}_U$ is an isomorphism $\mathcal{V}^{\lambda_P} \cong (\mathcal{V}_U)^{\lambda_U}$, for example when $\mathcal{V}$ is admissible, we therefore have

$$(\mathcal{V}_U)^{\lambda_U} \cong \tilde{\boldsymbol{j}}_\delta[\lambda_P]^*(\mathcal{V}^{\lambda_P})$$

as $e_U * \mathcal{H}(M) * e_U$-module. Composing with the Morita equivalence between $e_U * \mathcal{H}(M) * e_U$ and $\mathcal{H}(M, \lambda_U)$ and that between $e_P * \mathcal{H}(G) * e_P$ and $\mathcal{H}(G, \lambda_P)$, we get a natural isomorphism of $\mathcal{H}(M, \lambda_U)$-modules

$$\mathrm{Mod}_{\lambda_U}(\boldsymbol{r}_U(\mathcal{V})) \cong (\boldsymbol{j}_\delta[\lambda_P])^*(\mathrm{Mod}_{\lambda_P}(\mathcal{V})), \quad \mathcal{V} \in |\mathfrak{A}\mathfrak{d}_\lambda(G)|.$$

Defining $\boldsymbol{j}_{\delta_P}[\lambda]$ by (7.6.7), this is the same as

$$\mathrm{Mod}_{\lambda_U}(\boldsymbol{r}_U(\mathcal{V})) \cong (\boldsymbol{j}_{\delta_P}[\lambda])^*(\mathrm{Mod}_\lambda(\mathcal{V})), \quad \mathcal{V} \in |\mathfrak{A}\mathfrak{d}_\lambda(G)|.$$

This map $\boldsymbol{j}_{\delta_P}[\lambda]$ then satisfies *(i)* and *(ii)* of the theorem (except possibly the uniqueness).

For the uniqueness property, it is enough to show that $\boldsymbol{j}_{\delta_P}[\lambda_P]$ is uniquely determined by (7.6.11) and (7.6.2)*(i)*. This amounts to showing that if χ is an unramified quasicharacter of M, which does not agree with δ_P on $\boldsymbol{D}(\mathfrak{B})$, then there exists $\mathcal{V} \in |\mathfrak{A}\mathfrak{d}_\lambda(G)|$ such that

$$\boldsymbol{j}_\chi[\lambda_P]^*(\mathrm{Mod}_{\lambda_P}(\mathcal{V})) \not\cong \boldsymbol{j}_{\delta_P}[\lambda_P]^*(\mathcal{V}).$$

To produce such a $\mathcal{V}$, we take first a trivial module $\mathcal{M}$ over the affine Hecke algebra $\mathcal{H}(e, q_E^f)$, i.e. $\mathcal{M}$ has $\mathbb{C}$-dimension 1, and the generators $[s_i]$, $1 \le i \le e-1$, (in the notation of (5.4.6)) all act as multiplication by q_E^f. Then we let $\mathcal{V}$ correspond to $\mathcal{M}$ under some support-preserving algebra isomorphism $\mathcal{H}(e, q_E^f) \cong \mathcal{H}(G, \lambda)$. (7.5.12) then implies that $\mathcal{V}$ has the desired property.

To get the second diagram, we observe that if we use an isomorphism $\boldsymbol{\Psi} : \mathcal{H}(e, q_E^f) \xrightarrow{\approx} \mathcal{H}(G, \lambda)$ (as in (5.6.6)) to transport the picture to the affine Hecke algebra, then $\boldsymbol{\Psi}^{-1}(\boldsymbol{j}_{\delta_P}[\lambda](\mathcal{H}(M, \lambda_U))$ is the subalgebra $\mathcal{A}$ of $\mathcal{H}(e, q_E^f)$ generated by the elements $[\boldsymbol{a}]$, $\boldsymbol{a} \in \boldsymbol{D}^-(\mathfrak{B})$, and their inverses. (Here we use the elements $\boldsymbol{w}$ of $\widetilde{\boldsymbol{W}}(\mathfrak{B})$ to parametrise the canonical basis $[\boldsymbol{w}]$ of $\mathcal{H}(e, q_E^f)$, as in (5.5).) It follows from **[MW]** (I.3.1) that there is a finite subset $\boldsymbol{T}$ of $\widetilde{\boldsymbol{W}}(\mathfrak{B})$ such that

$$\mathcal{H}(e, q_E^f) = \sum_{\boldsymbol{t} \in \boldsymbol{T}} \mathcal{A}.[\boldsymbol{t}].$$

(In fact, we could take $\boldsymbol{T} = \boldsymbol{W}_0(\mathfrak{B})$.) It follows that the functor

$$(\boldsymbol{j}_{\delta_P}[\lambda])_* : \mathcal{H}(M, \lambda_U)\text{-}\mathfrak{Mod} \to \mathcal{H}(G, \lambda)\text{-}\mathfrak{Mod}$$

preserves finite-dimensionality, i.e. it takes $\mathcal{H}(M, \lambda_U)\text{-}\mathfrak{Mod}_\mathrm{f}$ to $\mathcal{H}(G, \lambda)$-$\mathfrak{Mod}_\mathrm{f}$. Thus $((\boldsymbol{j}_{\delta_P}[\lambda])^*, (\boldsymbol{j}_{\delta_P}[\lambda])_*)$ is an adjoint pair of functors between these categories. However, the functor $\boldsymbol{r}_U$ has an adjoint, namely Ind, so the commutativity of our diagram follows from the uniqueness properties of adjoint functors. ■

(7.6.15) Remark: We also have the functor ${}_*\boldsymbol{j}_\delta[\lambda] : \mathcal{H}(M, \lambda_U)\text{-}\mathfrak{Mod}_\mathrm{f} \to \mathcal{H}(G, \lambda)\text{-}\mathfrak{Mod}_\mathrm{f}$ and $({}_*\boldsymbol{j}_\delta[\lambda], \boldsymbol{j}_\delta^*[\lambda])$ is an adjoint pair. The corresponding functor $\mathfrak{A}\mathfrak{d}_{\lambda_U}(M) \to \mathfrak{A}\mathfrak{d}_\lambda(G)$, according to **[Cr]** Th.1.4, is

$$(\boldsymbol{\sigma}, \mathcal{W}) \mapsto \mathrm{Ind}(\boldsymbol{\sigma} \otimes \delta_P^{-1}), \quad (\boldsymbol{\sigma}, \mathcal{W}) \in |\mathfrak{A}\mathfrak{d}_{\lambda_U}(M)|.$$

We now choose to use a more concrete realisation of our Hecke algebra maps. We take an unramified field extension K/E as in (5.5.14). We assume that our parabolic subgroup P is defined by a K-decomposition of the underlying F-vector space V, subordinate to the $\mathfrak{o}_F$-order $\mathfrak{A}$, $[\mathfrak{A}, n, 0, \beta]$ being a simple stratum defining our simple type (J, λ). We set

(7.6.16) *(i)* $C = \mathrm{End}_K(V)$;
(ii) $\mathcal{I} = J \cap C^\times$.

Thus $C^\times \cong GL(e, K)$ and

(7.6.17) *(i) $\mathcal{I}$ is an Iwahori subgroup of $C^\times$;*
(ii) $(\mathcal{I}, \mathbf{1}_\mathcal{I})$ is a simple type in $C^\times$ (where $\mathbf{1}_\mathcal{I}$ is the trivial character of $\mathcal{I}$);
(iii) $Q = P \cap C^\times$ is a Borel subgroup of $C^\times$ subordinate to $(\mathcal{I}, \mathbf{1}_\mathcal{I})$;
(iv) $T = M \cap C^\times$ is a split maximal torus in $C^\times$ and $Q = T.(U \cap C^\times)$;
(v) $\widetilde{\boldsymbol{W}}(\mathfrak{B})$ is the affine Weyl group of $C^\times$ relative to T, $\mathcal{I}$, and the prime element π_E of K.

In the context of $(C^\times, (\mathcal{I}, \mathbf{1}_\mathcal{I}))$, the analogue $(\mathbf{1}_\mathcal{I})_Q$ of the representation λ_P is $\mathbf{1}_\mathcal{I}$ itself, and $(\mathbf{1}_\mathcal{I})_{U \cap C^\times} = \mathbf{1}_{T \cap \mathcal{I}}$ (always using $\mathbf{1}$ to denote a trivial character). Further, $\mathcal{I} \cap T \cong (\mathfrak{o}_K^\times)^e$. Just as in (7.6.2), (7.6.5), we get an algebra homomorphism

$$j_\xi[\mathbf{1}_\mathcal{I}] : \mathcal{H}(T, \mathbf{1}_{T \cap \mathcal{I}}) \to \mathcal{H}(C^\times, \mathbf{1}_\mathcal{I}),$$

for any unramified quasicharacter ξ of T.

The standard special case of (5.6.6) gives a support-preserving algebra isomorphism $\mathcal{H}(e, q_E^f) \cong \mathcal{H}(C^\times, \mathbf{1}_\mathcal{I})$. In fact, there is a canonical choice here:— define $\mathcal{H}(C^\times, \mathbf{1}_\mathcal{I})$ relative to the Haar measure μ_C on $C^\times$ for which $\mu_C(\mathcal{I}) = 1$, and map $[\boldsymbol{w}]$ to the characteristic function of $\mathcal{I}\boldsymbol{w}\mathcal{I}$, $\boldsymbol{w} \in \widetilde{\boldsymbol{W}}(\mathfrak{B})$. Composing, we get an algebra isomorphism

$$\textbf{(7.6.18)} \qquad \boldsymbol{\Psi} : \mathcal{H}(C^\times, \mathbf{1}_\mathcal{I}) \xrightarrow{\approx} \mathcal{H}(G, \lambda)$$

which preserves support of functions on $\widetilde{\boldsymbol{W}}(\mathfrak{B})$.

The maximal simple type associated to $(\mathcal{I}, \mathbf{1}_\mathcal{I})$ is the type $(\mathfrak{o}_K^\times, \mathbf{1})$ in $GL(1, K)$. As a special case of (7.6.18), we have a support-preserving algebra isomorphism $\boldsymbol{\Psi}_1 : \mathcal{H}(K^\times, \mathbf{1}_{\mathfrak{o}_K^\times}) \to \mathcal{H}(G', \lambda')$, where (J', λ') denotes the maximal type in G' associated to (J, λ). The affine Weyl group which supports these Hecke algebras is the cyclic group generated by π_E. We can identify

$$\mathcal{H}(T, \mathbf{1}_{\mathcal{I} \cap T}) = \mathcal{H}(K^\times, \mathbf{1}_{\mathfrak{o}_K^\times}) \otimes \ldots \otimes \mathcal{H}(K^\times, \mathbf{1}_{\mathfrak{o}_K^\times})$$

(e tensor factors) and likewise

$$\mathcal{H}(M, \lambda_U) = \mathcal{H}(G', \lambda') \otimes \ldots \otimes \mathcal{H}(G', \lambda').$$

Thus $\boldsymbol{\Psi}_1$ induces an algebra isomorphism $\mathcal{H}(T, 1_{\mathcal{I}\cap T}) \cong \mathcal{H}(M, \lambda_U)$. However, for our present purposes, we have to be very careful about how we make the identifications implicit in this.

The group $\widetilde{\boldsymbol{W}}(\mathfrak{B})$ acts on M by conjugation. This action normalises the subgroup $J \cap M$, and fixes the representation λ_U up to equivalence. Thus, for $\boldsymbol{w} \in \widetilde{\boldsymbol{W}}(\mathfrak{B})$, there is an automorphism $\boldsymbol{\alpha_w}$ of the representation space W of λ_U such that $\boldsymbol{\alpha_w} \circ \lambda_U(x) = \lambda_U(\boldsymbol{w}x\boldsymbol{w}^{-1}) \circ \boldsymbol{\alpha_w}$, $x \in J \cap M$, and this $\boldsymbol{\alpha_w}$ is uniquely determined up to scalar factor. We can then define an action of $\widetilde{\boldsymbol{W}}(\mathfrak{B})$ on $\mathcal{H}(M, \lambda_U)$ by

$$\boldsymbol{w} \cdot \phi : x \mapsto \boldsymbol{\alpha_w} \circ \phi(\boldsymbol{w}^{-1} x \boldsymbol{w}) \circ \boldsymbol{\alpha_w}^{-1}, \quad x \in M,\ \phi \in \mathcal{H}(M, \lambda_U),\ \boldsymbol{w} \in \widetilde{\boldsymbol{W}}(\mathfrak{B}).$$

This action preserves the algebra structure of $\mathcal{H}(M, \lambda_U)$. Moreover, the subgroup $\boldsymbol{D}(\mathfrak{B})$ acts trivially here:— acting on $\phi \in \mathcal{H}(M, \lambda_U)$ by an element $\boldsymbol{a} \in \boldsymbol{D}(\mathfrak{B})$ is the same as conjugating ϕ by a function $\phi_{\boldsymbol{a}} \in \mathcal{H}(M, \lambda_U)$ with support $(J \cap M)\boldsymbol{a}$, and the algebra $\mathcal{H}(M, \lambda_U)$ is commutative. In other words, this is effectively the obvious action of the permutation subgroup $\boldsymbol{W}_0(\mathfrak{B})$ of $\widetilde{\boldsymbol{W}}(\mathfrak{B})$.

We can similarly define an action of the group $\widetilde{\boldsymbol{W}}(\mathfrak{B})$ on the algebra $\mathcal{H}(T, \mathbf{1}_{T\cap\mathcal{I}})$. With this set up, we now have

(7.6.19) *Given a support-preserving algebra isomorphism*

$$\boldsymbol{\Psi}_1 : \mathcal{H}(K^\times, \mathbf{1}_{\mathfrak{o}_K^\times}) \xrightarrow{\approx} \mathcal{H}(G', \lambda'),$$

there is a unique support-preserving algebra homomorphism

$$\boldsymbol{\Psi}_1^{\otimes e} : \mathcal{H}(T, \mathbf{1}_{T\cap\mathcal{I}}) \xrightarrow{\approx} \mathcal{H}(M, \lambda_U)$$

which agrees with $\boldsymbol{\Psi}_1$ on the first tensor factor and commutes with the actions of $\widetilde{\boldsymbol{W}}(\mathfrak{B})$ on the two algebras.

Now we return to the normalised induction functor $\mathrm{Ind}^u = \mathrm{Ind}^u_{G/P}$: $\sigma \mapsto \mathrm{Ind}(\delta_P^{1/2} \otimes \sigma)$. Abbreviating $\delta_P^{1/2} = \sqrt{\delta}$ in subscripts, (7.6.1) gives us a commutative diagram

$$\begin{array}{ccc} \mathfrak{Ad}_{\lambda_U} & \xrightarrow{\mathrm{Mod}_{\lambda_U}} & \mathcal{H}(M, \lambda_U)\text{-}\mathfrak{Mod}_{\mathrm{f}} \\ {\scriptstyle \mathrm{Ind}^u}\downarrow & & \downarrow{\scriptstyle (j_{\sqrt{\delta_P}}[\lambda])_*} \\ \mathfrak{Ad}_{\lambda}(G) & \xrightarrow{\mathrm{Mod}_{\lambda}} & \mathcal{H}(G, \lambda)\text{-}\mathfrak{Mod}_{\mathrm{f}} \end{array}$$

We have a similar situation relative to the type $(\mathcal{I}, \mathbf{1}_{\mathcal{I}})$.

(7.6.20) Theorem: *With the notation above, let* $\mathbf{\Psi}_1 : \mathcal{H}(K^\times, \mathbf{1}_{\mathfrak{o}_K^\times}) \to \mathcal{H}(G', \lambda')$ *be a support-preserving algebra isomorphism, and define* $\mathbf{\Psi}_1^{\otimes e}$ *as in* (7.6.19). *Then there is a unique support-preserving algebra isomorphism* $\mathbf{\Psi} : \mathcal{H}(C^\times, \mathbf{1}_{\mathcal{I}}) \to \mathcal{H}(G, \lambda)$ *such that the following diagram is commutative:*

$$\begin{array}{ccc} \mathcal{H}(T, \mathbf{1}_{\mathcal{I}\cap T}) & \xrightarrow{\mathbf{\Psi}_1^{\otimes e}} & \mathcal{H}(M, \lambda_U) \\ {\scriptstyle j_{\sqrt{\delta_Q}}[\mathbf{1}_{\mathcal{I}}]}\big\downarrow & & \big\downarrow{\scriptstyle j_{\sqrt{\delta_P}}[\lambda]} \\ \mathcal{H}(C^\times, \mathbf{1}_{\mathcal{I}}) & \xrightarrow{\mathbf{\Psi}} & \mathcal{H}(G, \lambda). \end{array}$$

As an immediate consequence, we have:

(7.6.21) Corollary: *With the notation of the theorem, write* $\mathfrak{Ad}(\mathbf{\Psi})$ *for the equivalence* $\mathfrak{Ad}_{\mathbf{1}_{\mathcal{I}}}(C^\times) \to \mathfrak{Ad}_\lambda(G)$ *induced by* $\mathbf{\Psi}$, *and define* $\mathfrak{Ad}(\Psi_1^{\otimes e})$ *likewise. Then the diagram*

$$\begin{array}{ccc} \mathfrak{Ad}_{\mathbf{1}_{\mathcal{I}\cap T}}(T) & \xrightarrow{\mathfrak{Ad}(\mathbf{\Psi}_1^{\otimes e})} & \mathfrak{Ad}_{\lambda_U}(M) \\ {\scriptstyle \mathrm{Ind}^u}\big\downarrow & & \big\downarrow{\scriptstyle \mathrm{Ind}^u} \\ \mathfrak{Ad}_{\mathbf{1}_{\mathcal{I}}}(C^\times) & \xrightarrow[\mathfrak{Ad}(\mathbf{\Psi})]{} & \mathfrak{Ad}_\lambda(G) \end{array}$$

commutes.

Proof of (7.6.20): The first step is to give a partial characterisation of the maps $\boldsymbol{j}_{\sqrt{\delta}}$ in terms of the canonical inner products on the various Hecke algebras. At first, we work only in the context of (J, λ). The conclusions there then apply, *mutatis mutandis*, to the special case $(\mathcal{I}, \mathbf{1}_{\mathcal{I}})$.

We define our algebras $\mathcal{H}(G, \lambda)$, $\mathcal{H}(G, \lambda_P)$ and $\mathcal{H}(M, \lambda_U)$ relative to Haar measures μ_G, μ_C such that $\mu_G(K) = \mu_M(J \cap M) = 1$, where $K = (J \cap P)H^1$ as always. Write $\boldsymbol{h}_G$, $\boldsymbol{h}_M$ for the canonical inner products on $\mathcal{H}(G, \lambda_P)$, $\mathcal{H}(M, \lambda_U)$ respectively. The isomorphism $\mathcal{H}(G, \lambda_P) \cong \mathcal{H}(G, \lambda)$, given by the relation $\lambda \cong \mathrm{Ind}(\lambda_P : K, J)$, transports $\boldsymbol{h}_G$ to the canonical inner product on $\mathcal{H}(G, \lambda)$ (see (4.3.3)). Further, the form $\boldsymbol{h}_M$ is fixed by the action of $\widetilde{\boldsymbol{W}}(\mathfrak{B})$ on $\mathcal{H}(M, \lambda_U)$:—

$$\boldsymbol{h}_M(\boldsymbol{w} \cdot \phi, \boldsymbol{w} \cdot \psi) = \boldsymbol{h}_M(\phi, \psi), \qquad \phi, \psi \in \mathcal{H}(M, \lambda_U),\ \boldsymbol{w} \in \widetilde{\boldsymbol{W}}(\mathfrak{B}).$$

(7.6.22) Lemma: *Let* $\phi \in \mathcal{H}(M, \lambda_U)$ *have support* $(J \cap M)\boldsymbol{a}$, *for some* $\boldsymbol{a} \in \boldsymbol{D}^-(\mathfrak{B})$. *Then*

$$\boldsymbol{h}_G(\boldsymbol{j}[\lambda_P]\phi, \boldsymbol{j}[\lambda_P]\phi) = \delta_P(\boldsymbol{a})\boldsymbol{h}_M(\phi, \phi).$$

Proof: Let W denote the representation space of λ_P (or λ_U), and write "bar" for the involution on $\mathrm{End}_{\mathbb{C}}(\check{W})$ induced by a K-invariant (or just $(J \cap M)$-invariant) positive definite hermitian form on $\check{W}$. Then, for $\phi \in \mathcal{H}(M, \lambda_U)$, $\boldsymbol{h}_M(\phi, \phi) = \mathrm{tr}_{\check{W}}(\phi * \overline{\phi}(1_M))$, $\overline{\phi}$ being the function $x \mapsto \overline{\phi(x^{-1})}$. If ϕ is supported on $(J \cap M)\boldsymbol{a}$, $\boldsymbol{a} \in \boldsymbol{D}(\mathfrak{B})$, then (7.6.3) gives us $\phi * \overline{\phi}(1) = \phi(\boldsymbol{a}) \circ \overline{\phi(\boldsymbol{a})}$.

Now we put $\Phi = \boldsymbol{j}[\lambda_P]\phi$, and compute $\boldsymbol{h}_G(\Phi, \Phi)$. The function Φ is supported on $K\boldsymbol{a}K$, and $\Phi(\boldsymbol{a}) = \phi(\boldsymbol{a})$. Moreover, $\overline{\Phi}$ is supported on $K\boldsymbol{a}^{-1}K$, and $\overline{\Phi}(\boldsymbol{a}^{-1}) = \overline{\Phi(\boldsymbol{a})}$. We compute the convolution $\Phi * \overline{\Phi}$ at 1, and find

$$\Phi * \overline{\Phi}(1) = \sum_z \check{\lambda}_P(z) \circ \Phi(\boldsymbol{a}) \circ \overline{\Phi(\boldsymbol{a})} \circ \overline{\check{\lambda}_P(z)},$$

where z ranges over $K/(K \cap {}^{\boldsymbol{a}}K)$. Now we apply the trace, and recall that $\overline{\check{\lambda}_P(z)} = \check{\lambda}_P(z^{-1})$. This gives us $\boldsymbol{h}_G(\Phi, \Phi) = \delta_P(\boldsymbol{a})\boldsymbol{h}_M(\phi, \phi)$, and the lemma follows. ∎

For $0 \le i \le e$, let $\boldsymbol{a}_i \in \boldsymbol{D}^-(\mathfrak{B})$ be that element whose first i diagonal blocks are π_E, with all others 1. Put $\boldsymbol{b}_i = \boldsymbol{a}_i \boldsymbol{a}_{i-1}^{-1}$, $1 \le i \le e$. Let f_1 denote the characteristic function of $(\mathcal{I} \cap T)\boldsymbol{a}_1$, and put $\phi_1 = \boldsymbol{\Psi}_1^{\otimes e}(f_1)$. Form the function

$$\phi_i = \prod_{j=1}^{i} \left(\Pi(\mathfrak{B})^{j-1} \cdot \phi_1\right),$$

where we write convolutions as products and "$\cdot$" denotes the action of $\widetilde{\boldsymbol{W}}(\mathfrak{B})$ on $\mathcal{H}(M, \lambda_U)$ as above. This product makes sense since $\mathcal{H}(M, \lambda_U)$ is a commutative algebra. Thus ϕ_i has support $(J \cap M)\boldsymbol{a}_i$. Indeed, $\phi_i = \boldsymbol{\Psi}_1^{\otimes e}(f_i)$, where f_i is the characteristic function of $(\mathcal{I} \cap T)\boldsymbol{a}_i$, since $\boldsymbol{\Psi}_1^{\otimes e}$ commutes with the actions of $\widetilde{\boldsymbol{W}}(\mathfrak{B})$. Also, we let ϕ_0 denote the identity element of $\mathcal{H}(M, \lambda_U)$.

We define functions

$$\vartheta_i \in \mathcal{H}(G, \lambda_P), \quad 1 \le i \le e.$$

The function ϑ_i is to have support $K\boldsymbol{b}_iK$, and $\vartheta_i(\boldsymbol{b}_i) = \phi_{i-1}(\boldsymbol{a}_{i-1})^{-1} \circ \phi_i(\boldsymbol{a}_i)$. This is possible since ((7.2.19)) the intertwining spaces for λ_P, λ_U supported by any element of $\boldsymbol{D}(\mathfrak{B})$ are identical. In particular,

$$\vartheta_1 = c\boldsymbol{j}_{\sqrt{\delta_P}}[\lambda_P]\phi_1, \quad \textit{for some } c > 0.$$

(7.6.23) Lemma: *Let $\nu \in \mathcal{H}(G, \lambda_P)$ have support $K\Pi(\mathfrak{B})K$. Then, writing convolutions in $\mathcal{H}(G, \lambda_P)$ as products, we have $\vartheta_i = c_i \nu^{1-i} \vartheta_1 \nu^{i-1}$, for some $c_i > 0$, $1 \le i \le e$.*

Proof: First we observe that (7.2.10) implies the existence of a function ν with the required properties.

Temporarily write ι for the canonical isomorphism $\mathcal{H}(G,\lambda_P) \xrightarrow{\approx} \mathcal{H}(G,\lambda)$ implied by (4.1.3) and the equivalence $\lambda \cong \mathrm{Ind}(\lambda_P)$. Then $\iota(\nu)$ has support $J\Pi(\mathfrak{B})$, and (5.6.8), (5.6.9) show that, for $j \in \mathbb{Z}$ and $\phi \in \mathcal{H}(G,\lambda)$, the function $\iota(\nu)^j\phi\iota(\nu)^{-j}$ has support $\Pi(\mathfrak{B})^j\mathrm{supp}(\phi)\Pi(\mathfrak{B})^{-j}$. Further,

$$\iota(\nu)^j\phi\iota(\nu)^{-j}(\Pi(\mathfrak{B})^j g\Pi(\mathfrak{B})^{-j}) = c\iota(\nu)(\Pi(\mathfrak{B}))^j \circ \phi(g) \circ \iota(\nu)(\Pi(\mathfrak{B}))^{-j},$$

for $g \in G$ and some $c > 0$.

We show that the analogue of this holds in $\mathcal{H}(G,\lambda_P)$, with suitable restrictions. We know ((7.2.19)) that every $\boldsymbol{d} \in \boldsymbol{D}(\mathfrak{B})$ intertwines λ_P. Take $\boldsymbol{d} \in \boldsymbol{D}(\mathfrak{B})$, $j \in \mathbb{Z}$, and let $\phi \in \mathcal{H}(G,\lambda_P)$ have support $K\boldsymbol{d}K$. Since $\Pi(\mathfrak{B})^j\boldsymbol{d}\Pi(\mathfrak{B})^{-j} \in \boldsymbol{D}(\mathfrak{B})$, this element intertwines λ_P. It follows that the function $\nu^j\phi\nu^{-j}$ has support $K\Pi(\mathfrak{B})^j\boldsymbol{d}\Pi(\mathfrak{B})^{-j}K$. To compute its value at $\Pi(\mathfrak{B})^j\boldsymbol{d}\Pi(\mathfrak{B})^{-j}$, we use the definition of ι given by the proof of (4.1.3). Write W for the representation space of λ, W_P for that of λ_P. We have a canonical decomposition $W = W_P \oplus W'$ of W as K-space. By (4.1.5) and (5.6.6), $\Pi(\mathfrak{B})^j$ intertwines λ_P with itself and no other component of $\lambda \mid K$. This implies that the operator $\iota(\nu)^j(\Pi(\mathfrak{B})^j) = c.(\iota(\nu)(\Pi(\mathfrak{B})))^j$ commutes with the projection $W \to W_P$ (where c is again some positive constant). This just says

$$\nu^j\phi\nu^{-j}(\Pi(\mathfrak{B})^j\boldsymbol{d}\Pi(\mathfrak{B})^{-j}) = c\nu(\Pi(\mathfrak{B}))^j\phi(\boldsymbol{d})\nu(\Pi(\mathfrak{B}))^{-j},$$

as required. Taking $\phi = \vartheta_1$, we get the assertion of the lemma. ∎

If we take any support-preserving isomorphism $\boldsymbol{\Psi} : \mathcal{H}(C^\times, 1_\mathcal{I}) \to \mathcal{H}(G,\lambda)$, the functions $\boldsymbol{\Psi j}[1_\mathcal{I}](f_1)$, $\boldsymbol{j}[\lambda](\phi_1)$ have support $J\boldsymbol{a}_1J$, and so differ by a scalar. Since $\boldsymbol{a}_1 = \Pi(\mathfrak{B})\boldsymbol{c}$, for the obvious cycle $\boldsymbol{c} \in \boldsymbol{W}_0(\mathfrak{B})$, we may choose $\boldsymbol{\Psi}$ to satisfy

$$\textbf{(7.6.24)} \qquad \boldsymbol{\Psi j}_{\sqrt{\delta_Q}}[\mathbf{1}_\mathcal{I}](f_1) = \boldsymbol{j}_{\sqrt{\delta_P}}[\lambda](\phi_1).$$

This certainly determines $\boldsymbol{\Psi}$ uniquely.

(7.6.25) Lemma: *Suppose $\boldsymbol{\Psi}_1$ is unitary, and define $\boldsymbol{\Psi}$ by* (7.6.24). *Then $\boldsymbol{\Psi}$ is unitary.*

Proof: Write $\tilde{\boldsymbol{h}}_G$, $\boldsymbol{h}_T$, $\boldsymbol{h}_C$ for the canonical inner products on $\mathcal{H}(G,\lambda)$, $\mathcal{H}(T, \mathbf{1}_{\mathcal{I}\cap T})$, $\mathcal{H}(C^\times, \mathbf{1}_\mathcal{I})$, respectively. We have

$$\tilde{\boldsymbol{h}}_G(\boldsymbol{j}_{\sqrt{\delta_P}}[\lambda]\phi_1, \boldsymbol{j}_{\sqrt{\delta_P}}[\lambda]\phi_1) = \boldsymbol{h}_G(\boldsymbol{j}_{\sqrt{\delta_P}}[\lambda_P]\phi_1, \boldsymbol{j}_{\sqrt{\delta_P}}[\lambda_P]\phi_1) = \boldsymbol{h}_M(\phi_1,\phi_1)$$

by (7.6.22), and this equals $\dim(\lambda_U)\boldsymbol{h}_T(f_1,f_1)$ since $\boldsymbol{\Psi}_1^{\otimes e}$ is unitary. However, $\boldsymbol{h}_T(f_1,f_1) = \boldsymbol{h}_C(\boldsymbol{j}_{\sqrt{\delta_Q}}[\mathbf{1}_\mathcal{I}]f_1, \boldsymbol{j}_{\sqrt{\delta_Q}}[\mathbf{1}_\mathcal{I}]f_1)$ by (7.6.22) again, and so

$$\boldsymbol{h}_C(\boldsymbol{j}_{\sqrt{\delta_Q}}[\mathbf{1}_\mathcal{I}]f_1, \boldsymbol{j}_{\sqrt{\delta_Q}}[\mathbf{1}_\mathcal{I}]f_1) = \dim(\lambda_U)\tilde{\boldsymbol{h}}_G(\boldsymbol{j}_{\sqrt{\delta_P}}[\lambda]\phi_1, \boldsymbol{j}_{\sqrt{\delta_P}}[\lambda_P]\phi_1),$$

and this is enough to show that $\boldsymbol{\Psi}$ is unitary. ■

(7.6.26) Lemma: *For $1 \leq i \leq e$, there is a constant $\gamma_i > 0$ such that*

$$\boldsymbol{\Psi} \boldsymbol{j}_{\sqrt{\delta_Q}}[\mathbf{1}_{\mathcal{I}}](f_i) = \gamma_i \boldsymbol{j}_{\sqrt{\delta_P}}[\lambda](\phi_i).$$

Before proving this lemma, we show how it implies the theorem. First we note that if the theorem holds for one choice of $\boldsymbol{\Psi}_1$, then it holds for any other, provided we define the associated map $\boldsymbol{\Psi}$ by (7.6.24). We may therefore assume that $\boldsymbol{\Psi}_1$ is unitary. Next, if $f \in \mathcal{H}(T, \mathbf{1}_{\mathcal{I} \cap T})$ has support $(\mathcal{I} \cap T)\boldsymbol{a}$, $\boldsymbol{a} \in \boldsymbol{D}^-(\mathfrak{B})$, then both $\boldsymbol{\Psi}(\boldsymbol{j}_{\sqrt{\delta_Q}}[\mathbf{1}_{\mathcal{I}}](f))$ and $\boldsymbol{j}[\lambda](\boldsymbol{\Psi}_1^{\otimes e}(f))$ have support $J\boldsymbol{a}J$. The maps $\boldsymbol{\Psi} \circ \boldsymbol{j}_{\sqrt{\delta_Q}}[\mathbf{1}_{\mathcal{I}}]$, $\boldsymbol{j}[\lambda] \circ \boldsymbol{\Psi}_1^{\otimes e}$, therefore differ by a support-preserving automorphism of $\mathcal{H}(M, \lambda_U)$. Such an automorphism is of the form τ_χ ((7.6.5)), for some unramified quasicharacter χ of M. Therefore there exists an unramified quasicharacter χ of M such that $\boldsymbol{\Psi} \circ \boldsymbol{j}_{\sqrt{\delta_Q}}[\mathbf{1}_{\mathcal{I}}] = \boldsymbol{j}_\chi[\lambda] \circ \boldsymbol{\Psi}_1^{\otimes e}$. Now (7.6.26) shows that χ can only take real positive values on $\boldsymbol{D}^-(\mathfrak{B})$. (7.6.22) shows that $\chi \mid \boldsymbol{D}^-(\mathfrak{B}) = \sqrt{\delta_P} \mid \boldsymbol{D}^-(\mathfrak{B})$, whence $\boldsymbol{j}_\chi[\lambda] = \boldsymbol{j}_{\sqrt{\delta_P}}[\lambda]$, as required for the theorem.

We prove (7.6.26). It is more convenient to work with the composite $\boldsymbol{\Psi}'$ of $\boldsymbol{\Psi}$ with the canonical isomorphism $\mathcal{H}(G, \lambda) \to \mathcal{H}(G, \lambda_P)$. There certainly exists $\gamma_i \in \mathbb{C}^\times$ such that $\boldsymbol{\Psi}' \boldsymbol{j}_{\sqrt{\delta_Q}}[\mathbf{1}_{\mathcal{I}}] f_i = \gamma_i \boldsymbol{j}_{\sqrt{\delta_P}}[\lambda_P]\phi_i$, so we have to show that γ_i is real and positive. We have arranged matters (in (7.6.24)) to give $\gamma_1 = 1$, and we work by induction on i.

Let g_i denote the characteristic function of $\mathcal{I}\boldsymbol{b}_i\mathcal{I}$. We have $c > 0$ such that $\vartheta_1 = c\boldsymbol{j}_{\sqrt{P_\delta}}[\lambda_P]\phi_1 = c\boldsymbol{\Psi}' \boldsymbol{j}_{\sqrt{\delta_Q}}[\mathbf{1}_{\mathcal{I}}] f_1 = c'\boldsymbol{\Psi}' g_1$, with $c' > 0$. (7.6.23) and the $\widetilde{\boldsymbol{W}}(\mathfrak{B})$-invariance properties together imply that $\vartheta_i = c_i \boldsymbol{\Psi}' g_i$, with some positive c_i (different from the one in (7.6.23)).

(7.6.27) Lemma: *We have*

$$\begin{aligned} \boldsymbol{h}_G(\boldsymbol{j}[\lambda_P]\phi_{i-1} * \vartheta_i, \boldsymbol{j}[\lambda_P]\phi_i) > 0, \\ \boldsymbol{h}_C(\boldsymbol{j}[\mathbf{1}_{\mathcal{I}}]f_{i-1} * g_i, \boldsymbol{j}[\mathbf{1}_{\mathcal{I}}]f_i) > 0, \end{aligned} \quad 1 \leq i \leq r.$$

Proof: The second statement is a special case of the first. When $i = 1$, the statement reduces to the assertion $\boldsymbol{h}_G(\boldsymbol{j}[\lambda_p]\phi_1, \boldsymbol{j}[\lambda_P]\phi_1) > 0$, which is immediate. In general, $\boldsymbol{j}[\lambda_P]\phi_i$ is supported on $K\boldsymbol{a}_iK$. We compute the value of $\boldsymbol{j}[\lambda_P]\phi_{i-1} * \vartheta_i$ at $\boldsymbol{a}_i$, and show it is a positive multiple of $\boldsymbol{j}[\lambda_P]\phi_i(\boldsymbol{a}_i)$. Passing to $\mathcal{H}(G, \lambda)$ and then to the affine Hecke algebra via (5.6.6), the result follows from (5.6.18).

If we write the convolution $\boldsymbol{j}[\lambda_P]\phi_{i-1} * \vartheta_i$ as a sum, in the usual way, the fact that $\boldsymbol{a}_{i-1} \in \boldsymbol{D}(\mathfrak{B})$ reduces us to

$$\begin{aligned} \boldsymbol{j}[\lambda_P]\phi_{i-1} * \vartheta_i(\boldsymbol{a}_i) &= c\sum_z \boldsymbol{j}[\lambda_P]\phi_{i-1}(z\boldsymbol{a}_{i-1}) \circ \vartheta_i(\boldsymbol{a}_{i-1}^{-1}z^{-1}\boldsymbol{a}_i) \\ &= c\sum_z \boldsymbol{j}[\lambda_P]\phi_{i-1}(\boldsymbol{a}_{i-1}) \circ \vartheta_i(\boldsymbol{a}_{i-1}^{-1}z^{-1}\boldsymbol{a}_i), \end{aligned}$$

where z ranges over $(J^1 \cap U)/^{\boldsymbol{a}_{i-1}}(J^1 \cap U)$ and $c > 0$. An element z contributes to this sum if and only if $z^{-1} \in \boldsymbol{a}_{i-1} K \boldsymbol{a}_{i-1}^{-1} \boldsymbol{a}_i K \boldsymbol{a}_i^{-1}$. Expanding, we have

$$\boldsymbol{a}_{i-1} K \boldsymbol{a}_{i-1}^{-1} \boldsymbol{a}_i K \boldsymbol{a}_i^{-1} =$$
$$({}^{\boldsymbol{a}_{i-1}}J^1 \cap U)(J \cap M)({}^{\boldsymbol{a}_{i-1}}H^1 \cap U^-)({}^{\boldsymbol{a}_i}H^1 \cap U^-)(J \cap M)({}^{\boldsymbol{a}_i}J^1 \cap U).$$

Since we have at least $z \in U$, this forces

$$z \in ({}^{\boldsymbol{a}_{i-1}}J^1 \cap U)({}^{\boldsymbol{a}_i}J^1 \cap U)$$

and our convolution therefore comes down to

$$c'' \boldsymbol{j}[\lambda_P]\phi_{i-1}(\boldsymbol{a}_{i-1})\vartheta_i(\boldsymbol{b}_i) = c'' \boldsymbol{j}[\lambda_P]\phi_i(\boldsymbol{a}_i),$$

with $c'' > 0$, as required. ■

(7.6.27) now implies $\boldsymbol{j}_{\sqrt{\delta_P}}[\lambda_P]\phi_{i-1} * \vartheta_i = \alpha \boldsymbol{j}_{\sqrt{\delta_P}}[\lambda]\phi_i + \psi$, where $\alpha > 0$ and $\operatorname{supp}(\psi) \cap K\boldsymbol{a}_i K = \emptyset$. Likewise, $\boldsymbol{j}_{\sqrt{\delta_Q}}[\mathbf{1}_{\mathcal{I}}]f_{i-1} * g_i = \beta \boldsymbol{j}_{\sqrt{\delta_Q}}[\mathbf{1}_{\mathcal{I}}]f_i + h$, where $\beta > 0$ and $\operatorname{supp}(h) \cap \mathcal{I}\boldsymbol{a}_i\mathcal{I} = \emptyset$. We now save notation by using "c" to denote a variable positive real. Thus

$$\begin{aligned}
\boldsymbol{\Psi}' \boldsymbol{j}_{\sqrt{\delta_Q}}[\mathbf{1}_{\mathcal{I}}]f_i &= c\boldsymbol{\Psi}'(\boldsymbol{j}_{\sqrt{\delta_Q}}[\mathbf{1}_{\mathcal{I}}]f_{i-1} * g_i - h) \\
&= c\boldsymbol{j}_{\sqrt{\delta_P}}[\lambda_P]\phi_{i-1} * \boldsymbol{\Psi}' g_i - c\boldsymbol{\Psi}' h \quad \textit{(by induction)} \\
&= c\boldsymbol{j}_{\sqrt{\delta_P}}[\lambda_P]\phi_{i-1} * \vartheta_i - c\boldsymbol{\Psi}' h \\
&= c\boldsymbol{j}_{\sqrt{\delta_P}}[\lambda_P]\phi_i + \theta,
\end{aligned}$$

for some function θ with $\operatorname{supp}(\theta) \cap K\boldsymbol{a}_i K = \emptyset$. It follows that $\theta = 0$, and $\boldsymbol{\Psi}' \boldsymbol{j}_{\sqrt{\delta_Q}}[\mathbf{1}_{\mathcal{I}}]f_i = c\boldsymbol{j}_{\sqrt{\delta_P}}[\lambda_P]\phi_i$, for some $c > 0$, as required.

This proves (7.6.26) and completes the proof of the theorem. ■

Remark: For many applications, it is unnecessary to identify precisely the map $\boldsymbol{\Psi}$ which makes the diagram of (7.6.20) commute. For, if we take another support-preserving isomorphism $\boldsymbol{\Psi}'$, we have

$$\mathfrak{Ad}(\boldsymbol{\Psi}') : \boldsymbol{\sigma} \mapsto \chi \circ \det \otimes \mathfrak{Ad}(\boldsymbol{\Psi})(\boldsymbol{\sigma}),$$

for some fixed unramified quasicharacter χ of $F^\times$, by (7.5.10). Such an operation has no significant effect on, for example, the structure of composition series.

(7.7) Discrete series and formal degree

We now show that our Hecke algebra isomorphisms, or rather the categorical equivalences which they induce, preserve discrete series and, moreover, lead to a strikingly simple relation on the formal degree.

7. Typical representations

Two points deserve emphasis at the beginning. Firstly, our results are essentially formal consequences of the definition of discrete series, and rely only on the fact that our Hecke algebra isomorphisms can be chosen to respect the canonical involutions and inner products on the algebras. Thus we describe, *ab initio*, the λ-typical part of the discrete series of G in terms of the $(C^\times, \mathbf{1}_\mathcal{I})$-typical part of the discrete series of the group $C^\times$, using the notation of (7.6.16/17). Of course, discrete series representations with an Iwahori-fixed vector are described in **[KL]**. Secondly, we shall later see (in (8.5.10)) that any discrete series representation of G must contain some simple type, so this section, as we allow (J, λ) to vary over all simple types in G, gives a complete determination of the discrete series of G.

We fix once for all a Haar measure μ_G (or dg) on G, one μ_F (or dy) on $F^\times$, and we take $\dot{\mu}_G$ ($d\dot{g}$) to be the unique Haar measure on $G/F^\times$ for which $dg = dy d\dot{g}$.

Let $(\pi, \mathcal{V})$ be a smooth irreducible representation of G whose central quasicharacter ω_π is *unitary*, i.e. $|\omega_\pi| = 1$. Recall that $(\pi, \mathcal{V})$ is square-integrable if

$$\int_{G/F^\times} |\langle \pi(g)v, \check{v}\rangle|^2 \, d\dot{g} < \infty, \qquad v \in \mathcal{V},\ \check{v} \in \check{\mathcal{V}}.$$

An arbitrary irreducible smooth representation $(\pi', \mathcal{V}')$ of G is *essentially square-integrable* (or *discrete series*) if $\pi' = \pi \otimes \chi \circ \det_A$, where π is square-integrable and χ is an unramified quasicharacter of $F^\times$.

We return to our simple type (J, λ) in G, and use the notation (7.6.16), (7.6.17). We fix a support-preserving algebra isomorphism Ψ : $\mathcal{H}(C^\times, \mathbf{1}_\mathcal{I}) \to \mathcal{H}(G, \lambda)$, and write $\mathfrak{Ad}_\Psi : \mathfrak{Ad}_{\mathbf{1}_\mathcal{I}}(C^\times) \xrightarrow{\approx} \mathfrak{Ad}_\lambda(G)$ for the equivalence of categories which it defines.

(7.7.1) Theorem: *Let $(\pi, \mathcal{V})$ be an irreducible smooth representation of G containing the simple type λ. Then $(\pi, \mathcal{V})$ is discrete series if and only if $\pi \cong \mathfrak{Ad}_\Psi(\sigma)$, for some discrete series representation $(\sigma, \mathcal{W}) \in |\mathfrak{Ad}_{\mathbf{1}_\mathcal{I}}(C^\times)|$.*

(7.7.2) Gloss: We actually prove a more precise statement. We know from (7.5.10) that changing Ψ just tensors $\mathfrak{Ad}_\Psi(\sigma)$ with $\chi \circ \det_A$, for some fixed unramified χ. Therefore, in the theorem, we could take Ψ unitary. We show, in this situation, that *π is square-integrable if and only if $\pi \cong \mathfrak{Ad}_\Psi(\sigma)$, where σ is square-integrable.*

Our first task in the proof is to characterise square-integrable representations in terms that our Hecke algebra machinery can recognise. Let $(\pi, \mathcal{V})$ be an irreducible representation of G containing λ with $\omega_\pi = \omega$

unitary. Thus $\lambda \mid \mathfrak{o}_F^\times$ is a multiple of $\omega \mid \mathfrak{o}_F^\times$, and we can form the representation $\lambda\omega$ of $JF^\times$, as in (6.1.1).

Write $\mathcal{H}(G,\omega)$ for the convolution algebra (relative to $\dot{\mu}_G$) of smooth functions $\phi : G \to \mathbb{C}$ which are compactly supported modulo $F^\times$ and satisfy $\phi(zg) = \omega(z)^{-1}\phi(g)$, $z \in F^\times$, $g \in G$. This is contained in the space $\mathcal{H}^2(G,\omega)$ of left and right smooth functions $\phi : G \to \mathbb{C}$, with the same property $\phi(zg) = \omega(z)^{-1}\phi(g)$, but which are absolutely square integrable modulo $F^\times$. We view $\mathcal{H}^2(G,\omega)$ as a $G \times G$-space, with G acting by left and right translation, and this is the same as viewing it, in the obvious way, as an $(\mathcal{H}(G,\omega), \mathcal{H}(G,\omega))$-bimodule. Then, immediately,

(7.7.3) $(\pi, \mathcal{V})$ *is square-integrable if and only if* $\mathcal{V}$ *is isomorphic to a left* $\mathcal{H}(G,\omega)$*-submodule (equivalently a left G-subspace) of* $\mathcal{H}^2(G,\omega)$.

We can set this up more precisely. Write "bar" for the usual involution $\overline{\phi} : x \mapsto \overline{\phi(x^{-1})}$ of $\mathcal{H}(G,\omega)$, and (,) for its inner product

$$(\phi, \psi) = \int_{G/F^\times} \phi(x)\overline{\psi(x)}\, d\dot{x} = \phi * \overline{\psi}(1), \qquad \phi, \psi \in \mathcal{H}(G,\omega).$$

Both the involution and the inner product extend to $\mathcal{H}^2(G,\omega)$, and they restrict to the corresponding objects on the subalgebra $e_{\lambda\omega} * \mathcal{H}(G,\omega) * e_{\lambda\omega}$. The completion of $e_{\lambda\omega} * \mathcal{H}(G,\omega) * e_{\lambda\omega}$ with respect to (,) is then $e_{\lambda\omega} * \mathcal{H}^2(G,\omega) * e_{\lambda\omega}$.

(7.7.4) Proposition: *Let* $(\pi, \mathcal{V})$ *be an irreducible smooth representation of G which contains the simple type* λ*, and whose central quasicharacter* $\omega = \omega_\pi$ *is unitary. The* $(\pi, \mathcal{V})$ *is square-integrable if and only if the space* $\mathcal{V}^\lambda = \mathcal{V}^{\lambda\omega}$ *is isomorphic to a left* $e_{\lambda\omega} * \mathcal{H}(G,\omega) * e_{\lambda\omega}$*-submodule of* $e_{\lambda\omega} * \mathcal{H}^2(G,\omega) * e_{\lambda\omega}$.

Proof: We have $\mathcal{V} = \mathcal{H}(G) * \mathcal{V}^\lambda$, by (7.5.6), and indeed $\mathcal{V} = \mathcal{H}(G,\omega) * \mathcal{V}^{\lambda\omega}$. If $\mathcal{V}^{\lambda\omega}$ is isomorphic to a $e_{\lambda\omega} * \mathcal{H}(G,\omega) * e_{\lambda\omega}$-submodule $\mathcal{M}$ of $e_{\lambda\omega} * \mathcal{H}^2(G,\omega) * e_{\lambda\omega}$, we have $\mathcal{V} \cong \mathcal{H}(G,\omega) * \mathcal{M}$, by (7.5.6). This is a left G-subspace of $\mathcal{H}^2(G,\omega)$, and π is therefore square-integrable. Conversely, suppose $(\pi, \mathcal{V})$ is square-integrable, and choose a nonzero $\check{v} \in (\check{\mathcal{V}})^{\check{\lambda}}$. We map $\mathcal{V}$ to $\mathcal{H}^2(G,\omega)$ by $v \mapsto \phi_v$, where ϕ_v is the function $\phi_v(g) = \langle \pi(g^{-1})v, \check{v}\rangle$, $g \in G$. Then certainly $\phi_v \in \mathcal{H}^2(G,\omega)$, and ϕ is an injective G-homomorphism (with G acting on $\mathcal{H}^2(G,\omega)$ by left translation). Moreover, since $\check{v} \in (\check{\mathcal{V}})^{\check{\lambda}} = (\check{\mathcal{V}})^{(\lambda\omega)^\vee}$, we have $\phi_v \in \mathcal{H}^2(G,\omega) * e_{\lambda\omega}$. Therefore ϕ maps $\mathcal{V}^{\lambda\omega}$ into $e_{\lambda\omega} * \mathcal{H}^2(G,\omega) * e_{\lambda\omega}$, as required. ∎

On the other hand, we can form the completion $\mathcal{H}^2(G,\lambda\omega)$ of the space $\mathcal{H}(G,\lambda\omega)$ with respect to its canonical inner product. Writing W for the representation space of λ, (4.3.4) implies that the algebra isomorphism Υ of (4.2.4) extends to an isomorphism on the completions:

$$\Upsilon^2 : \mathcal{H}^2(G,\lambda\omega) \otimes \mathrm{End}_{\mathbb{C}}(W) \xrightarrow{\approx} e_{\lambda\omega} * \mathcal{H}^2(G,\omega) * e_{\lambda\omega}.$$

Further, $\boldsymbol{\Upsilon}^2$ preserves the involutions inherited by these completions. We deduce:

(7.7.5) Corollary: *With the hypotheses of* (7.7.4), *the representation* $(\boldsymbol{\pi}, \mathcal{V})$ *is square-integrable if and only if* $\mathrm{Mod}_\lambda(\mathcal{V})$ *is isomorphic to an* $\mathcal{H}(G, \lambda\omega)$*-submodule of* $\mathcal{H}^2(G, \lambda\omega)$.

All of this discussion applies equally to the group $C^\times$, relative to the type $(\mathcal{I}, \mathbf{1}_\mathcal{I})$, except that we must replace $F^\times$ by $K^\times$.

We now fix a prime element π_F of F, and let $\phi_1 \in \mathcal{H}(G, \lambda)$ be the function with support $\pi_F J$ such that $\phi_1(\pi_F) = \mu_G(J)^{-1}\omega(\pi_F)^{-1}1_{\check{W}}$. Then we know from (6.1.5) that $\mathcal{H}(G, \lambda\omega)$ is the quotient of $\mathcal{H}(G, \lambda)$ by the ideal generated by the (central) element $\phi_1 - \mathrm{e}_\lambda$, where e_λ denotes the unit element of $\mathcal{H}(G, \lambda)$ for this choice of Haar measure. The quotient map P_ω is given by

$$P_\omega(\phi) = \mu_F(\mathfrak{o}_F^\times) \sum_{n=-\infty}^{\infty} \phi_1^n * \phi, \quad \phi \in \mathcal{H}(G, \lambda), \ \textit{or}$$

$$P_\omega(\phi)(g) = \mu_F(\mathfrak{o}_F^\times) \sum_{n=-\infty}^{\infty} \omega(\pi_F)^n \phi(\pi_F^n g), \quad g \in G,$$

(where ϕ_1^n is the n-th power ϕ_1^{*n} of ϕ_1 in $\mathcal{H}(G, \lambda)$). Since ϕ_1 is central in $\mathcal{H}(G, \lambda)$ and $\overline{\phi}_1 = \phi_1^{-1}$, it follows that P_ω is a surjective homomorphism of algebras with involution. Write $\boldsymbol{h}_\lambda$ for the canonical inner product on $\mathcal{H}(G, \lambda)$, $\boldsymbol{h}_{\lambda\omega}$ for that on $\mathcal{H}(G, \lambda\omega)$. We have $\boldsymbol{h}_{\lambda\omega}(P_\omega(\phi), P_\omega(\psi)) = \mathrm{tr}_{\check{W}}(P_\omega(\phi) * \overline{P_\omega(\psi)}(1)) = \mathrm{tr}_{\check{W}}(P_\omega(\phi * \overline{\psi})(1))$, $\phi, \psi \in \mathcal{H}(G, \lambda)$, just as before.

Now we go to the other side. We have

$$\boldsymbol{\Psi}^{-1}(\phi_1) = \psi_1 = \alpha\mu_C(\mathcal{I})^{-1}\psi_1',$$

say, where $\alpha \in \mathbb{C}^\times$ and ψ_1' denotes the characteristic function of $\mathcal{I}\pi_F$. We define a quasicharacter ω' of the group $F^\times\mathfrak{o}_K^\times$ by $\omega' \mid \mathfrak{o}_K^\times = 1$, $\omega'(\pi_F) = \alpha^{-1}$. Since $\boldsymbol{\Psi}$ is unitary, we have $|\omega'| = 1$. The reason for this definition is given by:

(7.7.6) Lemma: *Let* $(\boldsymbol{\pi}, \mathcal{V}) \in |\mathfrak{Ad}_\lambda(G)|$, *let* ω, ω', $\boldsymbol{\Psi}$, ϕ_1, ψ_1 *be as above. Let* $(\boldsymbol{\sigma}, \mathcal{W}) = \mathfrak{Ad}_{\boldsymbol{\Psi}}^{-1}(\boldsymbol{\pi}, \mathcal{V}) \in |\mathfrak{Ad}_{\mathbf{1}_\mathcal{I}}(C^\times)|$, $\mathcal{M} = \mathrm{Mod}_\lambda(\mathcal{V})$, $M = \mathrm{Mod}_{\mathbf{1}_\mathcal{I}}(\mathcal{W})$, *so that* $M = \boldsymbol{\Psi}^*\mathcal{M}$. *The following are equivalent:*

(i) $\omega_{\boldsymbol{\pi}} = \omega$;
(ii) $(\mathrm{e}_\lambda - \phi_1)\mathcal{M} = \{0\}$;
(iii) $(\mathrm{e}_{\mathbf{1}_\mathcal{I}} - \psi_1)M = \{0\}$;
(iv) $\omega_{\boldsymbol{\sigma}}|F^\times\mathfrak{o}_K^\times = \omega'$.

Proof: Trivial. ∎

We can form the character $\mathbf{1}_{\mathcal{I}}\omega'$ of $F^\times\mathcal{I}$. We write $\mathcal{F} = F^\times\mathfrak{o}_K^\times$, and choose Haar measures μ_C, $\mu_{\mathcal{F}}$, $\dot{\mu}_C$ on $C^\times$, $\mathcal{F}$, $C^\times/\mathcal{F}$ respectively, such that $\mu_C = \dot{\mu}_C\mu_{\mathcal{F}}$, and form the algebra $\mathcal{H}(C^\times, \mathbf{1}_{\mathcal{I}}\omega')$ relative to the measure $\dot{\mu}_C$. We have a surjective homomorphism $P_{\omega'} : \mathcal{H}(C^\times, \mathbf{1}_{\mathcal{I}}) \to \mathcal{H}(C^\times, \mathbf{1}_{\mathcal{I}}\omega')$ of algebras with involution given by

$$P_{\omega'}(\phi) = \mu_{\mathcal{F}}(\mathfrak{o}_K^\times) \sum_{n=-\infty}^{\infty} \psi_1^n * \phi, \quad \phi \in \mathcal{H}(C^\times, \mathbf{1}_{\mathcal{I}}), \text{ or}$$

$$P_{\omega'}(\phi)(y) = \mu_{\mathcal{F}}(\mathfrak{o}_K^\times) \sum_{n=-\infty}^{\infty} \omega(\pi_F)^n \phi(\pi_F^n y), \quad y \in C^\times,$$

with ψ_1 as above. There is a unique isomorphism $\mathbf{\Psi}_\omega$ of algebras with involution such that the following diagram commutes:

$$\begin{array}{ccc} \mathcal{H}(C^\times, \mathbf{1}_{\mathcal{I}}) & \xrightarrow{\mathbf{\Psi}} & \mathcal{H}(G, \lambda) \\ {\scriptstyle P_{\omega'}}\downarrow & & \downarrow{\scriptstyle P_\omega} \\ \mathcal{H}(C^\times, \mathbf{1}_{\mathcal{I}}\omega') & \xrightarrow{\mathbf{\Psi}_\omega} & \mathcal{H}(G, \lambda\omega) \end{array}$$

The algebra $\mathcal{H}(C^\times, \mathbf{1}_{\mathcal{I}}\omega')$ carries a canonical inner product $\boldsymbol{h}_{\mathbf{1}_{\mathcal{I}}\omega'}$ in the usual way, and we have the standard relation

$$\boldsymbol{h}_{\mathbf{1}_{\mathcal{I}}\omega'}(P_{\omega'}(\phi), P_{\omega'}(\psi)) = P_{\omega'}(\phi * \overline{\psi})(1), \qquad \phi, \psi \in \mathcal{H}(C^\times, \mathbf{1}_{\mathcal{I}}).$$

The commutativity of the diagram above therefore gives us, for $\phi, \psi \in \mathcal{H}(C^\times, \mathbf{1}_{\mathcal{I}})$, the relation

$$\begin{aligned} \boldsymbol{h}_{\lambda\omega}(P_\omega\mathbf{\Psi}(\phi), P_\omega\mathbf{\Psi}(\psi)) &= \mathrm{tr}_{\check{W}}(P_\omega\mathbf{\Psi}(\phi * \overline{\psi})(1)) \\ &= \mathrm{tr}_{\check{W}}(\mathbf{\Psi}_\omega P_{\omega'}(\phi * \overline{\psi})(1)). \end{aligned}$$

The function $P_{\omega'}(\phi*\overline{\psi})$ takes the form $\alpha e+\xi$, where e denotes the identity element of $\mathcal{H}(C^\times, \mathbf{1}_{\mathcal{I}}\omega')$, $\alpha \in \mathbb{C}$, and $\mathrm{supp}(\xi) \cap \mathcal{I}F^\times = \emptyset$. Thus $P_{\omega'}(\phi * \overline{\psi})(1) = \alpha\dot{\mu}_C(\mathcal{I}F^\times/\mathcal{F})^{-1}$. The isomorphism $\mathbf{\Psi}_\omega$ preserves support of functions, so $\mathbf{\Psi}_\omega P_{\omega'}(\phi * \overline{\psi}) = \alpha e_{\lambda\omega} + \mathbf{\Psi}_\omega(\xi)$, and the support of $\mathbf{\Psi}_\omega(\xi)$ does not meet $JF^\times$. Thus

$$\mathbf{\Psi}_\omega P_{\omega'}(\phi * \overline{\psi})(1) = \alpha\dot{\mu}_G(JF^\times/F^\times)^{-1}1_{\check{W}}.$$

We deduce

$$\begin{aligned} \boldsymbol{h}_{\lambda\omega}(P_\omega\mathbf{\Psi}(\phi), P_\omega\mathbf{\Psi}(\psi)) &= \alpha \dim(\lambda).\dot{\mu}_G(JF^\times/F^\times)^{-1} \\ &= \dot{\mu}_C(\mathcal{I}F^\times/\mathcal{F})\dot{\mu}_G(JF^\times/F^\times)^{-1} \dim(\lambda) P_{\omega'}(\phi * \overline{\psi})(1) \\ &= \dot{\mu}_C(\mathcal{I}F^\times/\mathcal{F})\dot{\mu}_G(JF^\times/F^\times)^{-1} \dim(\lambda)\boldsymbol{h}_{\mathbf{1}_{\mathcal{I}}\omega'}(P_{\omega'}(\phi), P_{\omega'}(\psi)) \end{aligned}$$

for $\phi, \psi \in \mathcal{H}(C^\times, \mathbf{1}_\mathcal{I})$. Since the maps P_ω, $P_{\omega'}$ are surjective, we can rewrite this in the form

(7.7.7)

$$\dot\mu_G(JF^\times/F^\times)\boldsymbol{h}_{\lambda\omega}(\boldsymbol{\Psi}_\omega\varphi, \boldsymbol{\Psi}_\omega\vartheta) = \dim(\lambda)\dot\mu_C(\mathcal{I}F^\times/\mathcal{F})\boldsymbol{h}_{\mathbf{1}_\mathcal{I}\omega'}(\varphi, \vartheta),$$

for $\varphi, \vartheta \in \mathcal{H}(C^\times, \mathbf{1}_\mathcal{I}\omega')$. That is, the isomorphism $\boldsymbol{\Psi}_\omega$ preserves inner products, up to a harmless constant factor. It therefore extends to an isomorphism $\boldsymbol{\Psi}_\omega^2 : \mathcal{H}^2(C^\times, \mathbf{1}_\mathcal{I}\omega') \to \mathcal{H}^2(G, \lambda\omega)$ on completions. It therefore establishes a bijection between $\mathcal{H}(G, \lambda\omega)$-submodules of $\mathcal{H}^2(G, \lambda\omega)$ and $\mathcal{H}(C^\times, \mathbf{1}_\mathcal{I}\omega')$-submodules of $\mathcal{H}^2(C^\times, \mathbf{1}_\mathcal{I}\omega')$. The algebra $\mathcal{H}(C^\times, \mathbf{1}_\mathcal{I}\omega')$ splits as a direct product of algebras

$$\mathcal{H}(C^\times, \mathbf{1}_\mathcal{I}\omega') = \prod_\Omega \mathcal{H}(C^\times, \mathbf{1}_\mathcal{I}\Omega),$$

where Ω ranges over the unramified quasicharacters of $K^\times$ such that $\Omega \mid F^\times\mathfrak{o}_K^\times = \omega'$, and the algebra $\mathcal{H}(C^\times, \mathbf{1}_\mathcal{I}\Omega)$ is formed relative to the measure $\dot\mu_C$. The inner product on $\mathcal{H}(C^\times, \mathbf{1}_\mathcal{I}\omega')$ is the orthogonal sum of the inner products on the $\mathcal{H}(C^\times, \mathbf{1}_\mathcal{I}\Omega)$, so this extends to the completions. A simple $\mathcal{H}(C^\times, \mathbf{1}_\mathcal{I}\omega')$-submodule of $\mathcal{H}^2(C^\times, \mathbf{1}_\mathcal{I}\omega')$ is thus the same as a simple $\mathcal{H}(C^\times, \mathbf{1}_\mathcal{I}\Omega)$-submodule of the completion $\mathcal{H}^2(C^\times, \mathbf{1}_\mathcal{I}\Omega)$, for some Ω. The theorem now follows from (7.7.5). ■

(7.7.8) Remark: One would customarily form the algebra $\mathcal{H}(C^\times, \mathbf{1}_\mathcal{I}\Omega)$ relative to a measure $\ddot\mu_C$ on $C^\times/K^\times$, rather than the measure $\dot\mu_C$ on $C^\times/\mathcal{F}$. However, given such a measure $\dot\mu_C$, there is a unique measure $\ddot\mu_C$ on $C^\times/K^\times$ such that

$$\int_{C^\times/\mathcal{F}} f(x)\,d\dot\mu_C(x) = \frac{1}{e'}\int_{C^\times/K^\times} \sum_{y\in K^\times/\mathcal{F}} f(yx)\,d\ddot\mu_C(x),$$

for any L^1-function f on $C^\times/\mathcal{F}$, with $e' = (K^\times : \mathcal{F}) = e(K|F) = e(E|F)$. The convolutions on the function space $\mathcal{H}(C^\times, \mathbf{1}_\mathcal{I}\Omega)$ induced by these two measures are therefore identical. Observe that we also have

$$\ddot\mu_C(\mathcal{I}K^\times/K^\times) = e'\dot\mu_C(\mathcal{I}F^\times/\mathcal{F}).$$

Let us continue in the same situation. Take $(\pi, \mathcal{V}) \in |\mathfrak{A}\mathfrak{d}_\lambda(G)|$ irreducible and square-integrable, with $\omega_\pi = \omega$. Recall that *the formal degree* $\boldsymbol{d}(\pi)$ *of* π (with respect to the Haar measure $\dot\mu_G$) is defined by

$$\int_{G/F^\times} \langle\pi(g)v_1, \check v_1\rangle\langle\pi(g^{-1})v_2, \check v_2\rangle\,d\dot g = \boldsymbol{d}(\pi)^{-1}\langle v_1, \check v_2\rangle\langle v_2, \check v_1\rangle,$$

for $v_i \in \mathcal{V}$, $\check{v}_i \in \check{\mathcal{V}}$. It will be convenient to work with the more invariant object

$$\delta(\pi) = \dot{\mu}_G(JF^\times/F^\times)d(\pi). \tag{7.7.9}$$

We can rewrite this definition. Let h be some positive definite, G-invariant hermitian form on $\mathcal{V}$. Then

$$\int_{G/F^\times} h(\pi(g)v_1, w_1)\overline{h(\pi(g)v_2, w_2)}\, d\dot{g}$$
$$= \delta(\pi)^{-1}\dot{\mu}_G(JF^\times/F^\times)h(v_1, v_2)\overline{h(w_1, w_2)},$$

for $v_i, w_i \in \mathcal{V}$. We observe that the form h is uniquely determined up to a positive constant factor, so this expression is certainly independent of h. We now view $\mathcal{V}$ as embedded in $\mathcal{H}^2(G, \omega)$ via an embedding of $\mathcal{V}^{\lambda\omega}$ in $e_{\lambda\omega} * \mathcal{H}^2(G, \omega) * e_{\lambda\omega}$. We can replace the vectors v_i, w_i by functions $\varphi_i, \vartheta_i \in \mathcal{H}^2(G, \omega) \cap \mathcal{V}$, and h by the canonical inner product $(\,,\,)$ on $\mathcal{H}^2(G, \omega)$. We then have $h(\pi(g)\varphi, \vartheta) = \varphi * \overline{\vartheta}(g^{-1})$, and the relation above is

$$(\varphi_1 * \overline{\vartheta}_1, \varphi_2 * \overline{\vartheta}_2) = \varphi_1 * \overline{\vartheta}_1 * \vartheta_2 * \overline{\varphi}_1(1)$$
$$= \delta(\pi)^{-1}\dot{\mu}_G(JF^\times/F^\times)\, \varphi_1 * \overline{\varphi}_2(1)\, \vartheta_2 * \overline{\vartheta}_1(1).$$

There is no harm in taking the functions $\varphi_i, \vartheta_i \in e_{\lambda\omega} * \mathcal{H}^2(G, \omega) * e_{\lambda\omega} \cap \mathcal{V}$. (Note that we could take them all equal and nonzero, to make both sides of this equation positive.) Now we use the map $\boldsymbol{\Upsilon}^2$ to identify $e_{\lambda\omega} * \mathcal{H}^2(G, \omega) * e_{\lambda\omega}$ with $\mathcal{H}^2(G, \lambda\omega) \otimes_{\mathbb{C}} \mathrm{End}_{\mathbb{C}}(W)$ as Hermitian spaces. Then $\mathcal{V}^{\lambda\omega} = \boldsymbol{\Upsilon}^2(\mathcal{M} \otimes \mathfrak{a})$, where $\mathcal{M}$ is a simple left $\mathcal{H}(G, \lambda\omega)$-submodule of $\mathcal{H}^2(G, \lambda\omega)$ and $\mathfrak{a}$ is simple left ideal of $\mathrm{End}_{\mathbb{C}}(W)$. Then (4.3.4) and the definition of the form on $\mathcal{H}(G, \lambda\omega) \otimes \mathrm{End}_{\mathbb{C}}(W)$ give

$$(\boldsymbol{\Upsilon}(\Phi \otimes x), \boldsymbol{\Upsilon}(\Psi \otimes x)) = \mathrm{tr}_W(x\bar{x})\boldsymbol{h}_{\lambda\omega}(\Phi, \Psi), \quad \Phi, \Psi \in \mathcal{H}(G, \lambda\omega),\ x \in \mathfrak{a},$$

and this relation extends to completions. Of course, we have $\mathrm{tr}_W(x\bar{x}) > 0$ provided $x \neq 0$. Further, $\mathrm{tr}_W((x\bar{x})^2) = (\mathrm{tr}_W(x\bar{x}))^2$. We now take some nonzero $x \in \mathfrak{a}$ and $\varphi_i = \Xi_i \otimes x$, $\vartheta_i = \Theta_i \otimes x$, for various $\Xi_i, \Theta_i \in \mathcal{M}$, $i = 1, 2$. The relation above gives

$$\boldsymbol{h}_{\lambda\omega}(\Xi_1 * \overline{\Theta}_1, \Xi_2 * \overline{\Theta}_2)$$
$$= \delta(\pi)^{-1}\dot{\mu}_G(JF^\times/F^\times)\boldsymbol{h}_{\lambda\omega}(\Xi_1, \Xi_2)\overline{\boldsymbol{h}_{\lambda\omega}(\Theta_1, \Theta_2)}, \quad \Xi_i, \Theta_i \in \mathcal{M}. \tag{7.7.10}$$

Now we go through the same computation on the other side. We take the irreducible square-integrable $(\boldsymbol{\sigma}, \mathcal{W}) \in |\mathfrak{A}\mathfrak{d}_{1_T}(C^\times)|$ which corresponds to

$(\pi, \mathcal{V})$ via the unitary algebra isomorphism Ψ. Let Ω be the central character of σ. The inner product $h_{1_{\mathcal{I}}\omega'}$ on $\mathcal{H}(C^\times, 1_{\mathcal{I}}\omega')$ restricts to the inner product $h_{1_{\mathcal{I}}\Omega}$ on $\mathcal{H}(C^\times, 1_{\mathcal{I}}\Omega)$ (which we form relative to the measure $\ddot{\mu}_C$ of (7.7.8)). If $M = \Psi_\omega^{-1}(\mathcal{M})$ is the subspace of $\mathcal{H}^2(C^\times, 1_{\mathcal{I}}\Omega)$ corresponding to σ, we have the relation

$$\begin{aligned} h_{1_{\mathcal{I}}\Omega}(\xi_1 * \overline{\theta}_1, \xi_2 * \overline{\theta}_2) &= \delta(\sigma)^{-1}\ddot{\mu}_C(\mathcal{I}K^\times/K^\times) h_{1_{\mathcal{I}}\Omega}(\xi_1, \xi_2)\overline{h_{1_{\mathcal{I}}\Omega}(\theta_1, \theta_2)} \\ &= \delta(\sigma)^{-1}e'\dot{\mu}_C(\mathcal{I}F^\times/\mathcal{F}) h_{1_{\mathcal{I}}\Omega}(\xi_1, \xi_2)\overline{h_{1_{\mathcal{I}}\Omega}(\theta_1, \theta_2)} \end{aligned}$$

for $\xi_i, \theta_i \in M$. We can now put $\Xi_i = \Psi_\omega(\xi_i)$, $\Theta_i = \Psi_\omega(\theta_i)$, and substitute in the relations (7.7.7), (7.7.10). This gives us the identity

$$\delta(\pi) = \delta(\sigma)\frac{\dim(\lambda)}{e'}.$$

Now we substitute back, and observe that the formal degree is unchanged by twisting with an unramified quasicharacter. We have proved:

(7.7.11) Corollary: *Let $(\pi, \mathcal{V}) \in |\mathfrak{A}\mathfrak{d}_\lambda(G)|$ be irreducible and discrete series. Let $\Psi : \mathcal{H}(C^\times 1_{\mathcal{I}}) \to \mathcal{H}(G, \lambda)$ be a support preserving algebra isomorphism, and let $(\sigma, \mathcal{W}) \in |\mathfrak{A}\mathfrak{d}_{1_{\mathcal{I}}}(C^\times)|$ be the discrete series representation of $C^\times$ corresponding to π via Ψ. Define the formal degrees $d(\pi)$, $d(\sigma)$ relative to measures $\dot{\mu}_G$ on $G/F^\times$, $\ddot{\mu}_C$ on $C^\times/K^\times$. Then*

$$d(\pi)\dot{\mu}_G(JF^\times/F^\times) = \frac{\dim(\lambda)}{e'}\, d(\sigma)\ddot{\mu}_C(\mathcal{I}K^\times/K^\times),$$

where $e' = e(K|F) = e(E|F)$.

One can compute these measures explicitly, with little difficulty. The formal degree formula of **[Wa]** is thus a special case of (7.7.11). In this connection, see also **[CMS]**.

8. ATYPICAL REPRESENTATIONS

We come to the final chapter with one readily apparent problem outstanding. Having completely classified those irreducible smooth representations of our group $G = \mathrm{Aut}_F(V) \cong GL(N, F)$ which contain some simple type, we must now deal with those "atypical" representations which do not. The most pressing requirement here is to show that any supercuspidal representation of G contains a simple type. In this aspect, the basic idea goes back to **[Bu1]** and **[K4]** via **[KM2]** but is vastly more general than these. The absence of a simple type in an irreducible representation π is reflected in the presence of an obstruction of a very definite kind. This leads to the notion of "split type", which we define in **(8.1)**. A split type is a generalised version of a split fundamental stratum (as in **(2.3)**), and plays the same role here as the split fundamental strata in **[K4]** (or the relatively split strata in **[KM2]**). The main result of **(8.1)** shows that an irreducible smooth representation of G which contains no simple type must contain a split type. The converse of this statement almost certainly holds, but that does not concern us here.

One of the main properties of a split type is that it defines a proper parabolic subgroup P of G such that any irreducible representation of G containing that split type has a nontrivial Jacquet module relative to P (and this Jacquet module contains a particular subrepresentation defined by the type). There are two rather different cases to be considered, so this argument occupies **(8.2)**, **(8.3)**.

This immediately implies that an irreducible supercuspidal representation of G cannot contain a split type, and must therefore contain a simple type. Thus (6.2.2), (6.2.3) give all irreducible supercuspidal representations of G explicitly as induced representations. It also extends the scope of the results of §**7**:— an irreducible representation of G contains a simple type if and only if its supercuspidal support consists of unramified twists of a single representation. Representations of this sort are then classified by Kazhdan-Lusztig invariants via the Hecke algebra isomorphisms of (5.6.6). We summarise these consequences in **(8.4)**.

We thus come to **(8.5)** knowing that an irreducible representation of G is atypical if and only if its supercuspidal support contains at least two representations which are not unramified twists of each other. The simplest of these are the composition factors of $\pi_1 \times \pi_2$, where the π_i are irreducible and supercuspidal, and further $\pi_2 \ncong \pi_1 \otimes \chi \circ \det$, for any unramified quasicharacter χ of $F^\times$. The techniques associated with split types enable us to show directly that $\pi_1 \times \pi_2$ is irreducible and so $\pi_1 \times \pi_2 \cong \pi_2 \times \pi_1$. From there, it is trivial to show that an atypical

π embeds in $\pi_1 \times \ldots \times \pi_r$, where the π_i are typical and have "strongly disjoint" supercuspidal supports. Casselman's theorem on the composition structure of Jacquet modules of induced representations then shows, via an easy counting argument, that $\pi_1 \times \ldots \times \pi_r$ is irreducible. Hence $\pi \cong \pi_1 \times \ldots \times \pi_r$, and, moreover, this expression is unique up to permutation of the typical factors, any permutation being permissible here. Thus we achieve a complete classification of the irreducible smooth representations of G.

We conclude by reproducing a simple argument (which surely cannot be original!) showing that a representation of G which is irreducibly induced from a proper parabolic subgroup of G cannot be discrete series. It follows that any discrete series representation of G contains a simple type, and hence is given by (7.7.1).

(8.1) Split types

We construct a family of irreducible representations of certain compact open subgroups of our group $G = \mathrm{Aut}_F(V)$, to be called "split types". These generalise the split fundamental strata of **[K4]** recalled in (2.3.3). They extend the idea of a split refinement of a simple stratum, as discussed in **(2.3)**, to the contexts of §§**3**,**5**. Their function in the overall structure is straightforward:— the presence of a split type in a smooth irreducible representation π of G realises the obstruction to the presence of a simple type in G.

There are a couple of apparently exceptional cases to be dealt with, although this diversity is more in the description than the substance. However, it does mean that we have to give a list of definitions. A split type in G will be a pair (K, ϑ), where K is a compact open subgroup of G and ϑ an irreducible smooth representation of K, of a sort to be specified. A split type (K, ϑ) has a "level". This is a pair (x, y) of rational numbers with $x \geq y$, which we use to distinguish the various cases. Not all pairs (x, y) can arise here.

As in §§**1-3**, we fix a continuous character ψ_F of the additive group of F, with conductor $\mathfrak{p}_F$. We set $\psi = \psi_A = \psi_F \circ \mathrm{tr}_{A/F}$, where $A = \mathrm{End}_F(V)$ as always. Also, for $a \in A$, $\psi_a = \psi_{A,a}$ denotes the function $x \mapsto \psi(a(x-1))$, $x \in A$ (or some restriction thereof).

(8.1.1) Definition: *A split type of level (x, x), $x > 0$, is a pair (K, ϑ) given as follows:*

(i) $[\mathfrak{A}, n, n-1, b]$ *is a split fundamental stratum in A;*
(ii) $n > 0$, $\gcd(n, e(\mathfrak{A})) = 1$, $x = n/e(\mathfrak{A})$;
(iii) $K = \boldsymbol{U}^n(\mathfrak{A})$, $\vartheta = \psi_b$.

(8.1.2) Definition: *A* split type of level $(0,0)$ *is a pair* (K,ϑ) *given as follows:*

(i) $\mathfrak{A}$ *is a hereditary* $\mathfrak{o}_F$*-order in* A*;*

(ii) $\boldsymbol{U}(\mathfrak{A})/\boldsymbol{U}^1(\mathfrak{A}) = \mathcal{G}_1 \times \ldots \times \mathcal{G}_e$, *where* $\mathcal{G}_i \cong GL(n_i, \mathrm{k}_F)$*;*

(iii) $K = \boldsymbol{U}(\mathfrak{A})$ *and* ϑ *is the inflation of an irreducible representation of* $\boldsymbol{U}(\mathfrak{A})/\boldsymbol{U}^1(\mathfrak{A})$ *of the form* $\xi_1 \otimes \ldots \otimes \xi_e$, *where* ξ_i *is a cuspidal irreducible representation of* $\mathcal{G}_i$, *and* $\xi_i \not\cong \xi_j$ *for some* $i \neq j$.

(8.1.3) Definition: *A* split type of level (x,y), $x > y > 0$, *is a pair* (K,ϑ) *given as follows:*

(i) $[\mathfrak{A}, n, m, \beta]$ *is a simple stratum in* A *with* $E = F[\beta]$, $B = \mathrm{End}_E(V)$, $\mathfrak{B} = \mathfrak{A} \cap B$, $e_\beta = e(\mathfrak{B}|\mathfrak{o}_E)$, $\gcd(m, e_\beta) = 1$, $x = n/e(\mathfrak{A})$, $y = m/e(\mathfrak{A})$*;*

(ii) $K = H^m(\beta, \mathfrak{A})$*;*

(iii) $\vartheta = \theta\psi_c$, *for some* $\theta \in \mathcal{C}(\mathfrak{A}, m-1, \beta)$ *and some* $c \in \mathfrak{P}^{-m}$ *(where* $\mathfrak{P} = \mathrm{rad}(\mathfrak{A})$*) such that the stratum* $[\mathfrak{B}, m, m-1, s_\beta(c)]$ *is split fundamental, where* s_β *denotes a tame corestriction on* A *relative to* E/F.

Thus, in this case, $\vartheta \mid H^{m+1}(\beta, \mathfrak{A})$ lies in $\mathcal{C}(\mathfrak{A}, m, \beta)$.

(8.1.4) Definition: *A* split type of level $(x,0)$, $x > 0$, *is a pair* (K,ϑ) *given as follows:*

(i) $[\mathfrak{A}, n, 0, \beta]$ *is a simple stratum in* A, *with* $x = n/e(\mathfrak{A})$, $E = F[\beta]$, $B = \mathrm{End}_E(V)$, $\mathfrak{B} = \mathfrak{A} \cap B$, $e_\beta = e(\mathfrak{B}|\mathfrak{o}_E)$*;*

(ii) $K = J(\beta, \mathfrak{A})$, *and* ϑ *is an irreducible representation of* K *of the form* $\kappa \otimes \xi$, *where:*

(a) $\kappa \mid H^1(\beta, \mathfrak{A})$ *contains some* $\theta \in \mathcal{C}(\mathfrak{A}, 0, \beta)$, $\kappa \mid J^1(\beta, \mathfrak{A})$ *is irreducible, and* κ *is a* β*-extension of* $\kappa \mid J^1(\beta, \mathfrak{A})$ *(in the sense of* (5.2.1)*);*

(b) ξ *is the inflation of a representation* $\xi_1 \otimes \ldots \otimes \xi_{e_\beta}$ *of the group* $J(\beta, \mathfrak{A})/J^1(\beta, \mathfrak{A}) = \boldsymbol{U}(\mathfrak{B})/\boldsymbol{U}^1(\mathfrak{B}) = \mathcal{G}_1 \times \ldots \times \mathcal{G}_{e_\beta}$, *where* $\mathcal{G}_i \cong GL(m_i, \mathrm{k}_E)$ *for various* $m_i \geq 1$, *the factors* ξ_i *are all cuspidal, and* $\xi_i \not\cong \xi_j$ *for some pair* i, j.

In practice, one can often handle (8.1.1) as a special case of (8.1.3), and (8.1.2) as a special case of (8.1.4).

(8.1.5) Theorem: *Let* π *be a smooth irreducible representation of* G, *and suppose that* π *contains no simple type. Then* π *contains a split type.*

Proof: Suppose first that π contains the trivial representation of $\boldsymbol{U}^1(\mathfrak{A})$, for some hereditary $\mathfrak{o}_F$-order $\mathfrak{A}$ in A. Choose $\mathfrak{A}$ minimal for this property, and write

$$\boldsymbol{U}(\mathfrak{A})/\boldsymbol{U}^1(\mathfrak{A}) = \mathcal{G}_1 \times \ldots \times \mathcal{G}_e, \quad e = e(\mathfrak{A}|\mathfrak{o}_F),$$

for groups $\mathcal{G}_i \cong GL(n_i, \mathrm{k}_F)$ and various integers $n_i \geq 1$. The representation $\pi \mid U(\mathfrak{A})$ then has an irreducible factor ϑ which is null on $U^1(\mathfrak{A})$, and hence of the form $\vartheta = \vartheta_1 \otimes \ldots \otimes \vartheta_e$, where ϑ_i is an irreducible representation of $\mathcal{G}_i$. The first step is to show that the minimality of $\mathfrak{A}$ implies that all of these factors ϑ_i are cuspidal, or, equivalently, that ϑ is a cuspidal representation of the reductive k_E-group $\mathcal{G} = U(\mathfrak{A})/U^1(\mathfrak{A})$. Suppose otherwise. There exists a proper parabolic subgroup $\mathcal{P}$ of $\mathcal{G}$ such that $\vartheta \mid \mathcal{P}$ has an irreducible component ϑ' which is trivial on the unipotent radical $\mathcal{U}$ of $\mathcal{P}$. There exists a hereditary $\mathfrak{o}_F$-order $\mathfrak{A}' \subsetneq \mathfrak{A}$ such that $\mathcal{P}$ is the image of $U(\mathfrak{A}')$ in $\mathcal{G}$. Moreover, we have $U(\mathfrak{A}) \supset U(\mathfrak{A}') \supset U^1(\mathfrak{A}') \supset U^1(\mathfrak{A})$, and $\mathcal{U}$ is the image $U^1(\mathfrak{A}')/U^1(\mathfrak{A})$ of $U^1(\mathfrak{A}')$ in $\mathcal{G}$. The representation ϑ' (inflated to $U(\mathfrak{A}')$) certainly occurs in π, and is trivial on $U^1(\mathfrak{A}')$. This contradiction implies that the ϑ_i are all cuspidal, as asserted.

If the ϑ_i are all equivalent to each other, we have $n_1 = n_2 = \ldots = n_e$, so the order $\mathfrak{A}$ is principal, the pair $(U(\mathfrak{A}), \vartheta)$ is a simple type in G, and ϑ occurs in π. Therefore there exist i, j such that $\vartheta_i \not\cong \vartheta_j$, and $(U(\mathfrak{A}), \vartheta)$ is a split type in G, of level $(0,0)$, which occurs in π.

We may therefore assume that π does not contain the trivial representation of $U^1(\mathfrak{A})$ for any hereditary $\mathfrak{o}_F$-order $\mathfrak{A}$ in A. Thus, as in **[Bu1]** Th. 2, π contains a fundamental stratum $[\mathfrak{A}, n, n-1, b]$ with $n \geq 1$ (or rather $\pi \mid U^n(\mathfrak{A})$ contains ψ_b). If the stratum $[\mathfrak{A}, n, n-1, b]$ is split fundamental, we can argue as in **[K4]** (a more general version of this argument is given below) to show that we can further impose the condition $\gcd(n, e(\mathfrak{A})) = 1$. In this case, π contains a split type of level (x, x), $x = n/e(\mathfrak{A}) > 0$. Therefore we may assume that the stratum $[\mathfrak{A}, n, n-1, b]$ is nonsplit fundamental. By (2.3.4), we may assume that it is simple.

We are therefore reduced to the case in which π does contain some simple stratum. Therefore there exists a simple stratum $[\mathfrak{A}, n, m, \beta]$ in A, with $m \geq 0$, and a character $\theta \in \mathcal{C}(\mathfrak{A}, m, \beta)$ such that $\pi \mid H^{m+1}(\beta, \mathfrak{A})$ contains θ. We choose such a pair $([\mathfrak{A}, n, m, \beta], \theta)$ for which the quantity $m/e(\mathfrak{A})$ is minimal.

The argument now splits into two further cases according to whether or not $m = 0$. We start with the easier one in which $m = 0$. Among those pairs $([\mathfrak{A}, n, 0, \beta], \theta)$, $\theta \in \mathcal{C}(\mathfrak{A}, 0, \beta)$, for which π contains θ, we choose one for which the order $\mathfrak{A}$ is minimal. By (5.1.1), there is a unique irreducible representation η of $J^1(\beta, \mathfrak{A})$ such that $\eta \mid H^1(\beta, \mathfrak{A})$ contains θ. We deduce that $\pi \mid J^1(\beta, \mathfrak{A})$ contains η. Further, $\pi \mid J(\beta, \mathfrak{A})$ contains an irreducible representation ϑ such that $\vartheta \mid J^1(\beta, \mathfrak{A})$ contains η. It follows from (5.2.2) that ϑ is of the form $\vartheta = \kappa \otimes \xi$, where κ is some β-extension of η and ξ is the inflation of some irreducible representation of $J(\beta, \mathfrak{A})/J^1(\beta, \mathfrak{A})$. We invoke our usual notation $E = F[\beta]$, $B =$

$\mathrm{End}_E(V)$, $\mathfrak{B} = \mathfrak{A} \cap B$. We identify $J(\beta, \mathfrak{A})/J^1(\beta, \mathfrak{A}) = \boldsymbol{U}(\mathfrak{B})/\boldsymbol{U}^1(\mathfrak{B}) = \mathcal{G}_1 \times \ldots \times \mathcal{G}_e$, where $e = e(\mathfrak{B}|\mathfrak{o}_E)$ and $\mathcal{G}_i \cong GL(m_i, \mathsf{k}_E)$ for various integers $m_i \geq 1$. We therefore have $\xi = \xi_1 \otimes \ldots \otimes \xi_e$, where ξ_i is some irreducible representation of $\mathcal{G}_i$.

The next step is to show that the minimality condition on $\mathfrak{A}$ implies that all of these representations ξ_i are cuspidal. We can then complete the proof of this case very quickly:— if the ξ_i are all equivalent, then $\mathfrak{A}$ is a principal order and $(J(\beta, \mathfrak{A}), \vartheta)$ is a simple type in $\boldsymbol{\pi}$, so at least two of the ξ_i are inequivalent, and $(J(\beta, \mathfrak{A}), \vartheta)$ is a split type of level $(n/e(\mathfrak{A}), 0)$ in $\boldsymbol{\pi}$, as required.

So, suppose some ξ_i is not cuspidal (equivalently, ξ is not a cuspidal representation of the reductive k_E-group $\mathcal{G} = \prod \mathcal{G}_i$). Therefore there is a proper parabolic subgroup $\mathcal{P}$ of $\mathcal{G}$, with unipotent radical $\mathcal{U}$, such that $\xi \mid \mathcal{U}$ contains the trivial character. We choose $\mathcal{P}$ minimal for this property, and then $\xi \mid \mathcal{P}$ contains an irreducible representation ξ' inflated from a cuspidal representation of $\mathcal{P}/\mathcal{U}$. There is a uniquely determined hereditary $\mathfrak{o}_E$-order $\mathfrak{B}'$ in B, with $\mathfrak{B}' \subset \mathfrak{B}$, such that $\mathcal{P}$ is the image of $\boldsymbol{U}(\mathfrak{B}')$ under the quotient map $\boldsymbol{U}(\mathfrak{B}) \to \mathcal{G}$. The unipotent radical $\mathcal{U}$ is then the image of $\boldsymbol{U}^1(\mathfrak{B}')$. Let $\mathfrak{A}'$ be the unique hereditary $\mathfrak{o}_F$-order in A such that $E^\times \subset \mathfrak{K}(\mathfrak{A}')$ and $\mathfrak{A}' \cap B = \mathfrak{B}'$. The containment $\mathfrak{B} \supset \mathfrak{B}'$ implies $\mathfrak{A} \supset \mathfrak{A}'$, and this containment is strict since $\mathcal{P}$ is a proper parabolic subgroup of $\mathcal{G}$. Moreover, $[\mathfrak{A}', n', 0, \beta]$ is a simple stratum with $n'/e(\mathfrak{A}') = n/e(\mathfrak{A})$. Let $\theta' \in \mathcal{C}(\mathfrak{A}', 0, \beta)$ correspond to θ via (3.6.2). Let η' be the unique irreducible representation of $J^1(\beta, \mathfrak{A}')$ containing θ'. We show that $\boldsymbol{\pi}$ contains η', hence also θ', and this will contradict our minimality condition on $\mathfrak{A}$.

Now, by construction, the representation $\xi' \mid \boldsymbol{U}^1(\mathfrak{B}')J^1(\beta, \mathfrak{A})$ is null, so $\boldsymbol{\pi}$ contains a multiple of $\kappa \mid \boldsymbol{U}^1(\mathfrak{B}')J^1(\beta, \mathfrak{A})$. Therefore we need only prove:

(8.1.6) Lemma: *The representations of $\boldsymbol{U}^1(\mathfrak{A}')$ induced by η' and $\kappa \mid \boldsymbol{U}^1(\mathfrak{B}')J^1(\beta, \mathfrak{A})$ are irreducible and equivalent to each other.*

Proof: By (5.2.14), there is a unique β-extension κ' of η' such that $\kappa \mid \boldsymbol{U}(\mathfrak{B}')J^1(\beta, \mathfrak{A})$ and κ' induce the same (irreducible) representation of $\boldsymbol{U}(\mathfrak{B}')\boldsymbol{U}^1(\mathfrak{A}')$. Consider the restriction of this induced representation to $\boldsymbol{U}^1(\mathfrak{A}')$. On the one hand, the Mackey restriction formula gives us

$$\begin{aligned}\mathrm{Ind}(\kappa \mid \boldsymbol{U}(\mathfrak{B}')J^1(\beta, \mathfrak{A}) : \boldsymbol{U}(\mathfrak{B}')J^1(\beta, \mathfrak{A}), \boldsymbol{U}(\mathfrak{B}')\boldsymbol{U}^1(\mathfrak{A}')) \mid \boldsymbol{U}^1(\mathfrak{A}') \\ = \mathrm{Ind}(\kappa \mid \boldsymbol{U}^1(\mathfrak{B}')J^1(\beta, \mathfrak{A}) : \boldsymbol{U}^1(\mathfrak{B}')J^1(\beta, \mathfrak{A}), \boldsymbol{U}^1(\mathfrak{A}')),\end{aligned}$$

while on the other,

$$\begin{aligned}\mathrm{Ind}(\kappa' : J(\beta, \mathfrak{A}'), \boldsymbol{U}(\mathfrak{B}')\boldsymbol{U}^1(\mathfrak{A}')) \mid \boldsymbol{U}^1(\mathfrak{A}') \\ = \mathrm{Ind}(\kappa' \mid J^1(\beta, \mathfrak{A}') : J^1(\beta, \mathfrak{A}'), \boldsymbol{U}^1(\mathfrak{A}')).\end{aligned}$$

Of course, $\kappa' \mid J^1(\beta, \mathfrak{A}') = \eta'$. (5.1.1) implies that η' induces irreducibly to $\boldsymbol{U}^1(\mathfrak{A}')$, and we have proved the lemma. ∎

Now we return to the theorem. We have a simple stratum $[\mathfrak{A}, n, m, \beta]$ in A, and a character $\theta \in \mathcal{C}(\mathfrak{A}, m, \beta)$ which occurs in $\boldsymbol{\pi}$, and such that $m/e(\mathfrak{A})$ is minimal for this property. We have moreover reduced to the case $m \geq 1$. Therefore $\boldsymbol{\pi}$ contains some irreducible representation ϑ of $H^m(\beta, \mathfrak{A})$ such that $\vartheta \mid H^{m+1}(\beta, \mathfrak{A})$ contains θ. However, θ extends to an abelian character $\tilde{\theta} \in \mathcal{C}(\mathfrak{A}, m-1, \beta)$ of $H^m(\beta, \mathfrak{A})$ by (3.3.21), and the quotient $H^m(\beta, \mathfrak{A})/H^{m+1}(\beta, \mathfrak{A})$ is abelian. Thus ϑ is one-dimensional, $\vartheta \mid H^{m+1}(\beta, \mathfrak{A}) = \theta$. We may therefore write

$$\vartheta = \tilde{\theta}\psi_c \mid H^m(\beta, \mathfrak{A}), \quad \textit{for some } c \in \mathfrak{P}^{-m}.$$

Here, $\mathfrak{P} = \mathrm{rad}(\mathfrak{A})$, as usual. The notations E, B, $\mathfrak{B}$ are as before. We also choose a continuous character ψ_E of the additive group of E with conductor $\mathfrak{p}_E$, and set $\psi_B = \psi_E \circ \mathrm{tr}_{B/E}$. We take a tame corestriction s on A relative to E/F such that (1.3.5) holds, namely

$$\psi_A(ab) = \psi_B(s(a)b), \quad a \in A,\ b \in B.$$

We now consider the stratum $[\mathfrak{B}, m, m-1, s(c)]$ in B.

(8.1.7) Proposition: *Let $\mathfrak{G}$ be an open subgroup of $\boldsymbol{U}^m(\mathfrak{A})$, and ρ an irreducible representation of $\mathfrak{G}$ such that $\rho \mid H^m(\beta, \mathfrak{A}) \cap \mathfrak{G}$ contains $\vartheta \mid H^m(\beta, \mathfrak{A}) \cap \mathfrak{G}$. Let $\boldsymbol{\pi}$ be an irreducible representation of G which contains ϑ. Then $\boldsymbol{\pi} \mid \mathfrak{G}$ contains ρ.*

Proof: We first need an induction property of the character ϑ.

(8.1.8) Lemma: *There is a unique irreducible representation τ of $\boldsymbol{U}^m(\mathfrak{A})$ such that $\tau \mid H^m(\beta, \mathfrak{A})$ contains ϑ.*

Proof: Abbreviate $J^t = J^t(\beta, \mathfrak{A})$, $H^t = H^t(\beta, \mathfrak{A})$, $t \geq 0$. The commutator group $[J^m, J^m]$ is contained in $H^{2m} \subset H^{m+1}$ by (3.1.15), so the pairing

$$\boldsymbol{k}_\vartheta : J^m/H^m \times J^m/H^m \to \mathbb{C}^\times,$$
$$\boldsymbol{k}_\vartheta(x, y) = \vartheta[x, y],$$

depends only on $\vartheta \mid H^{m+1} = \theta$. (3.4.1) therefore shows $\boldsymbol{k}_\vartheta$ is nondegenerate. It follows that there is a unique irreducible representation μ of J^m containing ϑ. Indeed, $\mathrm{Ind}(\vartheta : H^m, J^m)$ is a multiple of μ and $\mu \mid H^m$ is a multiple of ϑ. We show that the intertwining of μ in $\boldsymbol{U}^m(\mathfrak{A})$ is contained in J^m. It will follow that $\mathrm{Ind}(\mu : J^m, \boldsymbol{U}^m(\mathfrak{A}))$ is irreducible, and this will give the result.

The G-intertwining of μ is contained in that of ϑ, hence in that of $\vartheta \mid H^{m+1} = \theta$. By (3.3.2), the G-intertwining of θ is

$$\mathcal{S} = (1 + \mathfrak{Q}^{r-m}\mathfrak{N}(\beta) + \mathfrak{J}^{[\frac{r+1}{2}]})B^{\times}(1 + \mathfrak{Q}^{r-m}\mathfrak{N}(\beta) + \mathfrak{J}^{[\frac{r+1}{2}]}).$$

Here, we use the notation of §**3**:— $\mathfrak{Q} = \mathrm{rad}(\mathfrak{B})$, $r = -k_0(\beta, \mathfrak{A})$, $\mathfrak{N}(\beta) = \mathfrak{N}_{-r}(\beta, \mathfrak{A})$, $\mathfrak{J}^{[\frac{r+1}{2}]} = \mathfrak{J}^{[\frac{r+1}{2}]}(\beta, \mathfrak{A})$. We have to show that $\mathcal{S} \cap \boldsymbol{U}^m(\mathfrak{A}) \subset J^m$. Immediately, we have $\mathcal{S} \cap \boldsymbol{U}^m(\mathfrak{A}) = (1 + \mathfrak{Q}^{r-m}\mathfrak{N}(\beta) \cap \mathfrak{P}^m)J^m$, so we are reduced to showing

$$\textbf{(8.1.9)} \qquad \mathfrak{Q}^{r-m}\mathfrak{N}(\beta) \cap \mathfrak{P}^m \subset \mathfrak{J}^m(\beta, \mathfrak{A}).$$

If $r - m \geq [\frac{r+1}{2}]$, the (3.1.10)*(i)* shows that $\mathfrak{Q}^{r-m}\mathfrak{N}(\beta) \subset \mathfrak{J}(\beta, \mathfrak{A})$, so we may as well assume that $m \geq [r/2] + 1$. We work by induction along β. If β is minimal over F, so that $r = n$ or ∞, we have $\mathfrak{J}^m(\beta, \mathfrak{A}) = \mathfrak{P}^m$ and there is nothing to prove. In general, we choose a simple stratum $[\mathfrak{A}, n, r, \gamma]$ equivalent to $[\mathfrak{A}, n, r, \beta]$ and put $r_1 = -k_0(\gamma, \mathfrak{A})$. Then, by (3.1.3), $\mathfrak{Q}^{r-m}\mathfrak{N}(\beta) = \mathfrak{Q}_\gamma^{r-m}\mathfrak{N}_{-r}(\gamma, \mathfrak{A})$, where $\mathfrak{Q}_\gamma = \mathfrak{P} \cap \mathrm{End}_{F[\gamma]}(V)$. Also, since $m \geq [r/2] + 1$, we have $\mathfrak{J}^m(\beta, \mathfrak{A}) = \mathfrak{J}^m(\gamma, \mathfrak{A})$ (remarks preceding (3.1.9)). We expand, using (1.4.9), to get

$$\mathfrak{Q}_\gamma^{r-m}\mathfrak{N}_{-r}(\gamma, \mathfrak{A}) = \mathfrak{Q}_\gamma^{r-m} + \mathfrak{Q}_\gamma^{r_1-m}\mathfrak{N}_{-r_1}(\gamma).$$

Now, we certainly have $\mathfrak{Q}_\gamma^{r-m} \subset \mathfrak{J}(\gamma, \mathfrak{A})$, while $\mathfrak{Q}_\gamma^{r_1-m}\mathfrak{N}_{-r_1}(\gamma) \cap \mathfrak{P}^m \subset \mathfrak{J}(\gamma, \mathfrak{A})$ by induction. Thus $\mathfrak{Q}^{r-m}\mathfrak{N}(\beta) \cap \mathfrak{P}^m$ is contained in $\mathfrak{J}(\beta, \mathfrak{A})$, hence in $\mathfrak{J}^m(\beta, \mathfrak{A})$, as required. ■

By (8.1.8), the induced representation $\mathrm{Ind}(\vartheta : K, \boldsymbol{U}^m(\mathfrak{A}))$ is a multiple of τ. If an irreducible representation $\boldsymbol{\pi}$ of G contains ϑ, it therefore contains τ and also any irreducible component of $\mathrm{Ind}(\vartheta : K, \boldsymbol{U}^m(\mathfrak{A})) \mid \mathfrak{G}$. The Proposition now follows from Mackey's restriction formula. ■

We return to our stratum $[\mathfrak{B}, m, m-1, s(c)]$ in B. We show first that it is fundamental, by assuming the contrary and contradicting the minimality of $m/e(\mathfrak{A})$. Put $e_\beta = e(\mathfrak{B}|\mathfrak{o}_E)$. Since we assume $[\mathfrak{B}, m, m-1, s(c)]$ is not fundamental, we can invoke (2.3.1) (and the remark following it) to find a hereditary $\mathfrak{o}_E$-order $\mathfrak{B}'$ in B, with radical $\mathfrak{Q}'$, $e'_\beta = e(\mathfrak{B}'|\mathfrak{o}_E)$, and an integer m' such that

(8.1.10) *(i)* $s(c) + \mathfrak{Q}^{1-m} \subset \mathfrak{Q}'^{-m'}$;
(ii) $m'/e'_\beta < m/e_\beta$;
(iii) there exist hereditary $\mathfrak{o}_E$*-orders* $\mathfrak{B}_1, \mathfrak{B}_2$ *in* B *such that* $\mathfrak{B}_1 \supset \mathfrak{B}, \mathfrak{B}' \supset \mathfrak{B}_2$.

Let $\mathfrak{A}'$ be the hereditary $\mathfrak{o}_F$-order in A such that $E^\times \subset \mathfrak{K}(\mathfrak{A}')$ and $\mathfrak{A}' \cap B = \mathfrak{B}'$. Put $\mathfrak{P}' = \mathrm{rad}(\mathfrak{A}')$. Because of (8.1.10)*(iii)*, we may assume that $\mathfrak{A}$, $\mathfrak{A}'$ have a common "(W,E)-decomposition" $\mathfrak{A} = \mathfrak{A}(E) \otimes_{\mathfrak{o}_E} \mathfrak{B}$, $\mathfrak{A}' = \mathfrak{A}(E) \otimes_{\mathfrak{o}_E} \mathfrak{B}'$ (*cf.* **(1.2)**) for some embedding $A(E) \to A$. By (8.1.10)*(i)*, we have $\mathfrak{Q}^{1-m} \subset \mathfrak{Q}'^{-m'}$. Dualising, we get $\mathfrak{Q}'^{1+m'} \subset \mathfrak{Q}^m$. This implies $\mathfrak{A}(E) \otimes_{\mathfrak{o}_E} \mathfrak{Q}'^{1+m'} \subset \mathfrak{A}(E) \otimes \mathfrak{Q}^m$, and (1.2.10) now gives $\mathfrak{P}'^{1+m'} \subset \mathfrak{P}^m$, whence

$$\boldsymbol{U}^{m'+1}(\mathfrak{A}') \subset \boldsymbol{U}^m(\mathfrak{A}).$$

We show

(8.1.11) *There exists $\theta' \in \mathcal{C}(\mathfrak{A}', m', \beta)$ which occurs in $\boldsymbol{\pi}$.*

By (8.1.10)*(ii)*, this will contradict the minimality of $m/e(\mathfrak{A})$.

We need a commutator property.

(8.1.12) Lemma: *Let $x \in \mathfrak{Q}^{r-m}\mathfrak{N}(\beta)$. Then $1+x$ normalises H^m and*

$$\vartheta^{1+x} = \vartheta\psi_{a_\beta(x)}.$$

Proof: (3.1.9)*(i)* implies that $\mathfrak{H}^m(\beta, \mathfrak{A})$ is a module over the ring $\mathfrak{N}_{-m}(\beta)$. (1.4.9) gives $\mathfrak{Q}^{r-m}\mathfrak{N}(\beta, \mathfrak{A}) \subset \mathfrak{N}_{-m}(\beta, \mathfrak{A})$. The first assertion follows.

We have $r - m \geq 1$ and $c \in \mathfrak{P}^{-m}$, so certainly $\psi_c^{1+x} = \psi_c$ as a character of $H^m \subset \boldsymbol{U}^m(\mathfrak{A})$. We therefore have to compute $\tilde{\theta}^{1+x}$. If $m \leq \left[\frac{r}{2}\right]$, the character $\psi_{a_\beta(x)}$ is null on $\boldsymbol{U}^m(\mathfrak{B})$, and hence on $H^m = \boldsymbol{U}^m(\mathfrak{B})H^{m+1}$. On the other hand, $\mathfrak{Q}^{r-m}\mathfrak{N}(\beta) \subset \mathfrak{J}^{\left[\frac{r+1}{2}\right]}(\beta, \mathfrak{A})$, whence $\tilde{\theta}^{1+x} = \tilde{\theta}$, by (3.3.1).

We therefore assume that $m \geq \left[\frac{r}{2}\right] + 1$. We can now apply (3.2.11) to get $\vartheta^{1+x} = \vartheta\psi_{(1+x)^{-1}\beta(1+x)-\beta}$. However, a simple computation gives

$$(1+x)^{-1}\beta(1+x) - \beta \equiv a_\beta(x) \pmod{\mathfrak{P}^{1-m}},$$

and the lemma follows. ■

Now, the conditions $c \in \mathfrak{P}^{-m}$, $s(c) \in \mathfrak{Q}'^{-m'}$ imply that

$$c \in (\mathfrak{P}'^{-m'} + a_\beta(A)) \cap \mathfrak{P}^{-m}.$$

We assert

$$\textbf{(8.1.13)} \quad (\mathfrak{P}'^{-m'} + a_\beta(A)) \cap \mathfrak{P}^{-m} = \mathfrak{P}'^{-m'} \cap \mathfrak{P}^{-m} + a_\beta(A) \cap \mathfrak{P}^{-m}.$$

Containment in the direction $\supset$ is clear. Both sides have the same intersection with $\mathrm{Ker}(s) = a_\beta(A)$, so it is enough to show they have the same image under s. However, the fact that the orders $\mathfrak{A}$, $\mathfrak{A}'$ admit a common (W,E)-decomposition implies that the lattices $\mathfrak{P}^{-m}$, $\mathfrak{P}'^{-m'}$ and $\mathfrak{P}^{-m} \cap \mathfrak{P}'^{-m'}$ are $\mathrm{Ad}(E^\times)$-invariant $(\mathfrak{A}(E), \mathfrak{o}_E)$-bilattices, for some embedding $A(E) \to A$. Thus, by (1.3.12), these lattices are E-exact and $s(\mathfrak{P}^{-m} \cap \mathfrak{P}'^{-m'}) = s(\mathfrak{P}^{-m}) \cap s(\mathfrak{P}'^{-m'})$. We therefore get

$$s((\mathfrak{P}'^{-m'} + a_\beta(A)) \cap \mathfrak{P}^{-m}) \subset s(\mathfrak{P}'^{-m'}) \cap s(\mathfrak{P}^{-m}) = s(\mathfrak{P}'^{-m'} \cap \mathfrak{P}^{-m}),$$

and (8.1.13) holds. We have $a_\beta(A) \cap \mathfrak{P}^{-m} = a_\beta(\mathfrak{Q}^{r-m}\mathfrak{N}(\beta))$, by (1.4.10). We deduce that there exists $x \in \mathfrak{Q}^{r-m}\mathfrak{N}(\beta)$ such that $c - a_\beta(x) \in \mathfrak{P}'^{-m'} \cap \mathfrak{P}^{-m}$. The character ϑ^{1+x} of H^m certainly occurs in π, and, for this x, we have

$$\vartheta^{1+x} \mid H^m \cap H^{m'+1}(\beta, \mathfrak{A}') = \tilde{\theta} \mid H^m \cap H^{m'+1}(\beta, \mathfrak{A}').$$

It follows from (3.6.1) and (3.3.21) that there exists $\theta' \in \mathcal{C}(\mathfrak{A}', m', \beta)$ agreeing with $\tilde{\theta}$ on $H^m \cap H^{m'+1}(\beta, \mathfrak{A}')$. We have

$$H^{m'+1}(\beta, \mathfrak{A}') \subset \boldsymbol{U}^{m'+1}(\mathfrak{A}'),$$

which, we have already observed, is contained in $\boldsymbol{U}^m(\mathfrak{A})$. The desired assertion (8.1.11) now follows from (8.1.7).

We now know that our stratum $[\mathfrak{B}, m, m-1, s(c)]$ in B must be fundamental. We next show that it cannot be nonsplit fundamental. We adopt the same strategy (and much the same argument) as above:— we assume the contrary and contradict the minimality of $m/e(\mathfrak{A})$. We use (2.3.4) to get a simple stratum $[\mathfrak{B}', m', m'-1, \alpha]$ in B such that

(8.1.14) *(i)* $s(c) + \mathfrak{Q}^{1-m} \subset \alpha + \mathfrak{Q}'^{1-m'}$;
(ii) $m/e_\beta = m'/e'_\beta$;
(iii) the lattice-chain defining $\mathfrak{B}'$ contains that defining $\mathfrak{B}$.

Here we have put $\mathfrak{Q}' = \mathrm{rad}(\mathfrak{B}')$, $e'_\beta = e(\mathfrak{B}'|\mathfrak{o}_E)$. Again let $\mathfrak{A}'$ denote the hereditary $\mathfrak{o}_F$-order in A with $E^\times \subset \mathfrak{K}(\mathfrak{A}')$, $\mathfrak{A}' \cap B = \mathfrak{B}'$, and $\mathfrak{P}' = \mathrm{rad}(A')$. The stratum $[\mathfrak{A}', n', m'-1, \beta]$ is simple, with $n'/e(\mathfrak{A}') = n/e(\mathfrak{A})$. Condition *(iii)* implies that $\mathfrak{A}$, $\mathfrak{A}'$ have a common (W, E)-decomposition. Condition *(i)* implies

$$\mathfrak{Q}^{1-m} \subset \mathfrak{Q}'^{1-m'} \subset \mathfrak{Q}'^{-m'} \subset \mathfrak{Q}^{-m}.$$

We have an embedding $A(E) \to A$ giving the common (W, E)-decomposition of $\mathfrak{A}$, $\mathfrak{A}'$, and so

$$\mathfrak{A}(E) \otimes_{\mathfrak{o}_E} \mathfrak{Q}^{1-m} \subset \mathfrak{A}(E) \otimes_{\mathfrak{o}_E} \mathfrak{Q}'^{1-m'} \subset \mathfrak{A}(E) \otimes_{\mathfrak{o}_E} \mathfrak{Q}'^{-m'} \subset \mathfrak{A}(E) \otimes_{\mathfrak{o}_E} \mathfrak{Q}^{-m},$$

whence, via (1.2.10),

$$\mathfrak{P}^{1-m} \subset \mathfrak{P}'^{1-m'} \subset \mathfrak{P}'^{-m'} \subset \mathfrak{P}^{-m}.$$

Choose $a \in \mathfrak{P}'^{-m'}$ such that $s(a) = \alpha$ (where s is a tame corestriction on A relative to E/F). Thus $c - a \in \mathfrak{P}^{-m}$ and $s(c-a) \in \mathfrak{Q}'^{1-m'}$. Therefore there exists $x \in \mathfrak{Q}^{r-m}\mathfrak{N}(\beta)$ such that $c - a - a_\beta(x) \in \mathfrak{P}'^{1-m'}$. (8.1.12) gives us $\vartheta^{1-x} = \vartheta\psi_{-a_\beta(x)}$, as a character of $H^m = H^m(\beta, \mathfrak{A})$. Restricting, we get

$$\vartheta^{1-x} \mid H^m \cap H^{m'}(\beta, \mathfrak{A}') = (\tilde{\theta}\psi_a) \mid H^m \cap H^{m'}(\beta, \mathfrak{A}').$$

Now take $\theta' \in \mathcal{C}(\mathfrak{A}', m'-1, \beta)$ to agree with $\tilde{\theta}$ on $H^m \cap H^{m'}(\beta, \mathfrak{A}')$ (as we may, by (3.6.1) and (3.3.21)). Dualising the above containment $\mathfrak{P}^{1-m} \subset \mathfrak{P}'^{1-m'}$, we get $\mathfrak{P}'^{m'} \subset \mathfrak{P}^m$. Thus $H^{m'}(\beta, \mathfrak{A}') \subset \boldsymbol{U}^{m'}(\mathfrak{A}') \subset \boldsymbol{U}^m(\mathfrak{A})$, so (8.1.7) applies to show that π contains the character $\theta'\psi_a$ of $H^{m'}(\beta, \mathfrak{A}')$. However, by (2.2.8), the stratum $[\mathfrak{A}', n', m'-1, \beta + a]$ is equivalent to a simple stratum $[\mathfrak{A}', n', m'-1, \beta']$, and $\theta'\psi_a \in \mathcal{C}(\mathfrak{A}', m'-1, \beta')$ by (3.3.18). Now we observe that $(m'-1)/e(\mathfrak{A}') < m'/e(\mathfrak{A}') = m/e(\mathfrak{A})$, and we have our desired contradiction.

Before proceeding, it is worth exhibiting the result of this argument for future use:—

(8.1.15) *Let $[\mathfrak{A}, n, m, \beta]$, $m \geq 1$, be a simple stratum in A, and let ϑ be a character of $H^m(\beta, \mathfrak{A})$ of the form $\vartheta = \theta\psi_c$, such that*

(i) $\theta \in \mathcal{C}(\mathfrak{A}, m-1, \beta)$;

(ii) $c \in \mathfrak{P}^{-m}$ (where $\mathfrak{P} = \mathrm{rad}(\mathfrak{A})$);

(iii) the stratum $[\mathfrak{B}, m, m-1, s(c)]$ is nonsplit fundamental, where $E = F[\beta]$, $\mathfrak{B} = \mathfrak{A} \cap \mathrm{End}_E(V)$, and s is a tame corestriction on A relative to E/F.

Then there exists a simple stratum $[\mathfrak{A}', n', m', \beta']$ and $\theta' \in \mathcal{C}(\mathfrak{A}', m', \beta')$ such that $m'/e(\mathfrak{A}') < m/e(\mathfrak{A})$ and any irreducible representation of G which contains ϑ also contains θ'.

So, finally, we know that our stratum $[\mathfrak{B}, m, m-1, s(c)]$ is split fundamental. To finish the proof of the theorem, we just have to show that we can arrange the condition $\gcd(m, e_\beta) = 1$. Put $d = \gcd(m, e_\beta)$ and assume that $d > 1$. Let $\mathcal{L} = \{L_i : i \in \mathbb{Z}\}$ be the lattice chain which defines $\mathfrak{B}$, set $\mathcal{L}' = \{L_{id} : i \in \mathbb{Z}\}$, $\mathfrak{A}' = \mathrm{End}^0_{\mathfrak{o}_F}(\mathcal{L}')$, $\mathfrak{B}' = \mathfrak{A}' \cap B = \mathrm{End}^0_{\mathfrak{o}_E}(\mathcal{L}')$. Put $m' = m/d$, $e'_\beta = e_\beta/d = e(\mathfrak{B}'|\mathfrak{o}_E)$, $\mathfrak{P}' = \mathrm{rad}(\mathfrak{A}')$, $\mathfrak{Q}' = \mathrm{rad}(\mathfrak{B}')$. Then we have

$$\mathfrak{P}'^{1-m'} \subset \mathfrak{P}^{1-m} \subset \mathfrak{P}^{-m} \subset \mathfrak{P}'^{-m'},$$

$$\mathfrak{P}'^{1+m'} \subset \mathfrak{P}^{1+m} \subset \mathfrak{P}^{m} \subset \mathfrak{P}'^{m'}.$$

The stratum $[\mathfrak{A}', n', m', \beta]$ is again simple, with $n' = n/d$, $m' = m/d$. By (3.6.1), there is a unique character $\theta' \in \mathcal{C}(\mathfrak{A}', m', \beta)$ which agrees with $\tilde{\theta}$ on the intersection $H^{m+1}(\beta, \mathfrak{A}) \cap H^{m'+1}(\beta, \mathfrak{A}')$. Since $H^{m'+1}(\beta, \mathfrak{A}') \subset U^{m+1}(\mathfrak{A})$, (8.1.7) shows that π contains θ'. It therefore contains a character $\vartheta' = \tilde{\theta}'\psi_{c'}$ of $H^{m'}(\beta, \mathfrak{A}')$ with $\tilde{\theta}' \in \mathcal{C}(\mathfrak{A}', m'-1, \beta)$ extending θ' and $c' \in \mathfrak{P}'^{-m'}$. The stratum $[\mathfrak{B}', m', m'-1, s(c')]$ must be split fundamental:— otherwise we repeat the whole procedure and reduce $m/e(\mathfrak{A})$.

This completes the proof of (8.1.5). ■

(8.2) Jacquet module of a split type I: the positive level case

In this and the following section, we take a split type (K, ϑ) in G. We associate to it a proper parabolic subgroup P of G, with unipotent radical U, and show that if $(\pi, \mathcal{V})$ is a smooth irreducible representation of G containing (K, ϑ), then the Jacquet module $(\pi_U, \mathcal{V}_U)$ is nontrivial. We also accumulate some extra detail for later application.

In this section, we treat the case where (K, ϑ) has level (x, y), with $x \geq y > 0$, leaving the case $y = 0$ to the next section. In the case $x > y$, we assume that (K, ϑ) is defined by a simple stratum $[\mathfrak{A}, n, m, \beta]$, with $\vartheta = \theta\psi_c$, for some $\theta \in \mathcal{C}(\mathfrak{A}, m-1, \beta)$, $c \in \mathfrak{P}^{-m}$ (where $\mathfrak{P} = \mathrm{rad}(\mathfrak{A})$) just as in the definition (8.1.3). We use the notations E, B, $\mathfrak{B}$, $\mathfrak{Q}$, $s = s_\beta$ etc. introduced there and in the rest of **(8.1)**. In particular, the stratum $[\mathfrak{B}, m, m-1, s(c)]$ is split fundamental and $\gcd(m, e(\mathfrak{B}|\mathfrak{o}_E)) = 1$. We handle the case $x = y$ simultaneously by the simple stratagem of putting $E = F$, $B = A$, $n = m$ and $s =$ the identity map $A \to A$.

We also need some terminology along the lines of **(7.1)**. An ordered decomposition $V = V_1 \oplus V_2$ of V as a direct sum of two proper F-subspaces leads to a pair (P, P^-) of mutually opposite maximal proper parabolic subgroups of G. Explicitly, P is the stabiliser of the flag $V \supset V_1 \supset \{0\}$ and P^- that of $V \supset V_2 \supset \{0\}$. We write

(8.2.1) *(i)* $A_{ij} = \mathrm{Hom}_F(V_j, V_i)$, $i, j \in \{1, 2\}$;
(ii) $G_i = A_{ii}^\times = \mathrm{Aut}_F(V_i)$, $i = 1, 2$;
(iii) $P = MU$, $M = G_1 \times G_2$, $U = 1 + A_{12}$;
(iv) $P^- = MU^-$, $U^- = 1 + A_{21}$.

In particular, we have $A = \coprod_{i,j} A_{ij}$. If $x \in A$, we write x_{ij} for the component of x in A_{ij} relative to this decomposition.

If $\mathcal{G}$ is a closed subgroup of G, we say that $\mathcal{G}$ *has an Iwahori decomposition relative to* (P, P^-) if

(8.2.2) $$\mathcal{G} = (\mathcal{G} \cap U^-).(\mathcal{G} \cap M).(\mathcal{G} \cap U), \quad \text{and} \\ \mathcal{G} \cap M = \mathcal{G} \cap G_1 \times \mathcal{G} \cap G_2.$$

Before stating and proving the main result of this section, we need some more details concerning split fundamental strata. We temporarily return to the context of **(2.3)**, and let $[\mathfrak{A}, n, n-1, b]$ be a split fundamental stratum in A with $\gcd(n, e(\mathfrak{A})) = 1$. In (2.3.9–10), we recalled a procedure for constructing a decomposition $V = V_1 \oplus V_2$ of V which splits $[\mathfrak{A}, n, n-1, b]$ (in the sense of (2.3.8)). This involves forming the characteristic polynomial $\phi(X) \in \mathfrak{o}_F[X]$ of the element $b^e \pi_F^n$, where $e = e(\mathfrak{A}|\mathfrak{o}_F)$ and π_F is a prime element of F. One factors $\phi(X)$ in the form $\phi = fg$, where $f, g \in \mathfrak{o}_F[X]$ are monic and relatively prime when reduced mod $\mathfrak{p}_F$. One further insists that $f(0) \not\equiv 0 \pmod{\mathfrak{p}_F}$. Then one puts $V_1 = \mathrm{Ker}(f(b))$, $V_2 = \mathrm{Ker}(g(b))$. Clearly, one could choose the factor $f(X)$ so that its reduction mod $\mathfrak{p}_F$ is a power of an irreducible polynomial in $\mathsf{k}_F[X]$. Of course, $f(X) \pmod{\mathfrak{p}_F}$ is the characteristic polynomial of the stratum $[\mathfrak{A} \cap A_{11}, n, n-1, b_{11}]$, so we have

(8.2.3) *Let $[\mathfrak{A}, n, n-1, b]$ be a split fundamental stratum in A with $\gcd(n, e(\mathfrak{A})) = 1$. There exists a decomposition $V = V_1 \oplus V_2$ of V which splits $[\mathfrak{A}, n, n-1, b]$ and such that the component $[\mathfrak{A} \cap A_{11}, n, n-1, b_{11}]$ is nonsplit fundamental.*

Now let $V = V_1 \oplus V_2$ be a decomposition of V which splits $[\mathfrak{A}, n, n-1, b]$, as in (2.3.10) (we do not need the extra condition of (8.2.3) here). For our present purposes, it is important to note that the two parabolic subgroups P, P^- defined by this decomposition actually play symmetric roles, even though this does not appear in the definition and despite the fact that the spaces V_i play very different roles. Formally, this amounts to:—

(8.2.4) Proposition: *Let $[\mathfrak{A}, n, n-1, b]$ be a split fundamental stratum in A with $\gcd(n, e(\mathfrak{A}|\mathfrak{o}_F)) = 1$, and write $\mathfrak{P} = \mathrm{rad}(\mathfrak{A})$. Let $V = V_1 \oplus V_2$ be a decomposition of V which splits $[\mathfrak{A}, n, n-1, b]$ as in (2.3.10). Then:*

(i) $\{\mathfrak{P}^j \cap A_{21} : j \in \mathbb{Z}\}$ is a uniform $\mathfrak{o}_F$-lattice chain in A_{21} with period $e(\mathfrak{A}|\mathfrak{o}_F)$;

(ii) the map $\partial_b^- : A_{21} \to A_{21}$ given by $\partial_b^-(x) = b_{22}x - xb_{11}$ induces an isomorphism $\partial_b^- : \mathfrak{P}^j \cap A_{21} \xrightarrow{\approx} \mathfrak{P}^{j-n} \cap A_{21}$.

The proof is identical to the "opposite" case treated in **[K4]**, so we omit it. We will, of course, apply this in the context of $[\mathfrak{B}, m, m-1, s(c)]$ with base field E.

(8.2.5) Theorem: *Let (K, ϑ) be a split type in G of level (x, y), with $x \geq y > 0$. Use the notation above. There exists a decomposition $V = V_1 \oplus V_2$, where the V_i are proper E-subspaces of V, leading to parabolic subgroups P, P^- of G as in (8.2.1), with the following properties:*

(i) $\mathfrak{A}_1 = \mathfrak{A} \cap A_{11}$ is a principal $\mathfrak{o}_F$-order in A_{11} with $e(\mathfrak{A}_1|\mathfrak{o}_F) = e(\mathfrak{A}|\mathfrak{o}_F)$ and $E^\times \subset \mathfrak{K}(\mathfrak{A}_1)$ (where we view E as embedded in A_{11} via its

action on V_1);

(ii) the groups $H^t(\beta,\mathfrak{A})$, $J^t(\beta,\mathfrak{A})$ have Iwahori decompositions relative to P, P^-, and

$$\begin{aligned} H^t(\beta,\mathfrak{A})\cap G_1 &= H^t(\beta,\mathfrak{A}_1), \\ J^t(\beta,\mathfrak{A})\cap G_1 &= J^t(\beta,\mathfrak{A}_1), \end{aligned} \qquad t\geq 1;$$

(iii) the character $\theta_1 = \theta \mid H^m(\beta,\mathfrak{A}_1)$ lies in $\mathcal{C}(\mathfrak{A}_1, m-1,\beta)$ and corresponds to θ via (3.6.14), while the characters $\theta \mid H^m(\beta,\mathfrak{A})\cap U$, $\theta \mid H^m(\beta,\mathfrak{A})\cap U^-$ are both null;

(iv) if $(\pi,\mathcal{V})$ is an irreducible smooth representation of G which contains (K,ϑ), then the Jacquet module $(\pi_U,\mathcal{V}_U)$ (resp. $(\pi_{U^-},\mathcal{V}_{U^-})$) contains a nonzero vector v_+ (resp. v_-) such that

$$\begin{aligned} \pi_U(x)v_+ &= \vartheta(x)v_+, \\ \pi_{U^-}(x)v_- &= \vartheta(x)v_-, \end{aligned} \qquad x\in K\cap M.$$

Proof: We take the split fundamental stratum $[\mathfrak{B}, m, m-1, s(c)]$ in B, and apply (2.3.10), (8.2.3) to get an E-decomposition $V = V_1\oplus V_2$ of V which splits the stratum. If $\mathcal{L} = \{L_k : k\in\mathbb{Z}\}$ is the lattice chain defining $\mathfrak{B}$ (or $\mathfrak{A}$), we have $L_k = L_k\cap V_1 \oplus L_k\cap V_2$, $k\in\mathbb{Z}$, and the order $\mathfrak{B}_1 = \mathfrak{B}\cap A_{11} = \mathrm{End}^0_{\mathfrak{o}_E}(\{L_k\cap V_1\})$ is a principal $\mathfrak{o}_E$-order in $B_{11} = B\cap A_{11}$ with $e(\mathfrak{B}_1|\mathfrak{o}_E) = e(\mathfrak{B}|\mathfrak{o}_E)$. Restricting the base field to F, we get *(i)*.

The decomposition $L_k = L_k\cap V_1\oplus L_k\cap V_2$ implies that the projections $\mathbf{1}_i : V_1\oplus V_2 \to V_i$ both lie in $\mathfrak{B}$. It follows that, if $\mathfrak{M}$ is any $(\mathfrak{B},\mathfrak{B})$-bimodule in A, we have

$$\textbf{(8.2.6)}\qquad \begin{aligned} \mathfrak{M}\cap A_{ij} &= \mathbf{1}_i.\mathfrak{M}.\mathbf{1}_j, \qquad i,j\in\{1,2\}, \text{ and} \\ \mathfrak{M} &= \coprod_{i,j} \mathfrak{M}\cap A_{ij}. \end{aligned}$$

In this situation, we often abbreviate

$$\mathfrak{M}_{ij} = \mathfrak{M}\cap A_{ij} = \mathbf{1}_i.\mathfrak{M}.\mathbf{1}_j.$$

Likewise, we put

$$x_{ij} = \mathbf{1}_i.x.\mathbf{1}_j, \quad x\in A,$$

and regard this as a block matrix decomposition. In this notation, we record for future use the fact that (8.2.3) gives us:

(8.2.7) *The stratum $[\mathfrak{B}_1, m, m-1, s(c)_{11}]$ is nonsplit fundamental.*

The relations (8.2.6) apply, in particular, to the lattices $\mathfrak{H}^t(\beta,\mathfrak{A})$, $\mathfrak{J}^t(\beta,\mathfrak{A})$, $t \geq 0$. The Iwahori decompositions for the groups $H^t(\beta,\mathfrak{A})$, $J^t(\beta,\mathfrak{A})$, $t \geq 1$, follow immediately. The proof of the identities

$$\begin{aligned} H^t(\beta,\mathfrak{A}) \cap G_1 &= H^t(\beta,\mathfrak{A}_1), \\ J^t(\beta,\mathfrak{A}) \cap G_1 &= J^t(\beta,\mathfrak{A}_1), \end{aligned} \qquad t \geq 1,$$

is exactly similar to, but easier than, that of (7.1.12), so we omit it. The same applies to the proof of *(iii)*.

This brings us to the main part *(iv)*. It is convenient to start by adjusting our split type (K,ϑ). By definition, the decomposition $V = V_1 \oplus V_2$ "splits" the element $s(c)$, in the sense that $s(c)_{ij} = 0$ when $i \neq j$. This need not be true of c itself, but we show that we may impose this condition. Write $\mathfrak{M} = \mathfrak{Q}^{r-m}\mathfrak{N}(\beta) + \mathfrak{J}^{[\frac{r+1}{2}]}(\beta,\mathfrak{A})$, with $\mathfrak{Q} = \mathrm{rad}(\mathfrak{B})$, $r = -k_0(\beta,\mathfrak{A})$, $\mathfrak{N}(\beta) = \mathfrak{N}_{-r}(\beta,\mathfrak{A})$ as usual. Take $x \in \mathfrak{M}$. The arguments of (8.1.12) apply here to show that $1{+}x$ normalises $K = H^m(\beta,\mathfrak{A})$, and that $\vartheta^{1+x} = \theta\psi_{c'}$, where $c' = c + a_\beta(x)$. This character must occur in any irreducible representation π of G which contains ϑ. We have $c' \in \mathfrak{P}^{-m}$ and $s(c') = s(c)$. Thus (K,ϑ^{1+x}) is a split type in G, attached to the same strata $[\mathfrak{A}, n, m{-}1, \beta]$ and $[\mathfrak{B}, m, m{-}1, s(c)]$, and occurring in just the same irreducible representations of G, as (K,ϑ). The condition $s(c)_{ij} = 0$ implies $c_{ij} \in \mathfrak{P}^{-m} \cap A_{ij} \cap a_\beta(A)$, when $i \neq j$. The maps s and a_β are $(\mathfrak{B},\mathfrak{B})$-bimodule homomorphisms, so they respect the block decomposition of A. By (3.1.16) therefore, we may choose $x \in \mathfrak{M}$ so that the element $c' = c + a_\beta(x)$ of A is split. Altogether, we are justified in assuming henceforth that:

(8.2.8) *The decomposition $V = V_1 \oplus V_2$ splits the element c.*

We now need to describe a special element and some auxiliary subgroups of G. Write $e = e(\mathfrak{B}|\mathfrak{o}_E) = e(\mathfrak{B}_1|\mathfrak{o}_E)$. Since $[\mathfrak{B}_1, m, m{-}1, s(c)_{11}]$ is nonsplit fundamental, we have $s(c)_{11} \in \mathfrak{K}(\mathfrak{B}_1)$, and hence $s(c)_{11}\mathfrak{B}_1 = \mathfrak{Q}_1^{-m}$, where we write $\mathfrak{Q}_1 = \mathrm{rad}(\mathfrak{B}_1)$. By definition, we have $\gcd(m,e) = 1$, so there exist $q, t \in \mathbb{Z}$ such that $qe - tm = 1$. Choose a prime element π_E of E and put

$$\zeta_1 = \pi_E^q s(c)_{11}^t.$$

Then $\zeta_1 \in \mathfrak{K}(\mathfrak{B}_1)$ and $\zeta_1\mathfrak{B}_1 = \mathfrak{Q}_1$. Also define

$$\zeta = \begin{pmatrix} \zeta_1 & 0 \\ 0 & 1 \end{pmatrix} \qquad \textbf{(8.2.9)}$$

in the obvious block matrix notation. Then $\zeta \in \mathfrak{B}$, and, inspecting its action on the lattice chain $\mathcal{L}$ defining $\mathfrak{B}$, we see that $\zeta^{-1} \in \mathfrak{Q}^{-1}$.

(8.2.10) Lemma: *For $t \in \mathbb{Z}$, we have $(\mathfrak{Q}^t)_{21} = \mathfrak{B}_{21}\zeta_1^t$, $(\mathfrak{Q}^t)_{12} = \zeta_1^t\mathfrak{B}_{12}$, $(\mathfrak{P}^t)_{21} = \mathfrak{A}_{21}\zeta_1^t$, $(\mathfrak{P}^t)_{12} = \zeta_1^t\mathfrak{A}_{12}$.*

Proof: We have $\zeta_1^{-1} \in (\mathfrak{Q}^{-1})_{11}$, so $\zeta_1^{-t}(\mathfrak{Q}^t)_{12} \subset (\mathfrak{Q}^{-t})_{11}\mathfrak{Q}^t_{12} \subset \mathfrak{B}_{12}$. Therefore $(\mathfrak{Q}^t)_{12} \subset \zeta_1^t\mathfrak{B}_{12}$. From the definition of splitting, $\{(\mathfrak{Q}^t)_{12}\}$ is a uniform $\mathfrak{o}_E$-lattice chain in B_{12} of period $e(\mathfrak{B}|\mathfrak{o}_E)$. The same clearly applies to $\{\zeta_1^t\mathfrak{B}_{12}\}$, and these chains coincide in degree zero. They are therefore identical. This proves that $(\mathfrak{Q}^t)_{12} = \zeta_1^t\mathfrak{B}_{12}$, $t \in \mathbb{Z}$, and the other statements are similar. ■

(8.2.11) Lemma: *Let $\mathfrak{M}$ be any $(\mathfrak{B},\mathfrak{B})$-bimodule in A. Then $\zeta_1\mathfrak{M}_{12} = (\mathfrak{Q}\mathfrak{M})_{12}$ and $\mathfrak{M}_{21}\zeta_1 = (\mathfrak{M}\mathfrak{Q})_{21}$.*

Proof: Again we only need prove one of these statements, say the first. We have $(\mathfrak{Q}\mathfrak{M})_{12} = \mathfrak{Q}_{11}\mathfrak{M}_{12} + \mathfrak{Q}_{12}\mathfrak{M}_{22} = \zeta_1\mathfrak{B}_1\mathfrak{M}_{12} + \zeta_1\mathfrak{B}_{12}\mathfrak{M}_{22} \subset \zeta_1\mathfrak{M}_{12}$. On the other hand, the element

$$\zeta_0 = \begin{pmatrix} \zeta_1 & 0 \\ 0 & 0 \end{pmatrix}$$

lies in $\mathfrak{Q}$, so $\zeta_1\mathfrak{M}_{12} = (\zeta_0\mathfrak{M})_{12} \subset (\mathfrak{Q}\mathfrak{M})_{12}$, and the lemma follows. ■

(8.2.12) Proposition: *Write $K = 1 + \mathfrak{h}$, $\mathfrak{h} = \mathfrak{H}^m(\beta,\mathfrak{A})$. The set $1 + \mathfrak{h} + \mathfrak{h}^\zeta$ is a group, equal to KK^ζ. Moreover, $K \cap K^\zeta = 1 + \mathfrak{h} \cap \mathfrak{h}^\zeta$ is a normal subgroup of KK^ζ and*

$$KK^\zeta/K \cap K^\zeta \cong (\mathfrak{h} + \mathfrak{h}^\zeta)/(\mathfrak{h} \cap \mathfrak{h}^\zeta).$$

In particular, this quotient is abelian.

Proof: We shall prove below in (8.2.13) that

$$\mathfrak{h}.\mathfrak{h} \subset \mathfrak{Q}\mathfrak{h}.$$

We start by deducing the Proposition from this fact. First recall from (3.1.9) that we have

$$\mathfrak{Q}^m\mathfrak{h} = \mathfrak{h}\mathfrak{Q}^m, \quad m \in \mathbb{Z},$$

which considerably simplifies the computations. Further, since $\mathfrak{h}$ is a $\mathfrak{B}$-module, we have $\mathfrak{h} \supset \mathfrak{Q}\mathfrak{h}$, whence $\mathfrak{h} \subset \mathfrak{Q}^{-1}\mathfrak{h}$. We write out our various lattices in blocks, using (8.2.11). We have

$$\mathfrak{h}^\zeta = \begin{pmatrix} \mathfrak{h}_{11} & (\mathfrak{Q}^{-1}\mathfrak{h})_{12} \\ (\mathfrak{h}\mathfrak{Q})_{21} & \mathfrak{h}_{22} \end{pmatrix},$$

whence

$$\mathfrak{h} \cap \mathfrak{h}^\zeta = \begin{pmatrix} \mathfrak{h}_{11} & \mathfrak{h}_{12} \\ (\mathfrak{h}\mathfrak{Q})_{21} & \mathfrak{h}_{22} \end{pmatrix} \subset \mathfrak{Q}\mathfrak{h}.$$

All the assertions of the Proposition follow once we show

$$(\mathfrak{h}+\mathfrak{h}^{\varsigma})(\mathfrak{h}+\mathfrak{h}^{\varsigma}) \subset \mathfrak{h}\cap\mathfrak{h}^{\varsigma}.$$

Expanding this, the term $\mathfrak{h}.\mathfrak{h}$ is contained in $\mathfrak{Q}\mathfrak{h} \subset \mathfrak{h}\cap\mathfrak{h}^{\varsigma}$. Next, we have $\mathfrak{h}^{\varsigma}.\mathfrak{h}^{\varsigma} = (\mathfrak{h}.\mathfrak{h})^{\varsigma} \subset (\mathfrak{Q}\mathfrak{h})^{\varsigma}$ which, in block terms, looks like

$$(\mathfrak{Q}\mathfrak{h})^{\varsigma} = \begin{pmatrix} (\mathfrak{Q}\mathfrak{h})_{11} & \mathfrak{h}_{12} \\ (\mathfrak{h}\mathfrak{Q}^2)_{21} & (\mathfrak{Q}\mathfrak{h})_{22} \end{pmatrix} \subset \mathfrak{h}\cap\mathfrak{h}^{\varsigma}$$

as required.

This leaves the terms $\mathfrak{h}.\mathfrak{h}^{\varsigma}$, $\mathfrak{h}^{\varsigma}.\mathfrak{h}$. The arguments in the two cases are similar, so we just treat the first. Expanding in blocks, we get

$$\begin{aligned} \mathfrak{h}.\mathfrak{h}^{\varsigma} &= \begin{pmatrix} \mathfrak{h}_{11} & \mathfrak{h}_{12} \\ \mathfrak{h}_{21} & \mathfrak{h}_{22} \end{pmatrix} \begin{pmatrix} \mathfrak{h}_{11} & (\mathfrak{h}\mathfrak{Q}^{-1})_{12} \\ (\mathfrak{h}\mathfrak{Q})_{21} & \mathfrak{h}_{22} \end{pmatrix} \\ &\subset \begin{pmatrix} (\mathfrak{h}.\mathfrak{h})_{11} & (\mathfrak{h}.\mathfrak{h}\mathfrak{Q}^{-1})_{12} \\ (\mathfrak{h}.\mathfrak{h})_{21} & (\mathfrak{h}.\mathfrak{h}\mathfrak{Q}^{-1})_{22} \end{pmatrix} \\ &\subset \begin{pmatrix} (\mathfrak{Q}\mathfrak{h})_{11} & \mathfrak{h}_{12} \\ (\mathfrak{Q}\mathfrak{h})_{21} & \mathfrak{h}_{22} \end{pmatrix} \subset \mathfrak{h}\cap\mathfrak{h}^{\varsigma}, \end{aligned}$$

as required. It therefore remains only to prove:

(8.2.13) Lemma: *Let $t \geq 1$, and abbreviate $\mathfrak{H}^t = \mathfrak{H}^t(\beta,\mathfrak{A})$. Then $\mathfrak{H}^t.\mathfrak{H}^t$ is contained in $\mathfrak{Q}\mathfrak{H}^t$.*

Proof: We proceed by induction along β. If β is minimal over F, the result follows from a trivial computation starting from the definition of $\mathfrak{H}$. In general, we set $r = -k_0(\beta,\mathfrak{A})$ and choose a simple stratum $[\mathfrak{A},n,r,\gamma]$ equivalent to $[\mathfrak{A},n,m,\beta]$. Assume first that $t \geq [r/2]+1$. Then $\mathfrak{H}^t = \mathfrak{H}^t(\gamma,\mathfrak{A})$ (by (3.1.9)). (3.1.9) further gives us $\mathfrak{Q}\mathfrak{H}^t = \mathfrak{Q}\mathfrak{N}(\beta)\mathfrak{H}^t$. Writing $\mathfrak{Q}_\gamma = \mathfrak{P}\cap\mathrm{End}_{F[\gamma]}(V)$, these identities and (3.1.3) give us

$$\mathfrak{Q}\mathfrak{H}^t = \mathfrak{Q}_\gamma\mathfrak{N}_{-r}(\gamma)\mathfrak{H}^t = \mathfrak{Q}_\gamma\mathfrak{H}^t(\gamma,\mathfrak{A}).$$

By induction, we have $\mathfrak{H}^t.\mathfrak{H}^t = \mathfrak{H}^t(\gamma,\mathfrak{A})\mathfrak{H}^t(\gamma,\mathfrak{A}) \subset \mathfrak{Q}_\gamma\mathfrak{H}^t(\gamma,\mathfrak{A})$, so the assertion holds in this case.

If, on the other hand, $t \leq [r/2]$, we write $\mathfrak{H}^t = \mathfrak{Q}^t + \mathfrak{H}^{[r/2]+1}$, and

$$\mathfrak{H}^t.\mathfrak{H}^t \subset \mathfrak{Q}^{2t} + \mathfrak{Q}^t\mathfrak{H}^{[r/2]+1} + (\mathfrak{H}^{[r/2]+1})^2 \subset \mathfrak{Q}^{t+1} + \mathfrak{Q}\mathfrak{H}^{[r/2]+1}$$

by the first case, as desired. ■

This also completes the proof of (8.2.12). ■

We finally take an irreducible smooth representation $(\boldsymbol{\pi},\mathcal{V})$ of G which contains (K,ϑ) (assumed to satisfy (8.2.8)), and consider the

Jacquet module $(\pi_U, \mathcal{V}_U)$. Let $q : \mathcal{V} \to \mathcal{V}_U$ denote the quotient map. Since q certainly maps the ϑ-isotypic component $\mathcal{V}^\vartheta$ of $\mathcal{V}$ to the $(\vartheta \mid K \cap M)$-isotypic component of $\mathcal{V}_U$, it is enough to show that q is non-trivial on $\mathcal{V}^\vartheta$. We assume the contrary and seek a contradiction. Arguing as in **(7.3)** (see especially (7.3.3)), if $\mathcal{V}^\vartheta \subset \mathrm{Ker}(q)$, there exists a compact open subgroup U' of U such that

$$\int_{U'} \pi(x)\nu \, dx = 0, \quad \nu \in \mathcal{V}^\vartheta,$$

where dx denotes some Haar measure on U. For $j \in \mathbb{Z}$, put

$$M_j = 1 + \mathfrak{P}^j \cap A_{12}.$$

Then $\{M_j\}$ is a descending sequence of compact open subgroups of U with union U and intersection $\{1\}$. For j sufficiently large (e.g. $j \geq n+1$), we have $M_j \subset \mathrm{Ker}(\vartheta)$, so there exists $j \in \mathbb{Z}$ which is maximal for the property

$$\int_{M_{j-1}} \pi(x)\nu \, dx = 0, \quad \nu \in \mathcal{V}^\vartheta.$$

Now we record a general technique, which we will use repeatedly in the ensuing arguments.

(8.2.14) Lemma: *Suppose we are given a compact open subgroup $\mathcal{K}$ of G, an abelian character ξ of $\mathcal{K}$, a vector $\nu \in \mathcal{V}^\xi$, and a compact open subgroup $\mathcal{M}$ of U such that*

$$\int_{\mathcal{M}} \pi(x)\nu \, dx \neq 0.$$

Let $\mathcal{H} \supset \mathcal{K}$ be another compact open subgroup of G, and suppose that the following conditions hold:

(i) $\mathcal{K}$ is a normal subgroup of $\mathcal{H}$, the quotient $\mathcal{H}/\mathcal{K}$ is abelian, and ξ admits extension to an abelian character of $\mathcal{H}$;

(ii) given any characters ξ_1, ξ_2 of $\mathcal{H}$ such that $\xi_1 \mid \mathcal{K} = \xi_2 \mid \mathcal{K} = \xi$, there exists $g \in G$ which normalises $\mathcal{K}$, $\mathcal{H}$, $\mathcal{M}$ and such that $\xi_2 = \xi_1^g$. Then, given an extension ξ' of ξ to $\mathcal{H}$, there exists $\nu' \in \mathcal{V}^{\xi'}$ such that

$$\int_{\mathcal{M}} \pi(x)\nu' \, dx \neq 0.$$

Proof: Let $\mathcal{U}$ be the $\mathcal{H}$-subspace of $\mathcal{V}$ generated by ν. Then $\mathcal{U} = \coprod \mathcal{U}^\chi$, where χ ranges over the extensions of ξ to $\mathcal{H}$. This gives a decomposition

$$\nu = \sum_\chi \nu_\chi,$$

for $v_\chi \in \mathcal{U}^\chi$. For some χ, we get

$$\int_{\mathcal{M}} \pi(x) v_\chi \, dx \neq 0.$$

We have $\xi' = \chi^g$, for some $g \in G$ satisfying the conditions in *(ii)*. Then $v' = \pi(g^{-1}) v_\chi$ is the vector we require. ■

We now embark on a lengthy and elaborate refinement argument, starting from our given split type (K, ϑ). The first step is:

(8.2.15) Proposition: *There exists a unique character ϑ' of the group K^ζ such that $\vartheta' \mid (K^\zeta \cap U)$ is null and $\vartheta' \mid K \cap K^\zeta = \vartheta \mid K \cap K^\zeta$. Moreover, there exists $v' \in \mathcal{V}^{\vartheta'}$ such that*

$$\int_{M_{j-1}} \pi(x) v' \, dx \neq 0.$$

Proof: (8.2.11) implies that $K^\zeta = (K \cap K^\zeta)(K^\zeta \cap U)$, so the character ϑ' is uniquely determined, if it exists. However, we can easily write it down explicitly. We form $\theta^\zeta \in \mathcal{C}(\mathfrak{A}^\zeta, m-1, \beta)$, and put $\vartheta' = \theta^\zeta \psi_c$. This certainly agrees with $\vartheta = \theta \psi_c$ on $K \cap K^\zeta$, since the fact $\zeta \in B^\times$ implies that ζ intertwines θ. Moreover, ϑ' is trivial on $K^\zeta \cap U$ by (8.2.5)*(iii)* and (8.2.8). (Note that (8.2.8) implies $c \in (\mathfrak{P}^\zeta)^{-m}$, so ψ_c defines a character of $\boldsymbol{U}^m(\mathfrak{A}^\zeta) \supset K^\zeta$.)

We next require another, even more arcane, subgroup of K. With $\mathfrak{H}^t = \mathfrak{H}^t(\beta, \mathfrak{A})$, $\mathfrak{h} = \mathfrak{H}^m$ again, define

$$\widetilde{\mathfrak{h}} = \mathfrak{Q}^m + \mathfrak{H}_{11}^{1+m} + \mathfrak{h}_{12} + \mathfrak{h}_{21} + \mathfrak{h}_{22},$$
$$\widetilde{H} = 1 + \widetilde{\mathfrak{h}}.$$

The group $\widetilde{H}$ inherits an Iwahori decomposition from that of K. Immediately, we have

$$\widetilde{H} \cap \widetilde{H}^\zeta = (\widetilde{H} \cap \widetilde{H}^\zeta \cap U^-)(\widetilde{H} \cap M)(\widetilde{H} \cap \widetilde{H}^\zeta \cap U).$$

The element $\zeta_1 \in \mathfrak{K}(\mathfrak{B}_1)$ was constructed to commute with $s(c)_{11}$. Also, by *(iii)* of the Theorem, the character ϑ is trivial on the outer factors here. We deduce:

(8.2.16) *The element ζ intertwines the character $\vartheta \mid \widetilde{H}$.*

By the definition of j, there exists $v_1 \in \mathcal{V}^\vartheta$ such that

$$\int_{M_j} \pi(x) v_1 \, dx \neq 0.$$

Consider the element $v_2 = \pi(\zeta^{-1})v_1$. We have

$$\int_{M_{j-1}} \pi(x)v_2\,dx = \pi(\zeta^{-1}) \int_{M_{j-1}} \pi(\zeta x\zeta^{-1})v_1\,dx,$$

while $\zeta M_{j-1}\zeta^{-1} = M_j$, by (8.2.10). Therefore there is a constant $k > 0$ such that

$$\int_{M_{j-1}} \pi(x)v_2\,dx = k\pi(\zeta^{-1}) \int_{M_j} \pi(x)v_1\,dx \neq 0.$$

By (8.2.16), the vector v_2 transforms according to ϑ under the group $\widetilde{H}\cap\widetilde{H}^\zeta$, and (8.2.5)*(iii)* shows that it is fixed under the group $(K\cap U)^\zeta = 1+\zeta_1^{-1}\mathfrak{h}_{12}$

(8.2.17) Lemma: *Let χ be a character of $K\cap K^\zeta$ such that $\chi \mid \widetilde{H}\cap\widetilde{H}^\zeta = \vartheta \mid \widetilde{H}\cap\widetilde{H}^\zeta$. Write $\mathfrak{M} = \mathfrak{Q}^{r-m}\mathfrak{N}(\beta)+\mathfrak{J}^{[\frac{r+1}{2}]}(\beta,\mathfrak{A})$. There exists $x \in \mathfrak{M}_{11}$ such that $\chi = \vartheta^{1+x}$ on the group $K\cap K^\zeta$.*

Proof: First we describe these characters χ. We have $K\cap K^\zeta = (K\cap G_1)(\widetilde{H}\cap\widetilde{H}^\zeta)$, and the intersection of the factors is $1+\mathfrak{Q}_1+\mathfrak{H}_{11}^{m+1}$. Thus any of our characters χ is of the form $\vartheta\chi_1$, where χ_1 is a character of $(K\cap G_1)/(1+\mathfrak{Q}_1+\mathfrak{H}_{11}^{m+1})$, extended to $K\cap K^\zeta$ by making it trivial on the other blocks. However, by part *(ii)* of the theorem, we have $K\cap G_1 = H^m(\beta,\mathfrak{A}_1)$, and $\mathfrak{H}_{11}^{m+1} = \mathfrak{H}^{m+1}(\beta,\mathfrak{A}_1)$. The adjoint map a_β respects the block decomposition of A, so now we can invoke (3.1.16) to show that χ_1 is of the form $\chi_1 = \psi_{a_\beta(x)}$, for some $x \in \mathfrak{Q}_1^{r-m}\mathfrak{N}(\beta,\mathfrak{A}_1)+\mathfrak{J}^{[\frac{r+1}{2}]}(\beta,\mathfrak{A}_1)$.

Now we observe that $\mathfrak{J}^{[\frac{r+1}{2}]}(\beta,\mathfrak{A}_1) = \mathfrak{J}^{[\frac{r+1}{2}]}(\beta,\mathfrak{A})_{11}$, by *(ii)* of the theorem. Also, again since a_β respects the block decomposition, it follows easily from (2.1.1) that $\mathfrak{Q}_1^{r-m}\mathfrak{N}(\beta,\mathfrak{A}_1) = (\mathfrak{Q}^{r-m}\mathfrak{N}(\beta))_{11}$. In all, we have

$$\chi = \vartheta\psi_{a_\beta(x)},$$

for some $x \in \mathfrak{M}_{11}$. On the other hand, the argument of (8.1.12) applies to give $\vartheta^{1+x} = \vartheta\psi_{a_\beta(x)}$, as a character of K, for any $x \in \mathfrak{M}$. The lemma now follows. ■

Now we apply the technique of (8.2.14), but some extra care is needed. We let χ range over the characters of $K\cap K^\zeta$ which extend $\vartheta \mid K\cap K^\zeta$, and write

$$v_2 = \sum_\chi v_\chi, \quad v_\chi \in \mathcal{V}^\chi.$$

Each v_χ here is a linear combination of $\pi(K\cap K^\zeta)$-translates of v_2. However, the groups $K\cap K^\zeta$, $\widetilde{H}\cap\widetilde{H}^\zeta$ differ only in the $(1,1)$-block, so we can choose coset representatives for $(K \cap K^{\zeta})/(\widetilde{H} \cap \widetilde{H}^{\zeta})$ from $K \cap G_1 = K^{\zeta} \cap G_1$. These coset representatives all normalise $K^{\zeta} \cap U$, so all the v_{χ} are fixed by $K^{\zeta} \cap U$. The group $1 + \mathfrak{M}_{11}$ normalises M_{j-1}, so now (8.2.14) gives us a vector v' in the $\vartheta \mid K \cap K^{\zeta}$-isotypic subspace of $\mathcal{V}$ with

$$\int_{M_{j-1}} \pi(x)v'\,dx \neq 0,$$

and which is also fixed by $K^{\zeta} \cap U$. In other words, $v' \in \mathcal{V}^{\vartheta'}$, and we have proved the Proposition. ∎

Now we write $J = 1 + \mathfrak{j} = J^m(\beta, \mathfrak{A})$, and consider the group $(J^{\zeta} \cap K)K^{\zeta} = J^{\zeta} \cap (KK^{\zeta})$. We can simplify this: for all pairs (i, j) other than $(2, 1)$, we have $(\mathfrak{j}^{\zeta} \cap \mathfrak{h})_{ij} \subset (\mathfrak{h}^{\zeta})_{ij}$. However, the $(2, 1)$-block of $\mathfrak{j}^{\zeta}$ is $(\mathfrak{j}\mathfrak{Q})_{21}$, and $\mathfrak{j}\mathfrak{Q} \subset \mathfrak{h}$, by (3.1.13). Therefore

$$J^{\zeta} \cap KK^{\zeta} = (J^{\zeta} \cap U^-)K^{\zeta}.$$

(8.2.18) Proposition: *The character θ^{ζ} of K^{ζ} extends to an abelian character of $J^{\zeta} \cap KK^{\zeta}$, which is trivial on $J^{\zeta} \cap U^-$. In particular, this character agrees with θ on $J^{\zeta} \cap K$. Moreover,*

(i) any two characters of $J^{\zeta} \cap KK^{\zeta}$ which extend θ^{ζ} are conjugate under $J^{\zeta} \cap U$;

(ii) the character ϑ' of K^{ζ} extends to an abelian character of $J^{\zeta} \cap KK^{\zeta}$, and any two such extensions are conjugate under $J^{\zeta} \cap U$.

Proof: Extending θ^{ζ} to $J^{\zeta} \cap KK^{\zeta} = (J^{\zeta} \cap U^-)K^{\zeta}$ is essentially the same as extending θ to $(J \cap U^-)K$. However, using the same argument as in the proof of (7.2.3), the image of $J \cap U^-$ in J/K (equipped with its canonical alternating form $\boldsymbol{k}_{\theta}$, as in **(3.4)**) is a totally isotropic subspace. Therefore θ does extend, and we can surely make this extension trivial on $J \cap U^-$.

In the same way, $J \cap U$ corresponds to a totally isotropic subspace of J/K. Further, just as in (7.2.3), these subspaces corresponding to $J \cap U^-$, $J \cap U$, are orthogonal to the image of $J \cap (G_1 \times G_2)$. As in (7.2.4), elementary properties of alternating spaces now show that any two extensions of θ to $(J \cap U^-)K$ are conjugate under $J \cap U$. Conjugating by ζ, we get the first assertion and *(i)* of the proposition.

The function ψ_c defines a character of the group $\boldsymbol{U}^m(\mathfrak{A}^{\zeta})$, since $c \in (\mathfrak{P}^{\zeta})^{-m}$. Its restriction to $J^{\zeta} \cap KK^{\zeta}$ is then certainly invariant under conjugation by J^{ζ}. This gives *(ii)*. ∎

(8.2.18) gives us a unique character φ of $J^{\zeta} \cap KK^{\zeta}$ such that

(8.2.19) *(i)* $\varphi \mid K^{\zeta} = \vartheta'$;
(ii) $\varphi \mid J^{\zeta} \cap U^-$ *is trivial.*

The conjugacy property (8.2.18)*(ii)* allows us to apply (8.2.14) and conclude:

(8.2.20) *There exists a vector* $\mathrm{v}_3 \in \mathcal{V}$ *with the following properties:*
(i) $\pi(x)\mathrm{v}_3 = \varphi(x)\mathrm{v}_3$, *for all* $x \in J^\varsigma \cap KK^\varsigma$;
(ii)

$$\int_{M_{j-1}} \pi(u)\mathrm{v}_3\, du \neq 0.$$

Now we have to divide the argument into two cases. *Assume first that* $m \leq \left[\frac{r}{2}\right]$. We relax the conditions a little, and set $K_1 = H^{m+1}(\beta, \mathfrak{A})$. We form the group $(J^\varsigma \cap K)K_1^\varsigma \subset J^\varsigma \cap KK^\varsigma$. We also write φ for the restriction to $(J^\varsigma \cap K)K_1^\varsigma$ of the character φ given by (8.2.18). We have $\mathrm{v}_3 \in \mathcal{V}^\varphi$ such that

$$\int_{M_{j-1}} \pi(x)\mathrm{v}_3\, dx \neq 0.$$

We also have

$$\boldsymbol{U}^m(\mathfrak{A}) \supset KK_1^\varsigma \supset (J^\varsigma \cap K)K_1^\varsigma \supset K_1,$$

$m \geq 1$ and $\boldsymbol{U}^m(\mathfrak{B}) \subset \boldsymbol{U}^{m-1}(\mathfrak{B}^\varsigma)$, so we can form the group

$$\mathcal{K} = \boldsymbol{U}^m(\mathfrak{B})(J^\varsigma \cap K)K_1^\varsigma.$$

The group $\mathcal{K}$ normalises its subgroup $(J^\varsigma \cap K)K_1^\varsigma$, differs from it only in the $(2,1)$-block, and the quotient is abelian. Now let us write φ in the form $\varphi = \varphi_0\psi_c$. The character φ_0 extends to a character Φ_0 of $\mathcal{K}$ as follows. We have $\mathcal{K} = (\mathcal{K} \cap K)K_1^\varsigma$. We define Φ_0 by

$$\Phi_0 \mid (\mathcal{K} \cap K) = \theta \mid (\mathcal{K} \cap K),$$
$$\Phi_0 \mid K_1^\varsigma = \theta^\varsigma \mid K_1^\varsigma \ (\in \mathcal{C}(\mathfrak{A}^\varsigma, m, \beta)).$$

This needs some justification. The group $\boldsymbol{U}^m(\mathfrak{B})$ normalises the character φ_0. However, φ factors through the determinant $\det_B$ on $(J^\varsigma \cap K)K_1^\varsigma \cap B^\times$, so its restriction here certainly extends to $\boldsymbol{U}^m(\mathfrak{B})$. We use this extension to get a well-defined extension of φ_0 to $\mathcal{K}$, which then satisfies the conditions above.

We have $\mathcal{K} \subset \boldsymbol{U}^m(\mathfrak{A})$, so the function ψ_c certainly defines a character of $\mathcal{K}$. We put $\Phi = \Phi_0\psi_c$. This satisfies $\Phi \mid (J^\varsigma \cap K)K_1^\varsigma = \varphi$. Moreover, if Φ' is another extension of φ to $\mathcal{K}$, it is of the form $\Phi_0\psi_{c'}$, where $c' \in \mathfrak{P}^{-m}$ and $\psi_{c'} \mid (J^\varsigma \cap K)K_1^\varsigma = \psi_c \mid (J^\varsigma \cap K)K_1^\varsigma$. Thus the extension Φ' is uniquely determined by the quantity

$$\psi_{c'} \mid \boldsymbol{U}^m(\mathfrak{B}) = \psi_{B,s(c')},$$

where s is a tame corestriction on A relative to E/F. We can go further. We have observed that the groups $\mathcal{K}$, $(J^\zeta \cap K)K_1^\zeta$ differ only in the $(2,1)$-block. The characters $\psi_{B,s(c)}$, $\psi_{B,s(c')}$ must agree on $\boldsymbol{U}^{m+1}(\mathfrak{B}) \subset (J^\zeta \cap K)K_1^\zeta$. Altogether, we have

(8.2.21) *Let $\Phi' = \Phi_0\psi_{c'}$ be some extension of φ to $\mathcal{K}$. Then $\Phi' = \Phi$ if and only if*

$$s(c')_{12} \equiv s(c)_{12} \pmod{\mathfrak{Q}^{1-m}}.$$

(8.2.22) Lemma: *The group $1 + \mathfrak{B}_{12}$ normalises the groups $\mathcal{K}$, $(J^\zeta \cap K)K_1^\zeta$, and fixes the characters Φ_0, φ. Moreover, if Φ' is any extension of φ to $\mathcal{K}$, there exists $y \in \mathfrak{B}_{12}$ such that $\Phi' = \Phi^{1+y}$.*

Proof: The first assertion follows from the observation $\mathfrak{B}_{12} \subset \mathfrak{B} \cap \mathfrak{Q}^\zeta$. The same reason implies that $1 + \mathfrak{B}_{12}$ fixes Φ_0, and also ψ_c viewed as a character of $\boldsymbol{U}^m(\mathfrak{A}^\zeta)/\boldsymbol{U}^{m+1}(\mathfrak{A}^\zeta)$. It therefore fixes φ, since $(J^\zeta \cap K)K_1^\zeta \subset J^\zeta \subset \boldsymbol{U}^m(\mathfrak{A}^\zeta)$.

Now consider the character Φ^{1+y}, for $y \in \mathfrak{B}_{12}$. Since the element c is split ((8.2.8)), this is $\Phi_0\psi_c^{1+y} = \Phi_0\psi_{cy-yc}$. The element $cy - yc$ is confined to the $(1,2)$-block. When we apply the tame corestriction s, we get $s(cy-yc) = \partial_{s(c)}(y)$ (notation of (2.3.8)). Now we invoke the original hypothesis that $V = V_1 \oplus V_2$ splits the stratum $[\mathfrak{B}, m, m-1, s(c)]$. The lemma is implied by (8.2.21) and (2.3.8)*(iii)*. ■

Now we can apply (8.2.14) to get:

(8.2.23) *There exists $v_4 \in \mathcal{V}^\Phi$ such that*

$$\int_{M_{j-1}} \pi(x)v_4\,dx \neq 0.$$

The final step in the proof of this case is to show:

(8.2.24) Lemma: *The character Φ of $\mathcal{K}$ constructed above induces irreducibly to the group KK_1^ζ.*

Proof: We have to show that the KK_1^ζ-intertwining of Φ is contained in $\mathcal{K}$. This intertwining is contained in that of the character $\Phi \mid K_1^\zeta = \theta^\zeta \mid K_1^\zeta$. Since $m \leq \left[\frac{r}{2}\right]$, the G-intertwining of $\theta^\zeta \mid K_1^\zeta$ (this character is an element of $\mathcal{C}(\mathfrak{A}^\zeta, m, \beta)$) is

$$\left((1 + (\mathfrak{J}^{[\frac{r+1}{2}]})B^\times(1 + (\mathfrak{J}^{[\frac{r+1}{2}]})\right)^\zeta.$$

We therefore have to compute the intersection of this with KK_1^ζ. Its intersection with $\boldsymbol{U}^m(\mathfrak{A})$ (which contains KK_1^ζ) is $1 + \mathfrak{Q}^m + ((\mathfrak{J}^{[\frac{r+1}{2}]})^\zeta \cap$

$\mathfrak{P}^m$). All the lattices here decompose according to the block decomposition $A = \coprod A_{ij}$ of A, so we can compute the required intersection block by block. Our groups $\mathcal{K}$, KK_1^ζ differ only in the (2,1)-block, so we are reduced to showing that

$$(\mathfrak{J}^{[\frac{r+1}{2}]})_{21}^\zeta \cap (\mathfrak{P}^m)_{21} \subset \mathfrak{j}_{21}^\zeta + \mathfrak{Q}_{21}^m.$$

Here, we have written $\mathfrak{j} = \mathfrak{J}^m$. Using (8.2.11), this containment is implied by

$$\mathfrak{Q}\mathfrak{J}^{[\frac{r+1}{2}]} \cap \mathfrak{P}^m \subset \mathfrak{Q}\mathfrak{J}^m,$$

which is immediate, since $m \leq [\frac{r}{2}]$. ■

Let ρ denote the irreducible representation $\mathrm{Ind}(\Phi : \mathcal{K}, KK_1^\zeta)$. The restriction of ρ to K is a sum of abelian characters, among which must be our original character ϑ. The components of $\rho \mid K$ are all conjugate under KK_1^ζ, and therefore under $K_1^\zeta \cap U$, since $KK_1^\zeta = K(K_1^\zeta \cap U)$. Now we apply (8.2.14), (8.2.23) to get a vector $\mathrm{v}_5 \in \mathcal{V}^\vartheta$ such that

$$\int_{M_{j-1}} \pi(x)\mathrm{v}_5\, dx \neq 0.$$

This contradicts the definition of j, and proves the theorem in this case.

Next, *we assume that* $m \geq [\frac{r}{2}]+1$. Here we relax more dramatically to the group $K \cap K^\zeta$. (8.2.20) gives us a vector v_3, which transforms under $K \cap K^\zeta$ according to ϑ, and satisfies

$$\int_{M_{j-1}} \pi(x)\mathrm{v}_3\, dx \neq 0.$$

(8.2.25) Lemma: *The group $1+\mathfrak{B}_{12}$ normalises $K \cap K^\zeta$, $\boldsymbol{U}^m(\mathfrak{B})(K \cap K^\zeta)$, and acts transitively on the set of characters of $\boldsymbol{U}^m(\mathfrak{B})(K \cap K^\zeta)$ which agree with ϑ on $K \cap K^\zeta$. There exists $\mathrm{v}_4 \in \mathcal{V}$ such that*

$$\pi(x)\mathrm{v}_4 = \vartheta(x)\mathrm{v}_4, \quad x \in \boldsymbol{U}^m(\mathfrak{B})(K \cap K^\zeta),$$
$$\int_{M_{j-1}} \pi(x)\mathrm{v}_4\, dx \neq 0.$$

Proof: Since $1 + \mathfrak{B}_{12} \subset \boldsymbol{U}(\mathfrak{B}) \cap \boldsymbol{U}^1(\mathfrak{B}^\zeta)$, the first two assertions are immediate. Conjugation by $1 + \mathfrak{B}_{12}$ certainly fixes the character $\theta \mid \boldsymbol{U}^m(\mathfrak{B})(K \cap K^\zeta)$. The next assertion is proved in exactly the same manner as (8.2.22). The final assertion follows from (8.2.14). ■

Now we set

$$\mathfrak{M} = \mathfrak{Q}^{r-m+1}\mathfrak{N}(\beta) + \mathfrak{J}^{[\frac{r+1}{2}]}(\beta, \mathfrak{A}),$$

and consider the action of the group $1 + \mathfrak{M}_{12}^{\zeta} = 1 + (\mathfrak{Q}^{-1}\mathfrak{M})_{12}$. We are operating under the hypotheses $1 \leq m < r$, $m \geq \left[\frac{r}{2}\right] + 1$. These imply $r \geq 3$, $m \geq 2$.

(8.2.26) Lemma: *The group $1 + (\mathfrak{M}^{\zeta})_{12}$ normalises the groups K, $\boldsymbol{U}^m(\mathfrak{B})(K \cap K^{\zeta})$, and fixes the character $\vartheta \mid \boldsymbol{U}^m(\mathfrak{B})(K \cap K^{\zeta})$. It acts transitively on the set of characters of K which agree with ϑ on $\boldsymbol{U}^m(\mathfrak{B})(K \cap K^{\zeta})$.*

Proof: The groups K, K^{ζ}, $\boldsymbol{U}^m(\mathfrak{B})(K \cap K^{\zeta})$ all have Iwahori decompositions, and differ only in the (2,1)-block. The (2,1)-blocks of these groups are respectively $1 + \mathfrak{h}_{21}$, $1 + (\mathfrak{Q}\mathfrak{h})_{21}$, $1 + (\mathfrak{Q}^m + \mathfrak{Q}\mathfrak{h})_{21}$. Here, we write $\mathfrak{h} = \mathfrak{H}^m(\beta, \mathfrak{A})$, as before. Therefore the normalisation assertions of the lemma all follow from the containment relation

$$\mathfrak{M}\mathfrak{h} + \mathfrak{h}\mathfrak{M} \subset \mathfrak{Q}^m + \mathfrak{Q}\mathfrak{h}. \tag{8.2.27}$$

Symmetry reduces us to showing $\mathfrak{M}\mathfrak{h} \subset \mathfrak{Q}\mathfrak{h}$. Since $m \geq \left[\frac{r}{2}\right] + 1$, (3.1.9) implies $\mathfrak{Q}^{r-m}\mathfrak{N}(\beta)\mathfrak{h} \subset \mathfrak{h}$, whence $\mathfrak{Q}^{r-m+1}\mathfrak{N}(\beta)\mathfrak{h} \subset \mathfrak{Q}\mathfrak{h}$. We therefore need only show that

$$\mathfrak{J}^{[\frac{r+1}{2}]}(\beta, \mathfrak{A})\mathfrak{H}^m(\beta, \mathfrak{A}) \subset \mathfrak{Q}\mathfrak{H}^m(\beta, \mathfrak{A}), \quad m \geq 2. \tag{8.2.28}$$

We leave this for the moment, and prove the remaining statements. We have $\boldsymbol{U}^m(\mathfrak{B})(K \cap K^{\zeta}) \subset \boldsymbol{U}^{m-1}(\mathfrak{B}^{\zeta})H^m(\beta, \mathfrak{A}^{\zeta})$. On $\boldsymbol{U}^m(\mathfrak{B})(K \cap K^{\zeta})$, the character θ agrees with some simple character $\theta_1 \in \mathcal{C}(\mathfrak{A}^{\zeta}, m-1, \beta)$. (3.5.10) now implies that the whole group $1 + \mathfrak{M}^{\zeta}$ fixes θ_1 on $\boldsymbol{U}^{m-1}(\mathfrak{B}^{\zeta})K^{\zeta}$, so $1 + (\mathfrak{M}^{\zeta})_{12}$ certainly fixes $\theta \mid \boldsymbol{U}^m(\mathfrak{B})(K \cap K^{\zeta})$. The function ψ_c defines a character of $\boldsymbol{U}^m(\mathfrak{A})$ which is trivial on $\boldsymbol{U}^{m+1}(\mathfrak{A})$. We have $1 + \mathfrak{M}^{\zeta} \subset \boldsymbol{U}^2(\mathfrak{A}^{\zeta})$, whence $1 + (\mathfrak{M}^{\zeta})_{12} \subset \boldsymbol{U}^1(\mathfrak{A})$. It follows that the group $1 + (\mathfrak{M}^{\zeta})_{12}$ fixes the character ψ_c on $\boldsymbol{U}^m(\mathfrak{B})(K \cap K^{\zeta})$, and even on K.

In all, this shows that $1 + (\mathfrak{M}^{\zeta})_{12}$ fixes the character ϑ on the group $\boldsymbol{U}^m(\mathfrak{B})(K \cap K^{\zeta})$ and the character ψ_c on K. Now we compute the $(1 + (\mathfrak{M}^{\zeta})_{12})$-stabiliser (call it $\mathcal{S}$) of ϑ as a character of K. We have just seen that $\mathcal{S}$ is the $1 + (\mathfrak{M}^{\zeta})_{12}$-stabiliser of θ on K. We have $1 + \mathfrak{M}^{\zeta} \subset \boldsymbol{U}^1(\mathfrak{A})$, and the $\boldsymbol{U}^1(\mathfrak{A})$-stabiliser of θ is $\boldsymbol{U}^1(\mathfrak{B})(1 + \mathfrak{M})$, by (3.3.17). The required stabiliser is therefore $1 + ((\mathfrak{M}^{\zeta})_{12} \cap (\mathfrak{Q} + \mathfrak{M})_{12})$. The index of this is stabiliser in $1 + (\mathfrak{M}^{\zeta})_{12}$ is

$$(a_\beta((\mathfrak{M}^{\zeta})_{12}) : a_\beta(\mathfrak{M}_{12})) = (\zeta_1^{-1}a_\beta(\mathfrak{M}_{12}) : a_\beta(\mathfrak{M}_{12})),$$

which depends only on the automorphism $x \mapsto \zeta_1^{-1}x$ of $a_\beta(A_{12})$.

On the other hand, the index $(K : \boldsymbol{U}^m(\mathfrak{B})(K \cap K^\zeta))$ is equal to

$$(\mathfrak{h}_{21} : (\mathfrak{Q}^m)_{21} + \mathfrak{h}_{21}\zeta_1) = (a_\beta(\mathfrak{h}_{21}) : a_\beta(\mathfrak{h}_{21})\zeta_1).$$

This index only depends on the automorphism $y \mapsto y\zeta_1$ of $a_\beta(A_{21})$. However, this is dual to the automorphism $x \mapsto \zeta_1^{-1}x$ of $a_\beta(A_{12})$, so these two indices are equal. Therefore $1 + (\mathfrak{M}^\zeta)_{12}$ acts transitively on the characters of K which agree with ϑ on $\boldsymbol{U}^m(\mathfrak{B})(K \cap K^\zeta)$, as required.

Now we must prove (8.2.28). We have

$$\mathfrak{M} = \mathfrak{Q}^{r-m+1}\mathfrak{N}(\beta) + \mathfrak{J}^{[\frac{r+1}{2}]}(\beta, \mathfrak{A}),$$

where $r = -k_0(\beta, \mathfrak{A})$, and $2 \leq m < r$. We have to show that

$$\mathfrak{M}\mathfrak{H}^m(\beta, \mathfrak{A}) \subset \mathfrak{Q}\mathfrak{H}^m(\beta, \mathfrak{A}).$$

We proceed by induction along β. If β is minimal over F and $m \geq [\frac{n}{2}]+1$, the assertion amounts to $(\mathfrak{P}^{n-m+1} + \mathfrak{P}^{[\frac{n+1}{2}]})\mathfrak{P}^m \subset \mathfrak{P}^{m+1}$, which surely holds. If $m \leq [\frac{n}{2}]$, we have $\mathfrak{M} = \mathfrak{J}^{[\frac{n+1}{2}]}(\beta, \mathfrak{A}) = \mathfrak{P}^{[\frac{n+1}{2}]}$, and we have to show

$$\mathfrak{P}^{[\frac{n+1}{2}]}(\mathfrak{Q}^m + \mathfrak{P}^{[\frac{n}{2}]+1}) \subset \mathfrak{Q}^{m+1} + \mathfrak{P}^{[\frac{n}{2}]+2}.$$

This is immediate. In general, we choose a simple stratum $[\mathfrak{A}, n, r, \gamma]$ equivalent to $[\mathfrak{A}, n, r, \beta]$. We set $\mathfrak{Q}_\gamma = \mathfrak{P} \cap \mathrm{End}_{F[\gamma]}(V)$, $s = -k_0(\gamma, \mathfrak{A})$. Assume first that $m \geq [\frac{r}{2}] + 1$. Then $\mathfrak{H}^m(\beta, \mathfrak{A}) = \mathfrak{H}^m(\gamma, \mathfrak{A})$. (3.1.3) and (1.4.9) give $\mathfrak{Q}^{r-m+1}\mathfrak{N}(\beta) = \mathfrak{Q}_\gamma^{r-m+1} + \mathfrak{Q}_\gamma^{s-m+1}\mathfrak{N}(\gamma)$. Also, $\mathfrak{J}^{[\frac{r+1}{2}]}(\beta, \mathfrak{A})$ $= \mathfrak{Q}_\gamma^{[\frac{r+1}{2}]} + \mathfrak{J}^{[\frac{s+1}{2}]}(\gamma, \mathfrak{A})$. Therefore

$$\mathfrak{M} = \mathfrak{Q}_\gamma^{r-m+1} + \mathfrak{Q}^{s-m+1}\mathfrak{N}(\gamma) + \mathfrak{J}^{[\frac{s+1}{2}]}(\gamma, \mathfrak{A}).$$

(3.1.3), (3.1.9) give us $\mathfrak{Q}\mathfrak{H}^m(\beta, \mathfrak{A}) = \mathfrak{Q}_\gamma\mathfrak{H}^m(\gamma, \mathfrak{A})$, and the assertion follows immediately from induction. If, on the other hand, $m \leq [\frac{r}{2}]$, we have $\mathfrak{M} = \mathfrak{J}^{[\frac{r+1}{2}]}(\beta, \mathfrak{A}) = \mathfrak{J}^{[\frac{r+1}{2}]}(\gamma, \mathfrak{A})$, and the assertion follows as before.

We have now completed the proof of (8.2.26). ■

(8.2.25), (8.2.26) and (8.2.14) now give us a vector $v_5 \in \mathcal{V}^\vartheta$ such that

$$\int_{M_{j-1}} \pi(x)v_5\, dx \neq 0.$$

This contradicts the definition of j, so we have proved the theorem, for the parabolic subgroup P. To prove it for P^-, the same argument suffices, with the roles of P, P^-, and ζ, ζ^{-1} interchanged, and using (8.2.4) in place of (2.3.8). ■

(8.3) Jacquet module of a split type II: the level zero case

In this section, we complement the results of **(8.2)** by showing that if $(\pi, \mathcal{V})$ is an irreducible smooth representation of G which contains a split type (K, ϑ) of level $(x, 0)$, $x \geq 0$, then it has a nonzero Jacquet module relative to some proper parabolic subgroup of G.

So, let (K, ϑ) be such a split type. When $x > 0$, we use the notation of the definition (8.1.4). In particular, we have a simple stratum $[\mathfrak{A}, n, 0, \beta]$ in A, with $E = F[\beta]$, $B = \mathrm{End}_E(V)$, $\mathfrak{B} = \mathfrak{A} \cap B$, $e = e(\mathfrak{B}|\mathfrak{o}_E)$, a character $\theta \in \mathcal{C}(\mathfrak{A}, 0, \beta)$, hence the unique irreducible representation η of $J^1(\beta, \mathfrak{A})$ which contains θ, and an irreducible representation κ of $J(\beta, \mathfrak{A})$ which is a β-extension of η. In this setup, $K = J(\beta, \mathfrak{A})$, and $\vartheta = \kappa \otimes \sigma$, for some "inhomogeneous" cuspidal representation σ of the k_E-reductive group $J(\beta, \mathfrak{A})/J^1(\beta, \mathfrak{A})$ (inflated to $J(\beta, \mathfrak{A})$). We handle the case $x = 0$ simultaneously by setting $E = F$, $B = A$, θ, η, κ all trivial, so that $H^t(\beta, \mathfrak{A}) = J^t(\beta, \mathfrak{A}) = \boldsymbol{U}^t(\mathfrak{A})$, $t \geq 0$.

We will need the machinery developed in **(7.1)**, **(7.2)**. In particular, let P denote a parabolic subgroup of G which is subordinate to the stratum $[\mathfrak{A}, n, 0, \beta]$ (just to the order $\mathfrak{A}$ in the case $x = 0$), and minimal for this property. Since the order $\mathfrak{B}$ cannot be maximal in the present context, this means that P is a proper subgroup of G. As in (7.1.13), P is determined by an E-decomposition

$$V = \coprod_{i=1}^{e} V^{(i)}$$

of V. We use the associated notations $A^{(ij)}$, $B^{(ij)}$, $\mathfrak{A}^{(i)}$, $M, U, U^-, \dots$ of **(7.1)**. In this situation, we can form the representation κ_P of the group $H^1(\beta, \mathfrak{A})(J(\beta, \mathfrak{A}) \cap P)$ as in (7.2.12). The inflated representation σ of $J(\beta, \mathfrak{A})$ restricts irreducibly to this subgroup, giving us a representation $\vartheta_P = \kappa_P \otimes \sigma$, which is the natural representation of $H^1(\beta, \mathfrak{A})(J(\beta, \mathfrak{A}) \cap P)$ on the space of $J(\beta, \mathfrak{A}) \cap U$-fixed vectors in ϑ. We have the relation

$$\textbf{(8.3.1)} \qquad \vartheta = \mathrm{Ind}(\vartheta_P : (J \cap P)H^1, J),$$

as in (7.2.15), where we abbreviate $H^1(\beta, \mathfrak{A}) = H^1$, $J(\beta, \mathfrak{A}) = J$. We can also form

$$\vartheta_U = \vartheta_P \mid J \cap M.$$

This is irreducible, and decomposes

$$\textbf{(8.3.2)} \qquad \begin{aligned} \vartheta_U &= \vartheta^{(1)} \otimes \vartheta^{(2)} \otimes \dots \otimes \vartheta^{(e)}, \\ \vartheta^{(i)} &= \kappa^{(i)} \otimes \sigma^{(i)}, \end{aligned}$$

where $\kappa^{(i)}$ is given by (7.2.14), and $\sigma^{(i)}$ is the inflation of an irreducible cuspidal representation σ_i of $J(\beta, \mathfrak{A}^{(i)})/J^1(\beta, \mathfrak{A}^{(i)}) = \boldsymbol{U}(\mathfrak{B}^{(i)})/\boldsymbol{U}^1(\mathfrak{B}^{(i)}) \cong GL(t_i, \mathrm{k}_E)$, for integers $t_i \geq 1$. In our original setup, J/J^1 is the product of the $J(\beta, \mathfrak{A}^{(i)})/J^1(\beta, \mathfrak{A}^{(i)})$, and σ is inflated from $\sigma_1 \otimes \ldots \otimes \sigma_e$.

We observe in passing that $(J(\beta, \mathfrak{A}^{(i)}), \vartheta^{(i)})$ is a maximal simple type in $G^{(i)} = \mathrm{Aut}_F(V^{(i)})$.

(8.3.3) Theorem: *Let (K, ϑ) be a split type in G of level $(x, 0)$, $x \geq 0$. Use the notation above. Let $(\pi, \mathcal{V})$ be an irreducible smooth representation of G which contains ϑ. Then the Jacquet module $(\pi_U, \mathcal{V}_U)$ is nonzero, and contains the representation ϑ_U of $J(\beta, \mathfrak{A}) \cap M$.*

Proof: We start by stating, and deferring the proofs of, a couple of subsidiary results which facilitate manipulation of these split types. Via (8.3.2), our split type (K, ϑ) determines a vector of equivalence classes of maximal simple types

$$\mathbf{V}(K, \vartheta) = ((K^{(1)}, \vartheta^{(1)}), \ldots, (K^{(e)}, \vartheta^{(e)})).$$

In the notation of (8.3.2), $K^{(i)} = J(\beta, \mathfrak{A}^{(i)})$. Note that (K, ϑ) determines $\mathbf{V}(K, \vartheta)$ up to a cyclic permutation (i.e. up to the choice of the base lattice L_0 in the chain which defines $\mathfrak{B}$). We say that split types $(K, \vartheta), (K', \vartheta')$ are *associate* if the vectors $\mathbf{V}(K, \vartheta)$, $\mathbf{V}(K', \vartheta')$ differ only by a permutation of entries. Of course, this concept applies equally to simple types, but only in a rather trivial way.

(8.3.4) Proposition: *Let (K, ϑ), (K', ϑ') be associate split types in G, and let $(\pi, \mathcal{V})$ be an irreducible smooth representation of G. Then π contains ϑ if and only if it contains ϑ'.*

We prove this later. Next, with our split type (K, ϑ) as before, let $\mathfrak{B}'$ be a hereditary $\mathfrak{o}_E$-order in B with $\mathfrak{B}' \supset \mathfrak{B}$ (whence $\boldsymbol{U}(\mathfrak{B}') \supset \boldsymbol{U}(\mathfrak{B}) \supset \boldsymbol{U}^1(\mathfrak{B}) \supset \boldsymbol{U}^1(\mathfrak{B}')$). Let $\mathfrak{A}'$ be the unique hereditary $\mathfrak{o}_F$-order in A with $E^\times \subset \mathfrak{K}(\mathfrak{A}')$ and $\mathfrak{A}' \cap B = \mathfrak{B}'$. Thus we have a simple stratum $[\mathfrak{A}', n', 0, \beta]$, for some integer n'. Let $\theta' \in \mathcal{C}(\mathfrak{A}', 0, \beta)$ correspond to θ via (3.6.2). Let η' be the unique irreducible representation of $J^1(\beta, \mathfrak{A}')$ which contains θ'. Then, by (5.2.14), there exists a unique β-extension κ' of η' such that $\kappa' \mid \boldsymbol{U}(\mathfrak{B})J^1(\beta, \mathfrak{A}')$ and κ induce the same irreducible representation of $\boldsymbol{U}(\mathfrak{B})\boldsymbol{U}^1(\mathfrak{A})$. A *cover of* (K, ϑ) is then a representation ϑ' of $J(\beta, \mathfrak{A}')$ of the form

$$\vartheta' = \kappa' \otimes \sigma',$$

where σ' is the inflation of an irreducible component of

$$\mathrm{Ind}(\sigma : \boldsymbol{U}(\mathfrak{B})/\boldsymbol{U}^1(\mathfrak{B}'), \boldsymbol{U}(\mathfrak{B}')/\boldsymbol{U}^1(\mathfrak{B}')).$$

Observe here that $\boldsymbol{U}(\mathfrak{B})/\boldsymbol{U}^1(\mathfrak{B}')$ is a parabolic subgroup of the reductive k_E-group $\boldsymbol{U}(\mathfrak{B}')/\boldsymbol{U}^1(\mathfrak{B}')$ with unipotent radical $\boldsymbol{U}^1(\mathfrak{B})/\boldsymbol{U}^1(\mathfrak{B}')$.

(8.3.5) Proposition: *Let (K, ϑ) be a split type as above, and $\mathfrak{B}'$ a hereditary $\mathfrak{o}_E$-order in B which contains $\mathfrak{B}$. Let $(\boldsymbol{\pi}, \mathcal{V})$ be an irreducible smooth representation of G. Then $\boldsymbol{\pi}$ contains ϑ if and only if it contains some representation ϑ' of $J(\beta, \mathfrak{A}')$ which is a cover of ϑ.*

Remark: The concept of "cover" applies equally, and (8.3.5) holds, when (K, ϑ) is a *simple* type.

We again defer the proof, and proceed to the theorem. Write

$$\mathcal{G}^{(i)} = \boldsymbol{U}(\mathfrak{B}^{(i)})/\boldsymbol{U}^1(\mathfrak{B}^{(i)}) = J(\beta, \mathfrak{A}^{(i)})/J^1(\beta, \mathfrak{A}^{(i)}),$$

so that $\mathcal{G}^{(i)} \cong GL(t_i, \mathrm{k}_E)$, for various integers t_i. The factor σ_i of σ is thus an irreducible cuspidal representation of $\mathcal{G}^{(i)}$. (8.3.4) allows us to assume that the σ_i "occur in blocks". That is to say, we have a set of pairwise inequivalent cuspidal representations $\tau_1, \ldots, \tau_r$, $2 \leq r \leq e$, and integers $0 = m_0 < m_1 < \ldots < m_r = e$ such that

$$\text{(8.3.6)} \qquad \sigma_i \cong \tau_j, \quad m_{j-1} + 1 \leq i \leq m_j, \ 1 \leq j \leq r.$$

We proceed by induction on the integer r. In this context, we can regard the result (7.3.2) on simple types as the special case $r = 1$ of (8.3.3). We may therefore appeal to this as the first step of this induction.

Now let us choose a maximal $\mathfrak{o}_E$-order $\mathfrak{B}_{\mathrm{M}}$ containing $\mathfrak{B}$. Thus

$$\boldsymbol{U}(\mathfrak{B}_{\mathrm{M}})/\boldsymbol{U}^1(\mathfrak{B}_{\mathrm{M}}) \cong GL(R, \mathrm{k}_E).$$

If we call this group $\mathcal{G}$, then $\mathcal{P} = \boldsymbol{U}(\mathfrak{B})/\boldsymbol{U}^1(\mathfrak{B}_{\mathrm{M}})$ is a proper parabolic subgroup of $\mathcal{G}$ with Levi factor $\mathcal{M} = \mathcal{G}^{(1)} \times \ldots \times \mathcal{G}^{(e)}$ and unipotent radical $\mathcal{U} = \boldsymbol{U}^1(\mathfrak{B})/\boldsymbol{U}^1(\mathfrak{B}_{\mathrm{M}})$. We choose a maximal proper parabolic subgroup $\mathcal{P}'$ of $\mathcal{G}$, with Levi decomposition $\mathcal{P}' = \mathcal{M}'\mathcal{U}'$, where $\mathcal{M}' \supset \mathcal{M}$ and $\mathcal{U}' \subset \mathcal{U}$, and moreover $\mathcal{M} \cap \mathcal{P}' = \mathcal{G}^{(1)} \times \ldots \times \mathcal{G}^{(m_1)}$. Informally, this just means that we put all the τ_1-components of σ on the first block of $\mathcal{M}'$.

Now let $\mathfrak{B}'$ be the unique hereditary $\mathfrak{o}_E$-order between $\mathfrak{B}$ and $\mathfrak{B}_{\mathrm{M}}$ such that $\mathcal{P}'$ is the image of $\boldsymbol{U}(\mathfrak{B}')$ in $\mathcal{G}$. Let $\mathfrak{A}'$ denote the hereditary $\mathfrak{o}_F$-order defined by the same lattice chain as $\mathfrak{B}'$. According to (8.3.5), our representation $\boldsymbol{\pi}$ contains a representation $\vartheta' = \kappa' \otimes \sigma'$ of $J(\beta, \mathfrak{A}')$ which is a cover of (K, ϑ). In particular, if we view σ as a representation of $\boldsymbol{U}(\mathfrak{B})$, then σ' is a component of the representation of $\boldsymbol{U}(\mathfrak{B}')$ induced by σ. We note a useful consequence of this construction:

(8.3.7) Lemma: *Let $\mathfrak{B}''$ be a hereditary $\mathfrak{o}_E$-order which contains $\mathfrak{B}'$. Then the representation $\mathrm{Ind}(\sigma' : \boldsymbol{U}(\mathfrak{B}'), \boldsymbol{U}(\mathfrak{B}''))$ is irreducible.*

Proof: This follows immediately from basic properties of cuspidal representations, **[Sp]**. ■

Let $\{L'_k\}$ be the lattice chain which defines $\mathfrak{B}'$. Thus $\{L'_k\}$ has $\mathfrak{o}_E$-period 2. We may think of the order $\mathfrak{B}_M$ above as $\mathrm{End}_{\mathfrak{o}_E}(L'_0)$. We then get a decomposition

$$L'_0 = L'_{01} \oplus L'_{02}$$

of L'_0 as $\mathfrak{o}_E$-module, such that

$$L'_1 = \pi_e L'_{01} \oplus L'_{02}$$

and the Levi factor $\mathcal{M}'$ above is the image of $\mathrm{Aut}_{\mathfrak{o}_E}(L'_{01}) \times \mathrm{Aut}_{\mathfrak{o}_E}(L'_{02})$. (Here, π_E denotes some prime element of E.) Let W_i denote the E-span of L'_{0i}, $i = 1, 2$, so that $V = W_1 \oplus W_2$. This decomposition gives rise to a pair of opposite parabolic subgroups P', P'^{-} of G as in **(7.1)**. Indeed, we have a simple stratum of the form $[\mathfrak{A}', n', 0, \beta]$, for some integer n'. The parabolic subgroup P' is then subordinate to $[\mathfrak{A}', n', 0, \beta]$, and minimal for this property. We use our standard notation $P' = M'U'$, where $M' = G_1 \times G_2$, $G_i = \mathrm{Aut}_F(W_i)$, and U' is the unipotent radical of P'.

We can therefore form the representation $\vartheta'_{P'} = \kappa'_{P'} \otimes \sigma'$ of the group $H^1(\beta, \mathfrak{A}')(J(\beta, \mathfrak{A}') \cap P')$ and

$$\vartheta'_{U'} = \vartheta'_{P'} \mid (J(\beta, \mathfrak{A}') \cap M' = \kappa'_{U'} \otimes \sigma'.$$

As usual, $\vartheta'_{P'}$ is the natural representation of $H^1(\beta, \mathfrak{A}')(J(\beta, \mathfrak{A}') \cap P')$ on the space of $J(\beta, \mathfrak{A}') \cap U'$-fixed vectors in ϑ'. (7.2.15) implies the relation

(8.3.8) $$\vartheta' = \mathrm{Ind}(\vartheta'_{P'} : H^1(\beta, \mathfrak{A}')(J(\beta, \mathfrak{A}') \cap P'), J(\beta, \mathfrak{A}')).$$

We have a decomposition

$$\vartheta'_{U'} = \vartheta'_1 \otimes \vartheta'_2,$$

where ϑ'_i is a representation of $J(\beta, \mathfrak{A}') \cap G_i = J(\beta, \mathfrak{A}'_i)$ and $\mathfrak{A}'_i = \mathfrak{A}' \cap \mathrm{End}_F(W_i) = \mathrm{End}^0_{\mathfrak{o}_F}(\{\pi_E^k L'_{0i}\})$, in the notation of **(1.1)**. Moreover, ϑ'_1 is a cover of a simple type in G_1 whose underlying maximal simple type is $(J(\beta, \mathfrak{A}^{(1)}), \vartheta^{(1)})$, in our original notation after renumbering according to (8.3.6). On the other hand, ϑ'_2 is a cover of a split type, $\overline{\vartheta}'_2$, say, and

$$\mathbf{V}(\overline{\vartheta}'_2) = (\vartheta^{(m_1+1)}, \ldots, \vartheta^{(e)}),$$

again after we have renumbered the $\vartheta^{(i)}$ according to (8.3.6).

From now on, let us abbreviate

$$J'' = H^1(\beta, \mathfrak{A}')(J(\beta, \mathfrak{A}') \cap P'), \quad \vartheta'' = \vartheta'_{P'}.$$

The theorem now follows from induction, (8.3.5), (7.3.2) and:

(8.3.9) Lemma: *The quotient map $\mathcal{V} \to \mathcal{V}_{U'}$ induces an isomorphism between $\mathcal{V}^{\vartheta''}$ and the $\vartheta'_{U'}$-isotypic component of $\mathcal{V}_{U'}$.*

Proof: We shall use the techniques of (7.3.7), (7.4.5). We must therefore investigate the structure of the Hecke algebra $\mathcal{H}(G, \vartheta'')$.

We use the decomposition $V = W_1 \oplus W_2$ to impose a block matrix structure on A, as usual. We choose a prime element π_E of E as above, and form the element

$$\zeta = \begin{pmatrix} \pi_E & 0 \\ 0 & 1 \end{pmatrix}.$$

This element clearly normalises the representation $\vartheta'_{U'}$ and, as in (7.2.19), it intertwines ϑ''. Moreover, the intertwining space $I_\zeta(\vartheta'')$ has dimension one.

(8.3.10) Lemma: *Let ϕ be some nonzero element of $\mathcal{H}(G, \vartheta'')$ with support $J''\zeta J''$. Then ϕ is an invertible element of $\mathcal{H}(G, \vartheta'')$, and, for $k \in \mathbb{Z}$, the k-fold convolution ϕ^k of ϕ has support $J''\zeta^k J''$.*

Proof: Let $\psi \in \mathcal{H}(G, \vartheta'')$ be nonzero with support $J''\zeta^{-1}J''$. We compute $\phi * \psi$. The support of $\phi * \psi$ is contained in $J''\zeta J''\zeta^{-1}J''$. The Iwahori decomposition of J'' (relative to P' and its opposite $P'^- \supset P^-$) gives

$$\zeta J''\zeta^{-1} = ({}^\zeta J'' \cap U')(J'' \cap M')({}^\zeta J'' \cap U'^-) \subset J''({}^\zeta J'' \cap U'^-).$$

Thus the support of $\phi * \psi$ is contained in $J''({}^\zeta J'' \cap U'^-)J''$.

Let $x \in {}^\zeta J'' \cap U'^-$ lie in the support of $\phi * \psi$. Then x intertwines ϑ''. From the relation (8.3.8), x also intertwines ϑ'. By (5.3.2), there exists $y \in B^\times$ which intertwines σ' and such that $x \in J'yJ'$, where we abbreviate $J(\beta, \mathfrak{A}') = J'$.

Now we write $U'^- = 1 + \mathbb{N}^-$. We show there exists a maximal $\mathfrak{o}_E$-order $\mathfrak{B}_{\mathrm{M}}$ in B such that $\mathfrak{B}_{\mathrm{M}} \supset \mathfrak{B}'$, and $\mathfrak{B}' \cap \mathbb{N}^- \subset \pi_E\mathfrak{B}_{\mathrm{M}} = \mathrm{rad}(\mathfrak{B}_{\mathrm{M}})$. It is easiest to use pictures here. We choose an E-basis of V which is the union of $\mathfrak{o}_E$-bases of the lattices L'_{0j}, and use this to identify B with $GL(R, E)$. The order $\mathfrak{B}'$ is then of the form

$$\mathfrak{B}' = \begin{pmatrix} \mathfrak{o}_E & \mathfrak{p}_E \\ \mathfrak{o}_E & \mathfrak{o}_E \end{pmatrix},$$

for various sizes of blocks. The group P' consists of all upper triangular block matrices with the same block structure. In this setup, the order we require is

$$\mathfrak{B}_{\mathrm{M}} = \begin{pmatrix} \mathfrak{o}_E & \mathfrak{p}_E \\ \mathfrak{p}_E^{-1} & \mathfrak{o}_E \end{pmatrix}.$$

Let $\mathfrak{A}_{\mathrm{M}}$ be the hereditary $\mathfrak{o}_F$-order in A defined by the same lattice chain as $\mathfrak{B}_{\mathrm{M}}$. We have $\mathfrak{A}' \cap \mathrm{N}^- \subset \pi_E \mathfrak{A}_{\mathrm{M}} = \mathrm{rad}(\mathfrak{A}_{\mathrm{M}})$. By construction, our element x lies in $1 + \pi_E^{-1}\mathfrak{H}^1(\beta, \mathfrak{A}') \cap \mathrm{N}^- \subset \boldsymbol{U}(\mathfrak{A}_{\mathrm{M}})$. Thus $y \in \boldsymbol{U}(\mathfrak{A}_{\mathrm{M}}) \cap B^\times = \boldsymbol{U}(\mathfrak{B}_{\mathrm{M}})$. The fact that, by definition, y intertwines σ' implies, via (8.3.7), that $y \in \boldsymbol{U}(\mathfrak{B}')$. Therefore $x \in J' \cap U'^-$.

The restriction of ϑ'' to $J^1(\beta, \mathfrak{A}') \cap J''$ is a multiple of the representation $\kappa'_{P'} \mid J^1(\beta, \mathfrak{A}') \cap J''$, and therefore the element x must also intertwine this representation. (7.2.4)*(iv)* now implies $x \in H^1(\beta, \mathfrak{A}') \cap U'^- \subset J''$. We deduce that the support of $\phi * \psi$ is contained in J''.

We now compute:

$$\phi * \psi(1) = \int_G \phi(x)\psi(x^{-1})\, dx.$$

We write this integral as a sum, in the usual way, to get

$$\phi * \psi(1) = \mu(J'' \cap J''^{\zeta}) \sum_x \phi(\zeta x)\psi(x^{-1}\zeta^{-1}),$$

where x ranges over a set of coset representatives for $(J'' \cap J''^{\zeta})\backslash J''$. We can choose these representatives to lie in the group $J'' \cap U'^-$, on which the representation ϑ'' is trivial. We deduce that $\phi * \psi(1) = \mu(J'')\phi(\zeta)\psi(\zeta^{-1})$, which is an automorphism of the representation space. Since this value is an intertwining operator for ϑ'' supported at 1, it is therefore a multiple of the identity.

We now treat $\psi * \phi$ similarly, and conclude that ϕ is an invertible element of $\mathcal{H}(G, \vartheta'')$, with inverse $c\psi$, for some constant c.

Finally, we observe that for $k \in \mathbb{Z}$, $k \geq 0$, we have $J''\zeta^k J''\zeta J'' = J''\zeta^{k+1}J''$, $J''\zeta^{-k}J''\zeta^{-1}J'' = J''\zeta^{-(k+1)}J''$, from the Iwahori decompositions. The last assertion of the lemma is now given by a simple computation. ∎

The hypotheses of (7.4.5) now apply, where D is now the group generated by ζ. We deduce

(8.3.11) *For $k \in \mathbb{Z}$, the operator $\pi(e_{\vartheta''}) \circ \pi(\zeta^k) \circ \pi(e_{\vartheta''})$ is an automorphism of the space $\mathcal{V}^{\vartheta''}$.*

We can now apply (7.3.7), and this proves (8.3.9). ∎

This leaves us only with the task of proving (8.3.4), (8.3.5). We start with (8.3.5). Assume first that π contains ϑ, and let $\mathfrak{B}'$ be a hereditary $\mathfrak{o}_E$-order in B which contains $\mathfrak{B}$. Let $\mathfrak{A}'$ be the associated $\mathfrak{o}_F$-order in A. Let $\theta' \in \mathcal{C}(\mathfrak{A}', 0, \beta)$ correspond to θ, let η' be the irreducible representation of $J^1(\beta, \mathfrak{A}')$ containing θ'. (5.2.14) gives a unique β-extension κ' of η' such that $\kappa' \mid \boldsymbol{U}(\mathfrak{B})J^1(\beta, \mathfrak{A}')$ and κ induce the same (irreducible) representation of $\boldsymbol{U}(\mathfrak{B})\boldsymbol{U}^1(\mathfrak{A})$. In our representation $\vartheta = \kappa \otimes \sigma$ of $K = J(\beta, \mathfrak{A})$, the factor σ extends to a representation of $\boldsymbol{U}(\mathfrak{B})\boldsymbol{U}^1(\mathfrak{A})$ which is null on $\boldsymbol{U}^1(\mathfrak{A})$. The representation $\mathrm{Ind}(\vartheta : J(\beta, \mathfrak{A}), \boldsymbol{U}(\mathfrak{B})\boldsymbol{U}^1(\mathfrak{A})) = \mathrm{Ind}(\kappa) \otimes \sigma$ is still irreducible, by (5.3.2), and it is therefore contained in π. It follows that π contains the representation

$$\nu = (\kappa' \mid \boldsymbol{U}(\mathfrak{B})J^1(\beta, \mathfrak{A}')) \otimes (\sigma \mid \boldsymbol{U}(\mathfrak{B})J^1(\beta, \mathfrak{A}')).$$

Thus π contains some irreducible component of

$$\mathrm{Ind}(\nu : \boldsymbol{U}(\mathfrak{B})J^1(\beta, \mathfrak{A}'), J(\beta, \mathfrak{A}')) = \kappa' \otimes \mathrm{Ind}(\sigma),$$

that is, by (5.3.2), a representation of the form $\kappa' \otimes \sigma'$, where σ' occurs in $\mathrm{Ind}(\sigma : \boldsymbol{U}(\mathfrak{B}), \boldsymbol{U}(\mathfrak{B}'))$.

Now we prove the converse. Suppose we have a hereditary $\mathfrak{o}_E$-order $\mathfrak{B}'$ containing $\mathfrak{B}$ and defining a hereditary $\mathfrak{o}_F$-order $\mathfrak{A}'$ in A. Let ϑ' be a representation of $J(\beta, \mathfrak{A}')$ which covers ϑ and occurs in π. We write $\vartheta' = \kappa' \otimes \sigma'$, in the notation of the definition. We restrict this representation to $\boldsymbol{U}(\mathfrak{B})J^1(\beta, \mathfrak{A}')$. Here it takes the form

$$\vartheta' \mid \boldsymbol{U}(\mathfrak{B})J^1(\beta, \mathfrak{A}') = \kappa'' \otimes \sum_{i=1}^{s} \tau_i,$$

where $\kappa'' = \kappa' \mid \boldsymbol{U}(\mathfrak{B})J^1(\beta, \mathfrak{A}')$ and τ_i ranges over the inflations of the irreducible components of $\sigma' \mid \boldsymbol{U}(\mathfrak{B})/\boldsymbol{U}^1(\mathfrak{B}')$. Each of the representations $\kappa'' \otimes \tau_i$ occurs in π, and σ is among the τ_i. Thus π contains the irreducible representation

$$\mathrm{Ind}(\kappa'' \otimes \sigma : \boldsymbol{U}(\mathfrak{B})J^1(\beta, \mathfrak{A}'), \boldsymbol{U}(\mathfrak{B})\boldsymbol{U}^1(\mathfrak{A})) = \mathrm{Ind}(\kappa'') \otimes \sigma,$$

and also the restriction of this to $J(\beta, \mathfrak{A})$. This restriction contains ϑ, as required. ■

(Observe that, in line with our earlier claim, this proof also holds when (K, ϑ) is a simple type.)

Now we can prove (8.3.4). Suppose that (K,ϑ) occurs in π and put $\mathbf{V}(K,\vartheta) = ((K^{(i)},\vartheta^{(i)}), 1 \le i \le e)$. It is enough to prove that π contains any associate (K',ϑ') of (K,ϑ) such that, for some j,

$$(K'^{(i)},\vartheta'^{(i)}) = (K^{(i)},\vartheta^{(i)}), \qquad i \neq j, j+1,$$
$$(K'^{(j)},\vartheta'^{(j)}) = (K^{(j+1)},\vartheta^{(j+1)}),$$

with $(K^{(j)},\vartheta^{(j)})$, $(K^{(j+1)}\vartheta^{(j+1)})$ not equivalent. To do this, we write $\mathfrak{B} = \mathrm{End}^0_{\mathfrak{o}_E}(\{L_k\})$ as before. We define an $\mathfrak{o}_E$-lattice chain $\{L'_k\}$ in V, of period $e-1$, by

$$L'_i = \begin{cases} L_i, & 0 \le i \le j-1, \\ L_{i+1}, & j \le i \le e-1. \end{cases}$$

Put $\mathfrak{B}' = \mathrm{End}^0_{\mathfrak{o}_E}(\{L'_k\})$, $\mathfrak{A}' = \mathrm{End}^0_{\mathfrak{o}_F}(\{L'_k\})$. There is then a unique representation ϑ'' of $J(\beta,\mathfrak{A}')$ which is a cover of ϑ, since in these circumstances, σ induces irreducibly to $\boldsymbol{U}(\mathfrak{B}')$. This representation also covers ϑ', so the proposition follows from (8.3.5). ■

(8.4) The main theorems

We have now formally completed the proofs of several results already anticipated. It seems worthwhile to state them fully as established fact.

(8.4.1) Theorem: *Let π be an irreducible supercuspidal representation of $G = \mathrm{Aut}_F(V) \cong GL(N,F)$. There exists a simple type (J,λ) in G such that $\pi \mid J$ contains λ. Further,*

(i) the simple type (J,λ) is uniquely determined up to G-conjugacy;

(ii) if (J,λ) is given by a simple stratum $[\mathfrak{A},n,0,\beta]$ in $A = \mathrm{End}_F(V)$, with $E = F[\beta]$, there is a uniquely determined representation Λ of $E^\times J$ such that $\Lambda \mid J = \lambda$ and $\pi = \mathrm{Ind}(\Lambda)$.

(iii) If (J,λ) is of the form (5.5.10)(b), i.e. if $J = \boldsymbol{U}(\mathfrak{A})$ for some principal $\mathfrak{o}_F$-order $\mathfrak{A}$ and λ is trivial on $\boldsymbol{U}^1(\mathfrak{A})$, then there is a uniquely determined representation Λ of $F^\times \boldsymbol{U}(\mathfrak{A})$ such that $\Lambda \mid \boldsymbol{U}(\mathfrak{A}) = \lambda$ and $\pi = \mathrm{Ind}(\Lambda)$.

Proof: If π is supercuspidal, then all of its Jacquet modules are trivial. (8.2.5) and (8.3.3) now imply that π cannot contain a split type. Therefore, by (8.1.5), π must contain a simple type. Part *(ii)* is thus given by (6.2.2), and part *(i)* by (6.2.4). ■

Now let ρ, ρ' be irreducible supercuspidal representations of $G_0 = GL(R,F)$, for some $R \ge 1$. We say that ρ, ρ' are *inertially equivalent* if there exists an unramified quasicharacter χ of $F^\times$ such that

$$\rho' \cong \rho \otimes \chi \circ \det,$$

where det is the determinant map $G_0 \to F^\times$. We write $\mathfrak{I}(\rho)$ for the inertial equivalence class of ρ. From (6.2.3), we get:—

(8.4.2) Theorem: *Let π_1, π_2 be irreducible supercuspidal representations of G. The following are equivalent:*

(i) $\mathfrak{I}(\pi_1) = \mathfrak{I}(\pi_2)$;

(ii) π_1, π_2 contain the same simple type in G.

This completes the results of §7:—

(8.4.3) Theorem: *Let π be an irreducible smooth representation of the group $G \cong GL(N, F)$. The following are equivalent:*

(i) there exists a simple type (J, λ) in G such that $\pi \mid J$ contains λ;

(ii) there exists a positive divisor R of N and an irreducible supercuspidal representation ρ of $GL(R, F)$ such that the supercuspidal support of π consists of elements of $\mathfrak{I}(\rho)$.

If these conditions hold, the representation ρ contains the maximal simple type associated to (J, λ), and any other simple type contained in π is conjugate to (J, λ).

In other words, the representation π contains the simple type (J, λ) if and only if π embeds in (equivalently is a subquotient of) $\rho_1 \times \rho_2 \times \ldots \times \rho_{N/R}$, for various $\rho_i \in \mathfrak{I}(\rho)$, where ρ is any supercuspidal representation of $GL(R, F)$ which contains the maximal simple type associated to (J, λ). The problem of decomposing this induced representation $\rho_1 \times \rho_2 \times \ldots \times \rho_{N/R}$ is then reduced to the case where $R = 1$ and ρ is an unramified quasicharacter, by (7.6.21).

(8.5) Classification

We are now in a position to give a complete classification of the irreducible smooth representations of our group $G = GL(N, F)$. It turns out that all that is required, beyond the now complete results of §7 and the above analysis of split types, is Casselman's theorem **[Cs]**(6.3.5) which describes the composition structure of the Jacquet modules of induced representations. Casselman's theorem is valid for arbitrary reductive F-groups so, altogether, we get a classification theory which is independent of the methods of **[Z]**. Of course, the results are completely compatible with the Zelevinsky classification.

We introduce some terminology. Write $\mathrm{Irr}(G)$ for the set of equivalence classes of irreducible smooth representations of G. If $\pi \in \mathrm{Irr}(G)$ (standard abuse of notation), $\mathrm{supp}(\pi)$ denotes the supercuspidal support of π (see (7.3.11) *et seq.*). This is a "multiset" or "set with multiplicity": when we wish to refer to the underlying set, we write $\{\mathrm{supp}(\pi)\}$. We also write $\mathfrak{I}$-$\mathrm{supp}(\pi)$ for the multiset of inertial equivalence classes

$\mathfrak{I}(\pi')$, $\pi' \in \mathrm{supp}(\pi)$, together with the associated notation $\{\mathfrak{I}\text{-supp}(\pi)\}$ for the underlying set.

We review the situation. We can divide the set $\mathrm{Irr}(G)$ into two disjoint subsets

$$\mathrm{Irr}(G) = \mathrm{Irr}^{(t)}(G) \,\dot\cup\, \mathrm{Irr}^{(a)}(G).$$

The subset $\mathrm{Irr}^{(t)}(G)$ consists of equivalence classes of "typical" representations of G, i.e. those which contain some simple type in G. The elements of the complement $\mathrm{Irr}^{(a)}(G)$ are called "atypical". Invoking (7.3.17), we can further dissect the set of (equivalence classes of) typical representations:—

$$\mathrm{Irr}^{(t)}(G) = \dot{\bigcup_{(J,\lambda)}} \mathrm{Irr}_\lambda(G),$$

where (J, λ) runs over conjugacy classes of simple types in G, and $\mathrm{Irr}_\lambda(G)$ is the set of equivalence classes of representations π which contain λ. We can equally well describe this set $\mathrm{Irr}^{(t)}(G)$ in terms of supercuspidal support. A representation $\pi \in \mathrm{Irr}(G)$ is typical if and only if $\#\{\mathfrak{I}\text{-supp}(\pi)\} = 1$. We can also, given $\pi \in \mathrm{Irr}^{(t)}(G)$ reconstruct the simple type contained in π from the inertial equivalence class of supporting supercuspidals, via (8.4.1) and (7.2.17).

So, we are left with the atypical representations π, characterised by the property $\#\{\mathfrak{I}\text{-supp}(\pi)\} \geq 2$. The situation is remarkably straightforward (as one could predict from **[Z]**).

(8.5.1) Theorem: *(i) For $1 \leq i \leq r$, let $N_i \geq 1$ be an integer, $G_i = GL(N_i, F)$, and let $\pi_i \in \mathrm{Irr}^{(t)}(G_i)$. Suppose that $\{\mathfrak{I}\text{-supp}(\pi_i)\} \neq \{\mathfrak{I}\text{-supp}(\pi_j)\}$ when $i \neq j$. Put $N = \sum_i N_i$. The representation $\pi_1 \times \pi_2 \times \ldots \times \pi_r$ of $G = GL(N, F)$ is irreducible.*

(ii) Let N_i, π_i be as in (i). Suppose we are also given positive integers M_j, with $H_j = GL(M_j, F)$, $1 \leq j \leq s$, and representations $\sigma_j \in \mathrm{Irr}^{(t)}(H_j)$ with $\{\mathfrak{I}\text{-supp}(\sigma_j)\} \neq \{\mathfrak{I}\text{-supp}(\sigma_k)\}$ whenever $j \neq k$. Then

$$\pi_1 \times \pi_2 \times \ldots \times \pi_r \cong \sigma_1 \times \sigma_2 \times \ldots \times \sigma_s$$

if and only if $r = s$ and there is a permutation p of $\{1, 2, \ldots, r\}$ such that $\sigma_i \cong \pi_{p(i)}$ for $1 \leq i \leq r$.

(iii) Let $\pi \in \mathrm{Irr}(G)$. There exist integers $r \geq 1$, $N_i \geq 1$, $1 \leq i \leq r$, and representations $\pi_i \in \mathrm{Irr}^{(t)}(GL(N_i, F))$, such that $\sum_i N_i = N$ and

$$\pi \cong \pi_1 \times \pi_2 \times \ldots \times \pi_r.$$

Proof: This starts by dealing directly with a very special case.

(8.5.2) Proposition: *For $i = 1, 2$, let π_i be an irreducible supercuspidal representation of $G_i \cong GL(N_i, F)$, and suppose that $\mathfrak{I}(\pi_1) \neq \mathfrak{I}(\pi_2)$. Then $\pi_1 \times \pi_2 \cong \pi_2 \times \pi_1$, and this representation is irreducible.*

We defer the proof of this and show first how it implies the theorem. If τ_i is an irreducible supercuspidal representation of $G_i = GL(n_i, F)$, $1 \le i \le t$, then (8.5.2) implies that

$$\textbf{(8.5.3)}\quad \tau_1 \times \tau_2 \times \ldots \times \tau_t \cong \sigma_1^{(1)} \times \sigma_2^{(1)} \times \ldots \times \sigma_{r(1)}^{(1)} \times \sigma_1^{(2)} \times \ldots \times \sigma_{r(s)}^{(s)},$$

where the $\sigma_j^{(i)}$ are irreducible supercuspidal, $\mathfrak{I}(\sigma_j^{(i)}) = \mathfrak{I}(\sigma_k^{(i)})$ for $1 \le j, k \le r(i)$, $1 \le i \le s$, while $\mathfrak{I}(\sigma_j^{(i)}) \ne \mathfrak{I}(\sigma_h^{(k)})$ whenever $i \ne k$. Moreover, in this expression, we can arrange for the inertial equivalence classes $\mathfrak{I}(\sigma_j^{(i)})$ to appear in any preassigned order. Explicitly,

$$\textbf{(8.5.4)}\quad \tau_1 \times \tau_2 \times \ldots \times \tau_t \cong \sigma_1^{(p(1))} \times \ldots \times \sigma_{r(p(1))}^{(p(1))} \times \sigma_1^{(p(2))} \times \ldots \times \sigma_{r(p(s))}^{(p(s))},$$

for any permutation p of $\{1, 2, \ldots, s\}$.

Now suppose we are given $\pi \in \mathrm{Irr}^{(a)}(G)$. We can find irreducible supercuspidal representations $\sigma_j^{(i)}$ such that π embeds in the representation (8.5.3), which we henceforward call Σ. We examine the composition factors of Σ. Suppose that $\sigma_j^{(i)}$ is a representation of $G_i = GL(m_i, F)$, and put $N_i = r(i)m_i$. Then $\sum_i N_i = N$, as π is a representation of $GL(N, F)$. Let $\boldsymbol{B}$ denote the standard Borel subgroup of G, say $\boldsymbol{B} = \boldsymbol{TN}$, where $\boldsymbol{T}$ is the diagonal split maximal torus, $\boldsymbol{N}$ the group of upper triangular unipotent matrices in G. Let P be the standard parabolic subgroup (i.e. $P \supset \boldsymbol{B}$) of G with diagonal block sizes $m_1, m_1, \ldots, m_1, m_2, \ldots, m_s$, where m_i occurs $r(i)$ times. Let $P = MU$ be the standard Levi decomposition of P, so that $M \supset \boldsymbol{T}$, $U \subset \boldsymbol{N}$. Thus

$$\sigma = \bigotimes \sigma_j^{(i)}$$

is a representation of P or P/U (or M). Likewise, let $P' = M'U'$ be the standard parabolic whose block sizes are $N_1, N_2, \ldots N_s$. As in (7.6.8), δ denotes the usual module character, while $\boldsymbol{W}$ is the Weyl group of G relative to the torus $\boldsymbol{T}$. We put $\boldsymbol{W}(P) = \boldsymbol{W} \cap P$, and so on.

We write $M' = \prod G_i'$, $G_i' = GL(N_i, F)$, and

$$\Sigma^{(i)} = \sigma_1^{(i)} \times \ldots \times \sigma_{r(i)}^{(i)} = \mathrm{Ind}((\sigma_1^{(i)} \otimes \ldots \otimes \sigma_{r(i)}^{(i)})\delta_P^{1/2}\delta_{P'}^{-1/2} : P, P')$$

(although we sometimes think of the induction as going from $P \cap M'$ to M'). Thus

$$\Sigma = \Sigma^{(1)} \times \ldots \times \Sigma^{(s)}.$$

(8.5.5) Lemma: *Let π' be a composition factor of Σ. Then $\pi_{U'}$ has a composition factor equivalent to $(\pi^{(1)} \otimes \ldots \otimes \pi^{(s)})\delta_{P'}^{1/2}$, where $\pi^{(i)}$ is a composition factor of $\Sigma^{(i)}$, $1 \le i \le s$.*

Proof: To revert to the notation of (8.5.3), we can embed $\boldsymbol{\pi}'$ in a representation $\boldsymbol{\tau}_{q(1)} \times \ldots \times \boldsymbol{\tau}_{q(t)}$, for some permutation q of $\{1, 2, \ldots, t\}$ (see (7.3.11)). Applying (8.5.2) to this representation, we have $\boldsymbol{\pi}'$ embedded in $\boldsymbol{\Sigma}'^{(1)} \times \ldots \times \boldsymbol{\Sigma}'^{(s)}$, where

$$\boldsymbol{\Sigma}'^{(i)} = \boldsymbol{\sigma}^{(i)}_{p_i(1)} \times \ldots \times \boldsymbol{\sigma}^{(i)}_{p_i(r(i))},$$

for a permutation p_i of $\{1, 2, \ldots, r(i)\}$, $1 \le i \le s$. By Frobenius reciprocity, we have a nontrivial homomorphism

$$\boldsymbol{\pi}'_{U'} \to (\boldsymbol{\Sigma}'^{(1)} \otimes \ldots \otimes \boldsymbol{\Sigma}'^{(s)})\delta_{P'}^{1/2}.$$

The composition factors of $\boldsymbol{\Sigma}'^{(1)} \otimes \ldots \otimes \boldsymbol{\Sigma}'^{(s)}$ are of the form $\boldsymbol{\pi}^{(1)} \otimes \ldots \otimes \boldsymbol{\pi}^{(s)}$, where $\boldsymbol{\pi}^{(j)}$ is a composition factor of $\boldsymbol{\Sigma}'^{(j)}$, and hence of $\boldsymbol{\Sigma}^{(j)}$. ■

Now let us consider the composition length $\ell = \ell(\boldsymbol{\Sigma})$ of $\boldsymbol{\Sigma}$. By (8.5.5), we have

$$\ell(\boldsymbol{\Sigma}) \le m(\boldsymbol{\Sigma}),$$

where $m(\boldsymbol{\Sigma})$ = *the number of composition factors of* $\boldsymbol{\Sigma}_{U'}$ *of the form* $\boldsymbol{\pi}^{(1)} \otimes \ldots \otimes \boldsymbol{\pi}^{(s)}$, *such that* $\{\mathfrak{J}\text{-supp}(\boldsymbol{\pi}^{(i)})\} = \mathfrak{J}(\boldsymbol{\sigma}^{(i)}_j)$, $1 \le i \le s$.

We use **[Cs]**(6.3.5) to find the multiset of composition factors of $\boldsymbol{\Sigma}_{U'}$. Since we shall use this technique repeatedly, it is worth exhibiting it formally:

(8.5.6) *The multiset of composition factors of the representation* $\boldsymbol{\Sigma}_{U'}$ *is the union of the multisets of composition factors of the representations*

$$\Phi_{\boldsymbol{w}} = \mathrm{Ind}^u(\boldsymbol{\sigma}^{\boldsymbol{w}} : P^{\boldsymbol{w}} \cap M', M').\delta_{P'}^{1/2},$$

where $\boldsymbol{w}$ *ranges over* $\boldsymbol{W}(P)\backslash\boldsymbol{W}/\boldsymbol{W}(P')$, *subject to the condition* $M^{\boldsymbol{w}} \subset M'$.

Take a typical composition factor $\varphi_{\boldsymbol{w}}$ of this space $\Phi_{\boldsymbol{w}}$, say $\varphi_{\boldsymbol{w}} = \varphi_{\boldsymbol{w},1} \otimes \ldots \otimes \varphi_{\boldsymbol{w},s}$, and consider $\{\mathfrak{J}\text{-supp}(\varphi_{\boldsymbol{w},j})\}$. There will exist a j such that this set contains a class $\mathfrak{J}(\boldsymbol{\sigma}^{(k)}_l)$ with $k \ne j$, unless $\boldsymbol{w} \in \boldsymbol{W}(P')$. We deduce that $m(\boldsymbol{\Sigma})$ equals the composition length of

$$\mathrm{Ind}^u(\boldsymbol{\sigma} : P \cap M', M')\delta_{P'}^{1/2} = (\boldsymbol{\Sigma}^{(1)} \otimes \ldots \otimes \boldsymbol{\Sigma}^{(s)})\delta_{P'}^{1/2}.$$

Therefore

$$\ell(\boldsymbol{\Sigma}) \le m(\boldsymbol{\Sigma}) = \ell(\boldsymbol{\Sigma}^{(1)} \otimes \ldots \otimes \boldsymbol{\Sigma}^{(s)}) \le \ell(\boldsymbol{\Sigma}).$$

Thus we have equality throughout. In particular,

$$\ell(\boldsymbol{\Sigma}^{(1)} \otimes \ldots \otimes \boldsymbol{\Sigma}^{(s)}) = \prod \ell(\boldsymbol{\Sigma}^{(i)}) = \ell(\boldsymbol{\Sigma}).$$

It follows that, for any composition factor $\boldsymbol{\pi}^{(i)}$ of $\boldsymbol{\Sigma}^{(i)}$, the representation $\boldsymbol{\pi}^{(1)} \times \ldots \times \boldsymbol{\pi}^{(s)}$ is irreducible. This proves *(i)* and *(iii)* of the theorem. Moreover, this same counting argument shows that, given a composition factor $\boldsymbol{\pi}$ of $\boldsymbol{\Sigma}$, the representation $\boldsymbol{\pi}_{U'}$ has a unique factor of the form

$$(\boldsymbol{\pi}^{(1)} \otimes \ldots \otimes \boldsymbol{\pi}^{(s)})\delta_{P'}^{1/2},$$

and part *(ii)* follows immediately. ■

Now we have to prove (8.5.2). It is enough to show that $\boldsymbol{\pi}_1 \times \boldsymbol{\pi}_2$ is irreducible, since $\boldsymbol{\pi}_1 \times \boldsymbol{\pi}_2$ and $\boldsymbol{\pi}_2 \times \boldsymbol{\pi}_1$ have the same composition factors. The simplest estimates (the technique of (8.5.6) used above is more than adequate) show that $\boldsymbol{\pi}_1 \times \boldsymbol{\pi}_2$ has length at most 2. Assume, for a contradiction, that this length is 2, and let $\boldsymbol{\sigma}_1, \boldsymbol{\sigma}_2$ be the composition factors.

Put $N = N_1 + N_2$, $G = GL(N, F)$. Let P be the standard (upper triangular) parabolic subgroup of G with diagonal block sizes N_1, N_2, and let $P = MU$ its obvious Levi decomposition. If P' is any other proper parabolic subgroup of G with unipotent radical U', then the representation $(\boldsymbol{\pi}_1 \times \boldsymbol{\pi}_2)_{U'}$ is null unless P' is conjugate to P or its opposite (i.e. transpose) P^-: this again follows from (8.5.6).

Take first the case where $N_1 \neq N_2$, and consider $(\boldsymbol{\pi}_1 \times \boldsymbol{\pi}_2)_U$. By **[Cs]**(6.3.5) again, this is irreducible and equivalent to $(\boldsymbol{\pi}_1 \otimes \boldsymbol{\pi}_2)\delta_P^{1/2}$. Thus we may assume $(\boldsymbol{\sigma}_1)_U = (\boldsymbol{\pi}_1 \times \boldsymbol{\pi}_2)_U$, while $(\boldsymbol{\sigma}_2)_U = 0$. Similar considerations apply to P^-, so we have $(\boldsymbol{\sigma}_1)_{U^-} = 0$ and $(\boldsymbol{\sigma}_2)_{U^-} = (\boldsymbol{\pi}_1 \times \boldsymbol{\pi}_2)_{U^-}$. The representation $\boldsymbol{\sigma}_1$ is atypical since $\#\{\mathfrak{I}\text{-supp}(\boldsymbol{\sigma}_1)\} = 2$ so, by (8.1.5), it contains a split type (K, ϑ). Let (x, y), $x \geq y \geq 0$, be the level of (K, ϑ). If $y > 0$, (8.2.5)*(iv)* shows that $(\boldsymbol{\sigma}_1)_{U^-} \neq 0$, which is a contradiction. If $y = 0$, we get the same conclusion from (8.3.4) and (8.3.3). Thus (8.5.2) holds when $N_1 \neq N_2$.

Now assume that $N_1 = N_2$, so that the groups P, P^- are conjugate. Thus P is the only proper parabolic subgroup of G (up to conjugacy) which gives $(\boldsymbol{\pi}_1 \times \boldsymbol{\pi}_2)_U$ nonzero. This Jacquet module has length two, with composition factors $(\boldsymbol{\pi}_1 \otimes \boldsymbol{\pi}_2)\delta_P^{1/2}$ and $(\boldsymbol{\pi}_2 \otimes \boldsymbol{\pi}_1)\delta_P^{1/2}$. Let us number the factors of $\boldsymbol{\pi}_1 \times \boldsymbol{\pi}_2$ so that $(\boldsymbol{\sigma}_1)_U = (\boldsymbol{\pi}_1 \otimes \boldsymbol{\pi}_2)\delta_P^{1/2}$. Again, $\boldsymbol{\sigma}_1$ contains a split type (K, ϑ), say, of level (x, y), $x \geq y \geq 0$. We first treat the case where $y = 0$. By (8.3.4), after a conjugation, we can assume that P is subordinate to the type (K, ϑ). We write $\vartheta_U = \vartheta_1 \otimes \vartheta_2$, where ϑ_i is a representation of $K \cap G_i$. We deduce that $\boldsymbol{\pi}_1$ contains ϑ_1. On the other hand, the conjugation which takes P^- to P takes ϑ_{U^-} to $\vartheta_2 \otimes \vartheta_1$. We deduce that $\boldsymbol{\pi}_1$ contains both ϑ_1 and ϑ_2. However, the pairs $(K \cap G_i, \vartheta_i)$ are nonconjugate simple types in G_1, both occurring in the supercuspidal representation $\boldsymbol{\pi}_1$. This contradicts (6.2.4), and gives the proposition in this case.

This leaves only the case where $N_1 = N_2$ and $\boldsymbol{\sigma}_1$ contains no split type of level $(x, 0)$. We can therefore assume that (K, ϑ) is defined by a simple stratum $[\mathfrak{A}, n, m, \beta]$, with $y = m/e(\mathfrak{A})$, $K = H^m(\beta, \mathfrak{A})$, $\vartheta = \theta\psi_c$ with $\theta \in \mathcal{C}(\mathfrak{A}, m-1, \beta)$, $c \in \mathfrak{P}^{-m}$, where $\mathfrak{P} = \text{rad}(\mathfrak{A})$. We use the other notations of **(8.2)**:— $E = F[\beta]$, $B =$ the A-centraliser of E, $\mathfrak{B} = \mathfrak{A} \cap B$, and so on. We can further assume that (K, ϑ) has been chosen to minimise the quantity $m/e(\mathfrak{A})$. Then:

(8.5.7) Lemma: *If $[\mathfrak{A}', n', m', \beta']$ is a simple stratum in the matrix ring $A = \mathbb{M}(N, F)$ and $\theta' \in \mathcal{C}(\mathfrak{A}', m', \beta')$ occurs in $\boldsymbol{\sigma}_1$, then $m'/e(\mathfrak{A}') \geq m/e(\mathfrak{A})$.*

Proof: For, suppose we have such a stratum $[\mathfrak{A}', n', m', \beta']$, and a character $\theta' \in \mathcal{C}(\mathfrak{A}', m', \beta')$ which occurs in $\boldsymbol{\pi}$, with the condition $m'/e(\mathfrak{A}') < m/e(\mathfrak{A})$. We choose such a stratum to minimise the non-negative quantity $m'/e(\mathfrak{A}')$. If we can achieve $m' = 0$ here, then the proof of (8.5.1) shows that $\boldsymbol{\pi}$ contains either a simple type or a split type of level $(x, 0)$, where $x \geq 0$. We have excluded both of these possibilities. Otherwise, the refinement process in the proof of (8.1.5) produces a split type in $\boldsymbol{\pi}$ of level (x', y'), with $y' \leq m'/e(\mathfrak{A}')$. This contradicts the choice of (K, ϑ). ∎

Returning to (K, ϑ) as above, we know that the stratum $[\mathfrak{B}, m, m-1, s(c)]$ is split fundamental, where s is a tame corestriction on A relative to E/F, and $\gcd(m, e(\mathfrak{B})) = 1$. We apply the procedure of **(8.2)** to get a splitting, and hence a parabolic subgroup P' of G relative to which the representation $\boldsymbol{\sigma}_1$ has a nontrivial Jacquet module. In present circumstances, P' must be conjugate to P. We can arrange that $P' = P$. We can also assume that the element c is split relative to the splitting defining P, as in (8.2.8), and that the "first component" of $[\mathfrak{B}, m, m-1, s(c)]$ is nonsplit fundamental, as in (8.2.3). We write $\vartheta \mid K \cap M$ in the form $\vartheta_1 \otimes \vartheta_2$. Arguing as in the last case, on the basis of (8.2.5)*(iv)*, we see that $\boldsymbol{\pi}_1$ and $\boldsymbol{\pi}_2$ both contain the character ϑ_1. By (8.2.7) and (8.1.15), we have

(8.5.8) *There exists a simple stratum $[\mathfrak{A}_1, n_1, m_1, \beta_1]$ in the algebra $A_1 = \mathbb{M}(N_1, F)$, and a character $\theta_1 \in \mathcal{C}(\mathfrak{A}_1, m_1, \beta_1)$ such that*

(i) $m_1/e(\mathfrak{A}_1) < m/e(\mathfrak{A})$, and

(ii) θ_1 occurs in both $\boldsymbol{\pi}_1$ and $\boldsymbol{\pi}_2$.

The character $\theta_1 \otimes \theta_1$ thus occurs in $(\boldsymbol{\sigma}_1)_U$.

We now construct an order $\overline{\mathfrak{A}}$ in the algebra $A = \mathbb{M}(N, F)$, with $N = N_1 + N_2 = 2N_1$ as above. Let $\mathcal{L}^1 = \{L_k^1 : k \in \mathbb{Z}\}$ be the lattice chain defining $\mathfrak{A}_1$. We define a new lattice chain $\mathcal{L} = \{L_k\}$ by

$$L_{2k} = L_k^1 \oplus L_k^1, \quad L_{2k+1} = L_k^1 \oplus L_{k+1}^1,$$

for $k \in \mathbb{Z}$. Let $\overline{\mathfrak{A}} = \mathrm{End}_{\mathfrak{o}_F}^0(\mathcal{L})$. Then $e(\overline{\mathfrak{A}}) = 2e(\mathfrak{A}_1)$. The stratum $[\overline{\mathfrak{A}}, 2n_1, 2m_1, \beta_1]$ is simple. Let $\overline{\theta} \in \mathcal{C}(\overline{\mathfrak{A}}, 2m_1, \beta_1)$ be the character corresponding to θ_1 via (3.6.14). Our original parabolic subgroup P is subordinate to $[\overline{\mathfrak{A}}, 2n_1, 2m_1, \beta_1]$, and $\overline{\theta} \mid H^{2m_1+1}(\beta_1, \overline{\mathfrak{A}}) \cap M = \theta_1 \otimes \theta_1$. Now we need another variation on part of (7.3.2).

(8.5.9) Lemma: *With the notation above, let $(\boldsymbol{\pi}, \mathcal{V})$ be an irreducible smooth representation of G. The quotient map $\mathcal{V} \to \mathcal{V}_U$ maps $\mathcal{V}^{\overline{\theta}}$ onto the $\theta_1 \otimes \theta_1$-isotypic component of $\mathcal{V}_U$.*

Proof: We take an Iwahori decomposition of $H^{2m_1+1}(\beta_1, \overline{\mathfrak{A}})$, as in **(7.1)**. Let π_1 be a prime element of the field $E_1 = F[\beta_1]$, and form the matrix

$$\varpi = \begin{pmatrix} \pi_1 & 0 \\ 0 & 1 \end{pmatrix}.$$

(Here, we are thinking of P as being in upper triangular form.) Then $\boldsymbol{\pi}_U(\varpi^k)$ acts as an automorphism of the $\theta_1 \otimes \theta_1$-isotypic subspace of $\mathcal{V}_U$, for any $k \in \mathbb{Z}$. The proof is now identical to (although much easier than) the surjectivity part of the proof of (7.3.2) (i.e. (7.3.5) *et seq.*). We therefore omit the details. ■

In particular, we see that our representation $\boldsymbol{\sigma}_1$ contains the character $\overline{\theta}$, contrary to (8.5.7).

This completes the proof of (8.5.2), and hence that of (8.5.1). ■

There is one remaining loose end to be tied.

(8.5.10) Proposition: *Let $(\boldsymbol{\pi}, \mathcal{V})$ be an irreducible smooth representation of $G = \mathrm{Aut}_F(V)$, and suppose that there exists a proper parabolic subgroup P of G, with unipotent radical U, together with an irreducible smooth representation $\boldsymbol{\sigma}$ of P/U such that $\boldsymbol{\pi} \cong \mathrm{Ind}(\boldsymbol{\sigma})$. Then $\boldsymbol{\pi}$ is not discrete series.*

(8.5.11) Corollary: *Any discrete series representation of G is typical, and hence described by* (7.7.1).

The corollary follows immediately from (8.5.1)*(iii)*. To prove the proposition, we just have to show that any representation of G, irreducible or not, induced from an irreducible representation of a proper parabolic subgroup and with unitary central quasicharacter, has a matrix coefficient which is not absolutely square-integrable mod centre.

To do this, let P be a proper parabolic subgroup of G, with unipotent radical U and Levi decomposition $P = MU$. Let $(\boldsymbol{\sigma}, \mathcal{W})$ be a smooth irreducible representation of M or P/U, and put

$$(\boldsymbol{\pi}, \mathcal{V}) = \mathrm{Ind}^u(\boldsymbol{\sigma} : P, G).$$

Then $\boldsymbol{\pi}$ has a central quasicharacter, which we may adjust to be unitary. We choose a compact open subgroup K of G, with an Iwahori decomposition relative to P and its opposite, and small enough to ensure that $\mathcal{W}$ has a nonzero element w fixed by $\boldsymbol{\sigma}(K \cap M)$. We define a function $\phi \in \mathcal{V}$ to have support PK and

$$\phi(pk) = \boldsymbol{\sigma}(p)w, \qquad p \in P,\ k \in K.$$

We can likewise choose a $(K \cap M)$-fixed vector $\check{w} \in \check{\mathcal{W}}$, with $\langle w, \check{w} \rangle_{\mathcal{W}}$ nonzero, to define a function $\psi \in \check{\mathcal{V}}$ supported on PK. We consider the coefficient function of $\boldsymbol{\pi}$ defined by the vectors ϕ, ψ. This is the function

$$\Phi : g \mapsto \int_{P\backslash G} \langle \phi(hg), \psi(h) \rangle \, d\ddot{h}$$

for a semi-invariant measure $d\ddot{h}$ on $P\backslash G$ (see **[S]** (1.7.8) for details). This function is constant on cosets of the group K. Let ω denote the central quasicharacter of $\boldsymbol{\sigma}$. It is easy to produce an element z of the centre of M satisfying

(i) the set $|\omega(z^n)|$ is bounded below, away from zero, as n tends to ∞;

(ii) the cosets $z^n F^\times .K$ are pairwise disjoint, for $n \geq 0$.

The coefficient Φ of $\boldsymbol{\pi}$ above is then not absolutely square-integrable mod $F^\times$ even when restricted to $\bigcup z^n F^\times K$. ■

REFERENCES

[BZ] **I. N. Bernstein & A. V. Zelevinsky,** *Induced representations of reductive p-adic groups.* Ann. Scient. Ec. Norm. Sup. (4) **10** (1977), 441–472.

[Bo] **A. Borel,** *Admissible representations of a semisimple group over a local field with vectors fixed under an Iwahori subgroup.* Invent. Math. **35** (1976), 233–259.

[BoCs] **A. Borel & W. Casselman (edd.),** *Automorphic forms, representations and L-functions.* Proc. Symposia in Pure Math. XXXIII parts 1 & 2, Amer. Math. Soc. (Providence RI), 1979.

[Br] **K. S. Brown,** *Cohomology of Groups.* Graduate Texts in Mathematics **87** (Springer, New York, 1982).

[Bu1] **C. J. Bushnell,** *Hereditary orders, Gauss sums and supercuspidal representations of GL_N.* J. reine angew. Math. **375/376** (1987), 184–210.

[Bu2] **C. J. Bushnell,** *Induced representations of locally profinite groups.* J. Alg. **134** (1990), 104–114.

[BF1] **C. J. Bushnell & A. Fröhlich,** *Gauss sums and p-adic division algebras.* Lecture Notes in Math. **987** (Springer, Berlin, 1983).

[BF2] **C. J. Bushnell & A. Fröhlich,** *Non-abelian congruence Gauss sums and p-adic simple algebras.* Proc. London Math. Soc. (3) **50** (1985), 207–264.

[BK] **C. J. Bushnell & P. C. Kutzko,** *The admissible dual of GL_N via restriction to compact open subgroups.* Harmonic analysis on reductive groups (W. Barker & P. Sally edd.). Progress in Math. **101**, Birkhauser (Boston 1991), 89–99.

[Ca] **H. Carayol,** *Représentations cuspidales du groupe linéaire.* Ann. Scient. Ec. Norm. Sup. (4) **17** (1984), 191–225.

[Cr] **P. Cartier,** *Representations of $\mathfrak{p}$-adic groups: a survey.* **[BoCs]**, pp. 111–156.

[Cs] **W. Casselman,** *Introduction to the theory of admissible representations of $\mathfrak{p}$-adic reductive groups.* Preprint.

[CMS] **L. Corwin, A. Moy & P. J. Sally Jr.,** *Degrees and formal degrees for division algebras and GL_n over a p-adic field.* Pacific J. Math. **141** (1990), 21–45.

[FM] **A. Fröhlich & A. W. McEvett,** *Forms over rings with involution.* J. Alg. **12** (1969), 79–104.

[GGP-S] **I. M. Gelfand, M. I. Graev & I. I. Pyatetskii-Shapiro,** *Representation theory and automorphic functions.* W. B. Saunders Company (Philadelphia 1969).

[GJ] **R. Godement & H. Jacquet,** *Zeta functions of simple algebras.* Lecture Notes in Math. **260** (Springer, Berlin 1972).

[HC] **Harish-Chandra,** *Harmonic analysis on reductive p-adic groups.* Harmonic Analysis on Homogeneous Spaces (C. C. Moore ed.), Proc. Symposia Pure Math. **XXVI**, American Math. Soc. (Providence 1973), 167–172.

[He] **G. Henniart,** *Représentations des groupes réductifs p-adiques.* Séminaire Bourbaki, 43ème année, 1990–91, n° 736.

[H1] **R. E. Howe,** *Tamely ramified supercuspidal representations of* GL_n. Pacific J. Math. **73** (1977), 437–460.

[H2] **R. E. Howe,** *Some qualitative results on the representation theory of* GL_n *over a p-adic field.* Pacific J. Math. **73** (1977), 479–538.

[HM1] **R. E. Howe & A. Moy,** *Harish-Chandra homomorphisms for p-adic groups.* CBMS Regional Conf. Series in Math. **59** (Amer. Math. Soc., Providence RI, 1985).

[HM2] **R. E. Howe & A. Moy,** *Minimal* K*-types for* $GL(n)$ *over a p-adic field.* Astérisque **171–172** (1989), 257–273.

[HM3] **R. E. Howe & A. Moy,** *Hecke algebra isomorphisms for* $GL(n)$ *over a p-adic field.* J. Alg. **131** (1990), 388–424.

[HL] **R. Howlett & G. Lehrer,** *Induced cuspidal representations and generalized Hecke rings.* Invent. Math. **58** (1980), 37–64.

[Iw] **N. Iwahori,** *Generalised Tits System (Bruhat Decomposition) on* $\mathfrak{p}$*-Adic Semisimple Groups.* Algebraic Groups and Discontinuous Subgroups (A. Borel & G. Mostow edd.), Proc. Symposia Pure Math. **IX** (American Mathematical Society, Providence, 1966), 71–83.

[IM] **N. Iwahori & H. Matsumoto,** *On some Bruhat decomposition and the structure of the Hecke rings of the* $\mathfrak{p}$*-adic Chevalley groups.* Publications Mathématiques de l'Institut des Hautes Études Scientifiques **25** (1965), 5–48.

[J1] **H. Jacquet,** *Représentations des groupes linéaires* $\mathfrak{p}$*-adiques.* Theory of group representations and harmonic analysis (CIME, II Ciclo, Montecatini Terme, 1970), (Edizioni Cremonese, Roma 1971), 119–220.

[J2] **H. Jacquet,** *Sur les représentations des groupes reductifs* $\mathfrak{p}$*-adiques.* C. R. Acad. Sci. Paris Sér. A **280** (1975), 1271–1272.

[JL] **H. Jacquet & R. P. Langlands,** *Automorphic forms on* $GL(2)$. Lecture Notes in Math. **114**, Springer (Berlin 1970).

[KL] **D. Kazhdan & G. Lusztig,** *Proof of the Deligne-Langlands conjecture for Hecke algebras.* Invent. Math. **87** (1987), 153–215.

[Ko] **H. Koch,** *Eisensteinsche Polynomfolgen und Arithmetik in Divisionsalgebren über lokalen Körpern.* Math. Nachr. **104** (1981), 239–251.

[K1] **P. C. Kutzko,** *On the supercuspidal representations of GL_2 II.* Amer. J. Math. **100** (1978), 705–716.

[K2] **P. C. Kutzko,** *On the restriction of supercuspidal representations to compact open subgroups.* Duke Math. J. **52** (1985), 753–764.

[K3] **P. C. Kutzko,** *On the supercuspidal representations of GL_n and other groups.* Proc. Internat. Congress of Mathematicians Berkeley 1986 (Amer. Math. Soc., Providence RI, 1987) 853–861.

[K4] **P. C. Kutzko,** *Towards a classification of the supercuspidal representations of GL_N.* J. London Math. Soc. (2) **37** (1988), 265–274.

[KM1] **P. C. Kutzko & D. C. Manderscheid,** *On intertwining operators for $GL_N(F)$, F a nonarchimedean local field.* Duke Math. J. **57** (1988), 275–293.

[KM2] **P. C. Kutzko & D. C. Manderscheid,** *On the supercuspidal representations of GL_N, N the product of two primes.* Ann. Scient. Ec. Norm. Sup. (4) **23** (1990), 39–88.

[KM2a] **P. C. Kutzko & D. C. Manderscheid,** *Correction to our paper, On the supercuspidal representations of GL_N, N the product of two primes.* Ann. Scient. Ec. Norm. Sup. (4) **24** (1991), 139-140.

[La1] **R. P. Langlands,** *Problems in the theory of automorphic forms. Lectures in modern analysis and applications III,* Lecture Notes in Math **170** (Springer, Berlin 1970), 18–86.

[La2] **R. P. Langlands,** *Base change for $GL(2)$.* Ann. Math. Studies **96**, Princeton University Press (Princeton, 1980).

[MW] **C. Moeglin & J.-L. Waldspurger,** *Sur l'involution de Zelévinski.* J. reine angew. Math. **372** (1986), 136–177.

[M1] **L. E. Morris,** *Some tamely ramified supercuspidal representations of symplectic groups.* Proc. London Math. Soc. (3) **63** (1991), 519–551.

[M2] **L. E. Morris,** *Tamely ramified supercuspidal representations of classical groups I: Filtrations.* Ann. Scient. Ec. Norm. Sup. (4) **24** (1991), 705–738.

[M3] **L. E. Morris,** *Tamely ramified supercuspidal representations of classical groups II: representation theory.* Ann. Scient. Ec. Norm. Sup. (to appear).

[M4] **L. E. Morris,** *Fundamental G-strata for p-adic classical groups.* Duke Math. J. **64** (1991), 501–553.

[Mo1] **A. Moy,** *Local constants and the tame Langlands correspondence.* Amer. J. Math. **108** (1986), 863–930.

[Mo2] **A. Moy,** *A conjecture on minimal K-types for GL_n over a p-adic field. Representation theory and number theory in connection with the local Langlands Conjecture* (J. Ritter, ed.), Contemporary Math. **86** (Amer. Math. Soc., Providence RI, 1989), 249–254.

[Mo3] **A. Moy,** *Representations of $U(2,1)$ over a p-adic field.* J. reine angew. Math. **372** (1986), 178–208.

[Mo4] **A. Moy,** *Representations of GSp_4 over a p-adic field I.* Comp. Math. **66** (1988), 237–284.

[Mo5] **A. Moy,** *Representations of GSp_4 over a p-adic field II.* Comp. Math. **66** (1988), 285–328.

[Re] **I. Reiner,** *Maximal orders.* (Academic Press, New York, 1975).

[Ro] **J. Rogawski,** *On modules over the Hecke algebra of a p-adic group.* Invent. Math. **79** (1985), 443–465.

[S] **A. J. Silberger,** *Introduction to harmonic analysis on reductive p-adic groups.* Mathematical Notes **23** (Princeton University Press, 1979).

[Sp] **T. A. Springer,** *Cusp forms for finite groups. Seminar on algebraic groups and related finite groups.* Lecture Notes in Math. **131** (Springer, Berlin, 1986), 97–120.

[T] **M. Tadić,** *On Jacquet modules of induced representations of p-adic symplectic groups.* Harmonic Analysis on Reductive Groups (W. Barker, P. Sally ed.), Progress in Math. **101** (Birkhauser, 1991), 305–314.

[Wa] **J.-L. Waldspurger,** *Algèbres de Hecke et induites de représentations cuspidales, pour $GL(N)$.* J. reine angew. Math. **370** (1986), 127–191.

[Z] **A. V. Zelevinsky,** *Induced representations of reductive p-adic groups II: On irreducible representations of $GL(n)$.* Ann. Scient. Ec. Norm. Sup. (4) **13** (1980), 165–210.

Index of notation and terminology

Non-alphabetic symbols